Ergebnisse der Mathematik und ihrer Grenzgebiete

Volume 42

3. Folge

A Series of Modern Surveys in Mathematics

Springer
Berlin
Heidelberg
New York
Barcelona
Hong Kong
London
Milan
Paris
Tokyo

Reinhardt Kiehl
Rainer Weissauer

Weil Conjectures, Perverse Sheaves and l'adic Fourier Transform

Springer

Reinhardt Kiehl
Institut für Mathematik und Informatik
Universität Mannheim
D7, 27
68159 Mannheim, Germany
e-mail: kiehl@math.uni-mannheim.de

Rainer Weissauer
Mathematisches Institut
Universität Heidelberg
Im Neuenheimer Feld 288
69120 Heidelberg, Germany
e-mail: weissauer@mathi.uni-heidelberg.de

Library of Congress Cataloging-in-Publication Data
Kiehl, Reinhardt.
Weil conjectures, perverse sheaves, and l-adic Fourier transform / Reinhardt Kiehl, Rainer Weissauer.
p. cm. – (Ergebnisse der Mathematik und ihrer Grenzgebiete; 3. Folge, v. 42)
Includes bibliographical references and index.
ISBN 3540414576 (alk. paper)
1. Weil conjectures. 2. Homology theory. 3. Sheaf theory. I. Weissauer, Rainer. II. Title. III. Ergebnisse der Mathematik und ihrer Grenzgebiete; 3. Folge, Bd. 42.
QA564 .K5 2001
516.3'52--dc21 2001031426

Mathematics Subject Classification (2000): 14-XX

ISSN 0071-1136
ISBN 3-540-41457-6 Springer-Verlag Berlin Heidelberg New York

Springer-Verlag Berlin Heidelberg New York
a member of BertelsmannSpringer Science+Business Media GmbH

http://www.springer.de

Printed in Germany

Typeset by the authors using a Springer TEX macro package.
Edited by Kurt Mattes, Heidelberg, using the MathTime fonts.
Printed on acid-free paper SPIN 10723113 44/3142LK - 5 4 3 2 1 0

Preface

The initial motivation for writing this book was given by N. Katz and his review of the book

> E. Freitag / R. Kiehl, Etale Cohomology and the Weil Conjecture
> Ergebnisse der Mathematik und ihrer Grenzgebiete, Springer-Verlag

in the bulletin of the AMS. In that review N. Katz remarks that it is especially the generalization of the original Weil conjectures, as given by P. Deligne in his fundamental paper "La Conjecture de Weil II", that had the most relevant applications in recent years. He continues:

> ... The book does not discuss Weil II at all, except for a two page summary (IV,5) of some of its main results near the end. Perhaps someday if the authors feel ambitious ...

Around that time we gave lectures in the *Arbeitsgemeinschaft Mannheim–Heidelberg* on Laumon's work, especially on his use of the Fourier transform for etale sheaves and his proof of the Weil conjectures. Therefore, we, that is one of the previous authors and the new author, decided to present these important and beautiful methods of Laumon in the form of this book. Pursuing this plan further the authors immediately felt that Deligne's work on the Weil conjectures was closely related to the sheaf theoretic theory of perverse sheaves. It seemed that only in this framework Deligne's results on global weights and his notion of purity of complexes obtain a satisfactory and final form. Therefore, it was desirable or even necessary for us to include the theory of middle perverse sheaves (as in asterisque 100) completely as a second main part in this *Ergebnisbericht*. The l-adic Fourier transform appears as a tool and proves to be a useful technique providing natural and simple proofs. This part of the book was also based on the lectures given in the *Arbeitsgemeinschaft Mannheim–Heidelberg*.

To round things off we present significant applications of these theories. For this purpose we included three chapters on the following topics: the Brylinski–Radon transform including a proof of the Hard Lefschetz Theorem, estimates for exponential sums reviewing the results of Katz and Laumon, and, finally, a chapter on the Springer representations of Weyl groups of semisimple algebraic groups. In these applications the l-adic Fourier transform always turns out to be of importance. So, looking back, it appears to us that in the course of writing this *Ergebnisbericht*, we were seemingly attracted by this elegant device, both by its vigour and its beauty.

The authors want to express thanks to all those who gave valuable comments or encouragement at the various stages of this project. Special thanks go to Dr. J. Ballmann, Dr. H. Baum, Dr. D. Fulea and Dr. U. Weselmann, to all whom we are indebted for their help during the final preparation of the manuscript. We also heartily thank the staff of Springer-Verlag for their friendly cooperation.

Reinhardt Kiehl
Rainer Weissauer

Table of Contents

Introduction 1

I. The General Weil Conjectures (Deligne's Theory of Weights) 5

I.1 Weil Sheaves 5
I.2 Weights 13
I.3 The Zariski Closure of Monodromy 25
I.4 Real Sheaves 33
I.5 Fourier Transform 38
I.6 Weil Conjectures (Curve Case) 45
I.7 The Weil Conjectures for a Morphism (General Case) 52
I.8 Some Linear Algebra 54
I.9 Refinements (Local Monodromy) 58

II. The Formalism of Derived Categories 67

II.1 Triangulated Categories 67
II.2 Abstract Truncations 74
II.3 The Core of a t-Structure 77
II.4 The Cohomology Functors 81
II.5 The Triangulated Category $D_c^b(X, \overline{\mathbb{Q}}_l)$ 86
II.6 The Standard t-Structure on $D_c^b(X, \mathfrak{o})$ 98
II.7 Relative Duality for Singular Morphisms 106
II.8 Duality for Smooth Morphisms 112
II.9 Relative Duality for Closed Embeddings 116
II.10 Proof of the Biduality Theorem 119
II.11 Cycle Classes 123
II.12 Mixed Complexes 129

III. Perverse Sheaves 135

III.1 Perverse Sheaves 135
III.2 The Smooth Case 137
III.3 Glueing 139
III.4 Open Embeddings 144
III.5 Intermediate Extensions 147
III.6 Affine Maps 153

III.7 Equidimensional Maps 156
III.8 Fourier Transform Revisited 159
III.9 Key Lemmas on Weights 161
III.10 Gabber's Theorem 167
III.11 Adjunction Properties 169
III.12 The Dictionary 173
III.13 Complements on Fourier Transform 177
III.14 Sections 181
III.15 Equivariant Perverse Sheaves 183
III.16 Kazhdan-Lusztig Polynomials 189

IV. Lefschetz Theory and the Brylinski–Radon Transform 203
IV.1 The Radon Transform 203
IV.2 Modified Radon Transforms 207
IV.3 The Universal Chern Class 215
IV.4 Hard Lefschetz Theorem 217
IV.5 Supplement: A Spectral Sequence 221

V. Trigonometric Sums 225
V.1 Introduction 225
V.2 Generalized Kloosterman Sums 226
V.3 Links with l-adic Cohomology 229
V.4 Deligne's Estimate 230
V.5 The Swan Conductor 231
V.6 The Ogg–Shafarevich–Grothendieck Theorem 236
V.7 The Main Lemma 237
V.8 The Relative Abhyankar Lemma 240
V.9 Proof of the Theorem of Katz 241
V.10 Uniform Estimates 244
V.11 An Application 246
Bibliography for Chapter V 248

VI. The Springer Representations 249
VI.1 Springer Representations of Weyl Groups of Semisimple Algebraic Groups 249
VI.2 The Flag Variety $\mathscr{B}$ 253
VI.3 The Nilpotent Variety $\mathscr{N}$ 256
VI.4 The Lie Algebra in Positive Characteristic 261
VI.5 Invariant Bilinear Forms on $\mathfrak{g}$ 263
VI.6 The Normalizer of $Lie(B)$ 264
VI.7 Regular Elements of the Lie Algebra $\mathfrak{g}$ 264
VI.8 Grothendieck's Simultaneous Resolution of Singularities 266
VI.9 The Galois Group W 269
VI.10 The Monodromy Complexes Φ and Φ' 272
VI.11 The Perverse Sheaf Ψ 276

VI.12 The Orbit Decomposition of Ψ 278
VI.13 Proof of Springer's Theorem 281
VI.14 A Second Approach 286
VI.15 The Comparison Theorem 290
VI.16 Regular Orbits 295
VI.17 W-actions on the Universal Springer Sheaf 301
VI.18 Finite Fields 310
VI.19 Determination of ε_T 317
Bibliography for Chapter VI 319

Appendix 323
A. $\overline{\mathbb{Q}}_l$-Sheaves 323
B. Bertini Theorem for Etale Sheaves 333
C. Kummer Extensions 336
D. Finiteness Theorems 338

Bibliography 355

Glossary 371

Index 373

Frequently Used Notation

Usually l denotes a prime number, $\mathbb{Z}_l$ the ring of l-adic integers, $\mathbb{Q}_l$ its quotient field of l-adic numbers. Finite extension fields of $\mathbb{Q}_l$ are usually denoted E. $\overline{\mathbb{Q}}_l$ denotes the algebraic closure of $\mathbb{Q}_l$ and $\tau : \overline{\mathbb{Q}}_l \to \mathbb{C}$ denotes a chosen isomorphism between the field $\overline{\mathbb{Q}}_l$ and the field $\mathbb{C}$ of complex numbers. For tensor products of $\overline{\mathbb{Q}}_l$-vector spaces we often write $\otimes$ instead of $\otimes_{\overline{\mathbb{Q}}_l}$, if confusion is unlikely from the context.

Usually κ denotes a finite field or an algebraically closed field which has a characteristic different from l. If κ is a finite field, k denotes its algebraic closure. $\mathbb{A}^n_S$ will denote the n-dimensional affine space over the base scheme S. Often we write $\mathbb{A}_S$ instead of $\mathbb{A}^1_S$. If the base scheme is fixed, we sometimes also write $\mathbb{A}^n$ instead of $\mathbb{A}^n_S$. Similar $\mathbb{P}^n$, respectively $\mathbb{P}^n_S$, denotes the n-dimensional projective space over the base scheme S.

Schemes of finite type over κ will be called finitely generated schemes over κ. They will be often called algebraic varieties or algebraic manifolds over κ (not following the usual notation!). If we write X_0 for an algebraic variety, the subscript 0 will be frequently used to indicate that X_0 is an algebraic variety over κ. Usually X will then denote the scheme $X_0 \times_{Spec(\kappa)} Spec(k) = X_0 \otimes_\kappa k$ over k, which is obtained by base field extension. A similar notation will be used for morphisms. Furthermore, $|X_0|$ denotes the set of closed points of the scheme X_0. Also if $\mathscr{G}_0$ is a sheaf on X_0, its pullback to X will be denoted $\mathscr{G}$. The Galois group $Gal(k/\kappa)$ acts on the field k (from the left), hence acts on $Spec(k)$ (from the right). Similarly for schemes X_0 over $Spec(\kappa)$ we get an induced (right) action of $Gal(k/\kappa)$ on $X = X_0 \times_{Spec(\kappa)} Spec(k)$ from its action on the second factor. For $\gamma \in Gal(k/\kappa)$ the corresponding automorphism of X will be denoted γ_X. For schemes X_0, Y_0 over $Spec(\kappa)$ the group $Gal(k/\kappa)$ acts on $Hom_{Spec(k)}(Y, X)$ via $f \mapsto f^\gamma = \gamma_X \circ f \circ \gamma_Y^{-1}$. For $Y_0 = Spec(\kappa)$, this defines a right action of $Gal(k/\kappa)$ on the set of k-valued points of X. We refer to these actions as Galois actions.

For a closed point x of the scheme X_0, $\kappa(x)$ will denote the residue field of x. This is a finite field. $N(x)$ will denote the number of its elements. A geometric point of X_0 over κ with values in a separable closed extension field k' of κ, which lies over the point $x \in |X_0|$, will usually be denoted $\overline{x}$. Any such $\overline{x}$ defines an embedding of the field $\kappa(x)$ into the field k: $\kappa \subset \kappa(x) \subset k$.

A morphism $f : X \to Y$ in the category of finitely generated schemes over κ is called compactifiable if it can be factorized in the form $f = \overline{f} \circ j$, where j is an open immersion and where $\overline{f}$ is a proper morphism. This is always the case for quasiprojective morphisms.

For the constant sheaf $\overline{\mathbb{Q}}_{l\,Y}$ on a scheme Y we often write δ_Y. We do this in particular in situations where $i : Y \hookrightarrow X$ is a closed embedding. Then we often write δ_Y instead of $i_*\overline{\mathbb{Q}}_{l\,Y}$, if there is no confusion possible. In complicated formulas we sometimes write $K\langle 2m\rangle$ instead of $K[2m](m)$, where $[2m]$ indicates a shift of a complex by $2m$ to the left, and (m) indicates an m-fold Tate twist. We sometimes write $i^*[n]K$ instead of $i^*(K)[n]$, if n has a meaning in terms of the morphism i (a relative dimension).

We freely use results of the book [FK]. In fact we will assume in this book most of the main results that were proven in [FK], without further mention of references. This includes, in particular, basic facts of etale cohomology theory, such as the proper base change theorem, the Poincaré duality theorem for smooth varieties over a field, and the Grothendieck trace formula, but besides this also some of the more elementary concepts, such as, for instance, the notion of etale sheaves. Because of its importance the reader will find a brief review of the statement of the Grothendieck trace formula at the beginning of the first section of Chap. I. Further information can be found also in the Appendix A of this book.

Introduction

This Ergebnisbericht is directed to a reader who is acquainted with the theory of *etale cohomology* to the extent, say, of the first two chapters of the book of E. Freitag / R. Kiehl

Etale Cohomology and the Weil Conjecture [FK].

In Chap. I of this book the theory of weights of l-adic sheaves and the proof of the generalized Weil conjecture for morphisms are described. Historically one of the first examples for the theory of weights is provided by the Gauss sums. If $\kappa = \mathbb{F}_q$ is a finite field with q elements and if $\psi : \kappa \to \mathbb{C}^*$ is a nontrivial additive character, the Gauss sums

$$\sum_{x \in \kappa} \chi(x)\psi(y \cdot x)$$

are the values of the Fourier transform of a nontrivial multiplicative character $\chi : \kappa^* \to \mathbb{C}^*$, where both the character and its Fourier transform are viewed as a function with support in κ^*. Whereas the character χ has absolute value one on its support, its Fourier transform has the absolute value $q^{1/2}$ on its support. This is the simplest case, illustrating how the Fourier transform shifts weights. Of course this is rather elementary in the context above. If a nontrivial function f and its Fourier $\widehat{f}$ on κ have, by chance, the same support and constant absolute values equal to c, respectively d, then $d = q^{1/2} \cdot c$. This is an immediate consequence of Plancherel's formula

$$\sum_{x \in \kappa} \widehat{f}(x)\overline{\widehat{h}}(x) = q \cdot \sum_{x \in \kappa} f(g)\overline{h}(x) \ .$$

In Chap. I it is shown that the basic idea of this argument can be used to prove the generalized Weil conjectures in the curve case. To make this work, the method has to be adapted to pertain to less rigid situations. Fortunately the Grothendieck fixed point formula gives all the help required. One derives from it mixedness and semicontinuouity statements, which finally allow to extend the kind of argument sketched above. The details of the arguments are carried through in the first seven sections of Chap. I. The early sections deal with some of the more elementary properties of weights; here the fundamental estimate for real sheaves proved by using the Rankin trick (Sect. 4) is an important step. The Fourier transform of a complex of l-adic sheaves and its remarkable properties form the content of Sect. 5. In the central Sect. 6 we finally give the proof of the Weil conjecture, essentially following Laumon. This fundamental result is the basis for the remaining part of Chap. I, which is then devoted to the proof of the generalized conjecture. In the last section of Chap. I on the local monodromy we already use derived categories and some of the material of the Chap. II. Although we might have postponed this, we preferred to have it included in Chap. I. The main sources for this chapter were

La conjecture de Weil II, by Deligne [De1],

and Laumon's work:

Transformation de Fourier, constants d'equations fonctionelles
et conjecture de Weil Conjecture [Lau].

In Chap. II the technical tools are provided; these are needed in the following chapter on perverse sheaves. In Sects. 3 and 4 of Chap. II abstract truncation structures are defined and their most important properties are studied. In particular, in Sect. 3 we describe the proof of the fact that the core of such a t-structure is an abelian category. In Sect. 7 the first important example of such a t-structure, namely the *standard* t-structure on the category $D_c^b(X, \mathbb{Z}_l)$, is constructed. The triangulated category $D_c^b(X, \mathbb{Z}_l)$ is not defined as the *derived category* of an abelian category, but rather as a projective limit of the derived categories $D_{ctf}^b(X, \mathbb{Z}/l^n\mathbb{Z})$. Its objects, therefore, are not complexes of l-adic sheaves in the usual sense. So the natural simple truncation operators $\tau_{\leq n}$ and $\tau_{\geq n}$ for complexes are not defined a priori. It is therefore necessary to define on $D_c^b(X, \mathbb{Z}_l)$ an abstract t-structure – called the *standard* t-structure – in a rather involved way to get a substitute for the non-existing naive truncation operators of complexes. The core of this triangulated category is isomorphic to the abelian category of l-adic sheaves. In Sects. 7–9, using the absolute Poincaré duality for smooth quasiprojective schemes over a field [FK, chap. II.1, theorem 1.13], we develop relative Poincaré duality for singular morphisms as far as this is needed in Chap. III.

After Chap. I, Chap. III is the second central chapter of this book. In this chapter we place the theory of global weights and the notation of purity of sheaves into the framework of the category of perverse sheaves. It is only in this generality that Deligne's theory of weights obtains its final form. First we develop the complete theory of (middle) perverse sheaves as in astérisque 100 [BBD]. Then this is supplemented by important theorems on the purity and weights of perverse sheaves. One of the main results is Gabber's theorem on the semisimplicity of pure complexes. At this point Fourier analysis again simplifies the picture. It significantly helps to prove the key lemmas in Sect. 10. In Sect. 12 we show how properties of Weil sheaf complexes are reflected by properties of certain functions related to these complexes, as already considered in the case of sheaves in Chap. I, Sect. 2. For the proof a variant of Deligne's Fourier transform appears. For this new "Fourier transform" the affine algebraic group $\mathbb{A}^n$ is replaced by another commutative algebraic group, a product $\mathbb{E}^n$ of elliptic curves. In the last section, as a first application of the theory of perverse sheaves, the Kazhdan–Lusztig polynomials – known for the remarkable role they play in the representation theory of groups of Lie type – are constructed using perverse sheaves.

The following chapters are devoted to applications of the theory.

In Chap. IV a new transformation for etale complexes, the Brylinski–Radon transform, is introduced and examined. It is related to Deligne's Fourier transform. In this context we prove the hard Lefschetz theorem.

One of the most beautiful applications of Grothendieck's trace formula and of Deligne's theory of weights are non-trivial estimates for trigonometric sums such as the well-known Kloosterman sums, which generalize the Gauss sums. Chapter V tries to convey the basic ideas of Katz and Laumon on the existence of uniform estimates for exponential sums of this type. Such estimates reflect a remarkable quality of the Deligne–Fourier transform. It behaves as if there exists a kind of global Fourier transformation on the affine space over $Spec(\mathbb{Z})$ from which we seemingly get the Fourier transform on the affine spaces over $Spec(\mathbb{Z}/p\mathbb{Z})$ – not uniquely determined – by reducing modulo p for almost all prime numbers p.

Chapter VI plays a special role and leads to further areas. It deals with the mysterious representations of Weyl groups of semisimple algebraic groups, discovered by Springer and named after him. Here we venture into the area of algebraic groups and of the representations of finite groups of Lie type. In Sects. 2–9 we report all the necessary background material, e.g. Lie algebras mod p, Grothendieck's simultaneous resolution of singularities etc. Proofs are carried out only where this is necessary and possible. The main content of Sects. 10–18 is Brylinski's construction of the Springer representations using perverse sheaves, using Gabber's decomposition theorem and in particular using the Fourier transform. Here it was our intention to present the ideas outlined by Brylinski in [49, §11 pp. 119–128] in greater detail, to make them available to a larger readership. Some other constructions are discussed as well, and their mutual relations and especially their relations to Springer's original construction are examined.

The appendices contain useful supplements. Appendix A represents a bridge between the book of Freitag/Kiehl [FK] and what is required for this *Ergebnisbericht*. In Freitag/Kiehl's book the theory of $\mathbb{Q}_l$-sheaves was developed. Without difficulty it is shown in Appendix A how the essential results on $\mathbb{Q}_l$-sheaves – e.g. Poincaré duality and Grothendieck's trace formula – carry over to the case of sheaves over a finite extension field E of the field $\mathbb{Q}_l$ of l-adic numbers. These are prerequisites for defining the l-adic Fourier transform. The headings of the Appendices B and C certainly speak for themselves. These appendices contain auxiliary results which are needed in Chap. I. For the convenience of the reader in Appendix D we finally present the proof of the finiteness theorems for the direct image functor in the case of non-proper morphisms. For the sake of completeness – it is not needed for this book – the finiteness theorems are proved also for the case of mixed characteristics, and furthermore the corresponding theorems for vanishing cycles and for nearby cycles are included, with short proofs.

I. The General Weil Conjectures (Deligne's Theory of Weights)

I.1 Weil Sheaves

Let κ be a finite field and k its algebraic closure. Fix a prime number l. The number q of elements of κ will always be assumed not to be divisible by the prime number l. The Galois group $Gal(k/\kappa)$ of k over κ contains the arithmetic Frobenius element $\sigma = \sigma_{k/\kappa}$, which acts on k as the automorphism

$$x \mapsto x^q \ .$$

This arithmetic Frobenius is a topological generator of the pro-cyclic Galois group $Gal(k/\kappa)$. Its inverse element F is called the geometric Frobenius automorphism. The dense cyclic subgroup of the Galois group generated by F is called the Weil group $W(k/\kappa)$ of κ. Under the map

$$\begin{array}{ll} W(k/\kappa) & \cong \mathbb{Z} \\ F & \mapsto 1 \end{array}$$

$W(k/\kappa)$ becomes canonically isomorphic to $\mathbb{Z}$, such that $Gal(k/\kappa) \cong \hat{\mathbb{Z}}$ becomes the profinite completion of its subgroup $W(k/\kappa)$. More generally for arbitrary schemes X over κ one has the morphism $\sigma_{X/\kappa} : X \to X$, which is the identity on the underlying space and is defined by $a \mapsto a^q$ on the structure sheaf. These morphisms $\sigma_{X/\kappa}$ are functorial in the category of schemes over κ.

Let X_0 be a finitely generated scheme X_0 over κ. Let X be the scheme $X_0 \times_{Spec(\kappa)} Spec(k) = X_0 \otimes_\kappa k$ over k, obtained by base field extension. Similar notation will be used for morphisms. Also if $\mathcal{G}_0$ is a sheaf on X_0, its pullback to X will be denoted $\mathcal{G}$. By its functoriality the morphism $\sigma_{X/\kappa}$ can be written in the form $(id_{X_0} \times \sigma_{k/\kappa}) \circ Fr_X$, where

$$Fr_X : X \to X$$

is now a morphism over k. It is called the **Frobenius endomorphism** of X.The morphism Fr_X is finite, hence proper. On the other hand the geometric Frobenius element $F \in W(k/\kappa) \subset Gal(k/\kappa)$ acts on the scheme via its Galois action, which is defined by $F_X = id_{X_0} \times F_{Spec(k)}$. The automorphism F_X is called the **Frobenius automorphism** of X. It is an automorphism over $Spec(\kappa)$ but not over $Spec(k)$. For schemes X_0, Y_0 over $Spec(\kappa)$ the Frobenius acts on the k-morphisms $Hom_{Spec(k)}(Y, X)$ by conjugation $f \mapsto f^F = F_X \circ f \circ F_Y^{-1}$. This defines an action of F on the k-valued points of X. This Galois action of F and the action of Fr_X on the set of k-valued points coincide. We write $X^F = X(\kappa)$ for the set of fixed points of F on the set $X(k)$ of k-valued points of X.

Example. For the affine space $X_0 = \mathbb{A}_0$, the action of F respectively Fr_X on $\mathbb{A}_0(k) = k$ is given by $k \ni a \mapsto F^{-1}(a) = \sigma(a) = a^q$.

Let $x \in |X_0|$ be some closed point of the scheme X_0. Let $\kappa(x)$ be the residue field of x and let $N(x)$ denote the number of elements of the finite field $\kappa(x)$. A geometric point of X_0 over κ with values in a separably closed extension field of k, which lies over the point $x \in |X_0|$, will usually be denoted $\overline{x}$. Any such $\overline{x}$ defines an embedding of the field $\kappa(x)$ into the field k: $\kappa \subset \kappa(x) \subset k$. Put

$$d(x) = [\kappa(x) : \kappa] .$$

One of the most important results of the theory of etale sheaves is the *Grothendieck trace formula* and the corresponding formula for the L-series of a $\overline{\mathbb{Q}}_l$-sheaf $\mathscr{G}_0$ on an algebraic scheme X_0 over κ. See for instance [FK] or [SGA5]. In [FK] this formula is proven in the context of etale $\mathbb{Q}_l$-sheaves. We also refer the reader to the Appendix A of this book for the precise definition of the notion of $\overline{\mathbb{Q}}_l$-sheaves. It is also explained there, how the results proved in [FK] extend to the case of $\overline{\mathbb{Q}}_l$-sheaves.

Let us now formulate this important result of Grothendieck. Let X_0 be an algebraic scheme and let $\mathscr{G}_0$ be an etale $\overline{\mathbb{Q}}_l$-sheaf on X_0. Then there exists a canonical isomorphism

$$Fr^*_{\mathscr{G}} : Fr^*_X(\mathscr{G}) \overset{\sim}{\longrightarrow} \mathscr{G} .$$

The existence of such an isomorphism can be reduced to the fact, that for an etale algebraic space $\pi : G_0 \to X_0$ over X_0 we have a diagram

$$\begin{array}{ccccc} & & \overset{Fr_G}{\curvearrowleft} & & \\ G & \longleftarrow & Fr^*_X(G) & \underset{\simeq}{\overset{Fr_{G/X}}{\longleftarrow}} & G \\ \downarrow{\scriptstyle \pi} & & \downarrow & & \downarrow{\scriptstyle \pi} \\ X & \underset{Fr_X}{\longleftarrow} & X & = & X , \end{array}$$

where the first square is cartesian and where $Fr_{G/X}$ is an isomorphism. So we have a canonical isomorphism $Fr^{-1}_{G/X} : Fr^*_X(G) \cong G$. This induces an isomorphism $Fr_{\mathscr{G}} : Fr^*_X(\mathscr{G}) \cong \mathscr{G}$ of the corresponding sheaves of sections.

Since the Frobenius endomorphism $Fr_X : X \to X$ is a proper map, there exists an induced morphism $H^i_c(X, \mathscr{G}) \to H^i_c(X, Fr^*_X(\mathscr{G}))$. Composed with the isomorphism provided by $Fr^*_{\mathscr{G}}$ we obtain an endomorphism of the cohomology group $H^i_c(X, \mathscr{G})$. So for an algebraic scheme X_0 and an etale $\overline{\mathbb{Q}}_l$-sheaf $\mathscr{G}_0$ on X_0 the Frobenius endomorphism of X induces a homomorphism

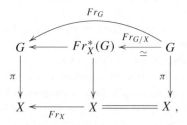

$$F : H_c^i(X, \mathcal{G}) \longrightarrow H_c^i(X, \mathcal{G})$$

of cohomology groups with compact support.

Let $x \in |X_0|$ be a closed point. Then the Galois group $Gal(k/\kappa(x))$, hence especially the geometric Frobenius substitution $F_x \in Gal(k/\kappa(x))$, acts $\overline{\mathbb{Q}}_l$-linearly on the stalk $\mathcal{G}_{0\overline{x}}$ of the sheaf $\mathcal{G}_0$ at the point $\overline{x}$

$$F_x : \mathcal{G}_{0\overline{x}} \to \mathcal{G}_{0\overline{x}} .$$

Up to isomorphism this action only depends on the closed point x and not on the choice of geometric point $\overline{x}$ over x.

With these notations the statement of the Grothendieck trace formula can be formulated as follows

Theorem 1.1 (Grothendieck) *The L-series $L(X_0, \mathcal{G}_0, t)$, defined for a $\overline{\mathbb{Q}}_l$-sheaf $\mathcal{G}_0$ on an algebraic scheme X_0 over κ by the formula*

$$L(X_0, \mathcal{G}_0, t) = \prod_{x \in |X_0|} det(1 - t^{d(x)} F_x, \mathcal{G}_{0\overline{x}})^{-1} ,$$

where $d(x)$ is the degree of the residue field $\kappa(x)$ over κ for the point x, satisfies

$$L(X_0, \mathcal{G}_0, t) = \prod_{i=0}^{2dimX} det(1 - tF, H_c^i(X, \mathcal{G}))^{(-1)^{(i+1)}} ,$$

Proof. See Verdier [308], or [FK], chap. II, §4 or Milne [249], chap. VI, §13. □

In the following we will use a generalization of the concept of $\overline{\mathbb{Q}}_l$-sheaves on an algebraic scheme X_0, namely the concept of Weil sheaves on X_0:

The Weil group $W(k/\kappa)$ operates on $X = X_0 \times_{Spec(\kappa)} Spec(k)$ via its right action on the second factor $Spec(k)$. In particular the Frobenius $F \in W(k/\kappa)$ acts on X by the Frobenius automorphism F_X. Now suppose $\mathcal{G}_0$ is a $\overline{\mathbb{Q}}_l$-sheaf on X_0. It extends to a sheaf $\mathcal{G}$ on X, which by extension of scalars is automatically a $W(k/\kappa)$-equivariant sheaf on X. Since $W(k/\kappa)$ is cyclic, this amounts to the existence of a canonical isomorphism $F_X^*(\mathcal{G}) \to \mathcal{G}$ of sheaves on X. So we are in the situation of the

Definition 1.2 *A* **Weil sheaf** *$\mathcal{G}_0$ on an algebraic scheme X_0 over κ is a $\overline{\mathbb{Q}}_l$-sheaf $\mathcal{G}$ on X together with an isomorphism*

$$F^* : F_X^*(\mathcal{G}) \cong \mathcal{G} ,$$

where F_X is the Frobenius automorphism $F_X : X \to X$. In other words $W(k/\kappa)$ acts on $(X, \mathcal{G})$. Formally the sheaf $\mathcal{G}$ on X will be called the pullback of the Weil *sheaf* $\mathcal{G}_0$ *on* X_0. *The Weil sheaf* $\mathcal{G}_0$ *is said to be smooth of rank* r, *if the pullback* $\mathcal{G}$ *on* X *is smooth of rank* r.

We state the following elementary properties of Weil sheaves:

1) The Weil sheaves on X_0 over κ can be made into the objects of an abelian category in the obvious way. This category contains the category of etale $\overline{\mathbb{Q}}_l$-sheaves on X_0 over κ as a full subcategory.

2) Let $\kappa' \subset \kappa$ be a subfield of κ. Every algebraic scheme X_0 over κ is also a scheme over κ'. The category of Weil sheaves does not change under restriction of scalars, i.e. the category of Weil sheaves on an algebraic scheme X_0 over κ is naturally equivalent to the category of Weil sheaves on X_0 over κ'. One uses the isomorphism

$$X_0 \otimes_{\kappa'} k \quad \cong \quad \coprod_{\alpha \in Gal(\kappa/\kappa')} X_0 \otimes_{\kappa_\alpha} k \,.$$

Here κ_α denotes the subfield

$$\alpha : \kappa \hookrightarrow k$$

of k.

3) For a morphism

$$f_0 : X_0 \to Y_0$$

one has natural functors, the functors of pullback and derived direct images, resp. derived images with compact support, between the categories of Weil sheaves.

Similar facts hold in related cases, e.g. in the case of the base change functors or the tensor product.

4) For Weil sheaves $\mathcal{G}_0$ on X_0 there is a map $F : H^i_c(X, \mathcal{G}) \to H^i_c(X, \mathcal{G})$, defined by the composition of the obvious map $H^i_c(X, \mathcal{G}) \to H^i_c(X, F^*_X(\mathcal{G}))$ and the map $H^i_c(X, F^*_X(\mathcal{G})) \cong H^i_c(X, \mathcal{G})$ induced by the imposed "structural" isomorphism $F^* : F^*_X(\mathcal{G}) \cong \mathcal{G}$. We remark, that in the case of an etale $\overline{\mathbb{Q}}_l$-sheaf this endomorphism F induced on cohomology groups coincides with the one defined earlier. See [SGA5], expose XV or [249], p. 292. Phrased in other words: The actions of Fr_X and the Galois action of the Frobenius induced both on geometric points and on cohomology with compact support coincide.

5) Let $\overline{x}$ be a geometric point of X_0 with values in k over the algebraic point $x \in X_0$. Then the Weil group $W(k/\kappa(x))$ acts on the stalk $\mathcal{G}_{0\overline{x}} = \mathcal{G}_{\overline{x}}$. Here $\kappa(x)$ is the residue field of x. Especially the geometric Frobenius element $F_x \in W(k/\kappa(x))$ operates on the stalk $\mathcal{G}_{0\overline{x}}$

$$F_x : \mathcal{G}_{0\overline{x}} \cong \mathcal{G}_{0\overline{x}} \,.$$

Now the L-series of the Weil sheaf $L(X_0, \mathcal{G}_0, t)$ can be defined as for etale sheaves (compare the formula in Theorem I.1.1)

Convention. If we use the notion of sheaves over algebraic varieties X_0, in particular in Chap. I of this book, then we usually mean Weil sheaves. The concept of Weil sheaves is a convenient but not essential tool, and will be of relevance especially in this first chapter. In particular we will not develop a theory of derived categories for

Weil sheaves later on. So from Chap. II on sheaf mainly is understood to mean etale $\overline{\mathbb{Q}}_l$-sheaf unless stated otherwise. In this chapter however will distinguish between Weil sheaves and etale $\mathbb{Q}_l$-sheaves only in special situations, and then mainly in order to emphasize their specific properties.

Grothendieck's result I.1.1 on L-series carries over to the case of Weil sheaves. This however uses a less obvious result about Weil sheaves:

To explain this property of Weil sheaves, let us start with the very useful description of smooth Weil sheaves in terms of representations. Let X_0 be a geometrically connected algebraic scheme over κ and $\overline{a}$ a geometric point of X_0 (over κ) in an extension field of k. The point $\overline{a}$ will be considered as a geometric point of $X = X_0 \otimes_\kappa k$. Then we have a natural exact sequence of fundamental groups

$$0 \longrightarrow \pi_1(X, \overline{a}) \longrightarrow \pi_1(X_0, \overline{a}) \longrightarrow Gal(k/\kappa) \longrightarrow 0\,.$$

The inverse image

$$W(X_0, \overline{a})$$

of $W(k/\kappa) \hookrightarrow Gal(k/\kappa)$ in $\pi_1(X_0, \overline{a})$ will be called the Weil group of X_0 attached to the base point $\overline{a}$. This group is a locally compact group. Its subgroup $\pi_1(X, \overline{a})$ is an open and closed normal subgroup in $W(X_0, \overline{a})$ with discrete factor group

$$\begin{array}{ccccc} W(X_0, \overline{a})/\pi_1(X, \overline{a}) \cong & W(k/\kappa) & \cong & \mathbb{Z} \\ & geom.Frob. & \mapsto & 1\,. \end{array}$$

The image of an element $u \in W(X_0, \overline{a})$ in $\mathbb{Z}$ is called the degree $deg(u)$ of u. This gives a short exact sequence

$$\begin{array}{ccccccccc} 0 & \longrightarrow & \pi_1(X, \overline{a}) & \longrightarrow & W(X_0, \overline{a}) & \longrightarrow & W(k/\kappa) & \longrightarrow & 0 \\ & & & & & \searrow & \downarrow \cong & & \\ & & & & & & \mathbb{Z} & & . \end{array}$$

Let $\mathscr{G}_0$ be an etale $\overline{\mathbb{Q}}_l$-sheaf. Then the fundamental group $\pi_1(X_0, \overline{a})$ operates continuously on its stalks $\mathscr{G}_{0\overline{a}} = \mathscr{G}_{\overline{a}}$; continuously means: There exists a finite dimensional extension field $E \subset \overline{\mathbb{Q}}_l$ of $\mathbb{Q}_l$ and a vector subspace W of $V = \mathscr{G}_{0\overline{a}}$ over E with $V = \overline{\mathbb{Q}}_l \otimes_E W$, such that $W(X_0, \overline{a})$ operates continuously on the finite dimensional E vector space W. The stalk functor

$$\mathscr{G}_0 \mapsto \mathscr{G}_{0\overline{a}} = \mathscr{G}_{\overline{a}}$$

defines an equivalence between the categories of smooth etale $\overline{\mathbb{Q}}_l$-sheaves on X_0 and the category of continuous representations of $\pi_1(X_0, \overline{a})$ on finite dimensional vector spaces V over $\overline{\mathbb{Q}}_l$. Remember X_0 was supposed to be geometrically connected. Similar also for X_0 replaced by X and $\pi_1(X_0, \overline{a})$ replaced by $\pi_1(X, \overline{a})$.

The corresponding statement for smooth Weil sheaves is the following

Proposition 1.3 *Let X_0 be an geometrically connected algebraic scheme over κ, let $\mathscr{G}_0$ be a Weil sheaf on X_0. The Weil group $W(X_0, \overline{a})$ operates in a natural way continuously on the stalk $\mathscr{G}_{0\overline{a}} = \mathscr{G}_{\overline{a}}$. The functor*

$$\mathscr{G}_0 \mapsto \mathscr{G}_{0\overline{a}}$$

gives rise to an equivalence between the category of smooth Weil sheaves on X_0 over κ and the category of continuous representations of the Weil group $W(X_0, \overline{a})$ on finite dimensional vector spaces V over $\overline{\mathbb{Q}}_l$.

Especially Weil sheaves of rank one on $Spec(\kappa)$ correspond to the characters

$$\phi : W(k/\kappa) \cong \mathbb{Z} \longrightarrow \overline{\mathbb{Q}}_l^{\;*} .$$

Such a character is completely determined by the image $b = \phi(F)$ of the geometric Frobenius element F. We write

$$\mathscr{L}_\phi \text{ or } \mathscr{L}_b$$

for this Weil sheaf, respectively its pullback from $Spec(\kappa)$ to X_0.

Theorem 1.4 *Let X_0 be an algebraic scheme over κ, and let $\mathscr{G}_0$ be a Weil sheaf on X_0. Then*

(1) Let X_0 be normal and geometrically connected over κ and let $\mathscr{G}_0$ be an irreducible smooth sheaf of rank n on X_0. Then $\mathscr{G}_0$ is an etale sheaf iff the determinant sheaf $\bigwedge^n \mathscr{G}_0$ is an etale sheaf. Especially one can always find an etale sheaf $\mathscr{F}_0$ and an element $b \in \overline{\mathbb{Q}}_l^{\;*}$ such that $\mathscr{G}_0$ is isomorphic to

$$\mathscr{F}_0 \otimes \mathscr{L}_b .$$

(2) In the general case there always exists a chain of subsheaves

$$0 = \mathscr{G}_0^{(0)} \subset \mathscr{G}_0^{(1)} \subset \ldots \subset \mathscr{G}_0^{(r)} = \mathscr{G}_0$$

with the following property:
Each factor sheaf is of the form

$$\mathscr{G}_0^{(j)} / \mathscr{G}_0^{(j-1)} \cong \mathscr{F}_0^{(j)} \otimes \mathscr{L}_{\psi_j} .$$

$\mathscr{F}_0^{(j)}$ is an etale sheaf on X_0 and $\mathscr{L}_{\psi_j}$ is a smooth Weil sheaf of rank 1 on X_0 attached to some character

$$\psi_j : W(k/\kappa) \to \overline{\mathbb{Q}}_l^{\;*} .$$

Proof. Assertion (2) follows from assertion (1) by Noetherian induction. It is therefore enough to prove the first part of assertion (1). For that we will have to appeal to a result proved later in §3, namely Theorem I.3.3 and its Corollary I.3.4 respectively Remark I.3.5 and Theorem I.3.1.

For the proof of assertion (1) we now assume, that X_0 is a normal geometrically irreducible algebraic scheme over κ. Let $\mathcal{G}_0$ be a smooth irreducible Weil sheaf of rank n on X_0, such that $\bigwedge^n \mathcal{G}_0$ is an etale sheaf on X_0. We attach to $\mathcal{G}_0$ an irreducible representation ρ of the Weil group $W(X_0, \overline{a})$ acting on a n-dimensional vectorspace V by Proposition I.1.3. Also we can find a finite field extension $E \subset \overline{\mathbb{Q}}_l$ of $\mathbb{Q}_l$ and a vectorspace W over E, such that $W \otimes_E \overline{\mathbb{Q}}_l = V$ and such that ρ induces a continuous representation of $W(X_0, \overline{a})$ on the E-vectorspace W.

Let $\mathfrak{o}$ be the valuation ring of E and let π be a generator of the maximal ideal of $\mathfrak{o}$. Let $\sigma \in Gl(W)$ be the image of an element in $W(X_0, \overline{a})$, whose image in $W(k/\kappa)$ generates the cyclic group $W(k/\kappa) \cong \mathbb{Z}$. Then

$$\rho\big(W(X_0, \overline{a})\big) = \bigcup_{j \in \mathbb{Z}} \rho\big(\pi_1(X, \overline{a})\big) \cdot \sigma^j .$$

By our assumption on the determinant sheaf it follows, that $det(\sigma)$ is a unit in $\mathfrak{o}$. We want to show, that furthermore the subset $\{\sigma^j | j \in \mathbb{Z}\}$ is bounded in $Gl(W)$ – or equivalently – that all eigenvalues of σ in a suitable extension field of E are units. Of course it is enough to verify this property for a suitable power σ^m, $m \neq 0$ of σ instead of σ itself. If this is true, then indeed the representation ρ extends continuously to a representation of the profinite group $\pi_1(X_0, \overline{a})$. This means, that $\mathcal{G}_0$ is an etale sheaf.

First assume, that V is an irreducible $\pi_1(X, \overline{a})$-module. Let G denote the Zariski closure of the group $\rho(\pi_1(X, \overline{a}))$ in $Gl(V)$. G is a semisimple algebraic subgroup of $Gl(V)$ defined over E (§3, Theorem I.3.3).

The group

$$G \cdot \rho(W(X_0, \overline{a})) = \bigcup_{j \in \mathbb{Z}} G \cdot \sigma^j$$

contains G as a normal subgroup. It is shown in §3, Theorem I.3.3, that for a suitable choice of E there exists a suitable power σ^m, $(m \geq 1)$ of σ, which can be written as a product

$$\sigma^m = g \cdot z$$

of an element $g \in G(E)$ and an element $z \in Gl(W)$ in the center of the group $G \cdot \rho(W(X_0, \overline{a}))$. But G is semisimple. Hence $det(g)$ has finite order. If we replace m by a suitable multiple, we can therefore assume

$$det(g) = 1$$

without restriction of generality for a suitable choice of the integer m. Hence

$$det(z) = det(\sigma)^m \in \mathfrak{o}^* .$$

The representation ρ was assumed to be irreducible. Therefore the element z lies in the center of $Gl(W)$. For our assertion on the eigenvalues of σ it is therefore enough to show that the eigenvalues of g are all units. This follows by a compactness argument. Of course

$$g\big(\rho(\pi_1(X,\overline{a}))\big)g^{-1} = \sigma^m\big(\rho(\pi_1(X,\overline{a}))\big)\sigma^{-m} = \rho\big(\pi_1(X,\overline{a})\big) .$$

Let

$$A \subset End(W)$$

be the $\mathfrak{o}$-submodule of the ring of endomorphisms of the vectorspace W generated by the compact subset $\rho(\pi_1(X,\overline{a}))$. The representation ρ was assumed to be irreducible with respect to the group $\pi_1(X,\overline{a})$. Therefore A defines a lattice in $End(W)$, i.e. a finite free $\mathfrak{o}$-module of full rank $n^2 = dim_E End(W)$, stable under conjugation by g.

$$gAg^{-1} = A .$$

In other words: The adjoint action of g stabilizes a full lattice. Therefore all eigenvalues of the adjoint matrix $Ad(\tau) \in Gl(End(W))$ are units. Let $\lambda_1, ..., \lambda_n$ denote the eigenvalues of g in E, after possibly replacing E by a suitable finite field extension of E. Then the eigenvalues of $Ad(g)$ satisfy

$$\lambda_i/\lambda_j \in \mathfrak{o}^* , \; i, j = 1, ..., n .$$

This together with the relation

$$\lambda_1 \cdots \lambda_n = det(g) = 1$$

proves, that all λ_i are units. This completes the proof for irreducible V.

In the general case V is not necessarily irreducible as a $\pi_1(X,\overline{a})$-module. By Remark I.3.5 the module V is a semisimple module. By §3, Corollary I.3.4 there exists a finite extension field κ' of κ of degree m, such that the pullback $\mathcal{G}_0'$ of the irreducible Weil sheaf $\mathcal{G}_0$ on $X_0' = X_0 \otimes_\kappa \kappa'$ has irreducible constituents, which are geometrically irreducible Weil sheaves on X_0' over κ'.

$W(X_0',\overline{a})$ is a normal subgroup of index m in $W(X_0,\overline{a})$ and

$$\rho\big(W(X_0,\overline{a})\big) = \bigcup_j \sigma^j \cdot \rho\big(W(X_0',\overline{a})\big)$$

$$\sigma^m \in \rho\big(W(X_0',\overline{a})\big) .$$

After a suitable field extension of E let U be a nonzero absolutely irreducible $W(X_0',\overline{a})$ submodule of W. Then U is also an absolutely irreducible $\pi_1(X',\overline{a})$ module.

$$W = \sum_j \sigma^j \cdot U .$$

On all vectorspaces $\sigma^j U$ the substitution σ^m has the same determinant. According to our assumption $det(\sigma)$ is a unit. Therefore the same holds for all determinants $det(\sigma^m|\sigma^j U)$. The vectorspaces $\sigma^j U$ are absolutely irreducible $W(X_0',\overline{a})$ modules, therefore also absolutely irreducible $\pi_1(X',\overline{a})$ modules.

As already shown, this forces the eigenvalues of the restrictions of σ^m to the subspaces $\sigma^j U$ to be units. Then all eigenvalues of σ^m on W, or equivalently all eigenvalues of σ on W are units. This completes the argument. □

Corollary 1.5 *Grothendieck's formula I.1.1 also holds for the L-series* $L(X_0, \mathscr{G}_0, t)$ *attached to a Weil sheaf* $\mathscr{G}_0$ *on* X_0.

Proof. The formula holds for etale sheaves $\mathscr{F}_0$ and therefore holds, by the replacement $t \mapsto tb$, for all twisted sheaves $\mathscr{F}_0 \otimes \mathscr{L}_\psi$. By additivity the formula extends to all Weil sheaves, using I.1.4. □

I.2 Weights

In this and in the following section we fix an isomorphism

$$\tau : \overline{\mathbb{Q}}_l \to \mathbb{C}$$

between the field $\overline{\mathbb{Q}}_l$ and the field of complex numbers.

Definition 2.1 *Let* X_0 *be an algebraic scheme over* κ, $\mathscr{G}_0$ *a sheaf on* X_0 *and* β *be a real number.*

(1) Choose a k valued geometric point $\overline{x}$ *for each closed point* $x \in |X_0|$. *Let* $\kappa(x)$ *be the residue field of* x. *The Weil group* $W(k/\kappa(x))$ *operates on the stalk* $\mathscr{G}_{0\overline{x}}$. $\mathscr{G}_0$ *is said to be* τ-**pure** *of weight* β, *if for all points* $x \in |X_0|$ *and all eigenvalues* $\alpha \in \overline{\mathbb{Q}}_l$ *of the geometric Frobenii*

$$F_x : \mathscr{G}_{0\overline{x}} \to \mathscr{G}_{0\overline{x}} \qquad F_x \in W(k/\kappa(x))$$

the following holds:

$$|\tau(\alpha)|^2 = N(x)^\beta \qquad N(x) = \#\kappa(x)\ .$$

(2) The sheaf $\mathscr{G}_0$ *is said to be* τ-**mixed**, *if there exists a finite filtration by subsheaves*

$$0 = \mathscr{G}_0^{(0)} \subset \mathscr{G}_0^{(1)} \subset \ldots \subset \mathscr{G}_0^{(r)} = \mathscr{G}_0\ ,$$

such that all factor sheaves $\mathscr{G}_0^{(j)}/\mathscr{G}_0^{(j-1)}$ *are* τ-*pure (of weight say* β_j*).*

(3) The sheaf $\mathscr{G}_0$ *is said to be (pointwise)* **pure** *of weight* β, *if for all isomorphisms* $\tau : \overline{\mathbb{Q}}_l \to \mathbb{C}$ *the sheaf* $\mathscr{G}_0$ *is* τ-*pure of weight* β.

(4) A sheaf $\mathscr{G}_0$ *is* **mixed**, *if there exists a finite filtration of* $\mathscr{G}_0$ *such that all successive factor sheaves are pure sheaves.*

Remark 2.2 (Twisting) Let $\mathscr{G}_0$ be a sheaf, which is τ-pure for all isomorphisms $\tau : \overline{\mathbb{Q}}_l \to \mathbb{C}$, but with weights depending on τ. Then there exists a pure sheaf $\mathscr{F}_0$ in the sense of I.2.1(3) and an element $b \in \overline{\mathbb{Q}}_l$ such that $\mathscr{G}_0$ is twisted of type

$$\mathscr{G}_0 \cong \mathscr{F}_0 \otimes \mathscr{L}_b \,.$$

Later in I.2.8 we will show:

Suppose $\mathscr{G}_0$ is τ-mixed for all $\tau : \overline{\mathbb{Q}}_l \cong \mathbb{C}$, then there exists a finite filtration of $\mathscr{G}_0$, whose successive factor sheaves are τ-pure for all $\tau : \overline{\mathbb{Q}}_l \cong \mathbb{C}$, hence can be derived from pure sheaves by a twist.

We list the following elementary

Permanence Properties

(1) Let $f_0 : X_0 \to Y_0$ be a morphism over κ, and $\mathscr{G}_0$ be a sheaf on Y_0. Then the pullback $f_0^*(\mathscr{G}_0)$ of $\mathscr{G}_0$ is τ-pure of weight β, if $\mathscr{G}_0$ is τ-pure of weight β. If f_0 is surjective, then $f_0^*(\mathscr{G}_0)$ is τ-pure of weight β iff $\mathscr{G}_0$ is τ-pure of weight β.
(2) Suppose $f_0 : X_0 \to Y_0$ is a finite morphism over κ, then the direct image $f_{0*}(\mathscr{G}_0)$ of a τ-pure sheaf $\mathscr{G}_0$ of weight β is τ-pure of weight β.
(3) Let X_0 be an algebraic scheme over κ, $\mathscr{G}_0$ a sheaf on X_0 and κ' a finite extension field of κ. Then $\mathscr{G}_0$ is τ-pure of weight β iff its inverse image on $X_0 \otimes_\kappa \kappa'$ is τ-pure of weight β.

Similar properties hold for pure sheaves. From (1)–(3) one derives permanence properties for τ-mixed sheaves in an obvious way. One exception should be noted. The statement

> "If the inverse image of a sheaf under a surjective morphism is τ-mixed, then the sheaf is τ-mixed itself"

is true in general only for finite morphisms.

Definition 2.3 *Let $\mathscr{G}_0$ be a sheaf on an algebraic scheme X_0. For fixed τ we define*

$$w(\mathscr{G}_0) = sup_{x\in|X_0|}\; sup_\alpha \; \frac{log\big(|\tau(\alpha)|^2\big)}{log\big(N(x)\big)}\,,$$

where α runs through all the eigenvalues of $F_x : \mathscr{G}_{0\overline{x}} \to \mathscr{G}_{0\overline{x}}$, if $\mathscr{G}_0$ is nontrivial. Here $\overline{x}$ always means a geometric point over x. For the trivial sheaf put $w(\mathscr{G}_0) = -\infty$.

Next we are interested in the radius of convergence for the complex valued L-series $\tau L(X_0, \mathscr{G}_0, t)$ attached to the sheaf $\mathscr{G}_0$. We will give an estimate in terms of the weights of the sheaf $\mathscr{G}_0$ in its stalks.

Lemma 2.4 *Let $\mathscr{G}_0$ be a sheaf on an algebraic scheme X_0 over κ with the property $w(\mathscr{G}_0) \leq \beta$: For all closed points $x \in |X_0|$ and corresponding geometric points $\overline{x}$ over x the eigenvalues α of*

$$F_x : \mathscr{G}_{0\overline{x}} \to \mathscr{G}_{0\overline{x}}$$

are bounded by

$$|\tau(\alpha)|^2 \leq N(x)^\beta = q^{d(x)\beta}$$
$$d(x) = [\kappa(x) : \kappa] .$$

Then the L-series $\tau L(X_0, \mathcal{G}_0, t)$ is a power series in t

$$\tau L(X_0, \mathcal{G}_0, t) = \prod_{x \in |X_0|} \tau det(1 - F_x t^{d(x)}, \mathcal{G}_{0\overline{x}})^{-1}$$

converging for all $|t| < q^{-\beta/2 - dim(X_0)}$. It has neither poles nor zeros in this region.

Proof. Without restriction of generality we can assume X_0 to be affine, reduced and irreducible. The Noether normalization theorem provides us with a finite morphism from X_0 to the affine space of dimension $dim(X_0)$. By a comparison with the affine space this gives us the following rough estimate for the number $A_n = \#X_0(\kappa_n)$ of geometric points over κ with values in the field κ_n of degree n over κ

$$A_n \leq C \cdot q^{n \cdot dim(X_0)} \qquad n = 1, 2, ..$$

for some constant C. From our assumptions the traces of

$$F_x^\nu : \mathcal{G}_{0\overline{x}} \to \mathcal{G}_{0\overline{x}}$$

can be estimated by

$$|\tau Tr(F_x^\nu)|^2 \leq r^2 \cdot q^{\nu d(x)\beta} ,$$

where $r = max_{x \in |X_0|}\big(dim_{\overline{\mathbb{Q}}_l} \mathcal{G}_{0\overline{x}}\big)$. Lemma I.2.4 follows, because this gives the following majorant power series

$$\sum_{n=1}^{\infty} r \cdot C \cdot q^{(dim(X_0)+\beta/2)n} \cdot t^{n-1}$$

for the logarithmic derivative

$$\frac{\tau L'(X_0, \mathcal{G}_0, t)}{L(X_0, \mathcal{G}_0, t)} = \sum_{n=1}^{\infty} \Big(\sum_{x \in |X_0|, d(x)|n} d(x) Tr(F_x^{n/d(x)}) \Big) \cdot t^{n-1} .$$

□

Lemma 2.5 *Let X_0 be a smooth irreducible curve over κ. Let U_0*

$$j_0 : U_0 \hookrightarrow X_0$$

be a nonempty open subset of X_0. Let $S_0 = X_0 \setminus U_0$ be the complement of U_0 in X_0. Let $\mathcal{G}_0$ be a sheaf on X_0, such that its restriction $j_0^*(\mathcal{G}_0)$ to U_0 is smooth and such that

$$H_S^0(X, \mathcal{G}) = 0 .$$

Finally assume, that for all points $x \in |U_0|$ and geometric points $\overline{x}$ over x and all eigenvalues of

$$F_x : \mathcal{G}_{0\overline{x}} \to \mathcal{G}_{0\overline{x}}$$

the following weight inequalities hold

$$|\tau(\alpha)|^2 \leq N(x)^\beta = q^{d(x)\beta} \quad , \quad d(x) = [\kappa(x) : \kappa] \, .$$

Then the corresponding inequalities hold for all points $s \in S_0$ in the complement, i.e. for all eigenvalues α of

$$F_s : \mathcal{G}_{0\overline{s}} \to \mathcal{G}_{0\overline{s}}$$

the following estimate holds

$$|\tau(\alpha)|^2 \leq N(s)^\beta \, .$$

Proof. 1) First of all we can replace κ by the algebraic closure of κ in the function field of X_0. We can therefore assume X_0 to be geometrically irreducible. By our assumptions the homomorphism

$$\mathcal{G}_0 \hookrightarrow j_{0*}(\mathcal{F}_0) \quad , \quad \mathcal{F}_0 = j_0^*(\mathcal{G}_0)$$

is injective. We may therefore assume $\mathcal{G}_0 = j_{0*}(\mathcal{F}_0)$ from the beginning. We can also assume X_0 to be affine, possibly by removing a point from U_0. Then the Grothendieck trace formula I.1.1 implies that

$$\tau L(X_0, \mathcal{G}_0, t) = \tau L(U_0, j_0^*(\mathcal{G}_0), t) \cdot \prod_{s \in |S_0|} \tau\Big(det(1 - F_s \cdot t^{d(s)}, \mathcal{G}_{0\overline{s}})\Big)^{-1}$$

$$= \frac{\tau det\Big(1 - Ft, H_c^1(X, \mathcal{G})\Big)}{\tau det\Big(1 - Ft, H_c^2(X, \mathcal{G})\Big)} \, .$$

2) For $x \in |U_0|$ the sheaf $\mathcal{F}_0$ defines a representation of the Weil group $W(U_0, \overline{x})$ on the vectorspace $V = \mathcal{F}_{0\overline{x}} = \mathcal{F}_{\overline{x}}$. Let $\mathcal{F}^\vee$ denote the dual sheaf of $\mathcal{F}$ and $W^\vee$ the dual of a $\overline{\mathbb{Q}}_l$-vectorspace W. Poincare duality on U implies

$$H_c^2(X, \mathcal{G}) = H_c^2(U, \mathcal{F}) = H^0(U, \mathcal{F}^\vee)^\vee(-1) = \Big(V_{\pi_1(U,\overline{x})}\Big)(-1) \, ,$$

where $V_{\pi_1(U,\overline{x})}$ denotes the largest factor space of V with trivial $\pi_1(U, \overline{x})$ action. Here (-1) is a Tate twist by the dual of the cyclotomic character.

Therefore the poles on the right hand side of Grothendieck's formula for the L-series, as stated above, are of the form

$$\tau(\alpha^{-1} q^{-1})$$

for eigenvalues αq of the geometric Frobenius substitution $F \in W(k/\kappa)$ acting on $V_{\pi_1(U,\overline{x})}(-1)$. Equivalently α is an eigenvalue of F on $V_{\pi_1(U,\overline{x})}$. The Frobenius homomorphism

$$F_x : \mathscr{F}_{0\overline{x}} = V \longrightarrow V$$

induces the homomorphism

$$F^{d(x)} : V_{\pi_1(U,\overline{x})} \to V_{\pi_1(U,\overline{x})}$$

on the factorspace $V_{\pi_1(U,\overline{x})}$ of V. Hence $\alpha^{d(x)}$ is also an eigenvalue of $F_x : \mathscr{F}_{0\overline{x}} \to \mathscr{F}_{0\overline{x}}$. Therefore by the Grothendieck trace formula the L-series $\tau L(X_0, \mathscr{G}_0, t)$ has no singularities in the region

$$|t| < q^{-\beta/2-1} .$$

3) The factor $\tau L(U_0, j_0^*(\mathscr{F}_0), t)$ of $\tau L(X_0, \mathscr{G}_0, t)$ converges for all $|t| < q^{-\beta/2-1}$ and does not have zeros in this region, by Lemma I.2.4. Furthermore $|S_0|$ is finite. Hence by 1) and 2) none of the factors $\tau det(1 - F_s \cdot t^{d(s)}, \mathscr{G}_{0\overline{s}})^{-1}$ for $s \in |S_0|$ has poles for

$$|t| < q^{-\beta/2-1} .$$

This implies the estimates

$$|\tau(\alpha)|^2 \leq q^{d(s)(\beta+2)} = N(s)^{\beta+2}$$

for all eigenvalues α of $F_s : \mathscr{G}_{0\overline{s}} \to \mathscr{G}_{0\overline{s}}$. Here $\mathscr{G}_{0\overline{s}}$ is one of the stalks for geometric points over $|S_0|$.

4) Now we apply these estimates also to the sheaves

$$j_{0*}(\mathscr{F}_0^{\otimes k}) \qquad k = 1, 2, \ldots .$$

If α is an eigenvalue of

$$F_s : j_{0*}(\mathscr{F}_0)_{\overline{s}} \longrightarrow j_{0*}(\mathscr{F}_0)_{\overline{s}} ,$$

then α^k is an eigenvalue of

$$F_s : \left(j_{0*}(\mathscr{F}_0)\right)_{\overline{s}}^{\otimes k} \longrightarrow \left(j_{0*}(\mathscr{F}_0)\right)_{\overline{s}}^{\otimes k} .$$

From the following Lemma I.2.6 we get the injectivity of the homomorphism

$$\left(j_{0*}(\mathscr{F}_0)\right)_{\overline{s}}^{\otimes k} \hookrightarrow \left(j_{0*}(\mathscr{F}_0^{\otimes k})\right)_{\overline{s}} .$$

This gives for all $k = 1, 2, 3, \ldots$ the upper estimates

$$|\tau(\alpha^k)|^2 \leq N(s)^{k\beta+2} ,$$

hence the improved estimate

$$|\tau(\alpha)|^2 \leq N(s)^{\beta+\frac{2}{k}}.$$

The claim of the lemma follows from the last inequality in the limit $k \to \infty$. □

Lemma 2.6 *We use the same notations as in Lemma I.2.5. Suppose $\mathcal{F}_0$ is smooth, given by the representation of the Weil group $W(X_0, \overline{x})$ on the stalk $V = \mathcal{G}_{0\overline{x}} = \mathcal{G}_{\overline{x}}$ for some $x \in |U_0|$. Let I_s be the ramification group of X in $s \in S$*

$$I_s \subset \pi_1(U, \overline{x}) \subset W(U_0, \overline{x}).$$

Then one has

$$\left(j_*(\mathcal{F})\right)_{\overline{s}} \cong V^{I_s},$$

hence

$$\left(j_*(\mathcal{F})\right)_{\overline{s}}^{\otimes k} = \left(V^{I_s}\right)^{\otimes k} \subset \left(V^{\otimes k}\right)^{I_s} = j_*\left(\mathcal{F}^{\otimes k}\right)_{\overline{s}}.$$

Proof. An immediate consequence of the definition of $\left(j_*(\mathcal{F})\right)_{\overline{s}}$. □

We have the following useful

Lemma 2.7 *Let X_0 be a normal, irreducible algebraic scheme over κ and $\mathcal{G}_0$ an irreducible smooth sheaf on X_0. Then the restriction of $\mathcal{G}_0$ to a nonempty open subscheme*

$$j_0 : U_0 \hookrightarrow X_0$$

of X_0 remains irreducible.

Proof. This is evident, because for any point $a \in |U_0|$ the homomorphism

$$\pi_1(U_0, \overline{a}) \to \pi_1(X_0, \overline{a}),$$

and therefore also the homomorphism

$$W(U_0, \overline{a}) \to W(X_0, \overline{a}),$$

is surjective. □

The next theorem reflects a fundamental property of smooth sheaves.

Theorem 2.8 (Semicontinuity of Weights) *Let $\mathcal{G}_0$ be a smooth sheaf on an algebraic scheme X_0 over κ and let*

$$j_0 : U_0 \hookrightarrow X_0$$

be the inclusion of an open dense subscheme of X_0. Then the following holds:

(1) $w(\mathcal{G}_0) = w(j_0^*(\mathcal{G}_0))$.
(2) If $j_0^*(\mathcal{G}_0)$ is τ-pure of weight β, then $\mathcal{G}_0$ is τ-pure of weight β.

(3) *Let X_0 be normal and irreducible, furthermore let $\mathcal{G}_0$ be irreducible and let $j_0^*(\mathcal{G}_0)$ be τ-mixed. Then $\mathcal{G}_0$ is τ-pure.*
(4) *Suppose X_0 is connected, $j_0^*(\mathcal{G}_0)$ is τ-mixed and $\mathcal{G}_0$ is τ-pure of weight β at a single point $x \in |X_0|$, i.e. all eigenvalues α of $F_x : \mathcal{G}_{0\overline{x}} \to \mathcal{G}_{0\overline{x}}$ satisfy $|\tau(\alpha)|^2 = N(x)^\beta$. Then $\mathcal{G}_0$ is τ-pure of weight β.*

Proof. We first prove (1). For that we may assume, that X_0 is irreducible and we can further replace X_0 by the normalization of its reduced subscheme X_{0red}. Nothing has to be proved for $dim(X_0) = 0$. The curve case was dealt with in Lemma I.2.5. The general case $dim(X_0) > 1$ on the other hand can be easily reduced to the curve case. For that observe, that every point $s \in |X_0| \setminus |U_0|$ can be connected to U_0 by a curve $Y_0 \subset X_0$, which has nonempty intersection with U_0. This completes the proof of (1).

Claim (2) follows from (1) applied to $\mathcal{G}_0$ and its dual sheaf.

Now consider (3): $j_0^*(\mathcal{G}_0)$ is τ-mixed. Any sheaf on U_0 becomes smooth on a suitable open nonempty subscheme. So by shrinking U_0 we can assume, that $j_0^*(\mathcal{G}_0)$ has a finite filtration by smooth subsheaves, whose successive quotient sheaves are τ-pure. But $j_0^*(\mathcal{G}_0)$ remains irreducible by Lemma I.2.7, hence is τ-pure. But assertion (2) then implies, that also $\mathcal{G}_0$ is τ-pure. This proves (3).

Assertion (4): It is enough to prove the assertion for each irreducible component of X_0, which contains the point x, and then to proceed by iteration. So we can assume X_0 to be irreducible and then even to be normal. Furthermore it is enough to proof the claim for all the irreducible constituents of $\mathcal{G}_0$, which allows us to assume $\mathcal{G}_0$ to be irreducible. This implies $j_0^*(\mathcal{G}_0)$ to be irreducible by Lemma I.2.7. Then $j_0^*(\mathcal{G}_0)$ is τ-pure by (3) and assertion (2) proves $\mathcal{G}_0$ to be τ-pure. The weight necessarily has to be β. □

Definition 2.9 *Let $\mathcal{G}_0$ be a sheaf on an algebraic scheme X_0 over κ. Then there is an open dense subscheme*

$$j_0 : U_0 \hookrightarrow X_0$$

such that the inverse image $j_0^*(\mathcal{G}_0)$ is smooth on U_0. We define

$$w_{gen}(\mathcal{G}_0) = w(j_0^*(\mathcal{G}_0)) .$$

Remark. This definition is independent of the choice of U_0 because of Theorem I.2.8(1).

Definition 2.10 (Real Sheaves) *Let $\mathcal{G}_0$ be a sheaf on an algebraic variety X_0. $\mathcal{G}_0$ is said to be τ-**real**, if for all closed points $a \in |X_0|$ and corresponding geometric points $\overline{a}$ over a, the characteristic polynomial*

$$\tau det(1 - F_a t, \mathcal{G}_{0\overline{a}})$$

of the geometric Frobenius

$$F_{\overline{a}} : \mathcal{G}_{0\overline{a}} \longrightarrow \mathcal{G}_{0\overline{a}}$$

has real coefficients, i.e. $\tau det(1 - F_a t, \mathcal{G}_{0\overline{a}}) \in \mathbb{R}[t] \subset \mathbb{C}[t]$.

We skip listing the obvious permanence properties of this notion.

Lemma 2.11 *Let $\mathcal{G}_0$ be a sheaf on an algebraic scheme. Assume $\mathcal{G}_0$ to be smooth and τ-pure of weight β. Then $\mathcal{G}_0$ is a direct summand of a smooth, τ-real and τ-pure sheaf $\mathcal{F}_0$ of weight β.*

Proof. Consider the dual sheaf $\mathcal{G}_0{}^\vee = \mathcal{H}om(\mathcal{G}_0, \overline{\mathbb{Q}}_l)$ of $\mathcal{G}_0$. The element $b \in \overline{\mathbb{Q}}_l$ with $\tau(b) = q^\beta$ gives the sheaf $\mathcal{F}_0 = (\mathcal{G}_0{}^\vee \otimes \mathcal{L}_b) \oplus \mathcal{G}_0$ wanted. □

2.12 The Functions $f^{\mathcal{G}_0}$. For the proof of the Weil conjectures an alternative characterization of the maximal weights of a τ-mixed sheaf is useful.

Let X_0 be an algebraic scheme over κ and $\mathcal{G}_0$ a sheaf on X_0. Let m be a natural number, we denote by κ_m the unique extension field of κ of degree m over κ in k. Let F_m denote the geometric Frobenius of $Gal(k/\kappa_m)$. The set $X_0(\kappa_m) = Hom_{Spec(\kappa)}(Spec(\kappa_m), X_0)$ of κ_m-valued points of X_0 is equal to the set of fixed points $X_0(k)^{F_m}$, where $X_0(k) = Hom_{Spec(\kappa)}(Spec(k), X_0)$ is the set of k-valued geometric points of X_0. For a sheaf $\mathcal{G}_0$ and an integer m we have the complex valued function

$$f^{\mathcal{G}_0} : X_0(\kappa_m) \to \mathbb{C}$$

defined by

$$f^{\mathcal{G}_0}(\overline{x}) = \tau Tr\left(F_x^{m/d(x)}, \mathcal{G}_{0\overline{x}})\right) = \tau Tr\left(F_m, \mathcal{G}_{0\overline{x}}\right).$$

Here $\overline{x} \in X_0(\kappa_m)$ denotes a geometric point over the closed point $x \in |X_0|$ relative to the structure map to κ. The trace is understood as the trace on the stalk $\mathcal{G}_{0\overline{x}}$. The trace value only depends on x and not on the geometric point $\overline{x}$ chosen over x. Of course these functions $f_m^{\mathcal{G}_0} = f^{\mathcal{G}_0}$ depend on m, though this dependence was dropped from the notation. Hopefully it will be clear from the context, which m is meant.

For any functions

$$f, g : X_0(\kappa_m) \to \mathbb{C}$$

one defines a scalar product $(f, g)_m$ and a norm $\|f\|_m$ by

$$(f, g)_m = \sum_{y \in X_0(\kappa_m)} f(y)\overline{g(y)} \qquad (X_0(\kappa_m) \subset X_0(k))$$

$$\|f\|_m^2 = (f, f)_m .$$

The logarithmic derivative of the function $L(X_0, \mathcal{G}_0, t)$ is given in terms of the functions $f^{\mathcal{G}_0}$ by

$$\frac{\tau L'(X_0, \mathcal{G}_0, t)}{\tau L(X_0, \mathcal{G}_0, t)} = \sum_{\nu=1}^{\infty}\Big(\sum_{x\in|X_0|, d(x)|\nu} d(x) Tr(F_x^{\nu/d(x)})\Big) \cdot t^{\nu-1}$$

$$= \sum_{n=1}^{\infty} (f^{\mathcal{G}_0}, 1)_n t^{n-1} .$$

Lemma I.2.4 respectively the method of proof used for Lemma I.2.4 gives an estimate

$$\| f^{\mathcal{G}_0}\|_m^2 \leq \Big(max_{x\in|X_0|}\Big(rang\ \mathcal{G}_{0\overline{x}}\Big)\Big)^2 \cdot \#\Big(X_0(\kappa_m)\Big) \cdot q^{m\cdot w(\mathcal{G}_0)}$$

$$\leq C \cdot q^{m\cdot(w(\mathcal{G}_0)+dim(X_0))} .$$

The constant C depends on $\mathcal{G}_0$ but not on m.

Lemma 2.13 *Let $\mathcal{G}_0$ be a sheaf on an algebraic scheme X_0 over κ. Then there is a constant C independent from m such that*

$$\| f^{\mathcal{G}_0}\|_m^2 \leq C \cdot q^{m\cdot(w(\mathcal{G}_0)+dim(X_0))} \qquad m = 1, 2, ...$$

Definition 2.14 *Let $\mathcal{G}_0$ be a sheaf on an algebraic scheme X_0 over κ. Using the preceding notations we put*

$$\|\mathcal{G}_0\| = sup\{\rho \,\Big|\, \limsup_m \frac{\| f^{\mathcal{G}_0}\|_m^2}{q^{m(\rho+dim(X_0))}} > 0\} .$$

There is a more direct characterization of $\|\mathcal{G}_0\|$, namely $q^{-\|\mathcal{G}_0\|-dim(X_0)}$ is the radius of convergence of the power series $\phi^{\mathcal{G}_0}(t)$ defined by

2.15 Definition

$$\phi^{\mathcal{G}_0}(t) = \sum_{n=1}^{\infty} \| f^{\mathcal{G}_0}\|_n^2 \cdot t^{n-1} = \sum_{n=1}^{\infty}\Big(\sum_{\overline{x}\in X_0(\kappa_n)} |\tau Tr(F_x^{n/d(x)})|^2\Big) \cdot t^{n-1} .$$

Here $\overline{x}$ is a geometric point in $X_0(\kappa_m) \subset X_0(k)$ over the closed point $x \in |X_0|$ and $F_x : \mathcal{G}_{0\overline{x}} \to \mathcal{G}_{0\overline{x}}$ is the geometric Frobenius homomorphism.

Theorem 2.16 (Radius of convergence) *Let $\mathcal{G}_0$ be a τ-mixed sheaf on an algebraic scheme X_0 of dimension $dim(X_0) = 1$. Let $j_0 : U_0 \hookrightarrow X_0$ be the open subscheme of all points of X_0, where X_0 has dimension 1. Then we have (with the convention $w_{gen}(j_0^*(\mathcal{G}_0)) = -\infty$ for $U_0 = \emptyset$)*

(1) $\|\mathcal{G}_0\| = max\big(w_{gen}(j_0^*(\mathcal{G}_0)), w(\mathcal{G}_0) - 1\big)$

(2) Assume X_0 *to be a smooth curve. If* $H_E^0(X, \mathcal{G}) = 0$ *holds for all closed subsets* E *of* X*, then*

$$\|\mathcal{G}_0\| = w(\mathcal{G}_0) .$$

Remark 2.17 Let us make some preliminary remarks concerning the radius of convergence of complex power series $P(t) = \sum_{\nu=0}^{\infty} a_\nu t^\nu$. From elementary function theory we get: If its radius of convergence R is finite and different from zero then the analytic function $P(t)$ has a singularity t_0 on the circle $\{t \in \mathbb{C} \mid |t| = R\}$; this means: There is no holomorphic continuation of the holomorphic function $P(t)$ to an open neighborhood of t_0 in $\mathbb{C}$. Another elementary observation is the following: Suppose the power series $P(t)$ is a finite or countable product of power series $P_n(t)$, such that each $P_n(t)$ has nonnegative coefficients and leading term 1. Then $P(t)$ is a majorant of $P_n(t)$ and the radius of convergence R of $P(t)$ is a lower bound for the radius of convergence of each of the power series $P_n(t)$.

Proof. We now prove I.2.16. We first observe, that the assertion (2) is a consequence of assertion (1) using Lemma I.2.5 (semicontinuity of weights).

In order to prove (1) we can assume X_0 to be reduced. For the proof we use the power series $\phi^{\mathcal{G}_0}(t)$ attached to a sheaf $\mathcal{G}_0$ as defined in I.2.15. We can assume X_0 to be connected, since $\phi^{\mathcal{G}_0}(t)$ is a sum of the corresponding power series attached to the restriction of $\mathcal{G}_0$ to the connected components. All these power series have non-negative coefficients. Therefore the radius of convergence of $\phi^{\mathcal{G}_0}(t)$ is the minimum of the radii of convergence of all these power series.

Case 1. $dim(X_0) = 0$.

Then we may suppose $s \in |X_0|$ is the single point in X_0. We view the stalk $V = \mathcal{G}_{0\overline{s}}$ of the sheaf $\mathcal{G}_0$ via $\tau : \overline{\mathbb{Q}}_l \cong \mathbb{C}$ as a complex vector space and choose a $\mathbb{C}$-basis. The map $F_s : V \to V$, now viewed as a $\mathbb{C}$-linear map, can then be described by a matrix A with entries in $\mathbb{C}$. Let $\overline{A}$ be the complex conjugate matrix. Then $\phi^{\mathcal{G}_0}(t)$ is the logarithmic derivative of the function

$$\frac{1}{det(E - A \otimes \overline{A} \cdot t^{d(s)})} .$$

Therefore the radius of convergence for $\phi^{\mathcal{G}_0}(t)$ is

$$\min_{\alpha,\beta} |\tau(\alpha)\tau(\beta)|^{-1/d(s)}$$

where α and β run through the set of eigenvalues of $F_s : \mathcal{G}_{0s} \to \mathcal{G}_{0s}$. The resulting value for the radius therefore is

$$q^{-w(\mathcal{G}_0)} .$$

Case 2. The case of a smooth sheaf $\mathcal{G}_0$ on a smooth affine curve $X_0 \neq \emptyset$.

We can assume X_0 to be connected and $\mathcal{G}_0 \neq 0$.

First we consider the case, where $\mathcal{G}_0$ is τ-pure of weight β. Let $b \in \overline{\mathbb{Q}}_l$ be such that $\tau(b) = q^\beta$. Let $\mathcal{G}_0^\vee$ be the dual of $\mathcal{G}_0$, then the "complex conjugate" sheaf $\overline{\mathcal{G}_0}$ is defined by

$$\overline{\mathcal{G}_0} = \mathcal{G}_0^\vee \otimes \mathcal{L}_b \qquad \text{see I.1.3, I.2.11 .}$$

Obviously the sheaf $\mathcal{G}_0 \otimes \overline{\mathcal{G}_0}$ is τ-real and the power series $\phi^{\mathcal{G}_0}(t)$ is the logarithmic derivative of the L-series

$$\tau L(X_0, \mathcal{G}_0 \otimes \overline{\mathcal{G}_0}, t) = \prod_{x \in |X_0|} \tau det(1 - F_x t^{d(x)}, \mathcal{G}_{0\overline{x}} \otimes \overline{\mathcal{G}_{0\overline{x}}})^{-1}$$

$$= \frac{\tau det\Big(1 - Ft, H_c^1(X, \mathcal{G} \otimes \overline{\mathcal{G}})\Big)}{\tau det\Big(1 - Ft, H_c^2(X, \mathcal{G} \otimes \overline{\mathcal{G}})\Big)} .$$

For this use $(f^{\mathcal{G}_0 \otimes \overline{\mathcal{G}_0}}(\overline{x}), 1)_n = |f^{\mathcal{G}_0}(\overline{x})|_n^2$. By our preliminary assumptions the sheaf $\mathcal{G}_0 \otimes \overline{\mathcal{G}_0}$ is τ-pure of weight 2β.

As in the proof of Theorem I.2.5 we can assume, that X_0 is geometrically irreducible after replacing κ by its algebraic closure in the function field of X_0. Similar, using Poincare duality as in the proof of I.2.5, the zeros α of $\tau det(1-Ft, H_c^2(X, \mathcal{G}_0 \otimes \overline{\mathcal{G}_0}))$, satisfy

$$|\alpha| = q^{-\beta-1} .$$

In particular any pole of the rational function $\tau L(X_0, \mathcal{G}_0 \otimes \overline{\mathcal{G}_0}, t)$ therefore satisfies $|\alpha| = q^{-\beta-1}$.

By definition of $\overline{\mathcal{G}_0}$ the local L-factors $\tau det(1 - F_x t^{d(x)}, \mathcal{G}_{0\overline{x}} \otimes \overline{\mathcal{G}_{0\overline{x}}})^{-1}$ attached to the smooth sheaf $\mathcal{G}_0$ are power series in t with nonnegative real coefficients and leading coefficient 1; since $\mathcal{G}_0 \neq 0$ they have a pole, say α. By the temporary purity assumption necessarily $|\alpha| = q^{-\beta}$. Using Remark I.2.17, therefore each local factor diverges as power series in t for $|t| > |\alpha| = q^{-\beta}$. Hence also the L-series $\tau L(X_0, \mathcal{G}_0 \otimes \overline{\mathcal{G}_0}, t)$ diverges for $|t| > q^{-\beta}$, by the second statement of Remark I.2.17. In particular the L-series must have at least one (real) pole (again by I.2.17). As already shown using the Grothendieck trace formula, any such pole has absolute value $q^{-\beta-1}$. Thus the L-series considered must have radius of convergence equal to $q^{-\beta-1}$. Lemma I.2.4 on the other hand – now applied with the weight 2β instead of β – implies, that the L-series also has no zeros in the domain $|t| < q^{-\beta-1}$. This implies, that the logarithmic derivative

$$\phi^{\mathcal{G}_0}(t) = \tau L'(X_0, \mathcal{G}_0 \otimes \overline{\mathcal{G}_0}, t) / \tau L(X_0, \mathcal{G}_0 \otimes \overline{\mathcal{G}_0}, t)$$

has radius of convergence exactly equal to

$$q^{-\beta-1} = q^{-w(\mathcal{G}_0)-1} .$$

The second characterization of $\|\mathcal{G}_0\|$ after I.2.14 implies $\|\mathcal{G}_0\| = w(\mathcal{G}_0)$. Of course, so far $\mathcal{G}_0$ was assumed to be τ-pure of weight $w(\mathcal{G}_0)$.

X_0 *being still a curve we now consider smooth τ-mixed sheaves* $\mathcal{G}_0$. We can replace $\mathcal{G}_0$ by its semisimplification. So without restriction of generality we can assume it to be semisimple. Using Theorem I.2.8(3) we can decompose such a sheaf $\mathcal{G}_0$ into a direct sum,

$$\mathcal{G}_0 = \mathcal{F}_0 \oplus \mathcal{H}_0 ,$$

where $\mathcal{F}_0$ is τ-pure of weight $\beta = w(\mathcal{G}_0)$ and $\mathcal{H}_0$ is τ-mixed of weight

$$w(\mathcal{H}_0) < \beta .$$

From the evident relation $f^{\mathcal{G}_0} = f^{\mathcal{F}_0} + f^{\mathcal{H}_0}$ we get

$$\phi^{\mathcal{G}_0}(t) = \phi^{\mathcal{F}_0}(t) + \sum_{n=1}^{\infty} 2Re(f^{\mathcal{F}_0}, f^{\mathcal{H}_0})_n t^{n-1} + \sum_{n=1}^{\infty} \| f^{\mathcal{H}_0} \|_n^2 t^{n-1} .$$

The power series $\phi^{\mathcal{F}_0}(t)$ has the precise radius of convergence $q^{-\beta-1}$, as already shown. The two other power series on the right have a larger radius of convergence, as follows from the inequalities

$$\| f^{\mathcal{H}_0} \|_n^2 \leq C_1 \cdot q^{n(w(\mathcal{H}_0)+1)}$$

$$|2Re(f^{\mathcal{F}_0}, f^{\mathcal{H}_0})_n| \leq 2\| f^{\mathcal{F}_0} \|_n \| f^{\mathcal{H}_0} \|_n \leq C_2 \cdot q^{n(\frac{w(\mathcal{F}_0)+w(\mathcal{H}_0)}{2}+1)}$$

deduced from Lemma I.2.13 and

$$w(\mathcal{H}_0) < (w(\mathcal{F}_0) + w(\mathcal{H}_0))/2 < \beta .$$

Therefore $\phi^{\mathcal{G}_0}(t)$ and $\phi^{\mathcal{F}_0}(t)$ have the same radius of convergence, namely $q^{-\beta-1}$.

Case 3. The general case:

Let X_0 be reduced. Then we can find an open affine smooth curve

$$h_0 : V_0 \hookrightarrow X_0 ,$$

such that the complement

$$i_0 : S_0 \hookrightarrow X_0$$

of V_0 in X_0 is finite and such that the sheaf $h_0^*(\mathcal{G}_0) = \mathcal{F}_0$ is smooth on V_0. Now consider $j_0 : U_0 \hookrightarrow X_0$ and the sheaf $j_0^*(\mathcal{G}_0)$ on U_0. From I.2.8(1) we obtain

$$V_0 \subset U_0$$

$$w_{gen}(j_0^*(\mathcal{G}_0)) = w(\mathcal{F}_0) .$$

Put $\mathcal{H}_0 = i_0^*(\mathcal{G}_0)$. Then

$$max\big(w(\mathcal{F}_0), w(\mathcal{H}_0)\big) = w(\mathcal{G}_0)$$
$$max\big(w(\mathcal{F}_0), w(\mathcal{H}_0) - 1\big) = max\big(w_{gen}(j_0^*(\mathcal{G}_0)), w(\mathcal{G}_0) - 1\big) .$$

From the exact sequence $0 \to h_{0!}(\mathscr{F}_0) \to \mathscr{G}_0 \to i_{0*}(\mathscr{H}_0) \to 0$ we obtain

$$\phi^{\mathscr{G}_0}(t) = \phi^{\mathscr{F}_0}(t) + \phi^{\mathscr{H}_0}(t) .$$

Now the coefficients of these power series are nonnegative. Therefore the radius of convergence of $\phi^{\mathscr{G}_0}(t)$ is the minimum of the corresponding radii of convergence for $\phi^{\mathscr{F}_0}(t)$ and $\phi^{\mathscr{H}_0}(t)$. We have already shown, that the radius of convergence for $\phi^{\mathscr{F}_0}(t)$ is $q^{-w(\mathscr{F}_0)-1}$. The radius of convergence for $\phi^{\mathscr{H}_0}(t)$ is $q^{-w(\mathscr{H}_0)}$. This implies

$$\|\mathscr{G}_0\| = max\Big(w(\mathscr{F}_0), w(\mathscr{H}_0) - 1\Big)$$

$$= max\Big(w_{gen}(j_0^*(\mathscr{G}_0)), w(\mathscr{G}_0) - 1\Big) .$$

□

I.3 The Zariski Closure of Monodromy

Theorem 3.1 *Let X_0 be a normal absolutely irreducible algebraic scheme over κ and let $\overline{a}$ be a geometric point. Let*

$$\chi : W(X_0, \overline{a}) \to \overline{\mathbb{Q}}_l^{\,*}$$

be a continuous character of the corresponding Weil group. Then the image $\chi(\pi_1(X, \overline{a}))$ of the geometric fundamental group is a finite group. Equivalently the following holds:

A suitable power χ^m, $m \geq 1$ of the character χ becomes trivial on the geometric fundamental group $\pi_1(X, \overline{a})$. In particular

$$\chi = \chi_1 \cdot \chi_2 ,$$

where χ_1 is a character of finite order and χ_2 factorizes over the quotient group $W(k/\kappa) \cong \mathbb{Z}$ of $W(X_0, \overline{a})$.

Corollary. Let $\mathscr{G}_0$ be a smooth sheaf of rank 1 on a normal geometrically irreducible algebraic scheme X_0 over κ. Then $\mathscr{G}_0$ is τ-pure.

More precisely: There is an element $b \in \overline{\mathbb{Q}}_l^{\,*}$ and an etale sheaf $\mathscr{F}_0$ attached to a finite order character of the Weil group of X_0, such that

$$\mathscr{G}_0 = \mathscr{F}_0 \otimes \mathscr{L}_b .$$

The τ-weight of $\mathscr{G}_0$ is $log(|\tau(b)|^2)/log(q)$.

Proof of I.3.1. Without restriction of generality one can make a finite base field extension and replace χ by a power χ^m, $m \neq 0$. χ being continuous means that we can further assume, that χ is E-valued

$$\chi : W(X_0, \overline{a}) \to E^*$$

for a suitable finite field extension $E \subset \overline{\mathbb{Q}}_l$ of $\mathbb{Q}_l$. Then one immediately reduces to the case, where the image $\chi(\pi_1(X, a))$ is a pro-l group. To proceed further, let us first consider the curve case.

First Case. X_0 is a smooth, projective geometrically irreducible curve.

Class field theory for function fields in one variable implies, that the image of $\pi_1(X, a)$ in the abelianized locally compact Weil group of X_0 is canonically isomorphic to the divisor class group Pic^0 of divisors of degree 0, which are rational over κ. It is enough to use the corresponding result for the l-parts of the groups under consideration. To derive it in this special case, one could use the theory of Kummer extensions (see Appendix C). The group Pic^0 is finite. In fact, Pic^0 is isomorphic to the finite group of κ-rational points of the Jacobi variety of X_0. Alternatively one may use finiteness of class numbers for global fields.

Now consider the higher dimensional case. For this let us make some preparations first. For an open nonempty subscheme U_0 of X_0 containing the base point a, the map

$$\pi_1(U, \overline{a}) \to \pi_1(X, \overline{a})$$

is surjective. Therefore we can assume X_0 to be smooth and quasiprojective. X_0 is an open dense subscheme of a normal projective scheme $\overline{X_0}$ over κ. So $\overline{X_0}$ is contained in some projective space $\mathbb{P}^N$ of dimension N over κ. Then there exists an open smooth subscheme Y_0 of $\overline{X_0}$ containing X_0

$$X_0 \subset Y_0 \subset \overline{X_0}\ ,$$

whose complement A_0 is of codimension at least 2 in $\overline{X_0}$ and such that the complement $D_0 = Y_0 \setminus X_0$ of X_0 in Y_0 is a smooth submanifold of pure codimension 1 in Y_0. After a base field extension the connected components

$$D_{10}, \ldots, D_{r0}$$

of D_0 can be assumed to be geometrically irreducible, such that each component $D_{\nu 0}$ has a κ-rational point, say a_ν.

The Second Case. Let $X_0 = Y_0$ and $dim(X_0) \geq 2$.

We choose a sufficiently general linear subspace L of codimension $dim(X) - 1$ of the projective space $\mathbb{P}^N$, such that $A \cap L$ is empty and such that $C = X \cap L$ is a smooth irreducible curve. After base field extension we can again assume L to be defined over κ.

$$L = L_0 \otimes_\kappa k\ ,\ C = C_0 \otimes_\kappa k\ .$$

By our choice of L and the assumption $X = Y$ the curve C_0 is even projective. From Bertini's theorem and Zariski's theorem on connectivity one deduces, that the homomorphism

$$\pi_1(C, \overline{a}) \to \pi_1(X, \overline{a})$$

is surjective. We can assume, that the base point is contained in C_0. After these preparations the second case follows from the first.

The Last Case is the General Case. Fix one of the components $D_{\nu 0}$ of D_0 where $\nu = 1, .., r$. Let $Y_0^{(\nu)}$ be the henselisation of Y_0 in the point a_ν, i.e. the spectrum of the henselisation of the local ring of Y_0 in the point a_ν. Then $Y_0^{(\nu)} \otimes_\kappa k = Y^{(\nu)}$ is the strict henselisation of $Y = Y_0 \otimes_\kappa k$ in the geometric point $\overline{a}_\nu \in Y_0(k)$ over a_ν. Let $X^{(\nu)}$ denote the inverse image of X in $Y^{(\nu)}$, similarly $X_0^{(\nu)}$ the inverse image of X_0 in $Y_o^{(\nu)}$ and $D^{(\nu)}$ the inverse image of D_ν in $X^{(\nu)}$. For the following considerations we can assume, that the base point $\overline{a}$ is the image of a geometric point of $X^{(\nu)}$ also denoted $\overline{a}$ for the sake of simplicity. The image of $\overline{a}$ in $X_0^{(\nu)}$ is also denoted $\overline{a}_\nu$. The inverse image of $W(k/\kappa) \subset Gal(k/\kappa)$ under the surjection $\pi_1(X_0^{(\nu)}, \overline{a}) \to Gal(k/\kappa)$ is called the Weil group $W(X_0^{(\nu)}, \overline{a})$ of $X_0^{(\nu)}$. We have the following natural commutative diagram with exact lines

$$\begin{array}{ccccccccc} 0 & \longrightarrow & \pi_1(X^{(\nu)}, \overline{a}) & \longrightarrow & W(X_0^{(\nu)}, \overline{a}) & \longrightarrow & W(k/\kappa) & \longrightarrow & 0 \\ & & \downarrow & & \downarrow & & \| & & \\ 0 & \longrightarrow & \pi_1(X, \overline{a}) & \longrightarrow & W(X_0, \overline{a}) & \longrightarrow & W(k/\kappa) & \longrightarrow & 0 \end{array}$$

The character χ induces characters

$$\chi_\nu^0 : W(X_0^{(\nu)}, \overline{a}) \to \overline{\mathbb{Q}}_l^*$$

$$\chi_\nu : W(X^{(\nu)}, \overline{a}) \to \overline{\mathbb{Q}}_l^* .$$

By assumption the image $\chi(\pi_1(X, \overline{a}))$ is a pro-l-group. Therefore χ_ν is tamely ramified along $D^{(\nu)}$ and factorizes over the l-part J of the tame ramification group along $D^{(\nu)}$. This group is isomorphic to $\mathbb{Z}_l$ after a choice of generator. For the arithmetic Frobenius element $\sigma \in W(k/\kappa)$ and some inverse image $\tilde{\sigma}$ in $W(X_0^{(\nu)}, \overline{a})$ we have $\tilde{\sigma} j \tilde{\sigma}^{-1} = j^q$ and

$$\chi_\nu(j) = \chi_\nu^0(j) = \chi_\nu^0(\tilde{\sigma} j \tilde{\sigma}^{-1}) = \chi_\nu(j)^q$$

for $j \in J$. Hence the character factors over the quotient group $\mathbb{Z}_l/(q-1)\mathbb{Z}_l$, a finite group with say m elements. Then the m-th power

$$\chi_\nu^m = 1$$

of χ_ν is the trivial character. We now repeat this consideration for all the other components of D_0. The theorem of the purity of branch points finally implies:

> χ^m is unramified in all points of D_0 and therefore factorizes over the Weil group $W(Y_0, \overline{a})$.

This reduces the general case to the second case already treated. Theorem I.3.1 is proved. □

For a smooth sheaf $\mathcal{G}_0$ on X_0 of rank n Theorem I.3.1 and its corollary imply, that its determinant sheaf, i.e. the highest exterior power $\bigwedge^n \mathcal{G}_0$ of $\mathcal{G}_0$, is τ-pure being a sheaf of rank 1.

Definition 3.2 (Determinant Weights) *Let X_0 be a geometrically connected normal algebraic scheme over κ and $\mathcal{G}_0$ be a smooth sheaf on X_0. For each irreducible constituent $\mathcal{F}_0$ of $\mathcal{G}_0$ we define the* **determinant weight** *of $\mathcal{G}_0$ with respect to τ and $\mathcal{F}_0$ to be*

$$\beta/n \quad , \qquad \beta = w(\bigwedge^n \mathcal{F}_0) \, ,$$

where n is the rank of $\mathcal{F}_0$ and β is the τ-weight of the determinant sheaf $\bigwedge^n \mathcal{F}_0$ of $\mathcal{F}_0$. A smooth sheaf $\mathcal{G}_0$ has finitely many such determinant weights (for fixed τ).

Let $\mathcal{G}_0$ be a smooth sheaf on a geometrically connected algebraic scheme X_0. Such a sheaf $\mathcal{G}_0$ is defined by a representation ρ of the Weil group $W(X_0, \overline{x})$ on a finite dimensional $\overline{\mathbb{Q}}_l$-vectorspace $V = \mathcal{G}_{0\overline{x}} = \mathcal{G}_{\overline{x}}$. There exists a finite extension field $E \subset \overline{\mathbb{Q}}_l$ of $\mathbb{Q}_l$ and an E-vectorspace $W \subset V$, such that $V = W \otimes_E \overline{\mathbb{Q}}_l$ and such that ρ defines a continuous representation $\rho : W(X_0, \overline{x}) \to Gl(W)$. The image $\rho(\pi_1(X, \overline{x}))$ of $\pi_1(X, \overline{x})$ in $Gl(W) \subset Gl(V)$ will be called the geometric monodromy group, whereas the image of $W(X_0, \overline{x})$ will be called the arithmetic monodromy group of the representation ρ or alternatively of the smooth sheaf $\mathcal{G}_0$.

For an extension κ' of the base field κ of degree m, the point $\overline{x}$ defines a geometric point of $X_0 \otimes_\kappa \kappa'$, again denoted $\overline{x}$. The Weil group $W(X_0 \otimes_\kappa \kappa', \overline{x})$ is canonically isomorphic to the inverse image in $W(X_0, \overline{x})$ of the unique subgroup of index m in $W(k/\kappa) \cong \mathbb{Z}$. Similarly if X_0' is a geometrically connected etale covering of degree n over X_0, then for a choice of a geometric point $\overline{x}'$ over $\overline{x}$ in X_0', the Weil group $W(X_0', \overline{x}')$ is canonically isomorphic to a subgroup of index n in $W(X_0, \overline{x})$ and $\pi_1(X', \overline{x}')$ is canonically isomorphic to the subgroup $W(X_0', \overline{x}') \cap \pi_1(X, \overline{x})$ of $\pi_1(X, \overline{x})$. This subgroup again has index n in $\pi_1(X, \overline{x})$. Conversely every subgroup of $W(X_0, \overline{x})$ with these properties arises from a geometrically connected etale covering of X_0 in this way.

$GL(V)$ is an algebraic group over $\overline{\mathbb{Q}}_l$, naturally defined over the field E by $V = W \otimes_E \overline{\mathbb{Q}}_l$. The Zariski closure G_{geom} of $\rho(\pi_1(X, \overline{x}))$ in $Gl(V)$ is a linear algebraic group over $\overline{\mathbb{Q}}_l$, which in fact is defined over the field E. Every element g in the Weil group $W(X_0, \overline{x})$ normalizes this group

$$\rho(g)\, G_{geom}\, \rho(g)^{-1} = G_{geom} \, .$$

This means that every element of the cyclic group $\langle \sigma \rangle \cong W(k/\kappa)$, where σ is some fixed element of degree $deg(\sigma) = 1$, defines an automorphism of the algebraic group G_{geom}, which is defined over the field E. Let G be the semidirect product of G_{geom} and the discrete group scheme $\mathbb{Z} \cong W(k/\kappa)$ defined over E via this action

$$G = G_{geom} \lhd W(k/\kappa) .$$

With these definitions G is Zariski locally a finitely generated group scheme over E, which contains G_{geom} as an open and closed subgroup scheme.

Let us arrange things into the following commutative diagram with exact horizontal lines:

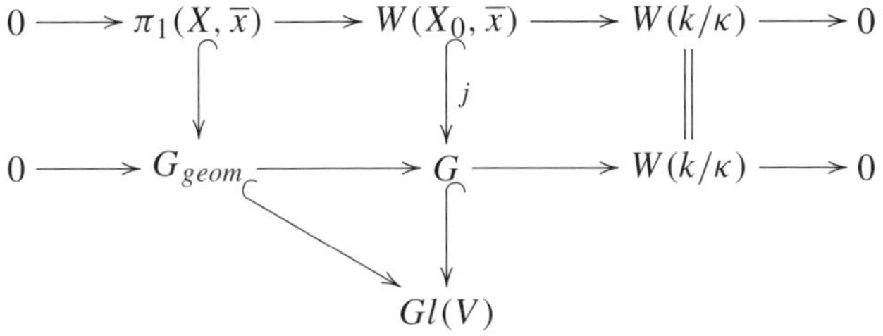

Further Notations. The connected component of G_{geom} will be denoted G^0_{geom}. For a field Ω between E and $\overline{\mathbb{Q}}_l$ let $G^0_{geom}(\Omega) \subset G_{geom}(\Omega) \subset G(\Omega)$ denote the groups of Ω rational points of G^0_{geom}, G_{geom}, G respectively.

Let $\mathcal{G}_0$ be a semisimple smooth sheaf, then the corresponding representation ρ of $W(X_0, \overline{x})$ on V and its restriction to the normal subgroup $\pi_1(X, \overline{x})$ are semisimple. This immediately implies, that also the representation of the algebraic group G_{geom} on W is semisimple. Hence G_{geom} has trivial unipotent radical. Thus G_{geom} is a – not necessarily connected – reductive linear algebraic group.

Theorem 3.3 (Grothendieck) *Let $\mathcal{G}_0$ be a geometrically semisimple smooth sheaf on a geometrically connected normal algebraic scheme over κ. Using the preceding notations we get:*

(1) G_{geom} respectively G^0_{geom} are semisimple algebraic groups.
(2) Let Z denote the center of the group $G(\overline{\mathbb{Q}}_l)$ of $\overline{\mathbb{Q}}_l$-valued points of G. Then the natural homomorphism

$$Z \to W(k/\kappa)$$

has finite kernel and cokernel. More precisely, Z contains a power of an element of degree 1.

Corollary 3.4 *If we replace κ by a suitable finite base field extension, the homomorphism*

$$Z \to W(k/\kappa)$$

becomes surjective.

Remark 3.5 Let $\mathcal{F}_0$ be a smooth, but not necessarily geometrically semisimple sheaf on X_0. All irreducible constituents $\mathcal{G}_0$ of $\mathcal{F}_0$ are semisimple, and therefore geometrically semisimple (see the remarks before Theorem I.3.3) We can therefore apply

Corollary I.3.4 to all the constituents $\mathscr{G}_0$. Then Schur's lemma easily implies: If we replace κ by a suitable finite base field extension κ', then all irreducible constituents of the inverse image of $\mathscr{F}_0$ on $X_0 \otimes_\kappa \kappa'$ become geometrically irreducible.

Proof of Theorem I.3.3(1). G_{geom} and G^0_{geom} are reductive algebraic groups, defined over $\overline{\mathbb{Q}}_l$. To prove I.3.3(1) we can replace κ by a finite extension field and X_0 by a geometrically connected etale covering, as indicated earlier. Such a replacement allows to assume $G^0_{geom} = G_{geom}$.

Let T be a maximal torus in the center of the now connected group $G_{geom} = G^0_{geom}$. Choose an element $\sigma \in W(X_0, \overline{x})$ of degree $deg(\sigma) = 1$. Conjugation by $\rho(\sigma) \in W(X_0, \overline{x})$ fixes G_{geom}, hence also the center of G_{geom}. Hence conjugation by $\rho(\sigma)$ fixes the maximal central torus T

$$\rho(\sigma) T \rho(\sigma)^{-1} = T .$$

The map $t \mapsto \rho(\sigma) t \rho(\sigma)^{-1}$ therefore permutes the finitely many characters of the representation of T on the vectorspace V. Since the underlying representation is faithful,

$$t \mapsto \rho(\sigma^m) t \rho(\sigma^m)^{-1}$$

is the identity map on T for some suitably chosen natural number m. There are only finitely many outer automorphisms of a reductive group, which induce the trivial automorphism on a maximal central torus T. Replacing m by a suitable multiple we may therefore assume, that there exists an element $g \in G_{geom}(\overline{\mathbb{Q}}_l)$ such that

$$\rho(\sigma^m) \cdot y \cdot \rho(\sigma^m)^{-1} = g^{-1} \cdot y \cdot g$$

holds for all $y \in G_{geom}(\overline{\mathbb{Q}}_l)$. Now we make a base field extension replacing κ by an extension κ_m of degree m. Then $\zeta = \rho(\sigma^m)$ becomes an element of degree $deg(\zeta) = 1$ in the Weil group $W(k/\kappa_m)$. This allows to assume $m = 1$ from now on. From above we obtain that $g\zeta$ commutes with all elements of $G_{geom}(\overline{\mathbb{Q}}_l)$, hence $g\zeta$ is in the center of $G(\overline{\mathbb{Q}}_l)$. In fact

$$G(\overline{\mathbb{Q}}_l) = \bigcup_{j \in \mathbb{Z}} G_{geom}(\overline{\mathbb{Q}}_l) \cdot (g\zeta)^j .$$

So the element $g\zeta$ of degree one (we are now over the extended base) defines a splitting

$$G(\overline{\mathbb{Q}}_l) \cong G_{geom}(\overline{\mathbb{Q}}_l) \times \mathbb{Z} .$$

This allows to define

$$\phi : W(X_0, \overline{x}) \to G_{geom}(\overline{\mathbb{Q}}_l)$$

as the composition of the natural map $j : W(X_0, \overline{x}) \to G(\overline{\mathbb{Q}}_l)$ and the projection $G(\overline{\mathbb{Q}}_l) \to G_{geom}(\overline{\mathbb{Q}}_l)$ onto the first factor of the direct product induced by the $g\zeta$-splitting.

If the connected group $G_{geom} = G^0_{geom}$ were not semisimple as stated in assertion (1), then there would exist a nontrivial character with values in the multiplicative group $\mathbb{G}_m$

$$\chi : G_{geom} \to \mathbb{G}_m \ .$$

Composition with $\phi : W(X_0, \overline{x}) \to G_{geom}(\overline{\mathbb{Q}}_l)$ induces a continuous character

$$\chi \circ \phi : W(X_0, \overline{x}) \to \mathbb{G}_m(\overline{\mathbb{Q}}_l) = \overline{\mathbb{Q}}_l^{\,*} \ .$$

Its restriction to the subgroup $\pi_1(X, \overline{x})$ would have Zariski dense image in $\mathbb{G}_m$. In particular, this image could not be finite contradicting Theorem I.3.1. This contradiction implies, that G_{geom} is a semisimple group. □

Proof of I.3.3(2). We are now again in the general situation, where G_{geom} need not necessarily be connected. To show I.3.3(2), it is enough to show that the image of the center Z of $G(\overline{\mathbb{Q}}_l)$ has finite index in $W(k/\kappa)$. The homomorphism $Z \to W(k/\kappa)$ has finite kernel, since G_{geom} is semisimple as shown above in the proof of I.3.3(1).

We choose an element ζ of $G(\overline{\mathbb{Q}}_l)$ of degree $deg(\zeta) = 1$. Any automorphism of a connected semisimple algebraic group defined over an algebraically closed field is a product of an inner automorphism and an automorphism of finite order. See [T], 1.5.6. Apply this for the connected group $G^0_{geom} \subset G_{geom}$. Hence ζ can be modified by an element in $G^0_{geom}(\overline{\mathbb{Q}}_l)$ – being still of degree 1 – such that for this modified choice of ζ a suitable power

$$z = \zeta^m, m \neq 0$$

commutes with all elements of $G^0_{geom}(\overline{\mathbb{Q}}_l)$, and in addition such that conjugation by z is trivial on the finite group $G_{geom}(\overline{\mathbb{Q}}_l)/G^0_{geom}(\overline{\mathbb{Q}}_l)$. If z is chosen in this way, then for any $g \in G_{geom}(\overline{\mathbb{Q}}_l)$

$$\phi_g(n) = g z^n g^{-1} z^{-n} \in G^0_{geom}(\overline{\mathbb{Q}}_l)$$

defines a cocycle of the cyclic group $\mathbb{Z} \cong \langle z \rangle$ generated by z with values in $G^0_{geom}(\overline{\mathbb{Q}}_l)$. This means $\phi_g(n+m) = \phi_g(n) z^n \phi_g(m) z^{-n}$. Since z acts trivially on $G^0_{geom}(\overline{\mathbb{Q}}_l)$, this map ϕ_g defines a group homomorphism

$$\phi_g : \mathbb{Z} \to G^0_{geom}(\overline{\mathbb{Q}}_l) \ .$$

$G^0_{geom}(\overline{\mathbb{Q}}_l)$ is normal in $G_{geom}(\overline{\mathbb{Q}}_l)$. The obvious relations

$$\phi_{(gg'g^{-1})g} = \phi_{gg'}(n) = \phi_g(n)$$

$$\phi_{g'g}(n) = g'\phi_g(n)(g')^{-1}$$

for all $g \in G_{geom}(\overline{\mathbb{Q}}_l)$ and all $g' \in G^0_{geom}(\overline{\mathbb{Q}}_l)$ therefore imply $\phi_g(n) = g'\phi_g(n)(g')^{-1}$. Hence $\phi_g(n)$ takes values in the center of the semisimple group $G^0_{geom}(\overline{\mathbb{Q}}_l)$. This center is a finite group of order say n. Then of course $\phi_g(n) = 0$ holds for all

$g \in G_{geom}(\overline{\mathbb{Q}}_l)$. Hence z^n commutes with all elements in $G_{geom}(\overline{\mathbb{Q}}_l)$. Since $G(\overline{\mathbb{Q}}_l)$ is generated by ζ and $G_{geom}(\overline{\mathbb{Q}}_l)$, this implies $z^n \in Z$. Since $deg(z^n) \neq 0$ the cokernel of the map $Z \to W(k/\kappa)$ is finite. This proves assertion I.3.3(2).

Corollary 3.6 *Let $\mathscr{G}_0$ be a semisimple smooth sheaf on a normal geometrically irreducible algebraic scheme X_0. Choose an element z from the center of $G(\overline{\mathbb{Q}}_l)$ with $deg(z) = m \neq 0$. Suppose $\alpha_1, .., \alpha_r$ are the eigenvalues of z on the representation space $V = \mathscr{G}_{0\overline{x}}$ with $|\tau(\alpha_i)|^2 = q^{m\beta_i}$. Then the numbers $\beta_1, .., \beta_r$ are the determinant weights of $\mathscr{G}_0$.*

Proof. By Schur's lemma z acts on each irreducible summand by multiplication with an eigenvalue. From the next remark one therefore easily reduces to the case, where $\mathscr{G}_0$ is irreducible. The claim then follows from Theorem I.3.1 applied to the determinant sheaves. □

Remark 3.7 Let the assumptions be as in Theorem I.3.3. Consider a representation of the group scheme G – the algebraized arithmetic monodromy group attached to V – on a finite dimensional $\overline{\mathbb{Q}}_l$ vectorspace V' defined over $\overline{\mathbb{Q}}_l$. This representation induces a continuous representation ρ' of $\pi_1(X_0, \overline{x})$ on V'. The image of G_{geom} in $Gl(V')$ is the Zariski closure of $\rho'(\pi_1(X, \overline{x}))$ in $Gl(V')$. Especially the restriction of ρ' to $\pi_1(X, \overline{x})$ is again semisimple. The induced map

$$G \to G'$$

of G to the group G' attached to (ρ', V') is surjective. This map therefore induces a homomorphism from the center Z of $G(\overline{\mathbb{Q}}_l)$ to the center Z' of $G'(\overline{\mathbb{Q}}_l)$. Assertion I.3.3(2) implies, that the image of Z in Z' has finite index in Z'.

The following properties of determinant weights follow from Corollary I.3.6 and the last statement of Remark I.3.7.

Theorem 3.8 *Let $f_0 : X_0' \to X_0$ be a morphism between normal geometrically irreducible schemes and let $\mathscr{F}_0$ and $\mathscr{G}_0$ be smooth sheaves on X_0. The image of X_0' is supposed to be dense in X_0. Then the following holds*

(1) $\mathscr{G}_0$ and $f_0^*(\mathscr{G}_0)$ have the same determinant weight with respect to a fixed τ.
(2) Suppose $\mathscr{G}_0$ resp. $\mathscr{F}_0$ has a single determinant weight α resp. β with respect to τ. Then $\mathscr{F}_0 \otimes \mathscr{G}_0$ has the single determinant weight $\alpha + \beta$ with respect to τ.
(3) For $\gamma \in \mathbb{R}$ let $r(\gamma)$ denote the sum of the ranks of all irreducible constituents of $\mathscr{F}_0$, which have the same determinant weight γ with respect to τ. Then the determinant weights of $\bigwedge^r \mathscr{F}_0$ are the numbers

$$\sum_{\gamma\in\mathbb{R}} n(\gamma)\gamma \qquad \textit{with} \qquad \sum_{\gamma\in\mathbb{R}} n(\gamma) = r \,,\ 0 \le n(\gamma) \le r(\gamma) \qquad n(\gamma) \in \mathbb{Z}\,.$$

Proof. For the proof one can replace $\mathscr{F}_0$ and $\mathscr{G}_0$ by their semisimplifications. (2) and (3) then follow from the Corollary I.3.6, using I.3.7 applied to the Zariski closure of the arithmetic monodromy group attached to $\mathscr{F}_0 \oplus \mathscr{G}_0$.

For (1) use, that the image of the geometric fundamental group of X' has finite index in the geometric fundamental group of X, hence contains $G^0_{geom}(\overline{\mathbb{Q}}_l)$. Choose $\zeta' \in G'(\overline{\mathbb{Q}}_l)$, such that $deg(\zeta') = 1$ and $(\zeta')^n \in Z'$ for some $n \in \mathbb{N}$ by I.3.3(2). The image of ζ' commutes with $G^0_{geom}(\overline{\mathbb{Q}}_l)$. The same argument as in I.3.3(2) shows, that some power commutes with $G_{geom}(\overline{\mathbb{Q}}_l)$. Therefore it is in Z, because $G(\overline{\mathbb{Q}}_l)$ is generated by $G_{geom}(\overline{\mathbb{Q}}_l)$ and the image of ζ'. Finally apply Corollary I.3.6. □

I.4 Real Sheaves

Let us recall the following basic facts on the cohomology groups of curves.

Let X_0 be a smooth geometrically irreducible curve over κ and $\mathscr{G}_0$ be a smooth sheaf on X_0. Fix a geometric point $\overline{x}$ over some closed point x in $|X_0|$. The Weil group $W = W(X_0, \overline{x})$ of X_0 acts on the $\overline{\mathbb{Q}}_l$-vectorspace $V = \mathscr{G}_{0\overline{x}} = \mathscr{G}_{\overline{x}}$. Consider

$$\pi = \pi_1(X, \overline{x}) \subset W(X_0, \overline{x}) \,.$$

If X_0 is affine, the cohomology group $H^0_c(X, \mathscr{G})$ vanishes. On the other hand $H^2_c(X, \mathscr{G})$ can be determined using Poincare duality. This allows to determine the eigenvalues of the geometric Frobenius element $F \in W(k/\kappa)$ on $H^2_c(X, \mathscr{G})$, since

$$H^2_c(X, \mathscr{G}) = V_\pi(-1),$$

where V_π is the largest factor space of V with trivial action of π. On V_π and on $V_\pi(-1)$ the Weil group $W(k/\kappa)$ still acts. The image of $F_x : \mathscr{G}_{0\overline{x}} = V \to V$ in $W(k/\kappa)$ is $F^{d(x)}$, where $d(x) = [\kappa(x) : \kappa]$. For every eigenvalue α of

$$F : V_\pi(-1) = H^2_c(X, \mathscr{G}) \to H^2_c(X, \mathscr{G})$$

the number αq^{-1} is an eigenvalue of $F : V_\pi \to V_\pi$ and $(\alpha q^{-1})^{d(x)}$ is thus an eigenvalue of $F_x : V = \mathscr{G}_{0\overline{x}} \to \mathscr{G}_{0\overline{x}}$. The module $M = V_\pi$ defines a smooth sheaf $\mathscr{F}_0$ on $Spec(\kappa)$. The pullback $\tilde{\mathscr{F}}_0$ to X_0 is the largest smooth quotient sheaf of $\mathscr{G}_0$, which becomes constant on $X = X_0 \otimes_\kappa k$. This representation theoretic characterization has the following consequence:

The determinant weights of $\mathscr{F}_0$ and $\tilde{\mathscr{F}}_0$ coincide and they are among the determinant weights of the sheaf $\mathscr{G}_0$.

This property of the determinant weights of $\mathscr{F}_0$ implies:

Lemma 4.1 *Let α be an eigenvalue of*

$$F : H^2_c(X, \mathscr{G}) \to H^2_c(X, \mathscr{G}) \,.$$

Then $log(|\tau(\alpha q^{-1})|^2)/log(q)$ is a determinant weight of $\mathscr{F}_0$, hence also of $\mathscr{G}_0$.

Proof. For an eigenvalue α of $F : H_c^2(X, \mathcal{G}) \to H_c^2(X, \mathcal{G})$ the number $log(|\tau(\alpha q^{-1})|^2)/log(q)$ is a determinant weight of $\mathcal{F}_0$, hence also of $\mathcal{G}_0$. Hence there exists a determinant weight β of $\mathcal{G}_0$, such that $|\tau(\alpha)|^2 = q^{\beta+2}$. □

Now suppose that $\mathcal{G}_0$ is also τ-real (Def. I.2.10). The logarithmic derivative with respect to t of $\tau det(1 - F_x t, \mathcal{G}_{0\overline{x}}^{\otimes k})^{-1}$ is

$$f(t) = \sum_{n=1}^{\infty} \tau(Tr(A^n)^k) \cdot t^{n-1},$$

where A is the Frobenius homomorphism

$$\mathcal{G}_{0\overline{x}} \to \mathcal{G}_{0\overline{x}}.$$

The identity

$$\tau det(1 - F_x \cdot t, \mathcal{G}_{0\overline{x}}^{\otimes k})^{-1} = e^{\int f(t)dt}$$

implies, that the sheaf $\mathcal{G}_0^{\otimes k}$ is again τ-real. In the even cases, i.e. if the integer k is replaced by $2k$, the coefficients of the power series $\tau det(1 - F_x t, \mathcal{G}_{0\overline{x}}^{\otimes 2k})^{-1}$ with respect to t are obviously *n*onnegative.

Lemma 4.2 (Rankin-Selberg Method) *Let $\mathcal{G}_0$ be a smooth sheaf on a geometrically irreducible smooth curve X_0 over κ. If $\mathcal{G}_0$ is τ-real, then all irreducible constituents of $\mathcal{G}_0$ are τ-pure. The τ-weights coincide with the determinant weights of the corresponding constituents.*

Proof. 1) We will first estimate the eigenvalues of

$$F_x : \mathcal{G}_{0\overline{x}} \to \mathcal{G}_{0\overline{x}} \qquad x \in |X_0|.$$

Suppose $\mathcal{G}_0 \neq 0$. Without restriction of generality we can assume the curve X_0 to be affine. Let β denote the largest determinant weight of $\mathcal{G}_0$ with respect to τ. Then $2k\beta$ is the largest determinant weight of $\mathcal{G}_0^{\otimes 2k}$ according to Theorem I.3.8(2). From our preliminary considerations in I.4.1 we find a lower bound

$$|t_0| \geq q^{-(2k\beta+2)/2}$$

for every zero $t = t_0$ of the polynomial $\tau det(1 - Ft, H_c^2(X, \mathcal{G}^{\otimes 2k}))$.

2) On the other hand, for every natural number k we have

$$\prod_{x \in |X_0|} \tau det(1 - F_x t^{d(x)}, \mathcal{G}_{0\overline{x}}^{\otimes 2k})^{-1} = \frac{\tau det(1 - Ft, H_c^1(X, \mathcal{G}^{\otimes 2k}))}{\tau det(1 - Ft, H_c^2(X, \mathcal{G}^{\otimes 2k}))}.$$

This shows, that the product expansion on the left hand side of the upper equation converges for all

$$|t| < q^{-(2k\beta+2)/2}.$$

Because all factors have nonnegative coefficients and the leading coefficients are equal to one, the same convergence holds true for each of the local L-factors. See Remark I.2.17. Hence for all $x \in |X_0|$ the polynomials

$$\tau det(1 - F_x t^{d(x)}, \mathcal{G}_{0\overline{x}}^{\otimes 2k})$$

are zero free for $|t| < q^{-(2k\beta+2)/2}$. Consider an eigenvalue α of $F_x : \mathcal{G}_{0\overline{x}} \to \mathcal{G}_{0\overline{x}}$ and the corresponding eigenvalue α^{2k} of

$$F_x : \mathcal{G}_{0\overline{x}}^{\otimes 2k} \to \mathcal{G}_{0\overline{x}}^{\otimes 2k} .$$

The above inequality implies

$$|\tau(\alpha)|^2 \leq q^{d(x)(2k\beta+2)/2k} = N(x)^{\beta+\frac{1}{k}} ,$$

which in the limit $k \to \infty$ gives

$$|\tau(\alpha)|^2 \leq N(x)^\beta .$$

To Put Things Together. For each eigenvalue α of

$$F_x : \mathcal{G}_{0\overline{x}} \to \mathcal{G}_{0\overline{x}}$$

the inequality

$$(*) \qquad |\tau(\alpha)|^2 \leq N(x)^\beta$$

holds, where β is the largest determinant weight of $\mathcal{G}_0$ with respect to τ.

3) Still one has to obtain further information on the irreducible constituents of $\mathcal{G}_0$. For this apply the last estimate $(*)$ to certain exterior powers of the sheaf $\mathcal{G}_0$. This finally proves Lemma I.4.2, as will be shown now:

Suppose the determinant weights of $\mathcal{G}_0$ being given in decreasing order

$$\gamma_1 > \gamma_2 > \ldots > \gamma_r .$$

We may assume $\mathcal{G}_0$ to be semisimple. For each determinant weight γ_i of $\mathcal{G}_0$ let then $\mathcal{G}_0(i)$ denote the direct sum of all irreducible constituents of $\mathcal{G}_0$ with determinant weight equal to γ_i with respect to τ.

$$\mathcal{G}_0 = \bigoplus_{i=1}^{r} \mathcal{G}_0(i) .$$

Let $r(i)$ denote the rank of $\mathcal{G}_0(i)$ and for any n with $0 \leq n < r$ put $N = \sum_{i=1}^{n} r(i)$. Then

$$\bigwedge^{N+1} \mathcal{G}_0 = \mathcal{G}_0(n+1) \bigotimes_{i=1}^{n} det(\mathcal{G}_0(i)) \oplus \ldots$$

has largest τ-determinant weight

$$\gamma_{n+1} + \sum_{i=1}^{n} r(i)\gamma_i$$

by I.3.8(3). A particular eigenvalue of

$$F_x : \bigwedge^{N+1} \mathcal{G}_0 \to \bigwedge^{N+1} \mathcal{G}_0$$

is

$$\alpha_i^{(n+1)} \prod_{i=1}^{n} \prod_{j=1}^{r(i)} \alpha_j^{(i)} ,$$

where $\alpha_1^{(i)}, \ldots, \alpha_{r(i)}^{(i)}$ denote the eigenvalues of $F_x : \mathcal{G}_0(i)_{\overline{x}} \to \mathcal{G}_0(i)_{\overline{x}}$ counted with multiplicities. The above estimate $(*)$ therefore shows

$$\left|\tau\left(\alpha_i^{(n+1)} \prod_{i=1}^{n} \prod_{j=1}^{r(i)} \alpha_j^{(i)}\right)\right|^2 \leq N(x)^{\gamma_{n+1}+\sum_{i=1}^{n} r(i)\gamma_i} .$$

This together with the determinant relation

$$\left|\tau\left(\prod_{i=1}^{n} \prod_{j=1}^{r(i)} \alpha_j^{(i)}\right)\right|^2 = N(x)^{\sum_{i=1}^{n} r(i)\gamma_i} ,$$

which follows from the definition of determinant weights, yields

$$|\tau(\alpha_i^{(n)})|^2 \leq N(x)^{\gamma_n} \qquad \text{for all } 1 \leq n \leq r\ ,\ \ 1 \leq i \leq r(n)\ .$$

The same argument applied to the dual sheaf of $\mathcal{G}_0$ proves the opposite inequalities and therefore completes the proof of Lemma I.4.2. □

Next we want to generalize I.4.2 from the curve case to higher dimensions and also to the case of more general sheaves. The idea is to use Bertini's theorem to reduce everything to the curve case I.4.2.

Theorem 4.3 *Let X be an algebraic scheme and $\mathcal{G}_0$ be a sheaf on X_0, which is τ-real. Then*

(1) $\mathcal{G}_0$ is τ-mixed
(2) Purity: Let X_0 be irreducible and normal and let $\mathcal{G}_0$ be smooth. Then the irreducible constituents of $\mathcal{G}_0$ are τ-pure. The τ-weight of each constituent coincides with the corresponding determinant weight.

Proof. Part (2) follows from part (1) and Theorem I.2.8(3).

We can assume, that X_0 is reduced. Let $j_0 : U_0 \to X_0$ be an open subscheme, $i_0 : S_0 \hookrightarrow X_0$ the closed complement. Then we have the exact sequence

$$0 \longrightarrow j_{0!}j_0^*(\mathcal{G}_0) \longrightarrow \mathcal{G}_0 \longrightarrow i_{0*}i_0^*(\mathcal{G}_0) \longrightarrow 0\ .$$

Hence $\mathcal{G}_0$ is mixed, provided $j_0^*(\mathcal{G}_0)$ and $i_0^*(\mathcal{G}_0)$ are mixed. This remark allows to use noetherian induction. It is therefore enough to show, that there exists a nonempty open subset

$$j_0 : U_0 \hookrightarrow X_0\ ,$$

such that $j_0^*(\mathcal{G}_0)$ is τ-mixed.

Next observe that we can freely use base field extension. Namely let κ' be a finite extension field of κ, put $X_0' = X_0 \otimes_\kappa \kappa'$ and let $\mathcal{G}_0'$ be the pullback of $\mathcal{G}_0$ to X_0'. If $\mathcal{G}_0'$ is τ-mixed, then also the direct image of $\mathcal{G}_0'$ onto X_0 is τ-mixed, which contains $\mathcal{G}_0$ as a τ-mixed subsheaf.

Using these preliminary remarks we can assume, that X_0 is a smooth irreducible affine algebraic scheme over κ and that $\mathcal{G}_0$ is a smooth sheaf on X_0. Replacing κ by its algebraic closure in the function field of X_0 we can also assume X_0 to be absolutely irreducible. Then by Lemma I.4.2 it is enough to consider the case

$$dim(X_0) > 1\ .$$

Furthermore using I.3.5 we can make all irreducible constituents of $\mathcal{G}_0$ geometrically irreducible by a finite base field extension. Now embed X_0 into some projective space $\mathbb{P}_0^N$ over κ

$$X_0 \subset \mathbb{P}_0^N\ .$$

Let $\mathcal{F}_0$ be one of the irreducible, geometrically irreducible constituents of $\mathcal{G}_0$. We have to show, that there exists a nonempty open subset U_0 of X_0, such that $\mathcal{F}_0$ becomes τ-pure when restricted to U_0. For that consider linear subspaces L of $\mathbb{P}^N$ of codimension $dim(X_0) - 1$, defined over k. They also arise as k-rational points of a certain Grassmannian G over k. Especially consider those L, where the intersection $C = L \cap X$ is a nonempty smooth irreducible curve over k and such that the restriction of the sheaf $\mathcal{F}$ from X to C is irreducible. According to the theorem of Bertini resp. the theorem of Bertini for geometrically irreducible sheaves (see Appendix B, Theorem 1) all linear spaces L from the k-rational points of a suitable open dense subset Ω of the Grassmannian G have these properties.

For the moment fix such an $L \in \Omega$. Then there is a finite extension field κ' of κ, such that C is defined over κ', i.e. that there exists a closed curve

$$C_0 \subset X_0 \otimes_\kappa \kappa'$$

with the property

$$C_0 \otimes_{\kappa'} k = C\ .$$

The pullback $\mathcal{F}_0'$ of $\mathcal{F}_0$ to C_0 is a constituent of the pullback $\mathcal{G}_0'$ of $\mathcal{G}_0$ to C_0, and it is geometrically irreducible according to our choice of Ω. The sheaf $\mathcal{G}_0'$ is still τ-real. According to Lemma I.4.2 (Rankin method) $\mathcal{F}_0'$ is a τ-pure sheaf on C_0. Again by I.4.2 the weights of $\mathcal{F}_0'$ coincide with the determinant weight of $\mathcal{F}_0'$, which coincides with the determinant weight β of $\mathcal{G}_0$ attached to $\mathcal{F}_0$. For any image point $\overline{x}$ of a geometrical point of $L \cap X$ in X_0 with values in k the stalk $\mathscr{G}_{0\overline{x}}$ is τ-pure of weight β.

Now we let $L \in \Omega$ vary. Then there exists an open nonempty subscheme U of X, such that every k-rational point of U is contained at least in one of the linear subspaces $L \in \mathbb{P}^N$ from Ω. But U is already defined over some finite extension field of κ. Intersecting U by its finitely many conjugates allows us to assume U already being defined over κ. Thus there exists a subscheme U_0 of X_0 such that

$$U = U_0 \otimes_\kappa k \qquad U_0 \neq \emptyset .$$

The restriction of the sheaf $\mathscr{F}_0$ to the open subscheme U_0 of X_0 is τ-pure of weight β, where β denotes the determinant weight of $\mathscr{F}_0$. This proves the claim. □

I.5 Fourier Transform

Let $\mathbb{A}_0$ denote the affine line over the field κ. For an extension field L of κ we have $\mathbb{A}_0(L) = L$. Now consider the Artin-Schreier covering

$$\begin{aligned} \wp_0 : \mathbb{A}_0 &\to \mathbb{A}_0 \\ x &\mapsto x^q - x \end{aligned}$$

of this affine line. It defines a geometrically irreducible, etale Galois covering of $\mathbb{A}_0$. The Galois group of this covering is canonically isomorphic to the additive group κ of the base field. It is a quotient group of the fundamental group $\pi_1(\mathbb{A}, \overline{x})$ for some fixed base point $\overline{x}$. So any character

$$\psi : \kappa \to \overline{\mathbb{Q}}_l^*$$

induces a character of this fundamental group and thus defines an etale sheaf

$$\mathscr{L}_0(\psi)$$

on $\mathbb{A}_0$ of rank one.

For any element x from κ, let ψ_x denote the character

$$\begin{aligned} \psi_x : \kappa &\to \overline{\mathbb{Q}}_l^* \\ y &\mapsto \psi(xy) \end{aligned}$$

Fix a nontrivial character ψ. Then $x \mapsto \psi_x$ identifies the additive group of the base field κ and the character group of the covering group of the Artin-Schreier extension defined by $\wp_0$.

Now let us consider the direct image sheaf $\wp_{0*}(\overline{\mathbb{Q}}_l)$ of the constant sheaf $\overline{\mathbb{Q}}_l$. One obtains:

$\wp_{0*}\overline{\mathbb{Q}}_l$ is a smooth sheaf on $\mathbb{A}_0$ of rank q. The underlying representation of the fundamental group is the natural representation on the group ring of factor group κ with coefficients in $\overline{\mathbb{Q}}_l$.

This implies

Lemma 5.1 *Let ψ be a nontrivial character of κ with values in $\overline{\mathbb{Q}}_l^{\,*}$. Then*

$$\wp_{0*}(\overline{\mathbb{Q}}_l) = \bigoplus_{x\in\kappa} \mathscr{L}_0(\psi_x)\,.$$

Apropo Base Field Extensions. For any natural number $n \geq 1$ consider the etale covering

$$\begin{aligned} \wp_0^{(n)} : \mathbb{A}_0 &\rightarrow \quad \mathbb{A}_0 \\ x &\mapsto x^{q^n} - x \end{aligned}\,.$$

In general this is not a Galois covering. But after a base field extension with the field κ_n of q^n elements it becomes the Artin-Schreier extension of the affine line $\mathbb{A}_0 \otimes_\kappa \kappa_n$ over κ_n. It then has the covering group κ_n. The morphism $\wp_0^{(n)}$ admits the following decomposition

$$\begin{aligned} \wp_0^{(n)} : \mathbb{A}_0 &\longrightarrow \mathbb{A}_0 \xrightarrow{\ \wp_0\ } \mathbb{A}_0\,, \\ x &\longrightarrow \textstyle\sum_{\nu=0}^{n-1} x^{q^\nu}\,. \end{aligned}$$

Over κ_n this decomposition induces a surjective homomorphism between the covering groups of the maps $\wp_0^{(n)}$ and $\wp_0 = \wp_0^{(1)}$, defined by the trace

$$\begin{aligned} Tr : \kappa_n &\rightarrow \quad \kappa \\ \sigma &\mapsto \textstyle\sum_{\nu=0}^{n-1} \sigma^{q^\nu} \end{aligned}\,.$$

Given a character $\psi : \kappa \rightarrow \overline{\mathbb{Q}}_l^{\,*}$ of κ and an element x from κ, the following holds true

$$\psi_x \circ Tr = \Big(\psi \circ Tr\Big)_x\,.$$

By abuse of notation the character

$$\psi \circ Tr : \kappa_n \rightarrow \overline{\mathbb{Q}}_l^{\,*}$$

will again be denoted ψ. Contrary to above, ψ_x is in general for elements x in κ_n only a character of κ_n. If ψ is nontrivial, then of course the induced character

$$\psi : \kappa_n \rightarrow \overline{\mathbb{Q}}_l^{\,*}$$

is again nontrivial. Therefore the construction of the sheaves $\mathscr{L}_0(\psi)$ and $\mathscr{L}_0(\psi_x)$ is compatible with finite base field extensions.

From the Leray spectral sequence obviously follows

$$0 = H^1(\mathbb{A}, \overline{\mathbb{Q}}_l) = H^1(\mathbb{A}, \wp_*(\overline{\mathbb{Q}}_l))$$

$$0 = H^1_c(\mathbb{A}, \overline{\mathbb{Q}}_l) = H^1_c(\mathbb{A}, \wp_*(\overline{\mathbb{Q}}_l)) .$$

Using Lemma I.5.1 and base field extension we get

Lemma 5.2 *For any character $\psi : \kappa \to \overline{\mathbb{Q}}_l^*$ and all elements $x \in k$ one has*

$$H^1(\mathbb{A}, \mathscr{L}(\psi_x)) = H^1_c(\mathbb{A}, \mathscr{L}(\psi_x)) = 0 .$$

Here we use the following convention. If x is an element of κ_n for some n, then $\mathscr{L}(\psi_x)$ will denote the pullback of the sheaf $\mathscr{L}_0(\psi_x)$ from $\mathbb{A}_0 \otimes_\kappa \kappa_n$ to $\mathbb{A}_0 \otimes_\kappa k$.

Definition 5.3 (Fourier Transform) *Let $\psi : \kappa \to \overline{\mathbb{Q}}_l^*$ be a nontrivial character. Consider the diagram*

$$\begin{array}{ccccc} & & \mathbb{A}_0 \times \mathbb{A}_0 & \xrightarrow{\quad m \quad} & \mathbb{A}_0 \qquad m(x,y) = xy \\ & \swarrow^{\pi^1} & & \searrow^{\pi^2} & \\ \mathbb{A}_0 & & & & \mathbb{A}_0 \end{array}$$

and define the functor Fourier transform

$$T_\psi : D^b_c(\mathbb{A}_0, \overline{\mathbb{Q}}_l) \to D^b_c(\mathbb{A}_0, \overline{\mathbb{Q}}_l)$$

by

$$T_\psi(K) = R\pi^1_{0!}\Big(\pi_0^{2*}(K) \otimes m^*(\mathscr{L}_0(\psi))\Big)[1] .$$

Here $K[1]^i = K^{i+1}$, hence [1] indicates a degree shift for the complex by 1 to the left.

Remarks to Definition I.5.3

(1) $D^b_c(\mathbb{A}_0, \overline{\mathbb{Q}}_l)$ is the "derived" category of bounded complexes of etale $\overline{\mathbb{Q}}_l$-sheaves on $\mathbb{A}_0$ with constructible cohomology sheaves. For details see Appendix A or Chap. II, in particular the section on the triangulated category $D^b_c(X, \overline{\mathbb{Q}}_l)$. Usually all results obtained for etale sheaves by Fourier transform in the following sections can be extended by Theorem I.1.4 to cover the case of Weil sheaves as well.

(2) For a sheaf K, i.e. a complex concentrated in degree 0, the Fourier transform $T_\psi(K)$ is a complex with cohomology concentrated at most in degree -1,0,1. This is due to the shift of degree of the complex to the left by [1], since the direct image complex $R\pi^1_{0!}(\mathscr{F})$ of a sheaf $\mathscr{F}$ has cohomology only in degrees 0,1,2.

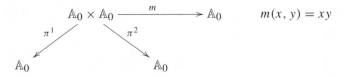

Similar to the later proof of Lemma I.5.10 one shows

Lemma 5.4 *For $a \in \kappa$ consider the morphism*

$$\begin{aligned} \lambda_0^a : \mathbb{A}_0 &\to \mathbb{A}_0 \\ x &\mapsto ax \end{aligned} .$$

Then for a character $\psi : \kappa \to \overline{\mathbb{Q}}_l^*$ the following holds

$$(\lambda_0^a)^*(\mathscr{L}_0(\psi)) = \mathscr{L}_0(\psi_a) .$$

After suitable base field extensions the base change theorem and this lemma imply

Theorem 5.5 *For a complex K_0 from $D_c^b(\mathbb{A}_0, \overline{\mathbb{Q}}_l)$ and a geometric point $a \in \mathbb{A}_0(k) = k$ one has*

$$\Big(T_\psi(K_0)\Big)_a = R\Gamma_c\Big(K \otimes \mathscr{L}(\psi_a)\Big)[1] .$$

Here K denotes the pullback of K_0 onto $\mathbb{A} = \mathbb{A}_0 \otimes_\kappa k$ and $\mathscr{L}(\psi_a)$ denotes the pullback of the sheaf $\mathscr{L}_0(\psi_a)$ defined over a suitable extension field κ_n containing a.

Let n be a fixed natural number. In I.2.12 we have attached functions

$$f^{\mathscr{G}_0} : X_0(\kappa_n) \to \mathbb{C}$$

to sheaves $\mathscr{G}_0$ on algebraic varieties X_0 over κ depending on n. This procedure can be extended to complexes $K_0 \in D_c^b(X_0, \overline{\mathbb{Q}}_l)$ by setting

$$f^{K_0} = \sum_\nu (-1)^\nu f^{\mathscr{H}^\nu(K_0)} ,$$

where $\mathscr{H}^\nu(K_0)$ is the ν-th cohomology sheaf of K_0.

Let x be an element from κ_n and $\alpha \in k$ be a solution of the Artin-Schreier equation $t^{q^n} - t = x$. The arithmetic Frobenius substitution of k over κ_n maps the solution α to the new solution $\beta = \alpha^{q^n}$. But $\beta - \alpha = x$. This shows, that the arithmetic Frobenius acts via translation by the element x, also viewed as the corresponding element x in the covering group κ_n. The geometric Frobenius element F_x in the covering group κ_n is therefore given by the element $-x \in \kappa_n$.

From this computation and Grothendieck's fixed point formula one derives, using Theorem I.5.5, the following

Lemma 5.6 *Let $n \geq 1$ be a natural number and K_0 be a complex from $D_c^b(\mathbb{A}_0, \overline{\mathbb{Q}}_l)$. The functions*

$$f^{T_\psi(K_0)} , \; f^{K_0} : \mathbb{A}_0(\kappa_n) = \kappa_n \to \mathbb{C}$$

are related by

$$f^{T_\psi(K_0)}(t) = -\sum_{x\in\kappa_n} f^{K_0}(x)\psi^{-1}(xt) \qquad (t \in \kappa_n)\,.$$

On the level of the functions $f : \kappa_n \to \mathbb{C}$ attached to complexes, the Fourier transform $T_\psi(K_0)$ of K_0 therefore corresponds to the usual Fourier transform on the finite abelian group κ_n with respect to the nondegenerate pairing $\psi^{-1}(xy)$, at least up to a sign. Remember the definition of L^2-norms given in §2. For these norms the Plancherel formula for the group κ_n implies

Theorem 5.7 (Plancherel Formula) *For a complex $K_0 \in D_c^b(\mathbb{A}_0, \overline{\mathbb{Q}}_l)$ and any natural number $n \geq 1$ one has*

$$\|f^{T_\psi(K_0)}\|_n \;=\; q^{n/2}\|f^{K_0}\|_n\,.$$

Of course these functions f^{K_0} depend on the choice of n as explained in §2. In other words, if one varies the natural number n, one obtains a whole collection of such functions.

Theorem 5.8 (Fourier Inversion) *For all K_0 in $D_c^b(\mathbb{A}_0, \overline{\mathbb{Q}}_l)$ one has*

$$\left(T_{\psi^{-1}} \circ T_\psi\right)(K_0) = K_0(-1)\,,$$

where $K_0(-1) = K_0 \otimes_{\overline{\mathbb{Q}}_l} \overline{\mathbb{Q}}_l(-1)$ denotes **Tate twist**.

The rest of this section is devoted to the proof of the Fourier inversion formula.

We start by computing the Fourier transform of the constant sheaf $\overline{\mathbb{Q}}_l$ on $\mathbb{A}_0$. The following facts are easily verified:

(1) From Lemma I.5.2 follows

$$H_c^1(\mathbb{A}, \mathscr{L}(\psi_x)) = 0$$

(2) $\mathbb{A}$ being affine implies

$$H_c^0(\mathbb{A}, \mathscr{L}(\psi_x)) = 0$$

(3) For ψ nontrivial ψ_x is trivial only for $x = 0$. Poincare duality therefore implies

$$H_c^2(\mathbb{A}, \mathscr{L}(\psi_x)) = \begin{cases} 0 & x \neq 0 \\ \overline{\mathbb{Q}}_l(-1) & x = 0 \end{cases}$$

Using Theorem I.5.5 these observations prove

Lemma 5.9 (Orthogonality Relations) *Let*

$$i_0 = \{0\} \hookrightarrow \mathbb{A}_0$$

be the embedding of the origin into the affine line and let $\delta_0 = i_{0*}(\overline{\mathbb{Q}}_l)$ *denote the sheaf concentrated at the origin with stalk equal to* $\overline{\mathbb{Q}}_l$. *Then*

$$T_\psi(\overline{\mathbb{Q}}_l) = \delta_0(-1)[-1] .$$

Equivalently this means in terms of the projection $\pi = \pi^2$

$$R\pi^2_!(m^*(\mathscr{L}_0(\psi)) = \delta_0(-1)[-2] .$$

This being said, we are ready to prove the inversion formula. For the proof we need a couple of maps and diagrams. The sign $\diamond$ will indicate cartesian squares. Consider

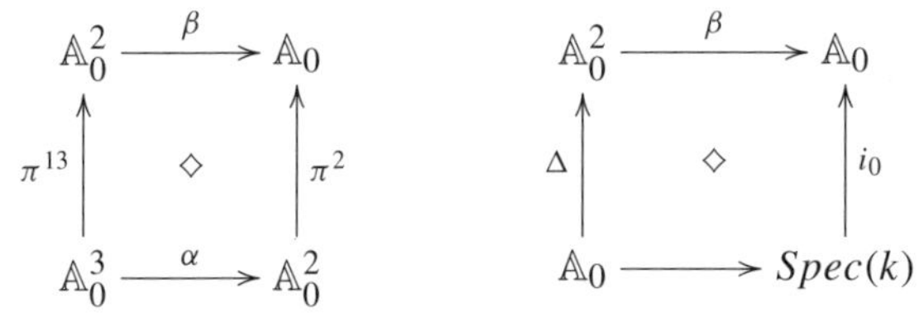

and the maps

$$\Delta(x) = (x, x) ,$$
$$\alpha(x, y, z) = (y, z - x) ,$$
$$\beta(x, y) = (y - x) .$$

Furthermore consider the diagram

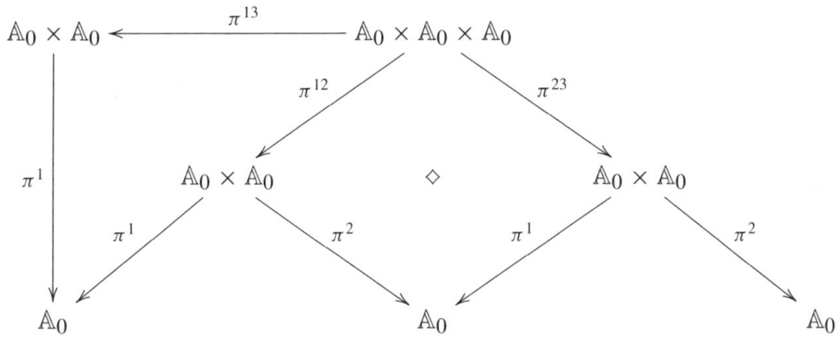

Here π^i resp. π^{ij} denotes projection on the i-th resp ij-th factor. In the diagram the middle square is cartesian and the morphisms are projections onto factors as indicated by the indexing. The essential step for the inversion formula is to prove the analog of the wellknown character formula $\sum_{y\in\kappa_n} \psi^{-1}(y(x-z)) = q^n\delta(x-z)$ for $x, z \in \kappa_n$. For that one uses, starting from the next Lemma I.5.10, the basechange theorem applied to the first cartesian square and then the orthogonality relations I.5.9. This gives

$$\begin{aligned}
R\pi_!^{13}&\Big(\pi^{12*}(m^*(\mathscr{L}_0(\psi^{-1}))) \otimes_{\overline{\mathbb{Q}}_l} \pi^{23*}(m^*(\mathscr{L}_0(\psi)))\Big) \\
&= R\pi_!^{13}\Big(\alpha^*(m^*(\mathscr{L}_0(\psi)))\Big) \\
&= \beta^*\Big(R\pi_!^2(m^*(\mathscr{L}_0(\psi)))\Big) \\
&= \beta^*\Big(\delta_0(-1)[-2]\Big) \\
&= \Delta_*(\overline{\mathbb{Q}}_l)(-1)[-2] \,.
\end{aligned}$$

The last equality uses base change in the second diagram. Next observe

$$\Delta_*(K_0) = \Delta_*(\overline{\mathbb{Q}}_l \otimes_{\overline{\mathbb{Q}}_l} \Delta^*\pi^{2*}(K_0)) = \Delta_*(\overline{\mathbb{Q}}_l) \otimes_{\overline{\mathbb{Q}}_l} \pi^{2*}(K_0) \,.$$

The base change theorem for the third cartesian diagram yields

$$\Big(T_{\psi^{-1}} \circ T_\psi\Big)(K_0)$$

$$= R\pi_!^1 \circ R\pi_!^{12}\Big(\pi^{12*}(m^*(\mathscr{L}_0(\psi^{-1}))) \otimes_{\overline{\mathbb{Q}}_l} \pi^{23*}(m^*(\mathscr{L}_0(\psi))) \otimes_{\overline{\mathbb{Q}}_l} \pi^{23*}(\pi^{2*}(K_0))\Big)[2] \,.$$

Now use $R\pi_!^1 \circ R\pi_!^{12} = R\pi_!^1 \circ R\pi_!^{13}$, $\pi^{23*} \circ \pi^{2*} = \pi^{13*} \circ \pi^{2*}$, the projection formula [FK], Proposition I.8.14 and the previous computations to obtain

$$\begin{aligned}
T_{\psi^{-1}} \circ T_\psi(K_0) &= R\pi_!^1\Big(\Delta_*(\overline{\mathbb{Q}}_l)(-1)[-2] \otimes_{\overline{\mathbb{Q}}_l} \pi^{2*}(K_0)\Big)[2] \\
&= R\pi_!^1\Big(\Delta_*(K_0)(-1)\Big) \\
&= K_0(-1) \,.
\end{aligned}$$

The Fourier inversion formula is proved. □

Lemma 5.10 *The following holds:*

$$\pi^{12*}\Big(m^*(\mathscr{L}_0(\psi^{-1}))\Big) \otimes_{\overline{\mathbb{Q}}_l} \pi^{23*}\Big(m^*(\mathscr{L}_0(\psi))\Big) = \alpha^*\Big(m^*(\mathscr{L}_0(\psi))\Big) \,.$$

Proof. We have $\mathbb{A}_0 = Spec(\kappa[s])$ and $\mathbb{A}_0 \times \mathbb{A}_0 \times \mathbb{A}_0 = Spec(\kappa[x, y, z])$ for certain variables s, x, y, z. We adjoin to $\kappa[x, y, z]$ the algebraic elements ε and η defined by the equations $\varepsilon^q - \varepsilon = xy$ and $\eta^q - \eta = yz$. The spectrum X_0 of the ring obtained by adjoining these two elements is an etale, irreducible Galois covering of $\mathbb{A}_0 \times \mathbb{A}_0 \times \mathbb{A}_0$ with covering group $\kappa \times \kappa$.

Consider the following commutative diagram

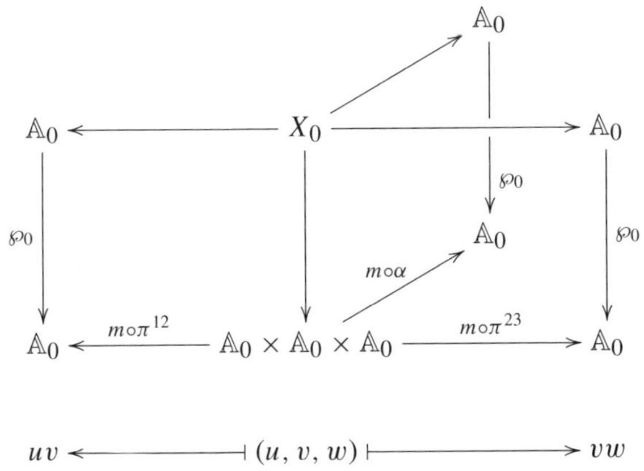

The three morphisms $X_0 \to \mathbb{A}_0$ in the diagram are best described by the corresponding homomorphisms of affine algebras, which are given by

$$s \mapsto \varepsilon\ , \quad s \mapsto \eta - \varepsilon\ , \quad s \mapsto \eta\ .$$

The induced homomorphisms of the covering groups

$$\kappa \times \kappa \to \kappa$$

are

$$(\sigma, \tau) \mapsto \sigma\ , \quad (\sigma, \tau) \mapsto \tau - \sigma\ , \quad (\sigma, \tau) \mapsto \tau\ .$$

The character $\psi : \kappa \to \overline{\mathbb{Q}}_l^{\,*}$ gives via composition with these three homomorphisms three characters $\psi_1, \psi_2, \psi_3 : \kappa \times \kappa \to \overline{\mathbb{Q}}_l^{\,*}$. These three characters correspond to the three smooth sheaves entering into Lemma I.5.10, namely

$$\begin{aligned}
\pi^{12*}(m^*(\mathscr{L}_0(\psi))) : &\qquad \psi_1\Big((\sigma, \tau)\Big) = \psi(\sigma) \\
\alpha^*(m^*(\mathscr{L}_0(\psi))) : &\qquad \psi_2\Big((\sigma, \tau)\Big) = \psi(\tau)/\psi(\sigma)\ . \\
\pi^{23*}(m^*(\mathscr{L}_0(\psi))) : &\qquad \psi_3\Big((\sigma, \tau)\Big) = \psi(\tau)
\end{aligned}$$

The assertion of the Lemma I.5.10 then boils down to the trivial statement $\psi_1^{-1}\psi_3 = \psi_2$. □

I.6 Weil Conjectures (Curve Case)

With all the necessary requisites at hand, we are now prepared to prove the fundamental

Theorem 6.1 (Deligne) *Let X_0 be a smooth, geometrically irreducible, projective curve over κ. Let*

$$j_0 : U_0 \hookrightarrow X_0$$

be an open nonempty subscheme. Let $\mathscr{F}_0$ be a smooth, τ-pure sheaf of weight w on U_0. Then the cohomology groups

$$H^i(X, j_*(\mathscr{F}_0))$$

are τ-pure of weight $w + i$ for $i = 0, 1, 2$.

One easily reduces the proof of this theorem to the case of etale sheaves. Therefore assume, that $\mathscr{F}_0$ is an etale sheaf.

Lemma 6.2 ([SGA4$\frac{1}{2}$] dualite, theorem 1.3, p. 156) *Let X be a smooth curve over a field, $j : U \hookrightarrow X$ an open dense subset and $\mathscr{F}_0$ a smooth, constructible $\overline{\mathbb{Q}}_l$-sheaf on U. Then*

$$R\mathscr{H}om(j_*(\mathscr{F}_0), \overline{\mathbb{Q}}_l) = j_*(\mathscr{H}om(\mathscr{F}_0, \overline{\mathbb{Q}}_l)) = j_*(\mathscr{F}_0^{\vee}) .$$

Theorem I.6.1 and Lemma I.6.2 are nowadays better understood in the context of the theory of perverse sheaves. We remark, that Theorem I.6.1 is a special case of more general purity statements proved in Chap. III. The statement of Lemma I.6.2 is a special case of the general results proved in Chap. III, §5. In fact, j_* turns out to be the intermediate extension of $\mathscr{F}_0$ in the sense of the theory of perverse sheaves. It should be emphasized, that a proof of I.6.2 will be given in Chap. III in the context of the theory of perverse sheaves. This proof is self contained and does not make use of the special case formulated in Lemma I.6.2 above. Therefore we skip the proof of I.6.2.

Proof of Theorem I.6.1. The case $i = 0$ is trivial. Because of Lemma I.6.2 and Poincare duality it is enough to treat the case $i = 1$ and to prove in that case: For every eigenvalue α of the Frobenius homomorphism

$$F : H^1(X, j_*(\mathscr{F}_0)) \to H^1(X, j_*(\mathscr{F}_0))$$

the following inequality holds

$$|\tau(\alpha)|^2 \leq q^{w+1} .$$

Let us first make some simplifications: Without restriction of generality we can make a base field extension, and one can furthermore replace U_0 by a smaller nonempty open subscheme $j_0' : U_0' \hookrightarrow U_0$, since

$$(j_0 \circ j_0')_*(j_0')^*(\mathscr{F}_0) = j_{0*}(\mathscr{F}_0) .$$

Especially we can assume U_0 to be affine. Then according to the Noether normalization theorem there exists a finite morphism

$$U_0 \to \mathbb{A}_0$$

to the affine line, which can be extended to a finite morphism

$$X_0 \to \mathbb{P}_0$$

of X_0 to the projective line $\mathbb{P}_0$ over κ. The direct image of $\mathscr{F}_0$ from U_0 onto $\mathbb{A}_0$ is a smooth sheaf on an open nonempty subset of $\mathbb{A}_0$. Therefore we can assume without restriction of generality, that $X_0 = \mathbb{P}_0$ and that U_0 is an open subscheme of the affine line $\mathbb{A}_0 \subset \mathbb{P}_0$.

The complement S_0 of U_0 in $\mathbb{P}_0$ is finite. The factor sheaf $\mathscr{H}_0 = j_{0*}(\mathscr{F}_0)/j_{0!}(\mathscr{F}_0)$ is concentrated on the finite set S_0, so $H^1(\mathbb{P}^1, \mathscr{H}_0)$ vanishes. Therefore the homomorphism on cohomology with compact support

$$H_c^1(U, \mathscr{F}) = H^1(\mathbb{P}^1, j_!(\mathscr{F})) \to H^1(\mathbb{P}^1, j_*(\mathscr{F}))$$

is surjective. It is therefore enough to prove the corresponding inequalities for the eigenvalues of the Frobenius homomorphism

$$H_c^1(U, \mathscr{F}) \to H_c^1(U, \mathscr{F}) .$$

Making a base field extension and after shrinking of U_0 we can assume, that $\mathbb{A}_0 \setminus U_0$ contains a κ-rational point s, where the sheaf $\mathscr{F}_0$ is unramified. One can interchange the points s and ∞ using a projective linear transformation. We can then assume, that $\mathscr{F}_0$ is unramified at the point $s = \infty$, i.e. it can be extended as a smooth sheaf to a neighborhood of the point ∞.

By Remark I.3.5 the irreducible constituents of $\mathscr{F}_0$ become geometrically irreducible after a finite base field extension. All constituents are then unramified at the point ∞. But $H_c^1(U, -)$ is a half exact functor. We can therefore assume $\mathscr{F}_0$ itself to be geometrically irreducible and unramified at the point ∞.

First we want to deal with the special case, where $\mathscr{F}_0$ is a geometrically constant sheaf, i.e. where the pullback $\mathscr{F}$ of $\mathscr{F}_0$ is a constant sheaf. Then again the direct image j_{0*} to $\mathbb{P}^1$ is geometrically constant. This implies the vanishing of the first cohomology

$$H^1(\mathbb{P}^1, j_*(\mathscr{F})) = 0 .$$

The factor sheaf $\mathscr{H}_0 = j_{0*}(\mathscr{F}_0)/j_{0!}(\mathscr{F}_0)$ is concentrated on the finite complement $S_0 = \mathbb{P}_0^1 \setminus U_0$. Therefore the long exact sequence attached to the short exact sequence

$$0 \to j_!(\mathscr{F}) \to j_*(\mathscr{F}) \to \mathscr{H} \to 0$$

yields a surjection

$$\prod_{s \in S} \left(j_{0*}(\mathscr{F}_0)\right)_{\overline{s}} = H^0(\mathbb{P}^1, \mathscr{H}) \quad \longrightarrow \quad H_c^1(U, \mathscr{F}) .$$

Semicontinuity of weights (Theorem I.2.8) gives us control over the weights of $\mathscr{H}$ and $H^0(\mathbb{P}^1, \mathscr{H})$ and therefore implies the estimates wanted.

We can assume now, in addition, that $\mathcal{F}_0$ is not geometrically constant.

At this point let us collect things together.

It is enough to consider the following situation: Let $j_0 : U_0 \hookrightarrow \mathbb{A}_0$ be an open nonempty subscheme of the affine line and $\mathcal{F}_0$ a smooth τ-pure sheaf of weight w on U_0. Let ρ be the corresponding representation of the fundamental group $\pi_1(U_0, \overline{a})$ on the finite dimensional $\overline{\mathbb{Q}}_l$-vectorspace V. Assume

(1) $\mathcal{F}_0$ is geometrically irreducible smooth etale $\mathbb{Q}_l$-sheaf, i.e. the representation $\rho\big|_{\pi_1(U,\overline{a})}$ is irreducible.
(2) $\mathcal{F}_0$ is geometrically nonconstant, i.e. the representation $\rho\big|_{\pi_1(U,\overline{a})}$ is nontrivial.
(3) $\mathcal{F}_0$ is unramified at the point ∞, i.e. ρ factorizes over a representation of $\pi_1(U_0 \cup \infty, \overline{a})$ on V.

To prove Theorem I.6.1 we have seen, that it will now completely suffice to prove in the situation above the

Claim 6.3 The eigenvalues α of the Frobenius homomorphism $F : H^1_c(U, \mathcal{F}) \to H^1_c(U, \mathcal{F})$ satisfy

$$|\tau(\alpha)|^2 \leq q^{w+1} .$$

The proof of claim I.6.3 is the main step of the proof of the Weil conjectures in the curve case. For the proof of I.6.3, following Laumon, we will use the Fourier transform as defined in §5. The Fourier transform $T_\psi(\mathcal{G}_0)$ of the sheaf

$$\mathcal{G}_0 = j_{0!}(\mathcal{F}_0)$$

will be shown to have the following properties

(a) $T_\psi(\mathcal{G}_0)$ is a sheaf, i.e. the complex $T_\psi(\mathcal{G}_0)$ is concentrated in degree 0.
(b) There are no sections with compact support, i.e. $H^0_c(\mathbb{A}, T_\psi(\mathcal{G}_0)) = 0$.
(c) The sheaf $T_\psi(\mathcal{G}_0)$ is a τ-mixed sheaf.

For the moment assume properties a)–c) to be true.

Then $T_\psi(\mathcal{G}_0)$ is a sheaf by (a). Hence we get from Theorem I.5.5

$$T_\psi(\mathcal{G}_0)_0 = H^1_c(\mathbb{A}, \mathcal{G}) = H^1_c(U, \mathcal{F}) .$$

Obviously $w = w(\mathcal{G}_0)$. Therefore the fundamental claim I.6.3 is equivalent to the statement, that the Fourier transform shifts upper weights at most by 1

$$\boxed{w(T_\psi(\mathcal{G}_0)) \leq w(\mathcal{G}_0) + 1 .}$$

To prove this recall: For a sheaf $\mathcal{K}_0$ on a smooth curve Y_0 we have defined the *upper weights*

$$w(\mathscr{K}_0) = \sup_{y\in|Y_0|} \sup_{\alpha} log(|\tau(\alpha)|^2)/log(N(y)) ,$$

where α runs over the eigenvalues of the Frobenii F_y, $y \in |Y_0|$ for all stalks of $\mathscr{K}_0$ (Definition I.2.3). We also defined the L^2-*norm*

$$\|\mathscr{K}_0\| = \sup \{\rho \mid \limsup_m \| f^{\mathscr{K}_0}\|_m^2 \cdot q^{-m(\rho+dim(Y_0))} > 0\}$$

(Definition I.2.14). For τ-mixed sheaves $\mathscr{K}_0$ on $\mathbb{A}_0$ with the property

$$H_c^0(\mathbb{A}, \mathscr{K}) = 0$$

we have shown in Theorem I.2.16(2) that

$$\boxed{\|\mathscr{K}_0\| = w(\mathscr{K}_0) .}$$

By the properties (a)-(c) stated above both $\mathscr{K}_0 = \mathscr{G}_0$ and $\mathscr{K}_0 = T_\psi(\mathscr{G}_0)$ are τ-mixed and satisfy $H_c^0(\mathbb{A}, \mathscr{K}) = 0$. So the last observation applied to both sheaves allows to restate the *Plancherel formula* (Theorem I.5.7)

$$\boxed{\|T_\psi(\mathscr{G}_0)\| = \|\mathscr{G}_0\| + 1}$$

in terms of the upper weights

$$w(T_\psi(\mathscr{G}_0)) = w(\mathscr{G}_0) + 1 .$$

This proves the desired inequality, hence claim I.6.3. □

In order to complete the proof of claim I.6.3 and Theorem I.6.1 we still have to verify the properties (a), (b) and (c) for the Fourier transform $T_\psi(\mathscr{G}_0) = T_\psi(j_{0!}\mathscr{F}_0))$ of the sheaf $\mathscr{G}_0 = j_{0!}\mathscr{F}_0$.

Property (a). Let $x \in \mathbb{A}_0(\kappa_n) = \kappa_n \subset \mathbb{A}(k) = k$ be a geometric point. We have to show (using Theorem I.5.5)

$$H_c^0(\mathbb{A}, j_!(\mathscr{F}) \otimes_{\overline{\mathbb{Q}}_l} \mathscr{L}(\psi_x)) = H_c^0(U, \mathscr{F} \otimes_{\overline{\mathbb{Q}}_l} \mathscr{L}(\psi_x)) = 0$$

$$H_c^2(\mathbb{A}, j_!(\mathscr{F}) \otimes_{\overline{\mathbb{Q}}_l} \mathscr{L}(\psi_x)) = H_c^2(U, \mathscr{F} \otimes_{\overline{\mathbb{Q}}_l} \mathscr{L}(\psi_x)) = 0 .$$

But $H_c^0(U, \mathscr{F} \otimes_{\overline{\mathbb{Q}}_l} \mathscr{L}(\psi_x))$ vanishes, because U is affine and $\mathscr{F}$ is smooth.

For the second assertion we consider the representation $\rho \otimes \psi_x$ on the vectorspace V attached to the sheaf $\mathscr{F}_0 \otimes_{\overline{\mathbb{Q}}_l} \mathscr{L}_0(\psi_x)$. Poincare duality implies

$$H_c^2(U, \mathscr{F} \otimes_{\overline{\mathbb{Q}}_l} \mathscr{L}(\psi_x)) = \big(V_{\rho\otimes\psi_x|\pi_1(U,\overline{a})}\big)(-1) ,$$

where $\left(V_{\rho\otimes\psi_x}\big|_{\pi_1(U,\overline{a})}\right)$ is the largest factor space of V with trivial action of $\pi_1(U,\overline{a})$ with respect to the representation $\rho\otimes\psi_x$. Suppose now $H_c^2(U,\mathscr{F}\otimes_{\overline{\mathbb{Q}}_l}\mathscr{L}(\psi_x))\neq 0$. This will lead to a contradiction. According to assumption (1) V is an irreducible $\pi_1(U,\overline{a})$ module. Therefore $\rho\otimes\psi_x$ would have to be the trivial representation of $\pi_1(U,\overline{a})$ on V. For $x=0$, hence $\rho\otimes\psi_x=\rho$, this contradicts assumption (2); $\mathscr{F}_0$ namely was supposed not to be geometrically constant. Let $x\neq 0$. Then the character ψ_x is nontrivial. On the other hand ρ is unramified at ∞ by assumption (3). $\rho\otimes\psi_x$ geometrically constant, hence unramified at ∞, would therefore imply, that the nontrivial character ψ_x is unramified at ∞, i.e. factorizes over

$$\pi_1(\mathbb{P}^1_0,\overline{a})=Gal(k/\kappa)\ .$$

This contradicts the geometric irreducibility of the Artin-Schreier covering (§5).

Property c) of the Fourier Transform $T_\psi(j_{0!}(\mathscr{F}_0))$. Let b be an element of $\overline{\mathbb{Q}}_l^{\,*}$ such that

$$\tau(b)=q^w\ .$$

Then, using the notation of Definition I.5.3, the Weil sheaf

$$\mathscr{H}_0=\pi^{2*}(j_{0!}(\mathscr{F}_0))\otimes_{\overline{\mathbb{Q}}_l}m^*(\mathscr{L}_0(\psi))\oplus\pi^{2*}(j_{0!}(\mathscr{F}_0^\vee))\otimes_{\overline{\mathbb{Q}}_l}m^*(\mathscr{L}_0(\psi^{-1}))\otimes_{\overline{\mathbb{Q}}_l}\mathscr{L}_b$$

is τ-real. (Compare Lemma I.2.11). But

$$\begin{aligned}&\mathrm{R}^\nu\pi^1_!\left(\pi^{2*}(j_{0!}(\mathscr{F}_0^\vee))\otimes_{\overline{\mathbb{Q}}_l}m^*(\mathscr{L}_0(\psi^{-1}))\otimes_{\overline{\mathbb{Q}}_l}\mathscr{L}_b\right)\\&=\mathrm{R}^\nu\pi^1_!\left(\pi^{2*}(j_{0!}(\mathscr{F}_0^\vee))\otimes_{\overline{\mathbb{Q}}_l}m^*(\mathscr{L}_0(\psi^{-1}))\right)\otimes_{\overline{\mathbb{Q}}_l}\mathscr{L}_b\\&=0\quad\text{for }\nu\neq 1.\end{aligned}$$

Namely, as well as $T_\psi(j_{0!}(\mathscr{F}_0))$, the complex $T_{\psi^{-1}}(j_{0!}(\mathscr{F}_0^\vee))$ is represented by a sheaf, i.e. its cohomology is concentrated in degree zero. For that use the same argument as for the proof of property a). Therefore

$$\mathrm{R}^\nu\pi^1_!(\mathscr{H}_0)=0\quad\text{for }\nu\neq 1\ .$$

$\mathscr{H}_0$ being τ-real, the Grothendieck formula for the L-series of a sheaf, applied to the fibers of the map π^1, shows that the Weil sheaf $\mathrm{R}^1\pi^1_!(\mathscr{H}_0)$ is again τ-real. Then according to Theorem I.4.3(1) $\mathrm{R}^1\pi^1_!(\mathscr{H}_0)$ is also τ-mixed. Therefore the direct summand $T_\psi(j_{0!}(\mathscr{F}_0))$ of $\mathrm{R}^1\pi^1_!(\mathscr{H}_0)$ is τ-mixed.

Property b) of the Fourier Transform $T_\psi(j_{0!}(\mathscr{F}_0))=\mathscr{H}_0$. The Fourier inversion formula I.5.8 implies

$$\begin{aligned}H^0_c(\mathbb{A},\mathscr{H})&=\mathscr{H}^{-1}(T_{\psi^{-1}}(\mathscr{H}_0))_0\\&=\mathscr{H}^{-1}(T_{\psi^{-1}}\circ T_\psi(j_{0!}(\mathscr{F}_0)))_0\end{aligned}$$

$$= \mathscr{H}^{-1}(j_{0!}(\mathscr{F}_0)(-1))_0 = 0 .$$

This is obvious, because $j_{0!}(\mathscr{F}_0)(-1)$ is a sheaf, i.e a complex with cohomology concentrated in degree 0.

All three properties of the Fourier transformed sheaf $T_\psi(j_{0!}(\mathscr{F}_0))$ have now been verified. So Theorem I.6.1 has been proved. □

Corollary 6.4 *Let $\mathscr{F}_0$ be a Weil sheaf on an algebraic scheme X_0 over κ. Assume*

$$dim(X_0) \leq 1 .$$

Let $\mathscr{F}_0$ be τ-mixed with highest weight equal to β. Then for all integers i $(= 0, 1, 2)$ the highest τ-weight β_i appearing in $H^i_c(X_0, \mathscr{F})$ satisfies

$$\beta_i \leq \beta + i .$$

Proof. We can assume X_0 to be reduced. By Theorem I.4.2 we can also restrict ourselves to the case of an etale sheaf $\mathscr{F}_0$.

Let us first discuss the case, when X_0 is a smooth curve and where $\mathscr{F}_0$ is a smooth sheaf. After some base field extension X_0 decomposes into geometrically irreducible connected components. We therefore can assume X_0 to be geometrically irreducible. The irreducible constituents of $\mathscr{F}_0$ are then τ-pure by Theorem I.2.8(3). On the other hand the functor H^i_c is half exact. So $\mathscr{F}_0$ can be taken to be τ-pure of weight β without loss of generality. X_0 is an open dense subscheme of a smooth projective curve $\overline{X}_0$ over κ.

$$j_0 : X_0 \hookrightarrow \overline{X}_0 .$$

Consider the short exact sequence

$$0 \longrightarrow j_{0!}(\mathscr{F}_0) \longrightarrow j_{0*}(\mathscr{F}_0) \longrightarrow \mathscr{H}_0 \longrightarrow 0 .$$

$\mathscr{H}_0$ is concentrated on the finite complement S_0 of X_0 in $\overline{X}_0$. Therefore

$$H^i(X, \mathscr{H}) = 0 \qquad i \geq 1 .$$

According to the semicontinuity of weights I.2.5 the τ-weights α of the stalks of $\mathscr{H}_0$ are controlled by the weight of $\mathscr{F}_0$

$$\alpha \leq \beta .$$

From Theorem I.6.1 and the long exact cohomology sequence one obtains the desired estimate for cohomology with compact support. The curve case has now been proven.

The general case. Everything is trivial in case $dim(X_0) = 0$. So let us assume $dim(X_0) > 0$. Then we can find an open smooth curve

$$j_0 : U_0 \hookrightarrow X_0$$

with finite complement

$$i_0 : S_0 \hookrightarrow X_0$$

such that $j_0^*(\mathscr{F}_0)$ is smooth on U_0.

The desired estimates were already shown for the sheaves $j_0^*(\mathscr{F}_0)$ and $i_0^*(\mathscr{F}_0)$. The corresponding estimate for $\mathscr{F}_0$ is derived from considering the long exact cohomology sequence attached to the short exact sequence

$$0 \longrightarrow j_{0!}(j_0^*(\mathscr{F}_0)) \longrightarrow \mathscr{F}_0 \longrightarrow i_{0*}(i_0^*(\mathscr{F}_0)) \longrightarrow 0\,.$$

□

I.7 The Weil Conjectures for a Morphism (General Case)

Theorem 7.1 (Deligne) *Suppose given a morphism*

$$f_0 : X_0 \to Y_0$$

of algebraic varieties over κ *and a* τ*-mixed sheaf* $\mathscr{F}_0$ *on* X_0*, whose largest* τ*-weight is* β*. Then the sheaves* $R^i f_{0!}(\mathscr{F}_0)$ *are* τ*-mixed for all* $i = 0, 1, 2, \ldots$ *Let* β_i *denote the largest* τ*-weight of* $R^i f_{0!}(\mathscr{F}_0)$*. Then*

$$\beta_i \leq \beta + i\,.$$

Remark 7.2 Let $\mathscr{F}_0$ be τ-mixed for all τ (with weights depending on τ), i.e. mixed in the sense of Definition I.2.1(4). Then all direct images $\mathrm{R}^i f_{0!}(\mathscr{F}_0)$ are τ-mixed for all τ. Using I.2.2 and Theorem I.2.8(3) one can show, that all these direct images have finite filtrations, whose factor sheaves are of the twisted form $\mathscr{G}_0 \otimes_{\overline{\mathbb{Q}}_l} \mathscr{L}_b$ for pure sheaves $\mathscr{G}_0$. (For the definition of $\mathscr{L}_b$ see Proposition I.1.3). Deligne furthermore proves [Del], that all direct images of a mixed sheaf are again mixed sheaves.

If $\mathscr{F}_0$ is a τ-pure sheaf of weight β, then Deligne proves [Del], that all the weights in $\mathrm{R}^i f_{0!}(\mathscr{F}_0)$ differ from $\beta + i$ only by an integer. We will discuss these refinements in §9 of this chapter, and in the section on mixed complexes in Chap. II.

The formulation in terms of mixed complexes and derived categories in Chap. II turns out to be the most flexible and the most elegant one, although the statements itself are immediate consequences of the corresponding statements for sheaves.

Using Poincare duality, one derives from Theorem I.7.1

Corollary 7.3 *If* $f_0 : X_0 \to Y_0$ *is a smooth and proper morphism and* $\mathscr{F}_0$ *a smooth* τ*-pure sheaf of weight* β*, then all the smooth sheaves* $R^i f_{0!}(\mathscr{F}_0)$ *are* τ*-pure of weight* $\beta + i$*.*

Proof of Theorem I.7.1. The statement of Theorem I.7.1 is proven by noetherian induction, which reduces everything to the special case of a smooth relative curve. In fact, this will be accomplished by the following reduction steps:

It is enough to prove the theorem after making a finite base field extension.

The assertion of the theorem is trivial for quasifinite morphisms f_0.

Suppose given a short exact sequence $0 \to \mathscr{F}_0 \to \mathscr{G}_0 \to \mathscr{H}_0 \to 0$. If the theorem is true for $\mathscr{F}_0$ and $\mathscr{H}_0$, then also for $\mathscr{G}_0$.

Let $U_0 \hookrightarrow X_0$ be an open subscheme and $S_0 \hookrightarrow X_0$ the closed complement. If the theorem is true for $\mathscr{F}_0 | U_0$ and the morphism $U_0 \to Y_0$ and also true for $\mathscr{F}_0 | S_0$ and the morphism $S_0 \to Y_0$, then it is true also for $\mathscr{F}_0$ and the morphism $X_0 \to Y_0$.

Let $f_0 = g_0 \circ h_0$ be a composition of two morphisms. If the theorem is true for g_0 and h_0, then it is true for f_0. This is a consequence of the spectral sequence for higher direct images with compact support.

One can replace X_0 and Y_0 by X_{0red} and Y_{0red}.

Using these remarks and noetherian induction one reduces the proof to the following special case $f_0 : X_0 \to Y_0$, where X_0 and Y_0 are smooth affine connected algebraic varieties and where f_0 is surjective and of relative dimension less or equal 1. Furthermore the sheaf $\mathscr{F}_0$ can be assumed to be smooth and irreducible, hence in particular τ-pure.

For a further reduction we use a simple result on curves: Let C be a finitely generated scheme over a field K with $dim(C) \leq 1$. Then there exists a finite field extension K' of K and an open subscheme U of the scheme $C' = \left(C \times_{Spec(K)} Spec(K')\right)_{red}$ with the following properties:

The complement of U in C' is finite and all connected components of U are geometrically irreducible smooth curves over K'. Mutatis mutandis this result carries over to the relative curve case $f_0 : X_0 \to Y_0$. Instead of the finite field extension one has to consider a finite ramified covering of a suitably small open nonempty subset of Y_0.

So it is finally enough to prove the theorem under the following further restrictions: Assume

$$f_0 : X_0 \to Y_0$$

is a smooth affine morphism, whose fibers are nonempty affine, geometrically irreducible smooth curves. Furthermore the sheaf $\mathscr{F}_0$ can assumed to be smooth and τ-pure of weight β.

The estimates for the τ-weights of the sheaves $R^i f_{0!}(\mathscr{F}_0)$ follow then via the basechange theorem from I.6.4, applied to the fibers of the morphism f_0. It only remains to show, that the sheaves $R^i f_{0!}(\mathscr{F}_0)$ are τ-mixed. For that it is enough to consider the cases $i = 1, 2$, because in all other cases the direct image vanishes, f_0 being affine. Lemma I.2.11 allows to assume, that $\mathscr{F}_0$ is τ-real. Because of Theorem I.4.3 it is then enough to show, that the sheaves $R^i f_{0!}(\mathscr{F}_0)$ are also τ-real for $i = 1$ and $i = 2$.

Let $y \in |Y_0|$ be a closed point and $\overline{y}$ a geometric point of Y_0 over y with values in k, let $\kappa(y) \subset k$ be the residue class field of y and let $C_0 = f_0^{-1}(y) = X_0 \times_{Y_0} Spec(\kappa(y))$ be the fiber of f_0 over the point y. Then C_0 is a smooth, affine, geometrically irreducible curve over $\kappa(y)$. The field $\kappa(y)$ contains $N(y) = q^{d(y)}$ elements. Let

$$C = C_0 \times_{\kappa(y)} k = X_0 \times_{Y_0} Spec(k) .$$

Then

$$(R^i f_{0!}(\mathscr{F}_0))_{\overline{y}} = H^i_c(C, \mathscr{F}) .$$

We have to show, that the polynomials $\tau det(1 - Ft, H^i_c(C, \mathscr{F}))$ are τ-real for $i = 1, 2$.

$\mathscr{F}_0$ τ-real implies by Grothendieck's formula for the L-series of a sheaf on the curve C_0, that the power series

$$\frac{\tau det(1 - Ft, H^1_c(C, \mathscr{F}))}{\tau det(1 - Ft, H^2_c(C, \mathscr{F}))}$$

has real coefficients in the variable t. The cohomology group $H^2_c(C.\mathscr{F})$ is easily computed via Poincare duality as in §4. Therefore the zeros α of the denominator satisfy

$$|\alpha| = N(y)^{-(\beta+2)/2} ,$$

because $\mathscr{F}_0|C_0$ is τ pure of weight β. According to I.6.4 the zeros α of the numerator satisfy

$$|\alpha| \geq N(y)^{-(\beta+1)/2} .$$

The polynomials in the numerator and denominator are therefore without common divisor. This implies, that both are real polynomials, because they have leading coefficients 1 and the quotient is real. This completes the proof of Theorem I.7.1. □

I.8 Some Linear Algebra

We consider the semidirect product G of the compact group $G_0 = \mathbb{Z}_l$ and the discrete cyclic group generated by an element σ

$$G = G_0 \rtimes \langle\sigma\rangle \quad , \quad G_0 = \mathbb{Z}_l ,$$

where we assume

$$\sigma t \sigma^{-1} = q \cdot t \quad , \quad t \in G_0$$

for some natural number $q \neq 1$ not divisible by the prime l. Let ρ be a continuous representation of the locally compact group G on a finite dimensional E-vectorspace V. Here $E \subset \overline{\mathbb{Q}}_l$ denotes a finite extension field of the field $\mathbb{Q}_l$. We define twisted representations $V(i)$ as the tensor product of the representation ρ with the character $\psi(\sigma^i t) = q^i$ of G.

Let the situation be as above. Then we have

Lemma 8.1 *For all $g \in G_0$ the matrix $\rho(g)$ is quasi-unipotent, i.e. all eigenvalues of this automorphism of V are roots of unity (whose order is bounded in terms of q and $dim_E(V)$).*

Proof. Let λ be an eigenvalue of $\rho(g)$. Then

$$\rho(\sigma)\rho(g)\rho(\sigma)^{-1} = \rho(g)^q .$$

Therefore λ^q is also an eigenvalue. Since V is finite dimensional, this implies that λ is a root of unity. Its order can be bounded in terms of q and the dimension $dim_E(V)$. □

Notation. Let G_0' be the subgroup of elements having unipotent action on V. Let a be an integer such that $a \cdot G_0 \subset G_0'$. Such an integer a exists by Lemma I.8.1.

Let t be a topological generator of the group G_0. Since $T^a = \rho(t)^a$ is unipotent on V, there exists an integer $b \leq dim_E(V)$, such that $(T^a - 1)^b = 0$. The nilpotent endomorphism

$$N = \text{"}a^{-1}log(T^a)\text{"}$$

defined by

$$N = -a^{-1}\sum_{n\geq 1}(1 - T^a)^n/n$$

is independent of the choice both of a.

In particular there exists a uniquely defined nilpotent endomorphism N of V, such that

$$\rho(\sigma)N\rho(\sigma)^{-1} = q \cdot N ,$$

and a subgroup G_0' of finite index a in $G_0 = \mathbb{Z}_l$, such that

$$\rho(t) = exp(tN) = \sum t^i N^i/i! \quad , \quad t \in G_0'$$

holds for all t in G_0'.

If we replace ρ by the representation $\rho \otimes \rho$ on the tensor product $V \otimes V$, the corresponding nilpotent endomorphism is $N \otimes id + id \otimes N$. Similarly the nilpotent endomorphism for the contragredient representation is $-{}^tN$ (minus the transposed endomorphism).

The σ-Decomposition. For $\alpha \in \overline{\mathbb{Q}}_l^*$ let $V_\alpha \subset V$ denote the subspace of all vectors $v \in V$, which are annihilated by $\left(\rho(\sigma) - \alpha^{-1}\right)^i$ for some integer $i \geq 1$. We have $V = \bigoplus_\alpha V_\alpha$. The equation $\left(\rho(\sigma) - q\alpha^{-1}\right)^i N = Nq^i\left(\rho(\sigma) - \alpha^{-1}\right)^i$ implies

$$NV_\alpha \subset V_{q^{-1}\alpha} .$$

For the representation $\rho \otimes \rho$ on $V \otimes V$ we have

$$V_\alpha \otimes V_\beta \subset V_{\alpha\beta} ,$$

since $\left((\rho(\sigma) - \alpha^{-1}) \otimes \rho(\sigma) + \alpha^{-1} \otimes (\rho(\sigma) - \beta^{-1})\right)^{2i}$ annihilates $v \otimes w$, provided $(\rho(\sigma) - \alpha^{-1})^i$ annihilates v and $(\rho(\sigma) - \beta^{-1})^i$ annihilates w.

The N-Filtration. For a nilpotent endomorphism N on the vectorspace V, there exists a unique finite increasing filtration $F_\bullet$ on V with the properties

$$N\big(F_i(V)\big) \subset F_{i-2}(V) \quad , \quad N^i : Gr_i(V) \cong Gr_{-i}(V) \quad i \geq 0 .$$

Here $Gr_i(V) = F_i(V)/F_{i-1}(V)$.

Existence and uniqueness of such a filtration is proved by induction on the minimal number m such that $N^{m+1} = 0$.

For $m = 0$ one endows V with the trivial filtration, i.e. $Gr_i(V) = 0$ for $i \neq 0$. In general put $F_m(V) = V$ and $F_{-m-1}(V) = 0$. This is forced, since $N^\nu = 0$ for $\nu \geq m+1$ has to induce an isomorphism between $Gr_\nu(V)$ and $Gr_{-\nu}(V)$. Therefore $Gr_\nu(V) = 0$ for $|\nu| \geq m+1$, hence

$$F_{-m-1}(V) = 0 \quad , \quad F_m(V) = V .$$

In order, that N^m induces an isomorphism between $Gr_m(V)$ and $Gr_{-m}(V)$, we have to put $F_{m-1}(V) = Ker(N^m)$ and $F_{-m}(V) = Im(N^m)$. But then $F_{m-1}(V)/F_{-m}(V)$ is annihilated by N^m. Hence a corresponding filtration uniquely exists on this quotient by the induction assumption. The inverse image of this filtration in $F_{m-1}(V)$ defines the desired filtration $F_\bullet(V)$ on V. □

Existence and uniqueness are obtained in the same way in the following general situation: Let **A** be a noetherian and artinian abelian category, let $V \in ob(\mathbf{A})$ and $N : V \to V$ be a morphism such that $N^{m+1} = 0$.

Let us return to our original situation. Consider the dual space V^*. It is clear that the filtration of the contragredient representation ρ^* on the dual space V^* is then given by $F_i(V^*) = F_{-i-1}(V)^\perp$.

The representation ρ respects the filtration $F_\bullet(V)$. We obtain an induced representation of G, actually of the factor group G/G_0', on $Gr_\bullet(V)$ such that $N^i : Gr_i(V)(i) \xrightarrow{\sim} Gr_{-i}(V)$. For simplicity from now on assume $G_0 = G_0'$.

Definition 8.2 *The primitive part $P_i(V)$ of $Gr_i(V)$ is defined to be the kernel of the induced homomorphism*

$$N : Gr_i(V) \to Gr_{i-2}(V) .$$

ρ induces a representation of $G/G_0 = \langle\sigma\rangle$ on $P_i(V)$.

Recall the definition of the twisted representations $V(i)$, where σ acts with an additional factor q^i. Then the following lemma is now more or less evident from the defining properties of the filtration, considered above.

Lemma 8.3

(1) $P_i(V) = 0$ for $i > 0$.
(2) $Gr_i(V) \cong \bigoplus_{j \geq |i|, j \equiv i(2)} P_{-j}(-\frac{1}{2}(j+i))$ as G/G_0-module.

(3) $Gr_i(V^*) \cong Gr_{-i}(V)^*$ *and* $Gr_{-i}(V) \cong Gr_i(V)(i)$ *as* G/G_0*-modules. Especially* $P_{-i}(V^*) \cong P_{-i}(V)^*(i)$.

(4) Let $i \geq 0$. *The map* $N^i : F_j(V) \to F_{j-2i}(V)$ *is surjective for all* $j \leq i$.

(5) For the induced filtration on $ker(N) \subset V$ *we have* $Gr_i(ker(N)) \cong P_i(V)$.

One obtains the following picture

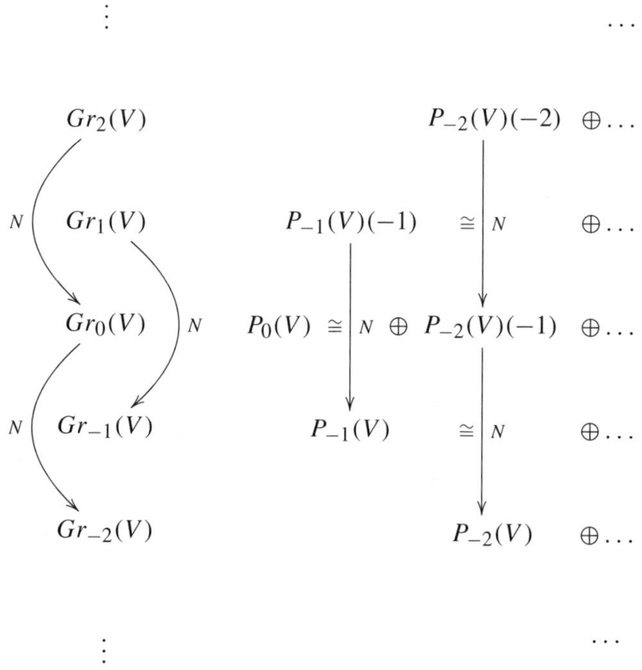

All vertical morphisms are isomorphisms induced by N. Each $Gr_i(V)$ is isomorphic to the direct sum of the rows of the diagram, i.e. $Gr_0(V) \cong P_0(V) \oplus P_{-2}(V)(-1) \oplus P_{-4}(V)(-2) \oplus \ldots$.

Proof of the Lemma. Part (1) is obvious from the fact that N^i defines an isomorphism between $Gr_i(V)$ and $Gr_{-i}(V)$ for $i \geq 0$. Part (2): It is enough to consider $Gr_{-i}(V)$ for $i \geq 0$. Proceed by induction with respect to i starting with $Gr_{-m-1}(V) = P_{-m-1}(V) = 0$ and $Gr_{-m}(V) = P_{-m}(V)$. Obviously $Gr_{-i}(V) = P_{-i}(V) \oplus N^{i+1}Gr_{i+2}(V)$. Apply the induction hypothesis to the right summand, which is isomorphic to $Gr_{-i-2}(V)$. The statement of part (3) is obvious. Part (4) is proved by passing to the associated graded, where the corresponding assertion is trivial. See the diagram above. (5) asserts, that $F_{-i}(V) \cap ker(N)$ maps onto $P_{-i}(V) = ker(N) : Gr_{-i}(V) \to Gr_{-i-2}(V)$. In other words, one has to show: An element $v \in F_{-i}(V)$ with $Nv \in F_{-i-4}(V)$ can be modified by an element $w \in F_{-i-2}(V)$, such that $N(v - w) = 0$. But this follows from property (4). □

Lemma 8.4 *For every eigenvalue α of σ^{-1} on $Ker(N)$ in V assume $|\tau(\alpha)| \leq 1$. Assume the corresponding statement also for the representations on the tensor products $V \otimes V$ resp. $V^* \otimes V^*$. Then every eigenvalue α of σ^{-1} on $Gr_i(V)$ has the property*

$$|\tau(\alpha)|^2 = q^i \ .$$

Proof. Using Lemma I.8.3(1)–(2) it is enough to consider eigenvectors of $\rho(\sigma^{-1})$, which lie in some $P_{-i}(V), i \geq 0$. By I.8.3(5) such a vector can be represented by a vector $v \in F_{-i}(V) \cap Ker(N)$. Projecting v onto V_α, we can furthermore assume

$$v \in V_\alpha \ \text{ and } \ N(v) = 0 \ .$$

By I.8.3(4) one can find a vector $w \in F_i(V)$ such that $N^i(w) = v$. Projecting onto $V_{q^i\alpha}$, we can assume

$$w \in V_{q^i\alpha} \ \text{ and } \ N^i(w) = v \ .$$

Then

$$u = \sum_{j=0}^{i} (-1)^j N^j(w) \otimes N^{i-j}(w)$$

is a nonvanishing eigenvector of σ^{-1} in $(V \otimes V)_{q^i\alpha^2}$. One immediately checks

$$(N \otimes id + id \otimes N)u = 0 \ .$$

The assumptions of Lemma I.8.4 imply $|\tau(q^i\alpha^2)| \leq 1$, hence $|\tau(\alpha)|^2 \leq q^{-i}$. The corresponding inequality holds for the eigenvalues of σ^{-1} on $P_{-i}(V^*) \cong P_{-i}(V)^*(i)$. $\alpha^{-1}q^{-i}$ is such an eigenvalue. Therefore $|\tau(\alpha^{-1}q^{-i})|^2 \leq q^{-i}$ or $|\tau(\alpha)|^2 \geq q^{-i}$. This proves the required equality $|\tau(\alpha)|^2 = q^{-i}$ on $P_{-i}(V)$. □

I.9 Refinements (Local Monodromy)

Let X_0 be a smooth geometrically irreducible curve over κ, $\bar{s}$ a geometric point of X_0 over κ with values in k, s the underlying closed point, let

$$j_0 : U_0 \hookrightarrow X_0$$

be the open complement of s in X_0 and $\mathscr{F}_0$ a τ-pure smooth sheaf on U of weight β.

In this situation the weights $w(j_{0*}(\mathscr{F}_0)_{\bar{s}}) \leq \beta$ of the direct image of $\mathscr{F}_0$ in the point of degeneration s can be estimated above by β, according to Lemma I.2.5. The linear algebra trick of Lemma I.8.4 of the last section will then imply refined properties for these weights.

Let K be the quotient field of the henselization of the local ring of X_0 in s, L the quotient field of the strict henselization with respect to s and $\overline{L}$ the separable closure of L. We look at the geometric point

$$\eta : Spec(\overline{L}) \to X_0 .$$

The Weil group $W(\overline{L}/K)$ is the inverse image of the Weil group $W(k/\kappa)$ with respect to the natural homomorphism

$$Gal(\overline{L}/K) \to Gal(k/\kappa) .$$

The ramification group I of X_0 in s is $Gal(\overline{L}/L)$. $W(\overline{L}/K)$ is the semidirect product of $Gal(\overline{L}/L)$ and the cyclic group $W(k/\kappa)$. The wild ramification group P is the p-Sylow group of $Gal(\overline{L}/L)$, p being the characteristic of k. It is a normal subgroup with factor group $Gal(\overline{L}/L)/P$, which is the tame ramification group

$$I^{tame} = I/P .$$

This tame ramification group is canonically isomorphic to

$$I^{tame} \cong \prod_{l \neq p, l \ prime} \mathbb{Z}_l(1)$$

or uncanonically isomorphic to

$$\prod_{l \neq p, l \ prime} \mathbb{Z}_l .$$

Its l-primary part is

$$\mathbb{Z}_l(1) \cong \mathbb{Z}_l .$$

The sheaf $\mathscr{F}_0$ is given by a continuous representation of $W(U_0, \eta)$ in the finite dimensional representation space V over $\overline{\mathbb{Q}}_l$. This defines a continuous representation of $Gal(\overline{L}/K)$ on V. The image of $Gal(\overline{L}/L)$ in $Gl(V)$ contains a pro-l subgroup of finite index. Passing to a finite covering of X_0, ramified in $\overline{s}$, allows to get rid of the wild ramification part and the tame part which is not pro-l, i.e. the representation of $Gal(\overline{L}/L)$ then factorizes over the pro-l part of the tame ramification group. In that case the representation of $W(\overline{L}/K)$ factorizes over the representation ρ of a semidirect product

$$G = \mathbb{Z}_l \rtimes \langle \sigma \rangle \quad , \quad W(k/\kappa(s)) = \langle \sigma \rangle .$$

Here σ denotes the arithmetic Frobenius element. We have for $t \in G_0 = \mathbb{Z}_l$

$$\sigma t \sigma^{-1} = qt \quad , \quad q = N(s) .$$

By some further modification of X_0 we can assume (see I.8.1), that there is a nilpotent endomorphism N of V such that

$$\rho(t) = exp(tN) \quad , \quad t \in G_0 = \mathbb{Z}_l .$$

Furthermore

$$j_{0*}(\mathscr{F}_0)_{\overline{s}} = V^{Gal(\overline{L}/L)} = ker(N) .$$

For the moment let us assume $\beta = 0$. Applying Lemma I.2.5 (weights are semi-continuous) to the sheaf $\mathscr{F}_0$, its dual and their tensor products, we can verify the assumptions of I.8.4. This determines the absolute values of the eigenvalues of σ^{-1} on the graded pieces of V with respect to the N-filtration. If $\beta \neq 0$, then twist $\mathscr{F}_0$ by a smooth sheaf on X_0 of rank 1 of τ-weight $-\beta$. This proves the following result

Lemma 9.1 *Let X_0 be a smooth geometrically irreducible curve over κ, S_0 a finite closed subset and $\mathscr{F}_0$ a smooth sheaf on $U_0 = X_0 \setminus S_0$. Let*

$$j_0 : U_0 \hookrightarrow X_0 .$$

$$\mathscr{H}_0 = j_{0*}(\mathscr{F}_0)/j_{0!}(\mathscr{F}_0) .$$

(1) Assume $\mathscr{F}_0$ is a τ-pure sheaf of weight β. Then the τ-weights of the τ-mixed sheaf $\mathscr{H}$ on the finite set S_0 differ from β only by an integer less or equal 0.
(2) Assume $\mathscr{F}_0$ to be pure (Definition I.2.1(3)). Then $\mathscr{H}_0$ is mixed (Definition I.2.1(4)).

By Theorem I.6.1 this implies

Corollary 9.2 *Suppose given a sheaf $\mathscr{F}_0$ on an algebraic scheme X_0 of dimension less or equal one. Then:*

(1) Let $\mathscr{F}_0$ be τ-pure of weight β. Then the τ-weights of $H^i_c(X, \mathscr{F})$ are of the form

$$\beta + i - n \quad , \quad n \in \mathbb{N},\ n \geqq 0 .$$

(2) Let $\mathscr{F}_0$ be mixed. Then $H^i_c(X, \mathscr{F})$ is mixed.

This corollary allows to strengthen Theorem I.7.1 to

Theorem 9.3 (Deligne) *Suppose*

$$f_0 : X_0 \to Y_0$$

is a morphism between algebraic varieties over κ and $\mathscr{F}_0$ is a sheaf on X_0, then the following holds:

(1) If $\mathscr{F}_0$ is τ-pure of weight β, then the τ-weights of $R^i f_{0!}(\mathscr{F}_0)$ are of the form

$$\beta + i - n \quad , \quad n \in \mathbb{N},\ n \geqq 0 .$$

(2) If $\mathscr{F}_0$ is mixed, then also $R^i f_{0!}(\mathscr{F}_0)$ is mixed.

Proof. One reduces the proof by noetherian induction – similar as for Theorem I.7.1 – to the case of a relative curve. Assertion (1) follows from Corollary I.9.2 and

Theorem I.7.1, applied to the fibers of the morphism f_0. For the proof of I.9.3(2) one uses Remark I.7.2. The sheaves $R^i f_{0!}(\mathscr{F}_0)$ are τ-mixed for all τ by Theorem I.7.1. Therefore each of theses sheaves has a finite filtration, such that all subquotients are of the form

$$\mathscr{G} \otimes \mathscr{L}_b$$

with τ-pure irreducible sheaves $\mathscr{G}_0$ and smooth sheaves $\mathscr{L}_b$ of rank 1. $\mathscr{L}_b$ is pullback of a smooth sheaf on $Spec(\kappa)$. See Proposition I.1.4. Applying Corollary I.9.2 to the fibers of the morphism one obtains for all nontrivial subquotient sheaves, that $\mathscr{L}_b$ is a pure sheaf. □

Permanence Properties. Our next aim is to describe the behaviour of τ-mixed (respectively mixed) sheaves with respect to the higher direct images in the case of a not necessarily proper map. (See also Deligne [Del], theorem 6.1.2).

Remark. In the following it is shown, that higher direct image sheaves of mixed sheaves are again mixed sheaves. Almost the same argument also implies, that the higher direct images of constructible sheaves with respect to morphisms between finitely generated schemes over a field κ are again constructible. (See [SGA $4\frac{1}{2}$]). The reader should be aware of the fact, that the constructibility for higher direct images is already used implicitly in the proof of results concerning mixedness. As the proof for both kinds of results are rather similar and in order not to repeat ourselves, we assume the result on constructibility to be known already. See also Appendix D. Under this hypothesis, let us now prove the corresponding result for mixedness.

Remark. A complex K_0 is called τ-mixed respectively mixed, if all its cohomology sheaves are τ-mixed respectively mixed sheaves.

Theorem 9.4 *Let $f_0 : X_0 \to Y_0$ be a morphism between algebraic schemes over κ and let K_0 be a τ-mixed respectively mixed complex on X_0. Then also the derived direct image complex $Rf_{0*}(K_0)$ in the derived category $D_c^b(X_0, \overline{\mathbb{Q}}_l)$ is a τ-mixed respectively mixed complex.*

Remark. A similar result is valid almost everywhere over $Spec(\mathbb{Z})$. In other words: the property of mixedness holds over a sufficiently small open nonempty subset of $Spec(\mathbb{Z})$.

For simplicity we restrict ourselves to the proof of the fact, that $\mathrm{R}f_{0*}K_0$ is mixed if K_0 is mixed. We can assume, and therefore will assume, that the underlying schemes are reduced.

The proof of Theorem I.9.4 is based on the following reduction to the case of an open immersion, using several reduction steps. This reduction uses that the Theorem I.9.4 already holds in the particular cases R1 and R2 mentioned below. To be more precise, let us make use of the followings auxiliary results:

R1 *The Proper Case.* The statement of Theorem I.9.4 obviously holds for proper maps f_0 by Theorem I.9.3(2), since then $\mathrm{R}^i f_{0!}(K_0) = \mathrm{R}^i f_{0*}(K_0)$.

R2 *The Smooth Case.* In this case Theorem I.9.4 follows from the relative Poincare duality Theorem II.7.1 by reduction to case R1. What we need is, that the duality map $\beta : \mathrm{R}f_*\mathrm{R}\mathcal{H}om(L, f^*(K)[2d](d)) \to \mathrm{R}\mathcal{H}om(\mathrm{R}f_!(L), K)$ defined in II §8 step 3) is an isomorphism. (For the notations and the proof of this see II §8). In fact we can apply this to cover Theorem I.9.4 in the following situation:

$$f_0 = g_0 \circ h_0 \,,$$

where g_0 is smooth and h_0 is purely inseparable and

$$K_0 = \mathcal{G}_0$$

is a sheaf such that $\mathcal{G}_0$ and the direct images $R^i f_!(\mathcal{G}_0^\vee)$ of the dual sheaf $\mathcal{G}_0^\vee$ are smooth.

R3 *Generic Points on the Base.* For a constructible sheaf $\mathcal{G}_0$ on X_0, there exists an open dense subscheme V_0 of Y_0 and an open dense subscheme U_0 of the inverse image of V_0 in X_0 such that for the induced map

$$\overline{f}_0 : U_0 \to V_0$$

and the restricted sheaf $\mathcal{G}_0|U_0$ the assumptions of R2 are satisfied. (This follows from the constructibility of the sheaves under consideration, since every finitely generated field extension is a purely inseparable extension of a separable finitely generated extension).

R4 *The Theorem is Local on Y_0.* So we can assume Y_0 to be affine. Using the spectral sequence attached to a finite open affine covering allows to reduce to the case, where X_0 is affine, hence quasiprojective. Furthermore this shows, that the map f_0 can be factorized into an open embedding followed by a proper map.

In view of Remark R4 – factorization into a proper and an open map – and Remark R1 Theorem I.9.4 is a consequence of the next lemma, since Theorem I.9.4 for an open embedding follows as the special case $S_0 = Spec(\kappa)$ of

Lemma 9.5 *Let $j_0 : X_0 \to Y_0$ be an open embedding*

$$\begin{array}{ccc} X_0 & \xrightarrow{\;j_0\;} & Y_0 \\ & \searrow \quad \swarrow & \\ & S_0 & \end{array}$$

over a finitely generated base scheme S_0 over κ and let $\mathcal{G}_0$ be a mixed sheaf on X_0. Then the restriction of the higher direct image sheaves $R^i j_{0*}(\mathcal{G}_0)$ to the open inverse image in Y_0 of some open dense subscheme S_0' of S_0, are mixed sheaves.

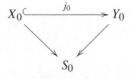

Lemma I.9.5 can be considered as a relative version of Theorem I.9.4 for open embeddings. It follows from the special case where $S_0 = Spec(\kappa)$. Its formulation has the advantage to allow an inductive proof. More precisely: It is obviously enough to proof the following partial statements (by induction on n).

Lemma 9.6 (n) *Same statement as in I.9.5 but under the additional assumption: Suppose that S_0 is an affine integral scheme, such that for the generic point η of S_0, the dimension of the fiber over η is less than n:*

$$dim(X_{0\eta}) \leq n \ .$$

Remark. If Lemma I.9.6(n) is true, we immediately get the assertion of Lemma I.9.6(n) also for locally closed embeddings. The assertion is local on Y_0, hence S_0 may be replaced by an open nonempty affine subscheme. So Y_0 can be assumed to be affine or, by embedding as an open subscheme of a projective scheme, then alternatively also to be projective over S_0. Under the latter assumption, the beginning of the induction – the case $n = 0$ – is obvious. Namely for $n = 0$ we can assume – possibly after shrinking S_0 – that X_0 is finite over S_0. Then j_0 is a finite, hence proper map, and we can apply Remark R1.

Lemma 9.7 *Let us make the assumptions of Lemma I.9.6(n). Then there exists, after shrinking S_0, an open subscheme V_0 of Y_0 which contains X_0, such that (a) the complement of V_0 in Y_0 is finite over S_0 (b) $Rj_{0*}(\mathscr{G}_0)|V_0$ is mixed on V_0.*

Proof. We may assume Y_0 to be affine. Embed Y_0 as a closed subscheme into some m-dimensional affine space over S_0

$$X_0 \overset{j_0}{\hookrightarrow} Y_0 \hookrightarrow \mathbb{A}_0 \times \cdots \times \mathbb{A}_0 \times S_0 \cong \mathbb{A}^m_{S_0} \ .$$

Consider the m projection maps

$$p_\nu : Y_0 \to \mathbb{A}_{S_0} \quad , \quad \nu = 1, .., m$$

to the affine line over S_0.

The dimension of the general fiber of X_0 over $\mathbb{A}_{S_0}$ with respect to the maps p_ν is smaller than n. Therefore by induction (applied in the case where S_0 is replaced by $\mathbb{A}_{S_0}$) there exist nonempty open subsets $W_{0\nu} \subset \mathbb{A}_{S_0}$, such that $\mathrm{R}j_{0*}(\mathscr{G}_0)$ is mixed on $V_{0\nu} = p_\nu^{-1}(W_{0\nu})$. Finally the union of X_0 together with the open subsets $V_{0\nu}$ is large in Y_0 in the following sense: Its complement is finite over S_0, after we further shrink S_0. Thus

$$V_0 = X_0 \cup V_{01} \cup ... \cup V_{0m}$$

has the desired properties, notably after replacing S_0 by an open dense subset. □

Proof of the Induction Step (of Lemma I.9.6(n)). Let $n > 0$. Assume Lemma I.9.6(m) holds for $m < n$.

Using Lemma I.9.7 we now give the proof of the induction step for Lemma I.9.6(n). This is done first in the following special case:

Induction Step (Smooth Case). Let us assume, that $\mathscr{G}_0$ and the map

$$f_{X_0} : X_0 \to S_0$$

satisfy the smoothness assumption of Remark R2. For the proof we can furthermore assume Y_0 to be projective over S_0. Apply Lemma I.9.7: By shrinking S_0, we find a large open subscheme

$$j_{V_0} : V_0 \hookrightarrow Y_0$$

of Y_0 containing X_0 with the properties mentioned in Lemma I.9.7. In particular, the complement Z_0

$$\begin{array}{ccccc} i_0 : Z_0 & \xrightarrow{i_0} & Y_0 & \xleftarrow{j_{V_0}} & V_0 \\ & & \uparrow{\scriptstyle j_0} & \nearrow & \\ & & X_0 & & \end{array}$$

of V_0 in Y_0 is finite over S_0. In the remaining part of the proof we use the language of derived categories to study the derived image complex

$$K = \mathrm{R}j_{0*}(\mathscr{G}_0)\ .$$

The reason is, that it is more convenient to work in this language, which avoids the use of spectral sequences. In particular one has a distinguished triangle

$$j_{V_0!}j_{V_0}^*(K) \to K \to i_{0*}i_0^*(K) \to \ .$$

For readers not acquainted with the language of derived categories, we refer to chapter II of this book. There he will find the most important notions – as for instance the notion of distinguished triangles; E.g. the third map above is a morphism in the derived category from the complex $i_{0*}i_0^*(K)$ to the translate $j_{V_0!}j_{V_0}^*(K)[1]$ of the complex $j_{V_0!}j_{V_0}^*(K)$. In the derived categories distinguished triangles are a substitute for exact sequences, which is a notion specific to abelian categories. For a distinguished triangle of complexes the long exact sequence of cohomology sheaves attached to it allows to conclude the following: Mixedness of any two of the three complexes of a distinguished triangle implies mixedness of all three complexes of the triangle.

To the distinguished triangle above we apply the direct image functor with respect to the morphism

$$f_{Y_0} : Y_0 \to S_0\ .$$

In this way we obtain a new distinguished triangle

$$\mathrm{R}(f_{Y_0}|V_0)_! j_{V_0}^*(K) \to \mathrm{R}f_{X_0*}(\mathscr{G}_0) \to \mathrm{R}(f_{Y_0}|Z_0)_* i_0^*(K) \to\ .$$

By construction of V_0 the left complex of the first distinguished triangle is mixed. According to Remark R1 also the left complex of the second distinguished triangle is mixed. The middle complex of the second distinguished triangle is mixed, by our smoothness assumptions using R2. Then also the third complex of the second distinguished triangle must be mixed. The restriction $f_{Y_0}|Z_0 : Z_0 \to S_0$ is a finite morphism. This implies that $i_{0*}i_0^*(K)$ is mixed. If we insert this information into the first distinguished triangle, we conclude that also the middle complex K of the first distinguished triangle has to be mixed. This proves the claim of Lemma I.9.6(n) in the "smooth" case.

General Case of the Induction Step. To attack the general case of Lemma I.9.6(n), we use Remark R3. According to it, there exists an open dense subscheme

$$j_{U_0} : U_0 \hookrightarrow X_0$$

after shrinking of S_0, such that the induced map

$$U_0 \to S_0$$

and the sheaf $\mathscr{G}_0|U_0$ satisfy the smoothness assumptions of Remark R2. Consider the complement

$$i_0 : Z_0 \to X_0$$

of U_0 in X_0. The cone of $\mathscr{G}_0 \to \mathrm{R}j_{U_0*}j_{U_0}^*(\mathscr{G}_0)$ has cohomology sheaves concentrated in $i_0(Z_0)$, hence is isomorphic to $i_{0*}(\Delta)$ for some complex Δ on Z_0. We obtain a distinguished triangle

$$\mathscr{G}_0 \to \mathrm{R}j_{U_0*}j_{U_0}^*(\mathscr{G}_0) \to i_{0*}(\Delta) \to \ .$$

Lemma I.9.6, already proved above in the "smooth" case, can be applied to the map $j_{U_0} : U_0 \to X_0$ and the sheaf $j_{U_0}^*(\mathscr{G}_0)$. Hence the complex $\mathrm{R}j_{U_0*}j_{U_0}^*(\mathscr{G}_0)$ in the middle, and of course also the complex $\mathscr{G}_0[1]$, of the distinguished triangle are mixed complexes. Hence Δ has to be a mixed complex on Z_0.

Now consider the direct image of the distinguished triangle under $\mathrm{R}j_{0*}$, obtained from the map $j_0 : X_0 \to Y_0$

$$\mathrm{R}j_{0*}(\mathscr{G}_0) \to \mathrm{R}(j_0 \circ j_{U_0})_* j_{U_0}^*(\mathscr{G}_0) \to \mathrm{R}(j_0 \circ i_0)_*(\Delta) \to \ .$$

Its middle complex is mixed, since the map $j_0 \circ j_{U_0} : U_0 \hookrightarrow Y_0$ and the complex $\mathscr{G}_0|U_0$ satisfy the smoothness assumptions of the special case already considered. Furthermore the dimension of the general fiber of Z_0 over S_0 is smaller than n. By our induction assumption therefore also the right complex $\mathrm{R}(j_0 \circ i_0)_*(\Delta)$ is mixed either. This implies, that the left complex

$$\mathrm{R}j_{0*}(\mathscr{G}_0)$$

is mixed. Using the cohomology sequence of this distinguished triangle completes the proof of Lemma I.9.6(n).

Since I.9.6 is proven, in particular Theorem I.9.4 follows. □

II. The Formalism of Derived Categories

II.1 Triangulated Categories

Let A be an abelian category. The derived category $D(A)$ of A is a quotient category of the abelian category $Kom(A)$ of complexes over A. The quotient category is defined by making quasiisomorphisms into isomorphisms and this allows to identify complexes with their resolutions. Recall, that a complex map $K' \to K$ is a quasiisomorphism, if the induced cohomology morphisms $H^\bullet(K') \to H^\bullet(K)$ are isomorphisms in all degrees. However, by taking this localization of the category $Kom(A)$, the notion of (short) exact sequences of complexes no longer exists and has to be replaced by the notion of distinguished triangles, which itself derives from the concept of mapping cones. The derived category thus becomes a triangulated category, a notion first introduced by Verdier in [Ver], [SGA4$\frac{1}{2}$].

Suppose D is an additive category with an additive automorphism of categories $\mathcal{T} : D \to D$, which is called the translation functor. By abuse of notation we write $X \in D$ to indicate that X is object of a category D. For $n \in \mathbb{Z}$ we usually write $\mathcal{T}^n(X) = X[n]$ resp. $\mathcal{T}^n(f) = f[n]$ both for objects X and morphisms f in D.

A triangle $T = (X, Y, Z, u, v, w)$ in D is a diagram in D

$$X \xrightarrow{u} Y \xrightarrow{v} Z \xrightarrow{w} X[1] .$$

Instead of writing such a diagram, we often use the abbreviated abusive way of writing (X, Y, Z) or $(X, Y, Z, *, *, *)$, if the underlying morphisms are understood from the context.

A morphism (f, g, h) between triangles is a commutative diagram

$$\begin{array}{ccccccc} X' & \xrightarrow{u'} & Y' & \xrightarrow{v'} & Z' & \xrightarrow{w'} & X'[1] \\ \downarrow{\scriptstyle f} & & \downarrow{\scriptstyle g} & & \downarrow{\scriptstyle h} & & \downarrow{\scriptstyle f[1]} \\ X & \xrightarrow{u} & Y & \xrightarrow{v} & Z & \xrightarrow{w} & X[1] . \end{array}$$

If

$$T = (X, Y, Z, u, v, w)$$

is a triangle, we call

$$rot(T) = (Z[-1], X, Y, -w[-1], u, v)$$

the rotated triangle. If $(f, g, h) : T' \to T$ is a morphism between triangles, then $rot(f, g, h) = (h[-1], f, g)$ is a morphism between the rotated triangles $rot(f, g, h) : rot(T') \to rot(T)$.

Definition 1.1 *A triangulated category is an additive category D with a translation functor $\mathscr{T}$ and a class of distinguished triangles satisfying the following axioms TR1, TR2, TR3 and TR4:*

TR1 (Rotation)

a) A triangle T is distinguished if and only if its rotated triangle $rot(T)$ is distinguished.
b) Triangles isomorphic to distinguished triangles are distinguished.

Especially $(id, -id, id)^* rot^3(T) = (X[-1], Y[-1], Z[-1], u[-1], v[-1], -w[-1])$ is distinguished, if $T = (X, Y, Z, u, v, w)$ is distinguished. This will be relevant for axiom TR4.

TR2 (Existence of cones)

a) Any morphism $u : X \to Y$ in D can be completed (not necessarily uniquely) to a distinguished triangle (X, Y, Z, u, v, w). Any such object Z will be called a cone or mapping cone for $u : X \to Y$.
b) The triangle $(X, X, 0, id_X, 0, 0)$ is distinguished.

TR3 (Morphisms). Any commutative diagram

$$\begin{array}{ccc} X' & \xrightarrow{u'} & Y' \\ \downarrow f & & \downarrow g \\ X & \xrightarrow{u} & Y \end{array}$$

can be extended (not necessarily uniquely) to a morphism (f, g, h) between given distinguished triangles (X', Y', Z', u', v', w') and (X, Y, Z, u, v, w).

Due to axiom TR1 one also has versions of the axioms TR2 and TR3, which are obtained by applying TR2 and TR3 to rotated triangles. Altogether this already has a number of consequences. The most important is the long exact Hom-sequence stated in Theorem II.1.3. We first discuss these consequences before we formulate the last axiom TR4 of triangulated categories.

Remark 1.2 Let (X, Y, Z, u, v, w) be a distinguished triangle. Then $v \circ u = 0$ holds. Use Axiom TR2b and Axiom TR3 with $(f, g) = (id_X, u)$ to deduce this from

$$\begin{array}{ccccccc} X & \xrightarrow{id_X} & X & \xrightarrow{0} & 0 & \xrightarrow{0} & X[1] \\ \Big\downarrow{\scriptstyle id_X} & & \Big\downarrow{\scriptstyle u} & & \Big\downarrow{\scriptstyle \exists h} & & \Big\downarrow{\scriptstyle id_X[1]} \\ X & \xrightarrow{u} & Y & \xrightarrow{v} & Z & \xrightarrow{w} & X[1]\,. \end{array}$$

Furthermore, any morphism $g : A \to Y$ satisfying $v \circ g = 0$ factors over X by a morphism $f : A \to X$. For this apply a rotated version of axiom TR3 to the pair $(g, h) = (g, 0)$:

$$\begin{array}{ccccccc} A & \xrightarrow{id} & A & \xrightarrow{0} & 0 & \xrightarrow{0} & A[1] \\ \Big\downarrow{\scriptstyle \exists f} & & \Big\downarrow{\scriptstyle g} & & \Big\downarrow{\scriptstyle 0} & & \Big\downarrow{\scriptstyle f[1]} \\ X & \xrightarrow{u} & Y & \xrightarrow{v} & Z & \xrightarrow{w} & X[1]\,. \end{array}$$

By repeated use of the rotation axiom this proves

Theorem 1.3 *Let (X, Y, Z, u, v, w) be a distinguished triangle in D. For any A in D the sequence*

$$\to Hom(A,X[i]) \xrightarrow{u[i]_*} Hom(A,Y[i]) \xrightarrow{v[i]_*} Hom(A,Z[i]) \xrightarrow{w[i]_*} Hom(A,X[i+1]) \xrightarrow{u[i+1]_*}$$

is a long exact sequence of abelian groups. similarly one gets a long exact Hom-sequence

$$\leftarrow Hom(X[i],B) \xleftarrow{u[i]^*} Hom(Y[i],B) \xleftarrow{v[i]^*} Hom(Z[i],B) \xleftarrow{w[i]^*} Hom(X[i+1],B) \xleftarrow{u[i+1]^*}$$

in the first variable.

Immediate consequences of the long exact Hom-sequence are

Corollary 1.4 *If (f, g, h) is a morphism between distinguished triangles and f and g are isomorphisms, then h is an isomorphism.*

Proof. This follows from the 5-lemma. Applying it to the two $Hom(A, .)$-sequences for (X', Y', Z') and (X, Y, Z) it shows $h_* : Hom(A, Z') \cong Hom(A, Z)$ for all A. Specializing to $A = Z', Z$ gives a right and left inverse to $h : Z' \to Z$. □

Corollary 1.5 *For a given morphism $u : X \to Y$ any two mapping cones Z_u are isomorphic. Two distinguished triangles (X, Y, Z_u, u, v, w) and (X, Y, Z_u, u, v', w') attached to $u : X \to Y$ with the same mapping cone Z_u are related by $v' = h^{-1} \circ v$, $w' = w \circ h$, with an isomorphism h of Z_u.*

Proof. Use Corollary II.1.4 and axiom TR3. □

Conversely, any object Z isomorphic by $h : Z \cong Z_u$ to a cone Z_u of $u : X \to Y$ is itself a cone, since the triangle (X, Y, Z, u, v', w') for $v' = h^{-1} \circ v$ and $w' = w \circ h$ is distinguished by axiom TR1b.

Remark (Rotated Version of Axiom TR2a). Any morphism $w : Z \to X[1]$ can be completed to a distinguished triangle (X, Y, Z, u, v, w). The resulting object Y is called an extension of Z by X attached to $w \in Hom(Z, X[1])$. As a variant of Corollary II.1.4 and II.1.5 any two extensions attached to w are isomorphic.

$$\begin{array}{ccccccc}
X & \xrightarrow{u'} & Y' & \xrightarrow{v'} & Z & \xrightarrow{w} & X[1] \\
\downarrow{\scriptstyle id} & & \downarrow{\scriptstyle \cong} & & \downarrow{\scriptstyle id} & & \downarrow{\scriptstyle id[1]} \\
X & \xrightarrow{u} & Y & \xrightarrow{v} & Z & \xrightarrow{w} & X[1]\,.
\end{array}$$

Corollary 1.6 *If (X, Y, Z, u, v, w) is a distinguished triangle and u is an isomorphism, then $Z \cong 0$. Conversely $Z \cong 0$ implies that u is an isomorphism.*

Proof. The long exact $Hom(Z, .)$-sequence implies $Hom(Z, Z) = 0$ and proves the first statement. The second statement again follows from the long exact Hom-sequence. □

Corollary 1.7 *If (X, Y, Z, u, v, w) is distinguished and $w = 0$, then $Y \cong Z \oplus X$ such that u and v correspond to the inclusion resp. projection map.*

Proof. Exercise! First find p with $p \circ u = id_X$, then find i with $id_Y = u \circ p + i \circ v$. Then $v \circ i = id_Z$ and finally $p \circ i = 0$ follow from the injectivity of the dual Hom-sequence at $X[1]$. □

So far we have discussed the axioms TR1-3 of a triangulated category and some trivial implications. We now formulate the remaining axiom TR4 for triangulated categories, the so called octaeder axiom. See also [140], especially for an explanation of the name.

Axiom TR4a (Composition Law for Mapping Cones). Suppose we are given morphisms $u : X \to Y$ and $v : Y \to Z$. For any choice of mapping cones C_u, C_v and $C_{v \circ u}$ with defining distinguished triangles

$$T_u = (X, Y, C_u, u, *, *)$$
$$T_v = (Y, Z, C_v, v, *, *)$$
$$T_{v \circ u} = (X, Z, C_{v \circ u}, v \circ u, *, *)\,,$$

there exists a distinguished triangle T relating these mapping cones

$$T = (C_u, C_{v \circ u}, C_v, \alpha, \beta, \gamma)\,,$$

which makes the following diagram commute

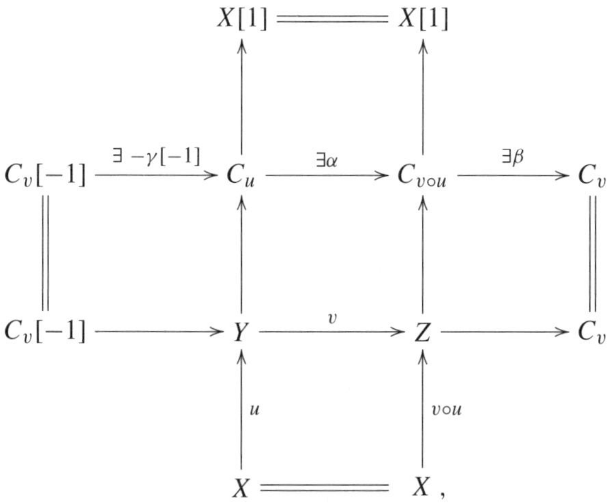

The two vertical lines of the diagram are defined by the morphisms of the two distinguished triangles T_u, T_{vou}, the two horizontal lines by the morphisms of the two rotated distinguished triangles $rot(T_v)$, $rot(T)$. Note, that although we assume the triangles T_u, T_v, T_{vou} to be chosen fixed, we preferred not to give names to all morphisms.

Axiom TR4b. For the full octaeder axiom one adds the further condition of commutativity for the diagram

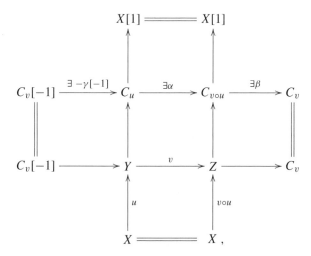

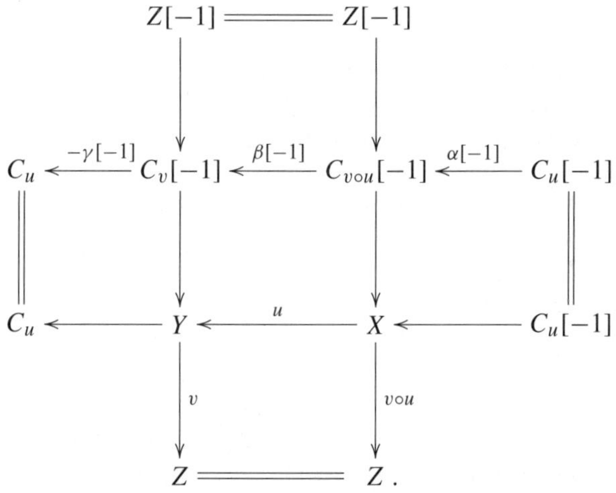

The two vertical lines of the diagram are defined by the morphisms of the distinguished triangles $rot^2(T_{vou})$, $rot^2(T_v)$, the two horizontal lines by the morphisms of the distinguished triangles $(id, -id, id)^*(rot^3(T))$ and $rot(T_u)$.

In the axioms of a triangulated category there is a certain redundancy: axiom TR3 is a consequence of axiom TR4a and TR4b. Consider maps f, g, u', u as in

TR3. Then axiom TR4a, which is a kind of pushout axiom, applied to $X' \xrightarrow{u'} Y' \xrightarrow{g} Y$ and axiom TR4b, which is a kind of pullback axiom, applied to $X' \xrightarrow{f} X \xrightarrow{u} Y$ together imply TR3. For $k := g \circ u' = u \circ f$ this defines morphisms $\alpha : C_{u'} \to C_k$ respectively $\beta' : C_k \to C_u$. Then $h = \beta' \circ \alpha$ defines a morphism (f, g, h) between the triangles $(X', Y', C_{u'})$ and (X, Y, C_u).

The Derived Categories $D(\mathbf{A})$

The most prominent example for a triangulated category is the derived category $D(\mathbf{A})$ of an abelian category $\mathbf{A}$, as mentioned in the introduction. It is the localization of the category of all complexes $Kom(\mathbf{A})$ over $\mathbf{A}$ by the class of quasiisomorphisms. A short review on this is in the appendix II of [FK], p. 292 and in chapter I of [140] (Hartshorne).

To verify the axioms TR1–4 for derived categories, it is useful to be aware of the fact that the localization functor is not injective on homomorphism groups. In particular homotopic complex maps become equal in the derived category. However this gives some extra freedom. So one can proceed in two steps: first pass to the category $K(\mathbf{A})$, whose morphisms are homotopy classes of complex maps, and then invert quasiisomorphisms. The axioms TR1–4 can be established already on the first level as properties of complexes up to homotopy. To invert quasiisomorphisms in the category $K(\mathbf{A})$ becomes much more convenient, because the class of quasiisomorphisms is a localizing class, i.e. it allows a calculus of fractions in the category $K(\mathbf{A})$. See [FK] A II.1. More details can be found in [Ver], [104], [140].

In the same way one can define the derived categories $D^+(\mathbf{A})$, $D^-(\mathbf{A})$, and $D^b(\mathbf{A})$ as the localization of the full subcategory of complexes which are bounded to the left, bounded to the right respectively are bounded. They can be embedded as full subcategories into $D(\mathbf{A})$. In particular for a morphism f in $Hom_{D^*(\mathbf{A})}(K, L)$ for $* \in \{+, -, b\}$ there exist quasiisomorphisms s, t, such that $f \circ s$ respectively $t \circ f$ are homotopic to complex maps.

We remark that the construction of $D(\mathbf{A})$ by localization gives set theoretic problems unless $\mathbf{A}$ is a small category or belongs to some given universe, since a priori $Hom_{D(\mathbf{A})}(X, Y)$ is not a set. However, let the category $\mathbf{A}$ satisfy the Grothendieck Tôhoku axioms ([113]). Then every – may be unbounded – complex K of $Kom(\mathbf{A})$ has a right resolution by a so called *K–injective* complex. Using this fact we can conclude that $D(\mathbf{A})$ is equivalent to the full subcategory of $Kom(\mathbf{A})$ given by all K–injective complexes. See [Spa], [Tar]. This implies, that $Hom_{D(\mathbf{A})}(X, Y)$ is a set in the case of an abelian Grothendieck category $\mathbf{A}$. See also the following remark for the case of the subcategory $D^+(\mathbf{A})$ of $D(\mathbf{A})$.

Remark 1.8 If K, L are in $D^+(\mathbf{A})$ and L is injective, i.e. all components L^ν of the complex L are injective objects of $\mathbf{A}$, then any morphism f in $Hom_{D^*(\mathbf{A})}(K, L)$ is represented by a complex map $p : K \to L$. The localizing property of quasiisomorphisms, which is formulated in [FK] A II 1(2), allows to reduce the proof of this statement to the case, where f is the inverse of a quasiisomorphism $u : L \to K$

(using fractions with left denominators we find u such that $u \circ f$ is a complex map). For the quasiisomorphism u it is enough to show, that there exists a complex map $p : K \to L$ which is a left inverse of u up to homotopy. Namely then (using calculus of fractions with right denominators) the morphism f and the complex map p

or $f : K \xleftarrow{u} L \xrightarrow{id} L$ and $p : K \xleftarrow{id} K \xrightarrow{p} L$ coincide in the derived category, since we have the commutative diagram

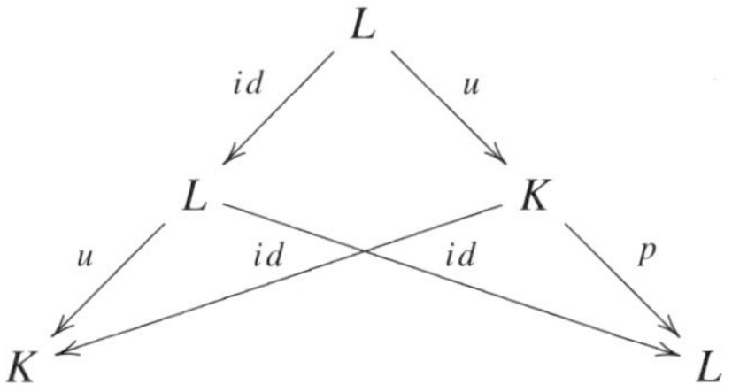

To construct p let $C = C_u$ be the cone in the category $K^+(\mathbf{A})$. Then we have a distinguished triangle $(L, K, C, u, *, *)$ and C is acyclic and bounded below. Using Corollary II.1.7 one reduces the construction of a homotopy left inverse p for u to the following fact: If C is acyclic and I is injective ($I = L[1]$) and both C and I are bounded from below, then any complex map $C \to I$ is homotopic to the zero map. This homotopy is constructed inductively using the extension property for injective objects. See e.g. [104], p. 180. So for $K, L \in D^+(\mathbf{A})$ and L injective, the natural map

$$Hom_{K^+(\mathbf{A})}(K, L) \ \to \ Hom_{D^+(\mathbf{A})}(K, L)$$

is surjective. It is also injective, since $u \circ f = 0$ for $f \in Hom_{K^+(\mathbf{A})}(K, L)$ and a quasiisomorphism $u : K' \to K$ implies $f = 0$. In fact $u^* : Hom_{K^+(\mathbf{A})}(K, L) \to Hom_{K^+(\mathbf{A})}(K', L)$ is injective. Since $Hom_{K^+(\mathbf{A})}(C_u, L)$ vanishes for the acyclic cone C_u of u, this follows from the long exact Hom-sequence for the triangulated category $K^+(\mathbf{A})$. Therefore the Hom-groups in the derived categories $D^+(\mathbf{A})$ and $D^b(\mathbf{A})$ can be computed in terms of the homotopy category $K(\mathbf{A})$ using injective resolutions in the second variable. This is convenient, provided the abelian category $\mathbf{A}$ has enough injective objects. See Verdier [SGA4$\frac{1}{2}$], p. 299 for further information.

Other examples of triangulated categories can be found in [104]. We will be mainly interested in the triangulated categories $D(X) = D_c^b(X, \overline{\mathbb{Q}}_l)$ for finitely generated schemes X over a finite or algebraically closed field. These categories are obtained as certain limits of derived categories. For further details on these categories the reader is referred to Appendix A and the corresponding section of this chapter.

Remark 1.9 The diagram of axiom TR4b is formally obtained from the diagram of axiom TR4a by replacing the direction of arrows (and renaming). This implies, that the notion of triangulated category is self dual: If D is a triangulated category, then also the opposite category D^{opp}, obtained by inverting arrows with the induced translation functor and induced distinguished triangles, is triangulated. Later we will use this in the proof of Corollary II.4.2. Nevertheless we mention, that the only information added by TR4b to TR4a is the commutativity of the middle square of the TR4b diagram. In other words TR4b is, modulo TR4a, equivalent to either one of the two statements:

TR4b′: $(u, id_Z, \beta) : T_{v\circ u} \to T_v$ is a morphism of distinguished triangles.
TR4b″: The two hidden ways, to go in the diagram of axiom TR4a over the upper right corner from $C_{v\circ u}$ to $Y[1]$, *anticommute.*

More precisely, Axiom TR4b″ states

$$(-i[-1]) \circ \beta = -(-u[1]) \circ k \,.$$

if $T_v = (*, *, *, v, *, i)$ and $T_{v\circ u} = (*, *, *, v \circ u, j, k)$ are the triangles chosen.

Remark 1.10 The isomorphism class of the distinguished triangle T, whose existence is imposed by axiom TR4, is determined (up to isomorphism) by the three cones $C_u, C_v, C_{v\circ u}$ and the morphism γ, according to Corollary II.1.5. However $-\gamma[-1]$ and therefore also γ is uniquely determined by the commutativity of the left square of diagram TR4a – as the composite of the given maps $C_v[-1] \to Y$ and $Y \to C_u$, appearing in the two distinguished triangles T_v and T_u. Thus the triangles $T_u, T_v, T_{v\circ u}$ determine T up to isomorphism.

II.2 Abstract Truncations

In the derived category $D(A)$ of an abelian category A one has full subcategories $D(A)^{\leq n}$ and $D(A)^{\geq m}$, consisting complexes with vanishing cohomology in degrees strictly larger than n resp. strictly smaller than m. By the process of truncation, a given complex can be split into two complexes, one of them in $D(A)^{\leq 0}$ and the other in $D(A)^{\geq 1}$. This has an abstract analog in an arbitrary triangulated category, motivated by the theory of D-modules and the Riemann-Hilbert correspondence. The notion of t-structures was first introduced in [BBD], inspired by the non obvious t-structures underlying perverse sheaves respectively holonomic D-modules with regular singularities.

Definition 2.1 *A t-structure in a triangulated category D consists of two strictly full subcategories $D^{\leq 0}$ and $D^{\geq 0}$ of D, such that with the definitions $D^{\leq n} = D^{\leq 0}[-n]$ and $D^{\geq n} = D^{\geq 0}[-n]$ we have*

(i) $Hom(D^{\leq 0}, D^{\geq 1}) = 0$.
(ii) $D^{\leq 0} \subset D^{\leq 1}$ *and* $D^{\geq 1} \subset D^{\geq 0}$.
(iii) For every object E in D there exists a distinguished triangle (A, E, B) with $A \in D^{\leq 0}$ and $B \in D^{\geq 1}$.

D is said to be bounded with respect to the t-structure, if every object of D is contained in some $D^{\geq a}$ and some $D^{\leq b}$ for certain integers a, b.

Truncation. Suppose we are given a t-structure.

In the situation of II.2.1(iii) we have $B \in D^{\geq 1}$ and $B[-1] \in D^{\geq 1}[-1] \subset D^{\geq 1}$ by property II.2.1(ii). Therefore the long exact Hom-sequence II.1.3 attached to the distinguished triangle $(A, E, B, u, *, *)$ together with the vanishing property II.2.1(i) of t-structures implies

$$u_* : Hom(X, A) \cong Hom(X, E) \quad , \qquad \text{for } X \in D^{\leq 0} .$$

This fundamental fact has the striking consequence that $u : A \to E$ is a universal morphism from $D^{\leq 0}$ to the given object $E \in D$. Every morphism from some $X \in D^{\leq 0}$ to E factors in a unique way over the morphism u

$$\begin{array}{ccc} A & \xrightarrow{\quad u \quad} & E \\ \uparrow {\scriptstyle \exists!} & \nearrow & \\ X \in D^{\leq 0} & & \end{array} .$$

This universal property characterizes the pair (A, u) uniquely up to isomorphism. We therefore write $A = \tau_{\leq 0}(E)$. The assignment $\tau_{\leq 0}$ is functorial in E. For every morphism $f : E' \to E$, the composite $f \circ u' : \tau_{\leq 0}(E') \to E$ factors through $u : \tau_{\leq 0}(E) \to E$ by a unique morphism $\tau_{\leq 0}(f)$. In other words:

$$\tau_{\leq 0} : D \to D^{\leq 0}$$

defines a functor, which is right adjoint to the inclusion functor of $D^{\leq 0}$ into D. The isomorphism u_* established above turns out to be the adjunction isomorphism.

Similarly, the assignment $E \mapsto B$ defines a functor $\tau_{\geq 1} : D \to D^{\geq 1}$, which is left adjoint to the inclusion of $D^{\geq 1} \subset D$. We will assume, after making some choices, that the functors $\tau_{\leq 0}, \tau_{\geq 1}$ are fixed from now on.

Resume. From the discussion of t-structures so far we see, that the distinguished truncation triangle of property (iii) for t-structures has now become the unique distinguished triangle

iii)′ $$\tau_{\leq 0}(E) \stackrel{ad_{\leq 0}}{\longrightarrow} E \stackrel{ad_{\geq 1}}{\longrightarrow} \tau_{\geq 1}(E) \to \tau_{\leq 0}(E)[1] .$$

The first two morphisms have become adjunction maps. They uniquely determine the third map of the triangle. For this recall Corollary II.1.5 and the fact, that $h = id_{\tau_{\geq 1}(E)}$ is the unique morphism $h : \tau_{\geq 1}(E) \to \tau_{\geq 1}(E)$ with the property $h \circ ad_{\geq 1} = ad_{\geq 1}$ because of the adjunction formula

$$Hom(E, Y) = Hom(\tau_{\geq 1}(E), Y) \quad , \qquad \text{for } Y \in D^{\geq 1} .$$

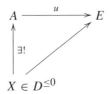

Similarly the isomorphism u_* from above gives the adjunction formula

$$Hom(X, \tau_{\leq 0}(E)) = Hom(X, E) \quad , \qquad \text{for } X \in D^{\leq 0} .$$

Properties of the truncation functors. Recursively define $\tau_{\leq n}$ for $n \in \mathbb{Z}$ by

$$(\tau_{\leq n+1}(X))[1] = \tau_{\leq n}(X[1])$$

or $\tau_{\leq n}(X) = (\tau_{\leq 0}(X[n])[-n]$. Then $\tau_{\leq n}$ is right adjoint to the inclusion functor $D^{\leq n} \subset D$. Therefore, by the obvious inclusion properties of the underlying categories

$$\tau_{\leq m} \circ \tau_{\leq n} = \tau_{\leq m} \qquad \text{for } \ m \leq n .$$

Lemma 2.2 (Orthogonality) *For objects $E \in D$ the following statements are equivalent*

(i) E is in $D^{\geq n+1}$.
(ii) $Hom(D^{\leq n}, E) = 0$.

Proof. One easily reduces to $n = 0$. Then one direction is the statement of property (i) for t-structures. For the converse direction (ii) $\Rightarrow$ (i) it is enough to show $\tau_{\leq 0}(E) = 0$ by Corollary II.1.6. But $\tau_{\leq 0}(E) = 0$ follows from the adjunction isomorphism $Hom(\tau_{\leq 0}(E), \tau_{\leq 0}(E)) = Hom(\tau_{\leq 0}(E), E) \subset Hom(D^{\leq 0}, E) = 0$. □

We also write $\tau_{>n}$ for $\tau_{\geq n+1}$ or $\tau_{<n}$ for $\tau_{\leq n-1}$.

Lemma 2.3 (Extensions) *Suppose (X, Y, Z) is a distinguished triangle. Then*

1) If X and Z are in $D^{\leq n}$, then also Y is in $D^{\leq n}$.
2) If Y is in $D^{\leq n}$ and if X is in $D^{\leq n+1} \supset D^{\leq n}$, then also Z is in $D^{\leq n}$.

Proof. Apply the long exact Hom-sequence II.1.3 and the criterion (ii) $\Rightarrow$ (i) of Lemma II.2.2. □

The statements of Lemma II.2.2 and II.2.3 have dual versions. We define

$$(\tau_{\geq m+1}(X))[1] = \tau_{\geq m}(X[1])$$

with $\tau_{\geq n} \circ \tau_{\geq m} = \tau_{\geq n}$ for $m \leq n$. Suppose (X, Y, Z) is distinguished. Again $X, Z \in D^{\geq m}$ implies $Y \in D^{\geq m}$ and $Y \in D^{\geq m}$, $Z \in D^{\geq m-1} \supset D^{\geq m}$ implies $X \in D^{\geq m}$.

Lemma 2.4 (Compatibility) *We have*

$$\tau_{\geq m}(D^{\leq n}) \subset D^{\leq n}$$

and similarly

$$\tau_{\leq n}(D^{\geq m}) \subset D^{\geq m} .$$

For $m > n$ we have $\tau_{\geq m}(D^{\leq n}) = \tau_{\leq n}(D^{\geq m}) = 0$.

Proof. We only proof the first statement. For $n < m$ and $X \in D^{\leq n}$ the natural map $X \to \tau_{\geq m}X$ is zero by II.2.1(i). Therefore the adjunction formula implies $Hom(\tau_{\geq m}X, \tau_{\geq m}X) = 0$, hence $id_{\tau_{\geq m}X} = 0$ or $\tau_{\geq m}X = 0$. So assume $m \leq n$. Consider the distinguished triangle $(\tau_{\leq m-1}X, X, \tau_{\geq m}X)$. For $X \in D^{\leq n}$ and $m \leq n$ we have $\tau_{\leq m-1}(X) \in D^{\leq n-1} \subset D^{\leq n+1}$. The claim now follows from the extension Lemma II.2.3. □

Exercise 2.5 Two t-structures α and β of a triangulated category D which are included in each other in the sense that ${}^{\alpha}D^{\leq n} \subset {}^{\beta}D^{\leq n}$, ${}^{\alpha}D^{\geq m} \subset {}^{\beta}D^{\geq m}$ necessarily coincide.

II.3 The Core of a t-Structure

In the derived category $D(A)$ attached to an abelian category A one can reconstruct A from the canonical t-structure by considering the full subcategory of complexes in $D(A)^{\leq 0} \cap D(A)^{\geq 0}$. This is the category of complexes with vanishing cohomology outside of degree zero. The natural functor $A \to D(A)^{\leq 0} \cap D(A)^{\geq 0}$ defines an equivalence of abelian categories. In general, a t-structure on a triangulated category D defines in a similar way an abelian category A, the core $Core(D) = Core_t(D)$. However, the relationship between D and $D(Core(D))$ is not clear in general[1].

Let D be a triangulated category with a given t-structure $D^{\leq 0}$ and $D^{\geq 0}$. Define the core $Core(D)$ to be the full subcategory $Core(D) = D^{\leq 0} \cap D^{\geq 0}$.

Theorem 3.1 *The core $Core(D) = D^{\leq 0} \cap D^{\geq 0}$ attached to a t-structure of a triangulated category D is an abelian category. A sequence in $Core(D)$*

$$0 \to X \overset{u}{\to} Y \overset{v}{\to} Z \to 0$$

is exact if and only if there exists a distinguished triangle (X, Y, Z, u, v, w) in D.

Proof. For a morphism $f : X \to Y$ between objects of $Core(D)$ choose a cone Z_f in D with distinguished triangle (X, Y, Z_f) and put

$$Ker_f := \tau_{\leq 0}\big(Z_f[-1]\big)$$

$$Koker_f := \tau_{\geq 0}Z_f .$$

Step 1. For $A \in D^{\leq 0}$ we have $Hom(A, \tau_{\leq 0}(Z_f[-1])) \cong Hom(A, Z_f[-1])$, the adjunction isomorphism. Furthermore $Hom(A, Y[-1]) \subset Hom(D^{\leq 0}, D^{\geq 1}) = 0$, and the long exact Hom-sequence gives the short exact sequence

[1] For the core of perverse sheaves considered in Chap. III this was clarified by Beilinson [Be1].

$$0 \to Hom(A, \tau_{\leq 0}(Z_f[-1])) \to Hom(A, X) \to Hom(A, Y) \, .$$

Thus $i : Ker_f \to X$ with $i : \tau_{\leq 0}(Z_f[-1]) \to Z_f[-1] \to X$ represents the kernel of f in $D^{\leq 0}$. Similarly $\tau_{\geq 0} Z_f$ with $\pi : Y \to Z_f \to \tau_{\geq 0} Z_f = Koker_f$ represents the kokernel of f in $D^{\geq 0}$.

Step 2. For the existence of kernels and kokernels in $Core(D)$ it remains to verify, that Ker_f, $Koker_f$ are objects contained in $Core(D)$. Indeed $Z_f \in D^{\leq 0} \cap D^{\geq -1}$, by the extension Lemma II.2.3 and its dual version. Therefore the compatibility Lemma II.2.4 implies that Ker_f, $Koker_f$ are in $Core(D)$.

Step 3 (Factorization Property). It remains to show, that $f : X \to Y$ decomposes into a composite of $a : X \to Z$ and $b : Z \to Y$ for some $Z \in Core(D)$ with induced isomorphisms $a_* : Z \cong Koker(i : Ker_f \to X)$ and $b_* : Z \cong Ker(\pi : Y \to Koker_f)$.

To find Z and the factorization $f = b \circ a$ consider the octaeder axiom attached to the composition of arrows defining $i : Ker_f \to X$, which gives the following commutative diagram

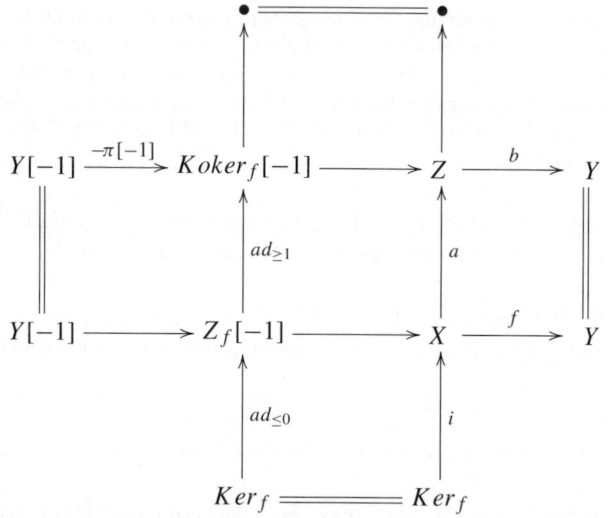

From $Y \in D^{\geq 0}$ and $Koker_f[-1] \in Core(D)[-1] \subset D^{\geq 1} \subset D^{\geq 0}$, we get $Z \in D^{\geq 0}$ using the extension Lemma II.2.3. Once more Lemma II.2.3, and $X, Ker_f \in Core(D) \subset D^{\leq 0}$, shows $Z \in D^{\leq 0}$. Thus Z is an object of $Core(D)$

$$Z \in Core(D) \, .$$

Step 4 (Remaining Verifications). The fact that Z is in $Core(D)$ implies, recalling the definition of kernel and kokernel in step 1,

$$b_* : Z \cong Ker(\pi : Y \to Koker_f)$$

$$a_* : Z \cong Koker(i : Ker_f \to X) .$$

The second isomorphism a_* comes from the distinguished triangle (Ker_f, X, Z) with Z as a cone for the morphism $i : Ker_f \to X$. Therefore $Koker_i = \tau_{\geq 0}(Z) = Z$.

The first isomorphism b_* comes from the triangle $(Y[-1], Koker_f[-1], Z)$, which makes Z into a cone for the morphism $\pi[-1] : Y[-1] \to Koker_f[-1]$. In fact, the commutativity of the left square of the diagram above shows that its upper map is $-\pi[-1]$ and we are allowed to modify the triangle by an isomorphic one. Therefore we get $Ker_\pi = \tau_{\leq 0}(Z[1][-1]) = Z$.

Step 5. The additivity axiom finally making $Core(D)$ into an abelian category is clearly satisfied, as $Core(D)$ is closed under extensions by Lemma II.2.3. □

Lemma 3.2 *For X, Z in $Core(D)$ we have*

$$Ext^1_{Core(D)}(Z, X) = Hom(Z, X[1]) .$$

Proof. Note that any distinguished triangle (X, Y, Z) satisfies $Y \in Core(D)$ by the assumptions and Lemma II.2.3 and its dual. Therefore the proof follows from the definition of $Core(D)$ and the remark after Corollary II.1.5. The identifying isomorphism is described as follows: Any extension of Z by X in the abelian category $Core(D)$ corresponds to an isomorphism class $[E]$ of an exact sequence in $Core(D)$

$$E : \qquad 0 \to X \to Y \to Z \to 0 .$$

By definition, such an exact sequence in the core arises from a distinguished triangle (X, Y, Z, u, v, w) in D. The isomorphism class $[E]$ of it is uniquely determined by $w \in Hom(Z, X[1])$. See Corollary II.1.5 and the remark thereafter. □

The corresponding statement for higher Ext-groups is satisfied for the derived categories $D(A)$ with their canonical t-structure. It might not be satisfied in general.

3.3 Thick Subcategories. A full triangulated subcategory C of a triangulated category D is called a **thick subcategory**, if objects $X, Y \in D$ must lie in C, provided there exists a morphism $f : X \to Y$ such that

1) f factors over an object in C
2) f has a cone $Cone_f \in C$.

If D is a triangulated category and if C is a thick triangulated subcategory, then the quotient category D/C is a triangulated category. See [SGA4$\frac{1}{2}$], p. 276ff.

Now let **A** be an abelian category. A **Serre subcategory B** or thick subcategory of the abelian category **A** is by definition a full subcategory, which is closed under taking subquotients and extensions. The quotient category **A**/**B** exists as an abelian category. It is obtained by inverting all morphisms, whose kernel and cokernel is in **B**. This class of morphisms allows a calculus of fraction. The quotient functor $\mathbf{A} \to \mathbf{A}/\mathbf{B}$ is an exact functor and maps objects of **B** to the zero object. Any functor from **A** into an abelian category with this property factorizes over the quotient functor $\mathbf{A} \to \mathbf{A}/\mathbf{B}$. See Gabriel [102], p. 364ff.

Examples. Suppose D is a triangulated category with t-structure and core $\mathbf{A} = Core(D)$. Suppose C is a thick subcategory of D, which is stable under the corresponding truncation functor $\tau_{\leq 0}$. Then $\mathbf{B} = C \cap Core(D)$ is a Serre subcategory of the abelian category **A**.

Suppose D is a triangulated category with t-structure. Suppose **B** is a Serre subcategory of $\mathbf{A} = Core(D)$. Then the full subcategory C defined by the objects X in D, whose cohomology objects $H^i(X)$ – as defined in the next section! – are in **B** for all $i \in \mathbb{Z}$, is a thick triangulated subcategory of D stable under the truncation functor $\tau_{\leq 0}$.

Now assume that **A** is a noetherian and artinian abelian category, and suppose **B** is a Serre subcategory of **A**. Then objects X of **A** have a unique maximal subobject X_s isomorphic to an object in **B**, and a unique (up to isomorphism) maximal quotient object X_q isomorphic to an object in **B**. The quotient ${}_rX = X/X_s$ is left reduced – i.e. has no nontrivial subobjects from **B** – and the kernel $X_r = Ker(X \to X_q)$ is right reduced – i.e. has no nontrivial quotient objects from **B**. The epimorphism $X \to {}_rX$ induces an isomorphism $({}_rX)_r \cong {}_r(X_r)$. For this consider the left vertical morphism of the next diagram

$$\begin{array}{ccccccccc}
0 & \longrightarrow & X_r & \longrightarrow & X & \longrightarrow & X_q & \longrightarrow & 0 \\
 & & \downarrow{\scriptstyle \exists!} & & \downarrow & & \downarrow{\scriptstyle \exists!} & & \\
0 & \longrightarrow & ({}_rX)_r & \longrightarrow & {}_rX & \longrightarrow & ({}_rX)_q & \longrightarrow & 0
\end{array}$$

Using the connecting morphism of the snake lemma applied to the diagram, it follows that the cokernel of the left vertical morphism is in **B**. Since $({}_rX)_r$ is right reduced, this cokernel therefore is zero. Thus $X_r \to ({}_rX)_r$ is an epimorphism. The kernel K of this epimorphism is in **B** (a subobject of X_s). Since ${}_rX$ has no subobjects in **B**, also $({}_rX)_r$ has no subobjects in **B**. Therefore K is the maximal subobject of X_r in **B**. Thus $({}_rX)_r \cong {}_r(X_r)$.

The object $({}_rX)_r \cong {}_r(X_r)$ is a reduced object, by which we mean that it has neither a nontrivial subobject nor a nontrivial quotient isomorphic to an object of **B**. Any object X becomes isomorphic in the quotient category $\mathbf{A}/\mathbf{B}$ to its reduced subquotient $({}_rX)_r$. The quotient category $\mathbf{A}/\mathbf{B}$ can be described up to equivalence in the following way: It has the same objects as **A**, but modified homomorphism groups such that – see [102]: objects X, Y of **A** become isomorphic in $\mathbf{A}/\mathbf{B}$ if and only if their reduced objects are isomorphic in **A**. For reduced objects X, Y one has the equality

$$Hom_{\mathbf{A}/\mathbf{B}}(X, Y) = Hom_{\mathbf{A}}(X, Y) .$$

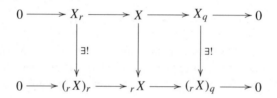

Hence $\mathbf{A}/\mathbf{B}$ is equivalent to the full subcategory of $\mathbf{A}$ defined by the reduced objects (this however is in general not an abelian subcategory of $\mathbf{A}$).

If Y is left reduced and X is arbitrary, then $Hom_{\mathbf{A}/\mathbf{B}}(X, Y) = Hom_{\mathbf{A}}(X_r, Y)$. The reader is advised to describe the general composition law!

II.4 The Cohomology Functors

It is a remarkable fact, that the derived category combines homological and cohomological properties. This property of derived categories carries over to triangulated categories with t-structure.

Let D be a triangulated category with a t-structure. We define

$$H^0(X) = \tau_{\leq 0}\tau_{\geq 0}X \in Core(D)\ .$$

More generally we define for $n \in \mathbb{Z}$ the **n-th cohomology functors**

$$H^n : D \to Core(D)$$

by $H^n(X) = H^0(X[n])$; and similar $H^n(u) = \tau_{\leq 0}\tau_{\geq 0}(u[n])$ for morphisms u. Note that

$$H^n(X)[-n] = \tau_{\leq n}\tau_{\geq n}X\ .$$

We will prove below, that for these functors and for a distinguished triangle (X, Y, Z) in the triangulated category D there exists a long exact cohomology sequence in the abelian category $Core(D)$.

To begin with let us assume $Z = \tau_{\geq 0}Z \in D^{\geq 0}$: Then $Z[-1] \in D^{\geq 1}$ is contained in $D^{\geq 0}$. Note that $\tau_{\leq 0}(Z[-1]) = 0$ by II.2.4. For a distinguished triangle (X, Y, Z, u, v, w) and any A in D we now get the following commutative diagram with exact columns and the exact horizontal row, using the long exact Hom-sequences (Theorem II.1.3)

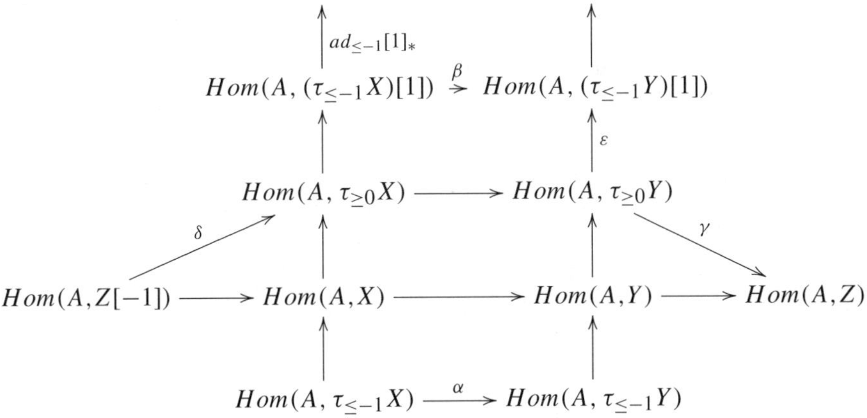

By our assumption on Z we get

a) γ as in the diagram exists! It is the map induced from

$$c : \tau_{\geq 0} Y \to Z\,,$$

where c is obtained from the factorization $Y \to \tau_{\geq 0} Y \to \tau_{\geq 0} Z = Z$.

b) $\tau_{\leq -1} X \xrightarrow{\sim} \tau_{\leq -1} Y$, thus α and β are isomorphisms.

The second statement follows by adjunction from $Hom(A, X) \cong Hom(A, Y)$ for all $A \in D^{\leq -1}$, using Theorem II.1.3 and $Hom(A, Z) = Hom(A, Z[-1]) = 0$ for $A \in D^{\leq -1}$ by II.2.1(i).

Claim. For $Z = \tau_{\geq 0} Z$ the sequence

$$(S) \qquad Hom(A, Z[-1]) \xrightarrow{\delta} Hom(A, \tau_{\geq 0} X) \to Hom(A, \tau_{\geq 0} Y) \xrightarrow{\gamma} Hom(A, \tau_{\geq 0} Z)$$

is exact for any object A of the triangulated category D.

By an easy diagram chase – which is left to the reader – this is reduced to the exactness of the horizontal and the two vertical lines of the diagram. Exactness at $Hom(A, \tau_{\geq 0} X)$ follows from observation b) above. Exactness at $Hom(A, \tau_{\geq 0} Y)$ uses

c) The following holds

$$\beta^{-1} \varepsilon\big(Ker(\gamma)\big) \subset Ker\big(ad_{\leq -1}[1]_*\big)\,.$$

Proof of c). Apply axiom TR3 of triangulated categories, which provides a morphism f and a commutative diagram (e inducing ε, c inducing γ)

$$\begin{array}{ccccccc}
Y & \longrightarrow & \tau_{\geq 0} Y & \xrightarrow{e} & (\tau_{\leq -1} Y)[1] & \xrightarrow{ad_{\leq -1}[1]} & Y[1] \\
\downarrow{\scriptstyle id} & & \downarrow{\scriptstyle c} & & \vdots\,{\scriptstyle \exists f} & & \downarrow{\scriptstyle id} \\
Y & \longrightarrow & Z & \longrightarrow & X[1] & \xrightarrow{u[1]} & Y[1]\,.
\end{array}$$

Note $f = adj \circ \tau_{\leq -2}(f)$. Apply $Hom(A, .)$ to this diagram, to obtain

$$\varepsilon\Big(Ker(\gamma)\Big) \subset Ker\Big(f_*\Big) := Ker\Big(Hom(A, (\tau_{\leq -1} Y)[1]) \to Hom(A, X[1])\Big)\,.$$

However under the isomorphism β^{-1} induced by the isomorphism $(\tau_{\leq -1} u)[1]$ the subgroup $Ker(f_*)$ maps to the kernel $Ker(ad_{\leq -1}[1]_*)$ of the morphism $ad_{\leq -1}[1] : (\tau_{\leq -1} X)[1] \to X[1]$, if the lower triangle of the following diagram commutes

$$\begin{array}{ccc}
X[1] & \xrightarrow{u[1]} & Y[1] \\
\uparrow{\scriptstyle ad_{\leq -1}[1]} & \nwarrow{\scriptstyle f} & \uparrow{\scriptstyle ad_{\leq -1}[1]} \\
(\tau_{\leq -1} X)[1] & \xrightarrow{(\tau_{\leq -1} u)[1]} & (\tau_{\leq -1} Y)[1]\,.
\end{array}$$

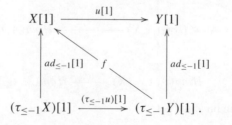

Commutativity of the diagram. The upper triangle and the outer square of this diagram commute by definition. Apply $\tau_{\leq -2}$, which collapses the diagram to its lower map. In particular $\tau_{\leq -2}(f)$ becomes an inverse of the isomorphism $(\tau_{\leq -1}u)[1]$ (fact b)). However f factors over the morphism $\tau_{\leq -2}(f)$ to $(\tau_{\leq -1}X)[1]$ by adjunction. Therefore the lower part of the diagram also commutes. □

Corollary 4.1 *Recall the exact sequence (S) of the last claim, which was proved for objects $Z \in D^{\geq 0}$. If we specialize to A varying in $Core(D)$, then it implies the exact sequence*

$$0 \to H^0(X) \to H^0(Y) \to H^0(Z) \qquad \text{in } Core(D)\,,$$

using the adjunction isomorphism $Hom(A, B) = Hom(A, \tau_{\leq 0}B)$ for $A \in Core(D)$, $B \in D$.

Corollary 4.2 *Dually we obtain for $H_0(X) = \tau_{\geq 0}\tau_{\leq 0}X \in Core(D)$ and distinguished triangles (X, Y, Z) in D with $X \in D^{\leq 0}$ the following exact sequence*

$$H_0(X) \to H_0(Y) \to H_0(Z) \to 0 \qquad \text{in } Core(D)\,.$$

The exact sequences II.4.1 and II.4.2 can be fitted together using

Theorem 4.3 *Let D be a triangulated category with t-structure. For the corresponding functors H_0 and H^0 there exists a functorial isomorphism $\delta : H_0(X) \xrightarrow{\sim} H^0(X)$.*

We postpone the proof of Theorem II.4.3 to the end of this section. Using this theorem, we will tacitly identify $H_0(X) = H^0(X)$. Then the two exact sequences in $Core(D)$, proved above, turn out to be special cases of

Theorem 4.4 *Let D be a triangulated category and let (X, Y, Z) be a distinguished triangle in D. The cohomological functors attached to a t-structure of D induce a long exact cohomology sequence in $Core(D)$*

$$\ldots \to H^{-1}(Z) \to H^0(X) \to H^0(Y) \to H^0(Z) \to H^1(X) \to \ldots\,.$$

Proof. Obviously one gets a complex. By the rotation axiom TR1 it is enough to show exactness at one place, say $H^0(Y)$. The proof of this reduces to the special cases II.4.1 and II.4.2 considered above. Namely apply the functor H^0 to the octaeder diagram TR4a (extended to the right)

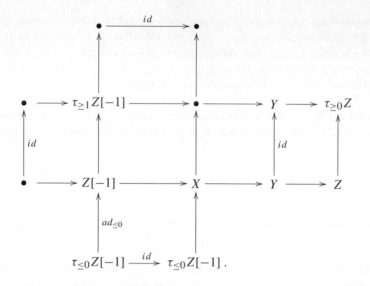

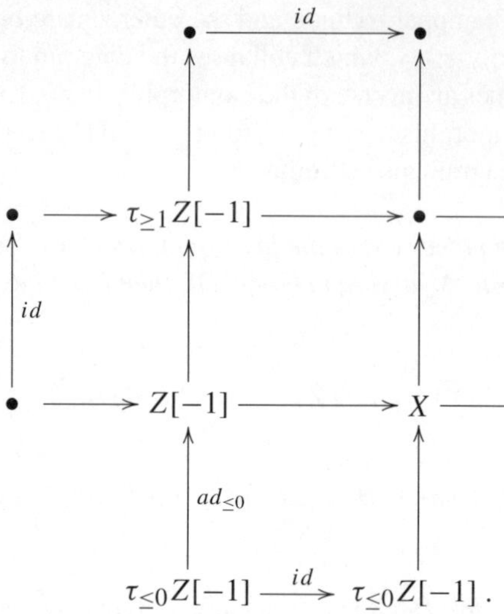

Now exactness at $H^0(Y)$ in the lower row can be deduced by a diagram chase from exactness at $H^0(Y)$ in the upper row. Except the trivial statement $H^0(Z) = H^0(\tau_{\geq 0} Z)$ this uses the fact, that $H^0(X) \to H^0(\bullet)$ is an epimorphism. This follows from II.4.2 for the distinguished triangle $(\tau_{\leq 0} Z[-1], X, \bullet)$ with $\tau_{\leq 0} Z[-1] \in D^{\leq 0}$. Finally exactness at $H^0(Y)$ in the upper row follows from II.4.1 applied to the distinguished triangle $(\bullet, Y, \tau_{\geq 0} Z)$ with $\tau_{\geq 0} Z \in D^{\geq 0}$. □

Exercise. If D is bounded with respect to the t-structure, then $X = 0$ resp. X is in $D^{\leq 0}$ or in $D^{\geq 1}$ iff $H^n(X) = 0$ for all n, resp. for all $n \geq 1$ resp. for all $n \leq 0$.

Proof of Theorem II.4.3. Suppose given integers $n \geq m$.

Apply $\tau_{\leq n} \to id \to \tau_{\geq n+1}$ to X and $\tau_{\geq m} X$, for $X \in D$. This gives two distinguished triangles T'' and T' together with a morphism $T'' \to T'$ induced by $(\beta, ad_{\geq m}, id)$ for some β, which is obtained from axiom TR3 by completing the left square of

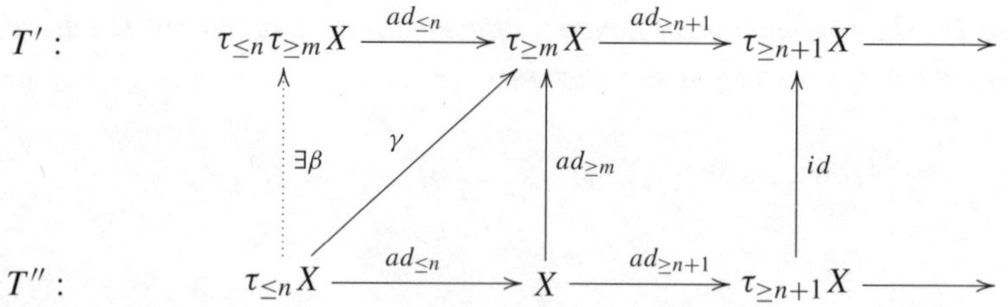

Here we used $ad_{\geq n+1} \circ ad_{\geq m} = ad_{\geq n+1}$.

On the other hand, starting from the lower square $ad_{\leq n} \circ ad_{\leq m-1} = ad_{\leq m-1}$ the octaeder diagram gives a morphism $T'' \to T$ of distinguished triangles:

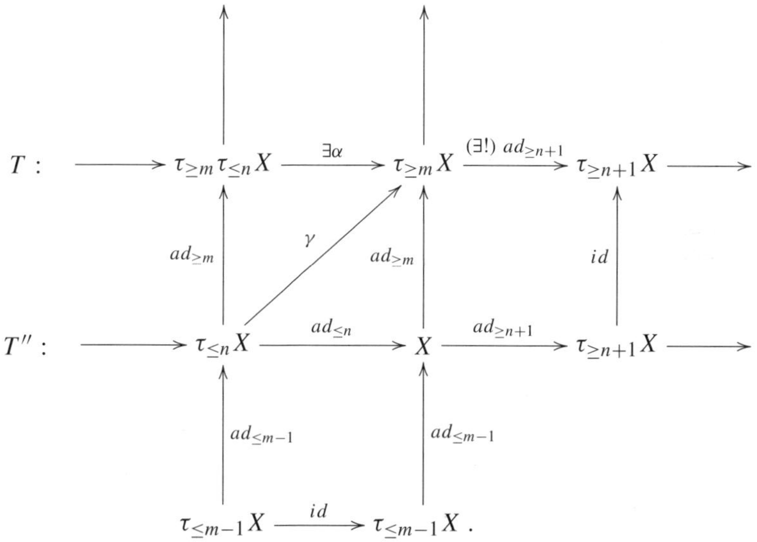

Actually the upper morphism in the right square, whose existence is provided by the octaeder axiom, is uniquely determined. It has to be $ad_{\geq n+1}$ by adjunction.

Now compare the distinguished triangles T' and T. Both triangles complete the morphism $ad_{\geq n+1} : \tau_{\geq m}X \to \tau_{\geq n+1}X$. Therefore, by Corollary II.1.4, there exists an isomorphism $(\delta, id, id) : T \xrightarrow{\sim} T'$. In particular we get an isomorphism

$$\delta : \tau_{\geq m}\tau_{\leq n}X \xrightarrow{\sim} \tau_{\leq n}\tau_{\geq m}X \ .$$

For $m = n = 0$ the existence of this isomorphism proves Theorem II.4.3, except that δ might not be functorial (since a priori it need not be uniquely defined). So the proof of Theorem II.4.3 will be completed by the following consideration.

Functoriality of δ. Put $\gamma = ad_{\geq m} \circ ad_{\leq n}$. This is a natural transformation.

Claim: α, β and δ (as chosen above) make the following two diagrams commutative and this characterizes them uniquely, making them into natural transformations as well:

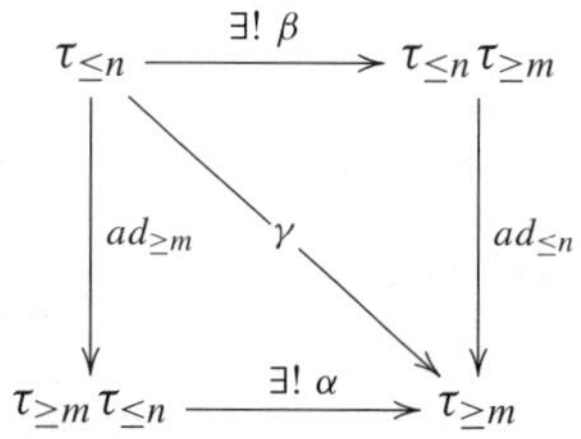

,
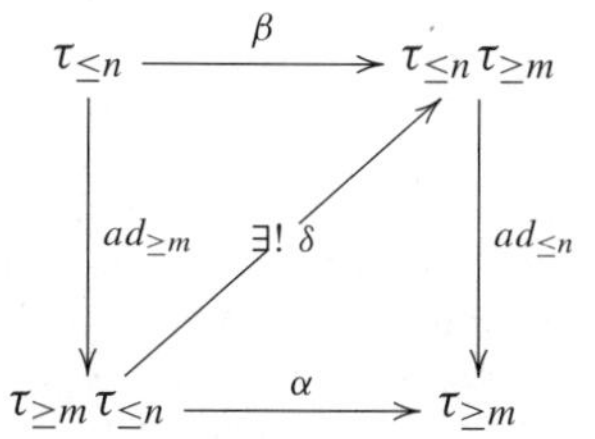

.

The proof of the claim is not difficult and we leave the details of its verification to the reader. One has to use the existence of the maps $T'' \to T'$ and $T'' \to T$ defined above (for the convenience of the reader we already have filled in γ into the left upper diagram) and the following sequence of isomorphisms (given by adjunction isomorphisms and Lemma II.2.4 for $m \leq n$)

$$\begin{array}{cccccc} \gamma \in & Hom(\tau_{\leq n}, \tau_{\geq m}) & \xleftarrow{\cong} & Hom(\tau_{\leq n}, \tau_{\leq n}\tau_{\geq m}) & \ni \beta \\ & \uparrow \cong & & & \\ \alpha \in & Hom(\tau_{\geq m}\tau_{\leq n}, \tau_{\geq m}) & \xleftarrow{\cong} & Hom(\tau_{\geq m}\tau_{\leq n}, \tau_{\leq n}\tau_{\geq m}) & \ni \delta\,. \end{array}$$

The horizontal maps are defined by the composition $\psi \mapsto ad_{\leq n} \circ \psi$, the vertical map by the composition $\phi \mapsto \phi \circ ad_{\geq m}$. Finally $\beta = \delta \circ ad_{\geq m}$ since $ad_{\leq n} \circ \beta = \gamma$ and $ad_{\leq n} \circ \delta \circ ad_{\geq m} = \alpha \circ ad_{\geq m} = \gamma$. □

We get as a

Corollary-Definition 4.5 (Assume $n \geq m$). The natural transformation δ defines an isomorphism $\delta : \tau_{\geq m}\tau_{\leq n} \cong \tau_{\leq n}\tau_{\geq m}$ and we obtain a functor

$$\tau_{[m,n]} = \tau_{\geq m}\tau_{\leq n} \cong \tau_{\leq n}\tau_{\geq m}\ ,$$

which is the identity on $D^{[m,n]} := D^{\leq n} \cap D^{\geq m}$.

Exercise 4.5 Let D be a triangulated category, bounded with respect to some t-structure. Suppose given $K \in Ob(D)$ and $N : K \to K$ a nilpotent morphism in D, i.e. $N^n = 0$ for some integer n. Then the following statements are equivalent:

(i) $K \in Core_t(D)$
(ii) The cone of $N : K \to K$ is in ${}^tD^{[-1,0]}(X)$.

Hint: Show $Cone(N^{2i}) \in {}^tD^{[a,b]}(X)$ if and only if $Cone(N^i) \in {}^tD^{[a,b]}(X)$, using the triangle $(Cone(N^i), Cone(N^{2i}), Cone(N^i))$, to reduce to the case $N = 0$.

II.5 The Triangulated Category $D_c^b(X, \overline{\mathbb{Q}}_l)$

The triangulated category of $\overline{\mathbb{Q}}_l$-sheaves is not a derived category in the original sense. It is obtained as a localization of a projective limit of derived categories, under certain finiteness assumptions. We remind the reader, that the triangulated category of $\overline{\mathbb{Q}}_l$-sheaves has already been used in Chap. I in order to define the Deligne-Fourier transform. It will play an even more important role in the following chapters of the book, in particular for the definition of the category of perverse sheaves. A natural way to construct such a triangulated category of $\overline{\mathbb{Q}}_l$-sheaves is to consider it as the direct limit of triangulated categories of E-sheaves, where E runs over all finite extension fields of the field $\mathbb{Q}_l$. However it is nontrivial, to define such a triangulated category of E-sheaves with good properties. A naive natural candidate to start from would be the derived category of the abelian category of π-adic sheaves introduced in this section. Unfortunately this abelian category of π-adic sheaves does not have sufficiently many injectives or even acyclic objects, in order to define the interesting functors $f_*(-)$ and $\mathcal{H}om(\mathcal{G}, -)$, ... as derived functors in the usual way. The authors of this book do not know, whether there is a mysterious way to define such derived functors with reasonable properties in a direct way on the derived category of π-adic sheaves or the derived category of the abelian category of constructible E-sheaves. On the other hand, Ekedahl [93] has defined a substitute of the derived category of the abelian category of π-adic sheaves. It is a triangulated category with all the good properties, and which allows to define the "derived" functors of all the usual functors in a reasonable way. In fact, Ekedahl's definition is technically rather involved. Fortunately, by the assumptions on the base scheme, which are made throughout this book, the definition considerably simplifies. Certain finiteness theorems then allow to define the desired category by a kind of projective limit, more precisely a projective limit of 2-categories. We start with the notion of π-adic sheaves.

The general reference for the following is [FK], chap. I, §12. See furthermore [SGA4$\frac{1}{2}$], rapport 4.5–4.8 and also Appendix A of this book.

In this section X will denote finitely generated schemes over a finite field or over an algebraically closed field.

Let E be a finite extension field of the field $\mathbb{Q}_l$ of l-adic numbers, let $\mathfrak{o}$ be the valuation ring of E and let π be a generator of the maximal ideal of $\mathfrak{o}$. Let

$$\mathfrak{o}_r = \mathfrak{o}/\pi^r\mathfrak{o} \quad , \quad r \geq 1 \, .$$

The prime number l is assumed to be invertible on X.

Excursion on π-Adic Sheaves

We freely use notions and results from [FK] in the following. In particular the notion of the **A-R category (Artin-Rees category)** will be of importance. However it will be necessary to generalize from the l-adic situation considered in [FK] to the π-adic case. The necessary modifications for this generalization are rather obvious. See also Appendix A.

Suppose

$$\mathscr{G} = (\mathscr{G}_r)_{r\geq 1}$$

is a projective system of constructible torsion $\mathfrak{o}$-module sheaves $\mathscr{G}_r$ on X, such that $\pi^r\mathscr{G}_r = 0$. Then $\mathscr{G}_r$ is a sheaf of $\mathfrak{o}_r$-modules.

The system $\mathscr{G}$ is called a **π-adic sheaf**, if the following holds

$$\mathscr{G}_r \otimes_{\mathfrak{o}_r} \mathfrak{o}_s = \mathscr{G}_s \qquad \text{for all} \quad r \geq s \, .$$

1. The projective system $\mathscr{G}$ is called **flat**, if all $\mathscr{G}_r$ have $\mathfrak{o}_r$-free stalks for all geometric points. The projective system $\mathscr{G}$ is called **smooth**[2], if all the sheaves $\mathscr{G}_r$ are locally constant sheaves. For a flat π-adic system $\mathscr{G}$ and integers $1 \leq s \leq r$ one has exact sequences of sheaves

$$0 \to \mathscr{G}_{r-s} \to \mathscr{G}_r \to \mathscr{G}_s \to 0 \, ,$$

where the left map is induced by multiplication with π^s and the right is induced by the natural quotient map $\mathscr{G}_r \to \mathscr{G}_r \otimes_{\mathfrak{o}_r} \mathfrak{o}_s \cong \mathscr{G}_s$.

[2] Called locally constant in [FK] I §12. In French one uses "lisse" or previously "constant tordue".

2. More generally the projective system $\mathscr{G}$ is called an **A-R π-adic sheaf**, if there is a π-adic sheaf $\mathscr{F} = (\mathscr{F}_r)$ in the sense above and a homomorphism of projective systems

$$\phi_r : \mathscr{F}_r \to \mathscr{G}_r \quad , \quad r = 1, 2, ..$$

such that the following associated projective systems

$$(Ker(\phi_r)) \quad , \quad (Koker(\phi_r))$$

are null systems. For that recall the notion of

3. **Null Systems.** A projective system

$$(\mathscr{N}_r)_{r \geq 1}$$

of constructible $\mathfrak{o}$-module sheaves, such that $\pi^r \mathscr{N}_r = 0$, is called a null system, if for some integer $s \geq 1$ the maps

$$\mathscr{N}_{s+r} \to \mathscr{N}_r$$

are zero for all r. If $\mathscr{F}$ and $\mathscr{G}$ are π-adic resp. A-R π-adic as above, they will become isomorphic in the A-R category.

The A-R Category. In the A-R category a null system becomes isomorphic to the zero object, which is represented by the π-adic zero sheaf. The Artin-Rees category is an abelian category. It is the quotient category of the abelian category of projective systems $(\mathscr{G}_r)_r$ of constructible sheaves of torsion $\mathfrak{o}$-modules $\mathscr{G}_r$, divided by the full exact subcategory of null systems. See [FK] p. 122.

Let $j : U \hookrightarrow X$ be an open embedding with closed complement Y. A projective system $\mathscr{G}$ of sheaves on X is A-R π-adic if and only if its restriction to U and Y is a A-R π-adic system. It is a A-R π-adic system if and only if it is etale locally a A-R π-adic system. See [FK], p. 124.

In the following we will often relax notation and address A-R π-adic sheaves $\mathscr{G}$ as π-adic sheaves. If a distinction is necessary we will stress that π-adic sheaves, as for example the sheaf $\mathscr{F}$ above, are π-adic sheaves in the "true sense".

The Derived Category $D^b_{ctf}(X, \mathfrak{o}_r)$ of Perfect Constructible Complexes of $\mathfrak{o}_r$-Modules on X

As already mentioned, the derived category of E-sheaves is not defined as the derived category of the abelian A-R category. The decisive step for its definition is to define the projective limit category $D^b_c(X, \mathfrak{o})$

$$D^b_c(X, \mathfrak{o}) = \text{"lim"} D^b_{ctf}(X, \mathfrak{o}_r) .$$

For this definition consider the transition functors

$$D^b_{ctf}(X, \mathfrak{o}_{r+1}) \longrightarrow D^b_{ctf}(X, \mathfrak{o}_r) \quad , \quad r = 1, 2, ...$$

$$K^\bullet \mapsto K^\bullet \otimes^L_{\mathfrak{o}_{r+1}} \mathfrak{o}_r .$$

To define the projective limit of categories correctly and to verify some of the main properties, it is useful to introduce a number of useful notions first.

Let the notation and assumptions be as above. Consider the bounded derived category of all etale $\mathfrak{o}_r$-module sheaves on X. Let then $D_c^b(X, \mathfrak{o}_r)$ denote the full triangulated subcategory of all objects, whose cohomology sheaves are constructible etale sheaves on X.

The category $D_{ctf}^b(X, \mathfrak{o}_r)$ is defined to be the full subcategory of $D_c^b(X, \mathfrak{o}_r)$ of those complexes, which are quasiisomorphic in $D_c^b(X, \mathfrak{o}_r)$ to a bounded $\mathfrak{o}_r$-**flat complex** $K^\bullet$, i.e. all components K^ν are flat $\mathfrak{o}_r$-sheaves and almost all of them are zero. In particular all cohomology sheaves of $K^\bullet$ are constructible. "**Bounded**" means: all K^ν vanish except finitely many. "**Cohomologically bounded**" means, that almost all cohomology sheaves vanish. The category $D_{ctf}^b(X, \mathfrak{o}_r)$ defines a triangulated subcategory of the derived category $D_c^b(X, \mathfrak{o}_r)$ of cohomologically bounded complexes of $\mathfrak{o}_r$-sheaves with constructible cohomology sheaves.

The larger category $D_c^b(X, \mathfrak{o}_r)$ has a natural t-structure. The underlying truncation operators

$$\tau_{\geq m} \quad , \quad \tau_{\leq n}$$

are defined as follows: $\tau_{\geq m} K^\bullet$ is obtained, if we replace K^ν by zero for $\nu < m$ and by $coker(d : K^{m-1} \to K^m)$ for $\nu = m$. Furthermore $(\tau_{\leq n} K)^\nu$ is obtained by replacing K^ν by 0 for $\nu > n$ and by $ker(d : K^n \to K^{n+1})$ for $\nu = n$. Using these truncation operators every complex in $D_c^b(X, \mathfrak{o}_r)$ turns out to be quasiisomorphic to a bounded complex.

This natural t-structure on $D_c^b(X, \mathfrak{o}_r)$ is not inherited by the triangulated full subcategory $D_{ctf}^b(X, \mathfrak{o}_r)$, because truncation does not preserve the $\mathfrak{o}_r$-flatness condition. Nevertheless in some special cases this is still true. See the technical remarks below, which the reader may skip on a first reading.

5.1 Technical Remarks

(1) For an $\mathfrak{o}_r$-module M – here we do not mean a sheaf! – the following assertions are equivalent:

(a) M is injective
(b) M is free
(c) M is projective
(d) M is flat
(e) $\pi^{r-s}M = Kern(\pi^s : M \to M)$ for all $1 \leq s \leq r$ ("divisibility").

Obviously (b) $\Longrightarrow$ (c) $\Longrightarrow$ (d) holds for modules over arbitrary rings Λ. Similarly, an injective module is always a direct factor of a cofree module. Hence in our case an injective module is a direct product of free modules, since $\mathfrak{o}_r \cong Hom(\mathfrak{o}_r, \mathbb{Q}/\mathbb{Z})$. Cofree or flat, hence (a),(b),(c),(d) all imply (e).

Assume (e). Consider lifts $v_i, i \in I$ of a basis of the quotient vector space $M \to M/\pi M$. Suppose $\sum_{i \in I} \lambda_i v_i = 0$. Choose $s \leq r$ maximal such that $\lambda_i \in \pi^s \mathfrak{o}_r$ for all i. Then $\lambda_i = \pi^s \mu_i$. If $s \neq r$ then $\pi^s \cdot w = 0$ for $w = \sum_i \mu_i \cdot v_i$ implies

$w \in \pi^{r-s}M$. This gives a contradiction mod πM. Hence the v_i are linear independent over $\mathfrak{o}_r$. Their span N satisfies $M \subset N + \pi M$, hence $M \subset N$. So $M = N$ is free.

Finally a free $\mathfrak{o}_r$-module M is "divisible" in the sense, that any morphism $N \to M$ of a cyclic module N into M can be extended to a morphism $N' \to M$ for any cyclic module N' containing N. A "divisible" $\mathfrak{o}_r$-module is injective (Zorn's lemma). □

Now some consequences of the facts stated above: Let

$$0 \longrightarrow K \longrightarrow L \longrightarrow M \longrightarrow 0$$

be an exact sequence of $\mathfrak{o}_r$-modules. If any two of the three modules satisfy (a)-(e), then also the third. This implies (by induction from below), that for any acyclic bounded below complex of injective $\mathfrak{o}_r$-modules all the modules $ker(d)$, for the differentials d of the complex, are again injective $\mathfrak{o}_r$-modules. The same is true for a acyclic bounded above complex of injective $\mathfrak{o}_r$-modules (induction from above).

(2) Suppose M, N are free $\mathfrak{o}_r$-modules and suppose $\phi : M \to N$ is an $\mathfrak{o}_r$-linear map. For any natural number $1 \leq s \leq r$ this induces maps $\phi_s : M_s \to N_s$ where $M_s = M \otimes_{\mathfrak{o}_r} \mathfrak{o}_s$, $N_s = N \otimes_{\mathfrak{o}_r} \mathfrak{o}_s$, $\phi_s = \phi \otimes_{\mathfrak{o}_r} \mathfrak{o}_s$. The map $\phi = \phi_r$ is injective iff ϕ_s is injective. One direction is clear from **(1a),(1b)**: If ϕ is injective, then it is an isomorphism onto a direct summand. This implies that ϕ_s is injective for all $s \leq r$. For the converse we can therefore assume $s = 1$ and use induction with respect to r. The claim follows from the snake lemma, which relates ϕ_r to ϕ_{r-1} and ϕ_1.

(3) For $1 \leq s \leq r$ an $\mathfrak{o}_s$-module can be viewed as an $\mathfrak{o}_r$-module via the surjective quotient map $\mathfrak{o}_r \to \mathfrak{o}_s$. In this sense, any $\mathfrak{o}_r$-module M is isomorphic to a direct sum of free $\mathfrak{o}_s$-modules N_s

$$M \cong \oplus_{1 \leq s \leq r} N_s \ .$$

Proof. Consider the $\mathfrak{o}_{r-1}$-module $M_{tor} = \{x \in M \mid \pi^{r-1} \cdot x = 0\}$. The quotient M/M_{tor} is an $\mathfrak{o}_1$-vectorspace. Lifting a basis gives $\mathfrak{o}_r$-linear independent elements, by a similar argument as in **(1)** above. They span a free submodule N of M. Since N is injective by **(1)** $M = N \oplus (M_{tor}/N \cap M_{tor})$. The claim follows now by induction. □

Let $(M_r)_{r \geq 1}$ be a projective system of finite $\mathfrak{o}_r$-modules M_r. Let $\mathbf{M}$ be the projective limit, which is an $\mathfrak{o}$-module. If the system is π**-adic**, i.e. if $M_{r+1} \otimes_{\mathfrak{o}_{r+1}} \mathfrak{o}_r \cong M_r$, then $\mathbf{M}$ is a finitely generated $\mathfrak{o}$-module (Nakayama's lemma) and $M_r = \mathbf{M}/\pi^r\mathbf{M}$ with the obvious transition maps. Let the system be not necessarily π-adic: Then

(a) $\mathbf{M}$ is flat (π-torsionfree) if the M_r are flat (π-torsionfree).
(b) $\mathbf{M}$ is finitely generated if $dim_{\mathfrak{o}_1}(M_r/\pi M_r) \leq const$ for some constant independent of r.

Proof of (b). The condition is inherited by subquotients. One therefore can assume that the transition maps are surjective. Then lift a primitive element. It remains primitive on each r-level. Divide by it and argue by induction on the length. □

(4) Let I be an injective sheaf of $\mathfrak{o}_r$-modules. Then all stalks are injective $\mathfrak{o}_r$-modules. In particular, I is an $\mathfrak{o}_r$-flat sheaf. Look at the stalks for the geometric points!

(5) Let $J^\bullet$ be a bounded (resp. bounded from below) complex with constructible cohomology sheaves, such that all J^ν are $\mathfrak{o}_r$-flat sheaves. Suppose given m such that

$$\mathscr{H}^\nu(J^\bullet) = 0 \quad , \quad \text{for } \nu < m \, .$$

Then $\tau_{\geq m}(J^\bullet)$ is again a bounded (resp. bounded from below), $\mathfrak{o}_r$-flat complex, which is quasiisomorphic to $J^\bullet$.

Proof. Recall that $\tau_{\geq m} J^\bullet$ is obtained from $J^\bullet$ by replacing J^ν by zero for $\nu < m$ and by $coker(d : J^{m-1} \to J^m)$ for $\nu = m$. Via induction from below $coker(d : J^{m-1} \to J^m)$ is $\mathfrak{o}_r$-flat because $\mathscr{H}^\nu(J^\bullet) = 0$ for $\nu < m$. Use **(1)**. Therefore $\tau_{\geq m} J^\bullet$ is a $\mathfrak{o}_r$-flat bounded complex. □

Let M be any $\mathfrak{o}_r$-module. Under the assumptions above $J^\bullet \otimes^L_{\mathfrak{o}_r} M = J^\bullet \otimes_{\mathfrak{o}_r} M$, because all J^ν_r are $\mathfrak{o}_r$-flat. Furthermore the natural map

$$J^\bullet \otimes_{\mathfrak{o}_r} M \to \tau_{\geq m}(J^\bullet) \otimes_{\mathfrak{o}_r} M$$

is a quasiisomorphism and induces an isomorphism

$$\tau_{\geq m}(J^\bullet \otimes_{\mathfrak{o}_r} M) \to \tau_{\geq m}(J^\bullet) \otimes_{\mathfrak{o}_r} M$$

(6) Let $I^\bullet$ be a complex, bounded from below, of injective sheaves I^ν over $\mathfrak{o}_r$ and suppose $I^\bullet$ is quasiisomorphic (i.e. isomorphic in the derived category) to a bounded $\mathfrak{o}_r$-flat constructible complex $J^\bullet$ (with respect to some honest morphism)

$$f : J^\bullet \to I^\bullet \, .$$

(For instance let $I^\bullet$ be an injective resolution of $J^\bullet$).

Choose an integer n such that

$$\mathscr{H}^\nu(I^\bullet) = 0 \quad , \quad \nu > n \, .$$

The stalks I^ν_x are injective $\mathfrak{o}_r$-modules for all geometric points x. By Remark (1) and (5) the mapping cone $C = C(f_x)$ has $\mathfrak{o}_r$-flat kernels $ker(d_C)$. Since f induces an isomorphism of stalk cohomologies, we have exact sequences

$$0 \longrightarrow Ker(d^\nu : I^\nu_x \to I^{\nu+1}_x) \longrightarrow Ker(d^\nu_C) \longrightarrow Im(d^\nu : J^\nu_x \to J^{\nu+1}_x) \longrightarrow 0$$

for all ν. Therefore $ker(d : I^n_x \to I^{n+1}_x)$ is $\mathfrak{o}_r$-injective, hence is a direct summand of I^n_x, since $Im(d^\nu : J^\nu_x \to J^{\nu+1}_x)$ is flat for $\nu \geq n$ by induction from above using $\mathscr{H}^\nu(J_x) = \mathscr{H}^\nu(I_x) = 0$ for $\nu > n$.

The truncated complex

$$K^\bullet = \tau_{\leq n} I^\bullet$$

is therefore an $\mathfrak{o}_r$-flat complex. Recall $K^\nu = I^\nu$ for $\nu < n$, $K^\nu = 0$ for $\nu > n$ and $K^n = ker(d : I^n \to I^{n+1})$. The $\mathfrak{o}_r$-flatness of all K^ν implies for an arbitrary $\mathfrak{o}_r$-module M that

$$K^\bullet \otimes_{\mathfrak{o}_r} M \cong I^\bullet \otimes_{\mathfrak{o}_r} M = I^\bullet \otimes^L_{\mathfrak{o}_r} M .$$

The first map is a quasiisomorphism. This is checked on stalks of geometric points using the $\mathfrak{o}_r$-flatness of all K^ν_x and I^ν_x. The second identity holds, because all I^ν_r are $\mathfrak{o}_r$-flat.

(7) Let $J^\bullet$ be a bounded complex of sheaves of flat $\mathfrak{o}_r$-modules with constructible cohomology sheaves. Suppose given integers $m \leq n$ such that

$$\mathscr{H}^\nu(J^\bullet) = 0 \quad \text{for} \quad \nu < m,\ \nu > n .$$

Then the truncated complex

$$L^\bullet = \tau_{\leq n}\tau_{\geq m} J^\bullet$$

is a complex, which is quasiisomorphic to the complex $J^\bullet$. By **(5)** and **(6)** the complex $L^\bullet$ is a complex with $\mathfrak{o}_r$-flat sheaves L^ν. Furthermore $L^\nu = 0$ for $\nu < m$ and $\nu > n$ and $\mathscr{H}^\nu(L^\bullet \otimes_{\mathfrak{o}_r} M) \cong \mathscr{H}^\nu(J^\bullet \otimes_{\mathfrak{o}_r} M)$ for all $\mathfrak{o}_r$-modules M.

(8) Let $K^\bullet$ be a bounded complex of $\mathfrak{o}_r$-flat sheaves with constructible cohomology. Suppose for some $1 \leq s \leq r$ that

$$\mathscr{H}^\nu(K^\bullet \otimes_{\mathfrak{o}_r} \mathfrak{o}_s) = 0 \quad \text{for all} \quad \nu < m,\ \nu > n .$$

Then the following holds

$$\mathscr{H}^\nu(K^\bullet) = 0 \quad \text{for all} \quad \nu < m,\ \nu > n .$$

Proof. Suppose $\mathscr{H}^\nu(K^\bullet) \neq 0$ for some ν. Let $\nu = M$ be the minimal choice. We have to show $M \geq m$. By **(7)** we may assume that $K^\nu = 0$ for $\nu < M$. Suppose $M < m$ and let d^M be the differential $d^M : K^M \to K^{M+1}$. By assumption the corresponding map

$$d^M \otimes_{\mathfrak{o}_r} \mathfrak{o}_s : \quad K^M \otimes_{\mathfrak{o}_r} \mathfrak{o}_s \to K^{M+1} \otimes_{\mathfrak{o}_r} \mathfrak{o}_s$$

is injective. Then d^M is injective by **(2)**. A contradiction! A similar but simpler argument applies for the $\nu > n$, where it is enough to use the right exactness of the tensor functor. □

(9) Let us close with the observation that for a bounded complex $K^\bullet$ of $\mathfrak{o}_r$-flat module sheaves with constructible cohomology sheaves such that $K^\nu = 0$ for $\nu < m$ or $\nu > n$, there exists a bounded, $\mathfrak{o}_r$-flat complex $L^\bullet$ in $D^b_{ctf}(X, \mathfrak{o}_r)$ with constructible sheaves L^ν and $L^\nu = 0$ for $\nu < m$ or $\nu > n$ and a complex map $L^\bullet \to K^\bullet$, which is a quasiisomorphism. This follows e.g. from [SGA4$\frac{1}{2}$] rapport Lemma 2.4, if we replace the infinite resolution $(K')^\bullet$ constructed there, by its truncation $L^\bullet = \tau_{\geq m}((K')^\bullet)$

using that $\tau_{\geq m}((K')^\bullet)^m = Cone((K')^\bullet \to K^\bullet)^{m-1}/Ker(d_{m-1})$ is flat by **(1)**. The same also holds for the ring $\Lambda = \mathfrak{o}$ itself.

End of Technical Remarks II.5.1. The last Remark (9) implies

Lemma 5.2 *Every complex in $D_{ctf}(X, \mathfrak{o}_r)$ can be represented by a* **perfect complex**, *i.e. a finite complex $K^\bullet$ of constructible $\mathfrak{o}_r$-flat sheaves K^ν.*

Lemma 5.3 *Let $u : K^\bullet \to J^\bullet$ be a morphism in $D_c^b(X, \mathfrak{o}_r)$. Suppose $K^\bullet$, $J^\bullet$ are bounded, $\mathfrak{o}_r$-flat with constructible cohomology. Then there exist bounded, $\mathfrak{o}_r$-flat complexes $\tilde{K}^\bullet$, $\tilde{J}^\bullet$ with constructible cohomology and maps of complexes $v : K^\bullet \to \tilde{J}^\bullet$ and $w : \tilde{K}^\bullet \to J^\bullet$, $j : J^\bullet \to \tilde{J}^\bullet$ and $k : \tilde{K}^\bullet \to K^\bullet$*

$$\begin{array}{ccc} \tilde{K}^\bullet & \xrightarrow{\;w\;} & J^\bullet \\ {\scriptstyle k}\downarrow & \overset{u}{\nearrow} & \downarrow{\scriptstyle j} \\ K^\bullet & \xrightarrow{\;v\;} & \tilde{J}^\bullet \end{array} ,$$

such that in the derived category $uk = w$ and $ju = v$ holds and j, k become isomorphisms in the derived category. The complex $\tilde{K}^\bullet$ can be chosen to be perfect.

Remark on Sizes. If K^ν, $J^\nu = 0$ for $\nu < m$ or $\nu > n$, the same can be achieved for $\tilde{K}^\nu$, $\tilde{J}^\nu$.

Proof. u is obtained by honest complex maps $u : K^\bullet \to I^\bullet \leftarrow J^\bullet$, where $I^\bullet$ is an injective resolution of $J^\bullet$. Use Remark II.1.8. Naive truncation $\tilde{J}^\bullet = \tau_{\leq n} I^\bullet$ for n – respecting the size of the finite complexes $K^\bullet$ and $J^\bullet$ – defines honest complex maps

$$v : K^\bullet \to \tilde{J}^\bullet$$
$$j : J^\bullet \to \tilde{J}^\bullet$$

such that $ju = v$ holds in the derived category, j being a quasiisomorphism. The assertion concerning $\tilde{J}^\bullet$ follows from **(6)** above.

The assertion concerning $\tilde{K}^\bullet$ is obtained, if we define $\tilde{K}^\bullet$ to be (the perfectification **(9)** of) the bounded flat complex

$$\tilde{K}^\bullet = Cone\big(v+j : \; K^\bullet \oplus J^\bullet \to \tilde{J}^\bullet\big)[-1] \, .$$

The maps k and w are the obvious projections. Then k is a quasiisomorphism and up to homotopy of complex maps the following is true

$$vk = wj \; .$$

See e.g. [104] p. 161. Therefore $uk = w$ holds in the derived category. □

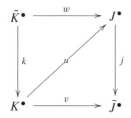

If sizes are to be respected, as claimed in the remark, one better replaces the complex $\tilde{K}^\bullet$ by the quasiisomorphic complex $\tau_{\leq n}\tilde{K}^\bullet$.

The Derived Category

$$D_c^b(X, \mathfrak{o}) = \text{“}\lim_{\overleftarrow{r}}\text{”}\ D_{ctf}^b(X, \mathfrak{o}_r)\ .$$

The field $\overline{\mathbb{Q}}_l$ is the direct limit of the finite extension fields E of the field $\mathbb{Q}_l$ of l-adic numbers. Each of the fields is obtained by localization from its ring $\mathfrak{o} = \mathfrak{o}_E$ of integers. The ring of integers $\mathfrak{o}$ is the projective limit of the finite self injective rings $\mathfrak{o}_r = \mathfrak{o}/\pi^r\mathfrak{o}$ for $r = 1, 2, ...$ Here π denotes a prime element in E. The construction of the triangulated categories $D_c^b(X, \overline{\mathbb{Q}}_l)$ follows the same pattern. Whereas direct limits and localization are exact functors, projective limits usually are not. So to construct $D_c^b(X, \overline{\mathbb{Q}}_l)$, the most critical step is the construction of a triangulated projective limit category $D_c^b(X, \mathfrak{o})$ from the system of derived categories $D_c^b(X, \mathfrak{o}_r)$, where $r = 1, 2, ...$ runs over all integers.

See also the references

Appendix A
[BBD], 2.2.14
Deligne [Del], 1.1 p. 149

The assumptions on X made at the beginning will imply, that this limit category $D_c^b(X, \mathfrak{o})$ exists and that it carries the structure of a triangulated category in a natural way.

The Problem. Suppose we have triangulated categories $(D_r)_{r\geq 1}$ with triangulated transition functors $F_{r+1} : D_{r+1} \to D_r$, i.e. compatible with shifts and preserving distinguished triangles. We would like to define objects of the limit category D as systems $A = (A_r, \phi_r)_{r\geq 1}$ of objects A_r and isomorphisms ϕ_{r+1} in the category D_r such that $\phi_{r+1} : F_{r+1}(A_{r+1}) \cong A_r$. A morphism $\psi : (A_r, \phi_r^A)_{r\geq 1} \to (B_r, \phi_r^B)_{r\geq 1}$ in D should be given by a system $\psi = (\psi_r)_{r\geq 1}$ of morphisms $\psi_r : A_r \to B_r$ in D_r, such that $\phi_{r+1}^B \circ F_{r+1}(\psi_{r+1}) = \psi_r \circ \phi_{r+1}^A$ holds for all r. In such a setting one could ask, whether D carries the structure of a triangulated category. The category D is obviously an additive category. Define the shift operator $\mathscr{T}(A) = A[1]$ by $(A_r, \phi_r)_{r\geq 1}[1] = (A_r[1], \phi_r[1])_{r\geq 1}$. To define distinguished triangles in D, the obvious attempt would be the following: A triangle (A, B, C, f, g, h) in D is distinguished if and only if the corresponding triangles $(A_r, B_r, C_r, f_r, g_r, h_r)$ of the system are distinguished triangles in the categories D_r (for all r). However, with this definition it is unfortunately not clear, how to verify for example axiom TR3. Only under the following strong finiteness assumption, it is easy to verify the axioms TR1–TR4 of triangulated categories.

Assumption. For all $r \geq 1$ the homomorphism groups

$$Hom_{D_r}(X, Y) \quad , \quad X, Y \in Ob(D_r)$$

are finite groups.

Under this assumption the limit category D is a triangulated category with the definitions as above. The proof is easy and left as an exercise for the reader. Just as an example, let us verify axiom TR3:

For given distinguished triangles (X, Y, Z, u, v, w) and (X', Y', Z', u', v', w') and morphisms $f : X' \to X$ and $g : Y' \to Y$, such that $g \circ u' = u \circ f$, we look for a morphism

$$h : Z' \to Z ,$$

such that (f, g, h) extends to a morphism of triangles. Let E_r be the set of morphisms $h_r : Z_r' \to Z_r$, such that (f_r, g_r, h_r) is morphism between the distinguished triangles $(X_r, Y_r, Z_r, u_r, v_r, w_r)$ and $(X_r', Y_r', Z_r', u_r', v_r', w_r')$. Then

(1) The set E_r is nonempty (axiom TR3 for D_r)
(2) The set E_r is a finite set (the finiteness assumption above)

The sets E_r obviously define a projective system

$$\to E_{r+1} \to E_r \to \ldots \to E_1 .$$

By (1) and (2) it follows, that the projective limit

$$E = \varprojlim_r E_r \neq \emptyset .$$

Any choice of $h \in E \subset Hom_D(Z', Z)$ now gives the required extension.

Let us now come back to our original problem. So we specialize to the case of the triangulated categories $D_r = D_c^b(X, \mathfrak{o}_r)$. The transition functors $F_{r+1} : D_{r+1} \to D_r$ are given by the tensor product $K_{r+1}^\bullet \mapsto K_{r+1}^\bullet \otimes_{\mathfrak{o}_{r+1}}^L \mathfrak{o}_r$. The construction above gives the desired limit category $D_c^b(X, \mathfrak{o})$, provided the finiteness assumption made above holds. In order that this finiteness condition holds, we now need the assumptions, made on the base field k. The relevant property is the following:

Let k' be a finite separable extension field of k and let G be the Galois group of the separable closure of K' over k'. Then we want, that the Galois cohomology groups

$$H^\nu(G, \mathbb{Z}/l\mathbb{Z})$$

are finite groups; here G is assumed to act trivially on the coefficients $\mathbb{Z}/l\mathbb{Z}$. Of course this is a strong assumption on the underlying base field k.

Let us assume that all the Galois cohomology groups $H^\nu(G, \mathbb{Z}/l\mathbb{Z})$ are finite groups. Under this assumption on the field k one can deduce from the finiteness theorems, proved by Deligne ([SGA4$\frac{1}{2}$], finitude; see also Appendix D of this book) that for schemes X finitely generated over k and perfect complexes $K^\bullet, L^\bullet \in D_{ctf}(X, \mathfrak{o}_r)$, the following

Theorem 5.4 (Finiteness Theorem) *Assume all Galois cohomology groups $H^\nu(G, \mathbb{Z}/l\mathbb{Z})$ are finite groups. Then the homomorphism groups*

$$Hom_{D(X, \mathfrak{o}_r)}(K^\bullet, L^\bullet)$$

are finite abelian groups.

There are other cases, where the analogous finiteness statements are valid. Since these cases are not relevant for this book, we only mention that the similar statement holds for schemes finitely generated over $\mathbb{Z}$, for schemes finitely generated over strictly henselian rings or over discrete valuation rings with finite residue fields. See Mazur [123].

Building on this finiteness theorem we can now define the category $D_c^b(X, \mathfrak{o})$ as above as the projective limit

$$D_c^b(X, \mathfrak{o}) = \text{“}\lim_r\text{”}\ D_{ctf}^b(X, \mathfrak{o}_r)$$

in down-to-earth terms.

Objects. First of all, an object of $D_c^b(X, \mathfrak{o})$ is a collection

$$K = K^\bullet = (K_r^\bullet)_{r\geq 1}$$

of complexes $K_r^\bullet$ in $D_{ctf}^b(X, \mathfrak{o}_r)$ together with quasiisomorphisms

$$\phi_{r+1} : K_{r+1}^\bullet \otimes_{\mathfrak{o}_{r+1}}^L \mathfrak{o}_r \cong K_r^\bullet$$

in the categories $D_c^b(X, o_r)$. The **ν-th cohomology sheaf** of $K^\bullet$ is by definition the induced projective system $\mathscr{H}^\nu(K^\bullet) = (\mathscr{H}^\nu(K_r^\bullet))_{r\geq 1}$.

Morphisms. They are given by compatible systems of morphisms for each r: For two objects of $D_c^b(X, \mathfrak{o})$ represented by projective systems $K^\bullet = (K_r^\bullet)_{r\geq 1}$ and $L^\bullet = (L_r^\bullet)_{r\geq 1}$ as above put

$$Hom_{D_c^b(X,\mathfrak{o})}(K^\bullet, L^\bullet) = \lim_{\overleftarrow{r}} Hom_{D_c^b(X,\mathfrak{o}_r)}(K_r^\bullet, L_r^\bullet)\ .$$

In other words, a homomorphism $\psi : K^\bullet \to L^\bullet$ in $Hom_{D_c^b(X,\mathfrak{o})}(K^\bullet, L^\bullet)$ is a family $\psi = (\psi_r)_{r\geq 1}$ of morphisms $\psi_r : K_r^\bullet \to L_r^\bullet$ in the derived categories $D_c^b(X, \mathfrak{o}_r)$, such that the following diagrams for $r = 1, 2...$ commute

$$\begin{array}{cccc} \psi_{r+1} \otimes_{\mathfrak{o}_{r+1}}^L \mathfrak{o}_r : & K_{r+1}^\bullet \otimes_{\mathfrak{o}_{r+1}}^L \mathfrak{o}_r & \longrightarrow & L_{r+1}^\bullet \otimes_{\mathfrak{o}_{r+1}}^L \mathfrak{o}_r \\ & \cong\downarrow \phi_{r+1}^K & & \cong\downarrow \phi_{r+1}^L \\ \psi_r : & K_r^\bullet & \longrightarrow & L_r^\bullet \end{array}$$

Note: A morphism $\psi = (\psi_r)_{r\geq 1}$ between two objects $K^\bullet = (K_r^\bullet, \phi_r^K)_{r\geq 1}$ and $L^\bullet = (L_r^\bullet, \phi_r^L)_{r\geq 1}$ is an isomorphism, if and only if all morphisms $\psi_r : K_r^\bullet \to L_r^\bullet$ are isomorphisms. As a consequence of the finiteness theorem formulated above, two objects $K^\bullet = (K_r^\bullet, \phi_r^K)_{r\geq 1}$ and $L^\bullet = (L_r^\bullet, \phi_r^L)_{r\geq 1}$ are isomorphic in the category $D_c^b(X, \mathfrak{o})$ if and only if the objects K_r and L_r are isomorphic in $D_c^b(X, \mathfrak{o}_r)$ for all r. In other words, the isomorphism class of an object $K^\bullet = (K_r^\bullet, \phi_r)_{r\geq 1}$ does not depend on the underlying transition isomorphisms ϕ_r, i.e. for any sequence of automorphisms $\tilde{\phi}_r$ in $Hom_{D_c^b(X,\mathfrak{o}_r)}(K_r^\bullet, K_r^\bullet)$, there exists a sequence of automorphisms ψ_r in $Hom_{D_c^b(X,\mathfrak{o}_r)}(K_r^\bullet, K_r^\bullet)$, such that

$$\phi_{r+1}^K \circ (\psi_{r+1} \otimes_{\mathfrak{o}_{r+1}}^L \mathfrak{o}_r) = \tilde{\phi}_r \circ \psi_r \circ \phi_{r+1}^K .$$

In fact, the obstruction to find such ψ_r is in $lim_r^1\ Aut_{D_c^b(X,\mathfrak{o}_r)}(K_r^\bullet, K_r^\bullet)$. See Jensen [163]. By the finiteness assumption, this lim^1-term vanishes.

In order to work with this definition of $D_c^b(X, \mathfrak{o})$, it is useful to make some preliminary remarks:

1. By assumption each $K_r^\bullet$ can be replaced by a quasiisomorphic bounded $\mathfrak{o}_r$-flat complex, whose bounds above and below – its size – may be controlled by the cohomology sheaves of $K_1^\bullet$ only (Remark (7) and (8)). So all the complexes $(K_r^\bullet)_{r\geq 1}$ can be chosen of uniform size, i.e. uniformly bounded.

2. Next one can use Remark (9) to replace each complex $K_r^\bullet$ by a quasiisomorphic perfect one, without altering the size.

3. So let us suppose that all complexes $K_r^\bullet$ are uniformly bounded perfect complexes. Flatness implies $K_{r+1}^\bullet \otimes_{\mathfrak{o}_{r+1}}^L \mathfrak{o}_r = K_{r+1}^\bullet \otimes_{\mathfrak{o}_{r+1}} \mathfrak{o}_r$. This gives induced transition morphisms

$$K_{r+1}^\bullet \to K_{r+1}^\bullet \otimes_{\mathfrak{o}_{r+1}} \mathfrak{o}_r \cong K_r^\bullet .$$

These morphisms, induced by ϕ_r, are only morphisms in the derived category $D_c^b(X, \mathfrak{o}_{r+1})$. One can choose perfect complexes $\tilde{K}_\nu^\bullet$, by modifying $K^\bullet$ up to quasi-isomorphism without changing the size, such that the transition maps are honest complex maps $\tilde{K}_{r+1}^\bullet \to \tilde{K}_r^\bullet$ instead of being just morphisms in the derived category. This is an immediate consequence of the Lemma II.5.3. We may therefore assume $\tilde{K}_r^\bullet = K_r^\bullet$, with "honest" transition maps

$$K_{r+1}^\bullet \to K_r^\bullet .$$

With the preparation 1.–3. above one can show

Lemma 5.5 *Let $K^\bullet = (K_r^\bullet)_{r\geq 1}$ be an object of $D_c^b(X, \mathfrak{o})$. Then its cohomology sheaves*

$$\mathcal{H}^\nu(K^\bullet) = (\mathcal{H}^\nu(K_r^\bullet))_{r\geq 1}$$

are A-R π-adic sheaves.

Remark. $\mathcal{H}^\nu(K^\bullet)$ is a π-adic sheaf in the relaxed way of speaking (see the comments on the A-R category).

Sketch of Proof of II.5.5. Over an open dense subscheme there exist locally constant sheaves $\mathcal{A}, \mathcal{B}$ of $\mathfrak{o}_1$-modules independent from $r \geq r_0$ such that

$$0 \longrightarrow \mathcal{A} \longrightarrow \mathcal{H}^\nu(K_{r+1}^\bullet) \longrightarrow \mathcal{H}^\nu(K_{r+1}^\bullet \otimes_{\mathfrak{o}_{r+1}} \mathfrak{o}_r) \longrightarrow \mathcal{B} \longrightarrow 0 .$$

Since

$$\mathscr{H}^\nu(K^\bullet_{r+1} \otimes_{\mathfrak{o}_{r+1}} \mathfrak{o}_r) \cong \mathscr{H}^\nu(K^\bullet_r) ,$$

$\mathscr{H}^\nu(K^\bullet)$ is smooth on an open dense subscheme $U \subset X$, which is shown by induction on r; for all r the sheaves $\mathscr{H}^\nu(K^\bullet_r)|U$ are locally constant.

The sheaves $\mathscr{A}$, $\mathscr{B}$ are constructed from the cohomology sequences of the distinguished triangles $(\pi^r K^\bullet_{r+1}, K^\bullet_{r+1}, K^\bullet_{r+1} \otimes_{\mathfrak{o}_{r+1}} \mathfrak{o}_r)$ as follows: They are obtained as constituents of $\mathscr{H}^\bullet(\pi^0 \cdot K^\bullet_1)$ using the isomorphisms

$$\mathscr{H}^\nu(\pi^{r-1} \cdot K^\bullet_r) \xrightarrow{\sim} \mathscr{H}^\nu(\pi^r \cdot K^\bullet_{r+1}) ,$$

that are induced from the π^{r-1} multiple of

$$K^\bullet_r \longleftarrow K^\bullet_{r+1} \otimes_{\mathfrak{o}_{r+1}} \mathfrak{o}_r \xrightarrow{id \otimes \pi} K^\bullet_{r+1} \otimes_{\mathfrak{o}_{r+1}} \mathfrak{o}_{r+1} = K^\bullet_{r+1} .$$

The left map is the transition map (a quasiisomorphism), the next map is multiplication by $id \otimes \pi$, which after multiplication by π^{r-1} becomes an isomorphism (divisibility of K^ν_{r+1}!). For the details confer [FK], lemma 12.14.

By noetherian induction one therefore reduces to the case, where X is the spectrum of a base field. This field may be replaced by a separably closed field. For this case see [FK], lemma 12.5, or the discussion in the next section below. For further properties of these cohomology sheaves see also Appendix A resp. [FK], I §12. □

Corollary 5.6 *$\mathscr{H}^\nu(K^\bullet)$ is a smooth π-adic sheaf on an open dense subscheme $U \subset X$.*

On one hand, as already observed, the triangulated categories $D^b_{ctf}(X, \mathfrak{o}_r)$ do not have natural t-structures by naive truncation of complexes. Therefore the naive truncations

$$(\tau_{\leq m} K^\bullet_r)_{r \geq 1}$$

do not yield an object in the category $D^b_c(X, \mathfrak{o})$. A natural t-structure on $D^b_c(X, \mathfrak{o})$ therefore a priori seems not to be present! On the other hand it is tempting to define a t-structure on $D^b_c(X, \mathfrak{o})$ by imposing vanishing conditions for the cohomology sheaves (considered in II.5.5). We will see, that this provides the required good substitute.

II.6 The Standard t-Structure on $D^b_c(X, \mathfrak{o})$

Let the situation be as in the last section. Our aim to define a t-structure on $D^b_c(X, \mathfrak{o})$, which will be called the standard t-structure.

We first deal with the case of a point

$$X_0 = Spec(k) \text{ , } k \text{ separably closed field.}$$

Then objects in $D^b_c(X, \mathfrak{o})$ can be represented by uniformly bounded projective systems $(K^\bullet_r)_{r\geq 1}$ of perfect complexes of finite $\mathfrak{o}_r$-modules with transition maps

$$K_{r+1} \to K_r \ .$$

(a) Let $\mathbf{K}^\bullet = \lim_s K^\bullet_s$ be the inverse limit complex. Since all the K^ν_s are finite groups (!) the lim^1-terms of such projective systems vanish (Mittag-Leffler condition). Therefore

$$H^\nu(\mathbf{K}^\bullet) \xrightarrow{\simeq} \varprojlim_s H^\nu(K^\bullet_s) \ .$$

The modules K^ν_r are free, so the transition maps $K^\bullet_{r+s} \otimes_{\mathfrak{o}_{r+s}} \mathfrak{o}_s \to K^\bullet_s$, obtained by composition, are quasiisomorphisms. Furthermore

$$0 \to K^\bullet_{r+s} \otimes_{\mathfrak{o}_{r+s}} \mathfrak{o}_s \xrightarrow{\pi^r} K^\bullet_{r+s} \to K^\bullet_{r+s} \otimes_{\mathfrak{o}_{r+s}} \mathfrak{o}_r \to 0$$

is exact. This holds for all $1 \leq r, s$. Take the cohomology sequence. The projective limit of these sequences for $s \to \infty$ is still exact by the finiteness assumptions made. This gives the exact sequences

$$0 \longrightarrow H^\nu(\mathbf{K}^\bullet))/\pi^r \longrightarrow H^\nu(K^\bullet_r) \longrightarrow H^{\nu+1}(\mathbf{K}^\bullet))[\pi^r] \longrightarrow 0 \ .$$

It will be called the **Tor-sequence**. Note $Tor^{\mathfrak{o}}_0(M, \mathfrak{o}/\pi^r\mathfrak{o}) \cong M/\pi^r$ and $Tor^{\mathfrak{o}}_1(M, \mathfrak{o}/\pi^r\mathfrak{o}) \cong M[\pi^r]$ for an $\mathfrak{o}$-module M.

(b) Now $dim_{\mathfrak{o}_1}(H^\nu(K^\bullet_r)/\pi) \leq dim_{\mathfrak{o}_1}(H^\nu(K^\bullet_1)) + dim_{\mathfrak{o}_1}(H^{\nu+1}(K^\bullet_1))$, since $dim_{\mathfrak{o}_1}(H^{\nu+1}(\mathbf{K}^\bullet)[\pi^r]/\pi) \leq dim_{\mathfrak{o}_1}(H^{\nu+1}(\mathbf{K}^\bullet)/\pi)$ and $H^\nu(\mathbf{K}^\bullet)/\pi \hookrightarrow H^\nu(K^\bullet_1)$, $H^{\nu+1}(\mathbf{K}^\bullet)/\pi \hookrightarrow H^{\nu+1}(K^\bullet_1)$. Thus the projective limits $H^\nu(\mathbf{K}^\bullet)$ are finitely generated $\mathfrak{o}$-modules by Remark II.5.1(3). Hence by the last sentence of Remark II.5.1(9) one can choose an $\mathfrak{o}$-perfectification $\mathbf{L}^\bullet \to \mathbf{K}^\bullet$ of the inverse limit complex $\mathbf{K}^\bullet$. In particular $H^\nu(\mathbf{L}^\bullet) \xrightarrow{\simeq} H^\nu(\mathbf{K}^\bullet)$. The composed map $\mathbf{L}^\bullet \to \mathbf{K}^\bullet \to K^\bullet_r$ induces the quasiisomorphisms $L^\bullet_r := \mathbf{L}^\bullet/\pi^r\mathbf{L}^\bullet \longrightarrow K^\bullet_r$ for all r (5-lemma)

$$\begin{array}{ccccccccc} 0 & \longrightarrow & H^\nu(\mathbf{L}^\bullet)/\pi^r & \longrightarrow & H^\nu(L^\bullet_r) & \longrightarrow & H^{\nu+1}(\mathbf{L}^\bullet)[\pi^r] & \longrightarrow & 0 \\ & & \downarrow \cong & & \downarrow & & \downarrow \cong & & \\ 0 & \longrightarrow & H^\nu(\mathbf{K}^\bullet)/\pi^r & \longrightarrow & H^\nu(K^\bullet_r) & \longrightarrow & H^{\nu+1}(\mathbf{K}^\bullet)[\pi^r] & \longrightarrow & 0 \end{array} .$$

We may therefore replace $(K^\bullet_r)_{r\geq 1}$ by the quasiisomorphic system

$$(L^\bullet_r)_{r\geq 1} = (\mathbf{L}^\bullet/\pi^r\mathbf{L}^\bullet)_{r\geq 1} \ .$$

(c) Let $\mathbf{A}$ be the abelian category of finitely generated $\mathfrak{o}$-modules. Let $D^b(\mathbf{A})$ be the derived category of $\mathbf{A}$ with bounded complexes. Any object in $D^b(\mathbf{A})$ can be represented by a perfect complex of $\mathfrak{o}$-modules.

Hence there exists a functor

$$D^b(\mathbf{A}) \to D^b_c(Spec(k), \mathfrak{o}) ,$$

which maps an $\mathfrak{o}$-perfect complex $\mathbf{L}^\bullet$ to the projective system $(\mathbf{L}^\bullet/\pi^r\mathbf{L}^\bullet)_{r\geq 1}$. Since a perfect complex is now its own projective resolution, any morphism between perfect complexes in $D^b(\mathbf{A})$ is induced by a complex map. Therefore isomorphic $\mathfrak{o}$-perfect complexes in $D^b(\mathbf{A})$ give isomorphic projective systems. This is true also in the converse direction by taking the projective limits. From the preceding discussion in **(b)** we therefore see, that this functor induces an equivalence of categories $D^b(\mathbf{A}) \cong D^b_c(Spec(k), \mathfrak{o})$.

Recall the equivalence of categories ([FK], 12.3) of the abelian category of A-R π-adic sheaves on $Spec(k)$ and $\mathbf{A}$, established by the functor $(F_r)_{r\geq 1} \mapsto \lim_r F_r$. The projective systems $N_r = (lim_s H^{\nu+1}(K_s^\bullet))[\pi^r]$ arising in the Tor-sequences define null systems; their objects become stable and the transition maps have an additional factor π. So the equivalence $D^b(\mathbf{A}) \cong D^b_c(Spec(k), \mathfrak{o})$ is compatible with the cohomology functors defined on both categories (see Lemma II.5.5).

Exercise. How do the natural truncation operators $\tau_{\leq n}, \tau_{\geq m}$ of the category $D^b(\mathbf{A})$ look like on the level of the category $D^b_c(Spec(k), \mathfrak{o})$?

Deligne's Truncation Operation

We now show, how things carry over from the special point case to the general case. The following two constructions of Deligne utilize that the homological dimension of a discrete valuation ring $\mathfrak{o}$ is one.

Let $(K_r^\bullet) \in D^b_c(X, \mathfrak{o})$ be given. For simplicity we assume, that the system $K_r^\bullet$ is uniformly bounded and that all the K_r^ν are constructible and $\mathfrak{o}_r$-flat and the transition morphisms $K_{r+1}^\bullet \to K_r^\bullet$ are complex maps. For

$$K_r^\bullet : \qquad \to \ldots \to K_r^{n-1} \to K_r^n \to K_r^{n+1} \to \ldots \to$$

let $\mathscr{K}_r^n \subset ker(K_r^n \to K_r^{n+1})$ be the subsheaf of those "elements", whose images in $\mathscr{H}^n(K_r^\bullet)$ are in the intersection of all images of the maps

$$\mathscr{H}^n(K_{r+s}^\bullet) \to \mathscr{H}^n(K_r^\bullet) \quad , \quad s \geq 0 .$$

This is well defined, and defines a constructible sheaf $\mathscr{K}_r^n$, since for the A-R π-adic sheaf

$$(\mathscr{H}^n(K_r^\bullet))_{r\geq 1}$$

the Mittag-Leffler condition is satisfied: There is a natural number t such that

$$Im(\mathscr{H}^n(K_{r+t}^\bullet) \to \mathscr{H}^n(K_r^\bullet)) = Im(\mathscr{H}^n(K_{r+s}^\bullet) \to \mathscr{H}^n(K_r^\bullet)) \quad , \quad s \geq t .$$

We then define $\tau_{\leq n}^{Del}(K_r^\bullet)$ to be the complex

$$\tau_{\leq n}^{Del}(K_r^\bullet) : \qquad \to \ldots \to K_r^{n-1} \to \mathscr{K}_r^n \to 0 \to 0 \to \ldots$$

and

$$\tau_{\leq n}^{Del} K^\bullet = (\tau_{\leq n}^{Del}(K_r^\bullet))_{r\geq 1} \,.$$

Lemma 6.1 *The sheaf $\mathcal{K}_r^n$ is an $\mathfrak{o}_r$-flat sheaf. The map $\tau_{\leq n}^{Del}(K_{r+1}^\bullet) \otimes_{\mathfrak{o}_{r+1}} \mathfrak{o}_r \to \tau_{\leq n}^{Del} K_r^\bullet$ is a quasiisomorphism, hence*

$$\tau_{\leq n}^{Del} K^\bullet \in D_c^b(X, \mathfrak{o}) \,.$$

We consider the homomorphism

$$\tau_{\leq n}^{Del} K^\bullet \to K^\bullet \,.$$

The induced homomorphism

$$\mathcal{H}^n(\tau_{\leq n}^{Del} K^\bullet) \to \mathcal{H}^n(K^\bullet)$$

is an isomorphism in the A-R category, i.e. kernel and cokernel of this homomorphism of projective systems are null systems.

How to prove this: Deligne's construction is compatible with taking stalk limits at geometric points. For an arbitrary geometric point $\overline{x}$ of X consider the projective system

$$((K_r^\bullet)_{\overline{x}})_{r\geq 1}$$

of $\mathfrak{o}_r$-modules. For this particular projective system the corresponding assertion was essentially proved in the last section. From this result for the geometric points $\overline{x}$ of X, all assertions made follow now immediately. □

Another Construction of Deligne. Let $\mathcal{G} = (\mathcal{G}_r)_{r\geq 1}$ be a π-adic sheaf. We will assume, that $\mathcal{G}$ is a π-adic sheaf on X in the honest sense. If $\mathcal{G}$ is flat, i.e. if all $\mathcal{G}_r$ are $\mathfrak{o}_r$-flat, then each $\mathcal{G}_r$ can be considered as a complex

$$0 \to \ldots \to 0 \to \mathcal{G}_r \to 0 \to \ldots \to$$

concentrated in degree 0. The projective system associated to these complexes, which is denoted $\tilde{\mathcal{G}}$ in the following, is contained in $D_c^b(X, \mathfrak{o})$

$$\tilde{\mathcal{G}} \in D_c^b(X, \mathfrak{o}) \,.$$

The same construction fails, if the given π-adic sheaf is not supposed to be flat any longer. Then instead we look for a complex $\tilde{\mathcal{G}} \in D_c^b(X, \mathfrak{o})$ such that

$$\mathcal{H}^n(\tilde{\mathcal{G}}) = \begin{cases} \mathcal{G} & n = 0 \\ 0 & n \neq 0 \end{cases}$$

– in the A-R category. This means for instance for $n \neq 0$ only that the projective system $(\mathcal{H}^n(\tilde{\mathcal{G}_r}))_{r\geq 1}$ is a null system. Such an object $\tilde{\mathcal{G}} \in D_c^b(X, \mathfrak{o})$ is provided by another construction of Deligne.

The Deligne Operator

Choose a natural number $t \geq 1$ such that π^t kills the torsion of $\mathscr{G}$, in other words such that in the A-R category

$$\mathscr{G}/Ker(\mathscr{G} \xrightarrow{\pi^t} \mathscr{G})$$

is flat. This means that the associated π-adic sheaf is flat by [FK], lemma 12.13. Let $s \geq t$ and choose a flat constructible resolution $\mathscr{F}^\bullet_{n+s}$ of $\mathscr{G}_{n+s}$ such that $\mathscr{F}^\bullet_{n+s} \in D^-_c(X, \mathfrak{o}_{n+s})$.

$$\mathscr{F}^\bullet_{n+s} \otimes_{\mathfrak{o}_{n+s}} \mathfrak{o}_n = \mathscr{G}_{n+s} \otimes^L_{\mathfrak{o}_{n+s}} \mathfrak{o}_n$$

Lemma 6.2 (The Deligne Operator $Del(\mathscr{G})$**)** *The complex* $\tau_{\geq -1}(\mathscr{F}^\bullet_{n+s} \otimes_{\mathfrak{o}_{n+s}} \mathfrak{o}_n)$ *is flat, in other words* $\tau_{\geq -1}(\mathscr{F}^\bullet_{n+s} \otimes_{\mathfrak{o}_{n+s}} \mathfrak{o}_n) \in D^b_{ctf}(X, \mathfrak{o}_n)$ *and is up to quasiisomorphism independent from the chosen* $s \geq t$. *This means*

$$Tor_1^{\mathfrak{o}_{n+s}}(\mathscr{F}^\bullet_{n+s}, \mathfrak{o}_n) \cong Tor_1^{\mathfrak{o}_{n+s'}}(\mathscr{F}^\bullet_{n+s'}, \mathfrak{o}_n) \quad , \quad s, s' \geq t \, .$$

Now study for fixed s and varying n the morphisms

$$\tau_{\geq -1}(\mathscr{F}^\bullet_{n+1+s}) \otimes_{\mathfrak{o}_{n+1+s}} \mathfrak{o}_n \to \tau_{\geq -1}(\mathscr{F}^\bullet_{n+s}) \otimes_{\mathfrak{o}_{n+s}} \mathfrak{o}_n$$

in the derived category. The projective system

$$Del(\mathscr{G}) = (\tau_{\geq -1}(\mathscr{F}^\bullet_{n+s} \otimes_{\mathfrak{o}_{n+s}} \mathfrak{o}_n))_{n \geq 1})$$

is an object of $D^b_c(X, \mathfrak{o})$. *Evidently* $Del(\mathscr{G})$ *is concentrated in degrees 0 and -1. We have*

$$\mathscr{H}^0(Del(\mathscr{G})) = \mathscr{G}$$

$$\mathscr{H}^{-1}(Del(\mathscr{G})) = 0 \, .$$

Concerning the Proof. By passage to the stalks the assertions of this lemma are reduced to the corresponding results for modules.

Suppose given a projective system of finite $\mathfrak{o}_r$-modules M_r such that

$$(M_r)_{r \geq 1}$$

is π-adic: $M_{r+1}/\pi^r M_{r+1} = M_r$.

Then $M = \lim_r M_r$ is a finitely generated $\mathfrak{o}$-module and we have

$$M_r = M/\pi^r M \, .$$

Without restriction of generality we can assume, that M is a torsion module and even of the special form

$$M = \mathfrak{o}/\pi^t \mathfrak{o} \, .$$

In particular $M_r = \mathfrak{o}/\pi^t\mathfrak{o}$ for all $r \geq t$. On the level $r = n + s$ for $s \geq t$ we can use the following $\mathfrak{o}_{n+s}$-flat resolution of the $\mathfrak{o}_{n+s}$-module M_{n+s}

$$\cdots \xrightarrow{\pi^t} \mathfrak{o}_{n+s} \xrightarrow{\pi^{n+s-t}} \mathfrak{o}_{n+s} \xrightarrow{\pi^t} \mathfrak{o}_{n+s} \longrightarrow 0 \longrightarrow \cdots \quad .$$

Using these complexes the proof of the lemma, in the special case of the projective system of modules

$$(M_r)_{r\geq 1}$$

under consideration, is a straight forward exercise. □

Short Notation. We define $D(X, \mathfrak{o}) = D_c^b(X, \mathfrak{o})$

Definition 6.3 *(The* **standard t-structure** *on* $D_c^b(X, \mathfrak{o})$*).*

$$D^{\leq 0}(X, \mathfrak{o}) = \{K^\bullet \in D(X, \mathfrak{o}) \mid \mathscr{H}^\nu(K^\bullet) = 0 \; \forall \nu > 0\}$$

$$D^{\geq 0}(X, \mathfrak{o}) = \{K^\bullet \in D(X, \mathfrak{o}) \mid \mathscr{H}^\nu(K^\bullet) = 0 \; \forall \nu < 0\} .$$

Theorem 6.4 *$D^{\leq 0}(X, \mathfrak{o})$ and $D^{\geq 0}(X, \mathfrak{o})$ define a t-structure on $D(X, \mathfrak{o})$. The core of this* **standard** *t-structure is the full subcategory of $D(X, \mathfrak{o})$*

$$Core(standard) = \{K^\bullet \in D(X, \mathfrak{o}) \mid \mathscr{H}^\nu(K^\bullet) = 0 \; \nu \neq 0\} .$$

The functor

$$Core(standard) \to \{\pi - adic\ sheaves\}$$

$$K^\bullet \mapsto \mathscr{H}^0(K^\bullet)$$

defines an equivalence of categories between the core of the standard t-structure and the abelian category of π-adic sheaves on X. The lower truncation operator

$${}^{st}\tau_{\leq n} \; := \; \tau_{\leq n}^{Del}$$

of this t-structure is the Deligne truncation operator defined in Lemma 6.1. The upper truncation

$${}^{st}\tau_{>n}$$

is obtained by completing the natural map

$$\tau_{\leq n}^{Del}(K^\bullet) \to K^\bullet$$

into a distinguished triangle.

Remark. We can and therefore will identify the cohomology objects of the standard t-structure with the "ordinary" π-adic sheaves. We identify ${}^{st}\tau_{\geq n}{}^{st}\tau_{\leq n}K^\bullet = {}^{st}\tau_{\leq n}{}^{st}\tau_{\geq n}K^\bullet = \mathscr{H}^n(K^\bullet)$. This being said we will later omit the index and write

$$^{st}\tau_{\geq n} = \tau_{\geq n} \quad , \quad ^{st}\tau_{\leq n} = \tau_{\leq n} \, .$$

We now give a sketch of the proof of Theorem II.6.4.

1) *First Step.* We have

$$K^\bullet = (K_r^\bullet) \in D^{\leq 0}(X, \mathfrak{o}) \Longleftrightarrow \mathscr{H}^\nu(K_r^\bullet) = 0 \; \nu \geq 1, r \geq 1$$

$$K^\bullet = (K_r^\bullet) \in D^{\geq 0}(X, \mathfrak{o}) \Longleftrightarrow \mathscr{H}^\nu(K_r^\bullet) = 0 \; \nu \leq -2, r \geq 1 \text{ and } \mathscr{H}^{-1}(K^\bullet) = 0 \, .$$

Considering the stalks, the proof is reduced to the case of the corresponding statement for modules. This case is an easy consequence of the Tor-sequence in subsection **(a)** at the beginning of this section.

2) *Second Step.* For $K^\bullet \in D^{\leq 0}(X, \mathfrak{o})$ and $L^\bullet \in D^{\geq 1}(X, \mathfrak{o})$ we have

Claim. $Hom_{D(X,\mathfrak{o})}(K^\bullet, L^\bullet) = 0$.

By step 1) and the remarks at the beginning we can assume that $K^\bullet = (K_r^\bullet)$ and $L^\bullet = (L_r^\bullet)$, such that $K_r^\bullet$ is uniformly bounded, constructible and $\mathfrak{o}_r$-flat and

$$K_r^\nu = 0 \quad , \quad \nu > 0 \, .$$

$L_r^\bullet$ is an injective complex such that

$$L_r^\nu = 0 \quad , \quad \nu < 0 \, .$$

Then a morphism

$$K^\bullet \to L^\bullet$$

is given by a homotopy compatible family of complex homomorphisms

$$K_r^\bullet \to L_r^\bullet \, ,$$

hence by a family of sheaf homomorphisms

$$\psi_r : K_r^\nu \to L_r^\nu \quad , \quad \nu = 0$$

such that

$$\psi_r(K_r^0) \subset Ker(L_r^0 \to L_r^1) = \mathscr{H}^0(L_r^\bullet) \, .$$

However $\mathscr{H}^0(L^\bullet) = 0$ implies, that the projective system $(\mathscr{H}^0(L_r^\bullet))_{r \geq 1}$ is a null system. Therefore for some (sufficiently large) natural number $t \geq 1$ the sheaf homomorphism

$$\mathscr{H}^0(L_{r+t}^\bullet) \longrightarrow \mathscr{H}^0(L_r^\bullet)$$

is the zero map for all $r \geq 1$.

By definition and flatness

$$K_{r+t}^\bullet \otimes_{\mathfrak{o}_{r+t}} \mathfrak{o}_r \longrightarrow K_r^\bullet$$

is a quasiisomorphism. By Remark II.5.1(6) we also have a quasiisomorphism

$$L_{r+t}^\bullet \otimes_{\mathfrak{o}_{r+t}} \mathfrak{o}_r \longrightarrow L_r^\bullet$$

The map induced by ψ_{r+t}

$$K_{r+t}^\bullet \otimes_{\mathfrak{o}_{r+t}} \mathfrak{o}_r \longrightarrow L_{r+t}^\bullet \otimes_{\mathfrak{o}_{r+t}} \mathfrak{o}_r$$

is the zero map! We may replace $K^\bullet$, $L^\bullet$ by isomorphic objects in the category $D_c^b(X, \mathfrak{o})$, e.g. by the shifted systems

$$\tilde{K}^\bullet = (K_{r+t}^\bullet \otimes_{\mathfrak{o}_{r+t}} \mathfrak{o}_r)_{r \geq 1}$$

$$\tilde{L}_{r+t}^\bullet = (L^\bullet \otimes_{\mathfrak{o}_{r+t}} \mathfrak{o}_r)_{r \geq 1} .$$

Hence the underlying morphism of the projective systems is shown to be zero.

3) *Third Step*. The *truncation axiom* for the standard t-structure is an immediate consequence of the properties of Deligne's operator $\tau_{\leq}^{Del}$ (Lemma II.6.1).

4) *Last Step*. A *right inverse functor* for the functor

$$Core(standard) \to \pi - \text{adic sheaves}$$

$$K^\bullet \mapsto \mathcal{H}^0(K^\bullet)$$

is the Deligne operator

$$Del(\mathcal{G})$$

(Lemma II.6.2). It is therefore enough to show the following: Let

$$h : K^\bullet \to L^\bullet$$

be a morphism between objects in the core of the standard t-structure, which induces the zero map $\mathcal{H}^0(h) = 0$

$$\mathcal{H}^0(h) : \mathcal{H}^0(K^\bullet) \to \mathcal{H}^0(L^\bullet)$$

– in the A-R category. Then we claim $h = 0$ in $D_c^b(X, \mathfrak{o})$.

First of all the functor

$$Core(standard) \ni K^\bullet \longmapsto \mathcal{H}^0(K^\bullet)$$

is an exact functor. This follows from the long exact cohomology sequence. Applied for $Ker(h)$, $Koker(h)$ it is therefore enough to show: A morphism

$$h : K = (K_r^\bullet) \; \to \; L = (L_r^\bullet)$$

in the abelian category $Core(standard)$, for which

$$\mathscr{H}^0(h) : \mathscr{H}^0(K) \to \mathscr{H}^0(L)$$

is an isomorphism – in the A-R category – is itself an isomorphism. For this we can assume that the complexes $K_r^\bullet$, $L_r^\bullet$ are bounded, $\mathfrak{o}_r$-flat and that the morphism h is represented by complex homomorphisms $h_r : K_r^\bullet \to L_r^\bullet$. We then have to show, that the morphisms h_r are quasiisomorphisms. Passage to the stalks, reduces to the case where X is replaced by a geometric point of X. The corresponding statement in this particular case of modules is an easy consequence of what was proved already. □

The category

$$D_c^b(X, E)$$

is deduced from $D_c^b(X, \mathfrak{o})$ by "localization", the category $D_c^b(X, \overline{\mathbb{Q}}_l)$ on the other hand by a direct limit

$$D_c^b(X, \overline{\mathbb{Q}}_l) = \text{“} \varinjlim_{E \subset \overline{\mathbb{Q}}_l} \text{”}\, D_c^b(X, E)$$

(See Appendix A).

In an evident way, the distinguished triangles of $D_c^b(X, \mathfrak{o})$ carry over to distinguished triangles and hence structures of triangulated categories for $D_c^b(X, E)$ respectively $D_c^b(X, \overline{\mathbb{Q}}_l)$; and similar for the t-structure. The corresponding truncation operators are again denoted ${}^{st}\tau_{\geq n}$, ${}^{st}\tau_{\leq n}$ or shorter

$${}^{st}\tau_{\geq n} = \tau_{\geq n} \qquad {}^{st}\tau_{\leq n} = \tau_{\leq n}\ .$$

Again for $K^\bullet \in D_c^b(D, \overline{\mathbb{Q}}_l)$ we identify

$$\tau_{\leq n}\tau_{\geq n} K^\bullet \ = \ \mathscr{H}^n(K^\bullet)$$

with a $\overline{\mathbb{Q}}_l$-sheaf on X. We often write

Short Notation. $D(X) = D_c^b(X, \overline{\mathbb{Q}}_l)$.

II.7 Relative Duality for Singular Morphisms

In the next Chap. III we will consider the theory of perverse sheaves in the sense of [BBD]. One of the cornerstones of this theory is the formalism of Poincare duality in its relative version for arbitrary, i.e. possibly singular morphisms between schemes. It gives rise to functors D and $f^!$, which have remarkable properties.

To begin with, we want to formulate the main results of this duality theory. These are formulated in terms of the derived categories $D_c^b(X, \overline{\mathbb{Q}}_l)$ of bounded complexes of $\overline{\mathbb{Q}}_l$-sheaves with constructible cohomology sheaves on a scheme X. For the purpose of this book it is enough to restrict oneself to the category of finitely generated schemes over a fixed finite field or algebraically closed field. This restriction allows to define the derived category $D_c^b(X, \overline{\mathbb{Q}}_l)$ in a simple and naive way. See also Chap. II §5 and Appendix A. The prime l will always be assumed to be different from the characteristic of the residue field. After stating the main results we present the main steps of the proofs.

Theorem 7.1 (Deligne [SGA4], exp. XVIII) *Let*

$$f : X \to S$$

be a compactifiable morphism between finitely generated schemes over a finite or algebraically closed base field. Then the functor

$$Rf_! : D_c^b(X, \overline{\mathbb{Q}}_l) \to D_c^b(S, \overline{\mathbb{Q}}_l)$$

admits a right adjoint triangulated functor

$$f^! : D_c^b(S, \overline{\mathbb{Q}}_l) \to D_c^b(X, \overline{\mathbb{Q}}_l) .$$

This means, that we have functorial isomorphisms

$$\boxed{Hom(K, f^!(L)) \xrightarrow{\simeq} Hom(Rf_!(K), L) .}$$

for all $K \in D_c^b(X, \overline{\mathbb{Q}}_l)$ and all $L \in D_c^b(S, \overline{\mathbb{Q}}_l)$. The functor $Hom(-,-)$ denotes the functor of homomorphisms in the derived categories $D_c^b(-, \overline{\mathbb{Q}}_l)$.

A slightly stronger statement – including the adjointness statements also for all etale schemes S' over S and X' over X – is the following sheafified version:

Relative Poincare Duality. There exists a functorial isomorphism

$$\boxed{\mathrm{R}f_*\mathrm{R}.\mathcal{H}om(K, f^!(L)) \xrightarrow{\simeq} \mathrm{R}.\mathcal{H}om(\mathrm{R}f_!(K), L) .}$$

At this point remember the similar, but comparatively trivial adjunction formula

$$\boxed{Hom(f^*K, L) \xrightarrow{\simeq} Hom(K, \mathrm{R}f_*(L))}$$

respectively

$$\mathrm{R}f_*\mathrm{R}.\mathcal{H}om(f^*K, L) \xrightarrow{\simeq} \mathrm{R}.\mathcal{H}om(K, \mathrm{R}f_*(L)) .$$

For the definition of the complexes $R.\mathcal{H}om(K,L)$ we refer to [FK], appendix AII, p. 300. Now Poincare duality in the singular case is formulated by means of the dualizing complex.

Definition 7.2 *Let*

$$f : X \to S = Spec(k)$$

be a finitely generated scheme over the fixed base field k - with the restrictions on k made at the beginning. The **dualizing complex** *of X is*

$$K_X = f^!(\overline{\mathbb{Q}}_{l\,S}) \in D_c^b(X, \overline{\mathbb{Q}}_l) .$$

The dualizing complex is compatible with base change for the ground field k. One defines the contravariant **dualizing functor** *by*

$$D_X(L) = R.\mathcal{H}om(L, K_X) .$$

We will often write $DL = D(L) = D_X(L)$, if the scheme X is fixed.

Corollary 7.3 (Poincare Duality) *Under the assumptions of Theorem II.7.1 for* $f : X \to S$ *and* $K \in D_c^b(X, \overline{\mathbb{Q}}_l)$ *the following holds*

$$\boxed{Rf_*(D_X(K)) = D_S(Rf_!(K)) .}$$

Proof. Apply relative Poincare duality for $L = K_S$ and use the functoriality $f^! K_S = K_X$ of the adjoint functor $f^!$. □

Theorem 7.4 (Biduality) *(Deligne, SGA4$\frac{1}{2}$, Theoreme de finitude)*
The natural functorial homomorphism

$$K \to D_X(D_X(K))$$

is a canonical isomorphism

$$\boxed{D_X \circ D_X = Id .}$$

Therefore the dualizing functor defines an anti-equivalence of categories

$$D_X : D_c^b(X, \overline{\mathbb{Q}}_l) \xrightarrow{\simeq} D_c^b(X, \overline{\mathbb{Q}}_l) .$$

$$Hom(K, L) \;=\; Hom(D_X(L), D_X(K)) .$$

We now collect some formulas, which will be frequently used later. They are immediate or formal consequences of the adjunction Theorem II.7.1, the biduality Theorem II.7.4 and the well known elementary tensor identity $f^*(K \otimes^L K') = f^*(K) \otimes^L f^*(K')$ and the R.$\mathcal{H}om$-formula

$$\mathrm{R}.\mathcal{H}om(K, \mathrm{R}.\mathcal{H}om(M, N)) = \mathrm{R}.\mathcal{H}om(K \otimes^L M, N)$$

for $K, M, N \in D_c^b(X, \overline{\mathbb{Q}}_l)$. These tensor identities are easily verified. The first is reduced to the corresponding statement for the stalks, since tensor products and pull back commute with passage to the stalks. The R.$\mathcal{H}om$-formula is reduced to the corresponding identity for the functors $\mathcal{H}om$ instead of the functors R.$\mathcal{H}om$. The remaining verification is then an obvious fact already on the level of presheaves. However for the reduction to the case of that $\mathcal{H}om$-functors - which is done by replacing the complex $N = N^\bullet$ by an injective resolution $I^\bullet$ - one has to use that for a complex $M \in D_c^b(X, \overline{\mathbb{Q}}_l)$ the complex $\mathrm{R}.\mathcal{H}om(M^\bullet, I^\bullet) = \mathcal{H}om(M^\bullet, I^\bullet)^\bullet$ has flabby, hence R.$\mathcal{H}om(K, .)$-acyclic components! The details of the argument are left as an exercise to the reader.

Corollary 7.5 *Suppose*

$$f : X \to S$$

is a morphism satisfying the assumptions of Theorem II.7.1. Then the following formulas hold:

a) $D \circ D = id$
b) $D \circ Rf_! = Rf_* \circ D$ *or alternatively* $D \circ Rf_! \circ D = Rf_*$
c) $D \circ Rf_* = Rf_! \circ D$ *or alternatively* $D \circ Rf_* \circ D = Rf_!$
d) $D \circ f^* = f^! \circ D$ *or alternatively* $D \circ f^! \circ D = f^*$
e) $D \circ f^! = f^* \circ D$ *or alternatively* $D \circ f^* \circ D = f^!$
f) $R.\mathcal{H}om(A, B) = D(A \otimes^L D(B))$
g) $Rf_!(A \otimes^L f^*B) = Rf_!A \otimes^L B$ *(Künneth type formula)*
h) $f^! R.\mathcal{H}om(A, B) = R.\mathcal{H}om(f^*(A), f^!(B))$

Proof. a) and b) are stated in II.7.3 and II.7.4. c) is an immediate consequence of a) and b). d) and e) follow from b) and c) and $Hom(K, L) = Hom(D_X(L), D_X(K))$ by the adjunction formulas stated in II.7.1. f) is the R.$\mathcal{H}om$-formula if we put $B = D(C)$ and use a), i.e. the biduality $D(B) = C$. For g) restate the relative Poincare duality theorem using f) in the form

$$\mathrm{R}f_* D(A \otimes^L Df^! C) = D(\mathrm{R}f_! A \otimes^L DC)$$

and use $Df^! C = f^* B$ for $DC = B$ applying e). Now use $D \circ \mathrm{R}f_* \circ D = \mathrm{R}f_!$. Statement h) finally is equivalent to

$$f^! D(A \otimes^L DB) = D(f^* A \otimes^L Df^! B)$$

via formula f). Now d) and a) reduce this to the obvious statement, that $f^*(A \otimes^L DB) = f^* A \otimes^L f^* DB$. □

We remark, that formula g) above was obtained purely formally as a consequence of the duality theorems. However, the reader should be aware of the fact that formula g) has to be proved independently at an early stage of the theory. See [FK], chap. I, proposition 8.14. This formula implies the Künneth formulas (see [FK]) and it is also one of the ingredients for the proof of the fundamental duality Theorem II.7.1. This will be explained in the next section.

After having stated the duality theorem, a general remark concerning the proofs of the statements of Theorem II.7.1 and II.7.4 is in order. These two theorems were formulated in the $\overline{\mathbb{Q}}_l$-adic setting, i.e. for the $\overline{\mathbb{Q}}_l$-adic derived categories. The usual techniques explained in Appendix A (projective limits, localization etc.) actually allow to reduce the statements made for $\overline{\mathbb{Q}}_l$-sheaves, to analogous statements for etale sheaves over finite commutative self injective rings. On that level the restrictions made on the base field – which had the purpose of defining the $\overline{\mathbb{Q}}_l$-derived categories – are no longer necessary. So for the rest of this section, which is devoted to the proof of the duality theorems, we can therefore completely ignore $\overline{\mathbb{Q}}_l$-sheaves.

As already explained, it is enough to consider the corresponding statements for etale sheaves of Λ-modules. Here Λ means a finite commutative self injective ring. It is enough to consider the case, where Λ is a finite factor ring of the ring of integers of a finite extension field $E \subset \overline{\mathbb{Q}}_l$ of the field $\mathbb{Q}_l$ of l-adic numbers. We will only consider etale sheaves of Λ-modules on noetherian schemes such that the prime l is invertible on Λ.

The proof of the duality Theorem II.7.1 in the absolute smooth case, i.e. in the case of a smooth morphism

$$f : X \to Spec(k)$$

from X to the spectrum of a field k, is contained in [FK]. See also the Appendix A of this book. Without restriction of generality we can assume k to be a separably or even algebraically closed field. Our intention for the following is to show how the general statement of Theorem II.7.1 – now in the case of Λ-sheaves – can be deduced from the special case of an absolutely smooth morphism, at least under the slight restriction that the morphism f can be extended to a smooth compactifiable morphism g. This includes – for instance – the case of a quasiprojective morphism f. This is a setting which is sufficiently general for our purposes. Secondly we will prove the biduality Theorem II.7.4 – always assuming, that the underlying scheme is finitely generated over the base field.

We begin by formulating two fundamental finiteness theorems. Both these finiteness theorems are supplements to a finiteness theorem proved in [FK] for proper morphisms.

Theorem 7.6 *Let*

$$f : X \to S$$

be a morphism between finitely generated schemes over an arbitrary base field. Then the direct image functor f_* has finite cohomological dimension independent from Λ.

Proof. This statement is an immediate consequence of M. Artin's theorem on the cohomological dimension of affine algebraic schemes over a separably closed field. See [SGA4], expose X or [FK], chap. I, theorem 9.1. □

Theorem 7.7 (Deligne, [SGA4$\frac{1}{2}$], finitude)
Let

$$f : X \to S$$

be a morphism between finitely generated schemes over a field and let $\mathscr{G}$ be a constructible sheaf on X. Then all higher direct images $R^\nu f_*(\mathscr{G})$ are constructible.

Proof. The proof is similar to the proof, which gives the permanence properties for mixed sheaves (Theorem I.9.4 at the end of Chap. I). We leave it to the reader to fill in the details of the argument. See also Appendix D. □

Remark. Be aware of the fact, that the argument for the proof of Theorem II.7.7 uses special cases of the Poincare duality theorem – namely the case of smooth morphisms considered in the next section!

The last theorem now has the following consequence

Corollary 7.8 (Deligne, [SGA4$\frac{1}{2}$], finitude corollary 1.6)
Let X be a finitely generated scheme over a field and let $\mathscr{F}$, $\mathscr{G}$ be constructible sheaves on X. Then also the Ext-sheaves

$$\mathscr{E}xt^\nu_\Lambda(\mathscr{F}, \mathscr{G})$$

are constructible.

Further consequences of II.7.6–II.7.8 are

Corollary 7.9 *Let the assumptions be the same as in II.7.7.*

(1) $R.\mathscr{H}om(-,-)$ defines a bifunctor

$$D_c^-(X,\Lambda) \times D_c^+(X,\Lambda) \to D_c^+(X,\Lambda)\,.$$

(2) The functors

$$R.\mathscr{H}om_\Lambda(-,-)\,,\ Rf_*\,,\ Rf_!$$

preserve the derived categories $D_c^b(-,\Lambda)$ and $D_{ctf}(-,\Lambda)$.

Recall that $D_c^+(X,\Lambda)$, $D_c^-(X,\Lambda)$ denote the full subcategories of $D_c(X,\Lambda)$ of complexes of Λ-module sheaves with constructible cohomology sheaves, which are bounded to the left respectively to the right. Similar $D_c^b(X,\Lambda)$ denotes the full subcategory of bounded complexes, $D_{ctf}(X,\Lambda)$ the full subcategory of complexes with finite Tor-dimension. For further details see also Appendix A and Appendix D.

II.8 Duality for Smooth Morphisms

After our preliminary statements made in the last section we now come to the proof of the first reduction step: Under the additional hypothesis, that $f : X \to S$ is a smooth morphism, we prove Theorem II.7.1 by reduction to the absolute smooth case, where f is smooth and S is the spectrum of a field. This remaining case is covered in [FK], chap. II, §1. Recall that all sheaves are now sheaves of Λ-modules.

Let

$$f : X \to S$$

be a smooth finitely generated compactifiable morphism between noetherian schemes. Further assumptions on f – except that l is invertible in Λ – will not be necessary.

1) We begin by assuming, that f has constant fiber dimension. All nonempty fibers of f are then equidimensional of dimension say d. Hence for all sheaves $\mathscr{F}$ we have

$$\mathrm{R}^\nu f_!(\mathscr{F}) = 0 \quad , \quad \nu > 2d \; .$$

For $\nu = 2d$ one has the functorial **trace map**

$$Tr_f : \; \mathrm{R}^{2d} f_!(\Lambda_X(d)) \to \Lambda_S \; .$$

See [FK], chap. II, theorem 1.6. This trace map has obvious permanence properties, which are stated in loc. cit. and which will subsequently be used without further mentioning. On the level of complexes the trace map may be viewed as a complex homomorphism

$$\mathrm{R}f_!(\Lambda_X[2d](d)) \to \Lambda_S$$

due to the vanishing of the higher direct images with compact support for $\nu > 2d$, stated above.

Recall that

1. the functor $f_!$ has finite cohomological dimension
2. $\mathrm{R}f_!(\Lambda_X(d)) \in D^b_{ctf}(S, \Lambda)$

Now the sought for functor $f^! : D(S, \Lambda) \to D(X, \Lambda)$ is in the first case 1) simply defined to be

Definition 8.1 *For smooth f put $f^!(K) = f^*(K)[2d](d)$.*

The proper base change theorem implies the projection formula [FK], chap. I, proposition 8.14

$$\mathrm{R}f_!(\Lambda_X(d) \otimes^L f^*(K)) = \mathrm{R}f_!(\Lambda_X(d)) \otimes^L K$$

for any complex $K \in D(S, \Lambda)$. From this projection formula we obtain a new "trace map" $\mathrm{R}f_!(f^!(K)) \to K$ via

$$\begin{array}{ccc}
Rf_!(f^!K)) & \longrightarrow & K \\
\Big\| & & \Big\uparrow {\scriptstyle Tr_f \otimes^L id_K} \\
Rf_!(\Lambda_X[2d](d] \otimes^L f^*(K)) & = & Rf_!(\Lambda_X[2d](d)) \otimes^L K
\end{array}$$

2) Suppose now f is smooth, but does not have constant fiber dimension. Then f decomposes into morphisms of constant fiber dimension

$$X = X_1 \cup .... \cup X_r \qquad \text{(disjoint decomposition)} .$$

The X_ν are open subschemes of X such that $f_\nu = f \mid X_\nu : X_\nu \to S$ has constant fiber dimension d_ν. We then define for a complex K in $D(S, \Lambda)$

$$f^!(K) = \bigoplus_\nu f_\nu^!(K) | X_\nu .$$

By a term by term addition we again obtain a trace map of the form

$$Tr_{f,K} : Rf_!(f^!(K)) \to K .$$

All results stated in [FK], chap. II, §1 remain true. They carry over from the case of constant fiber dimension - considered in loc. cit. - to the slightly more general case of nonconstant fiber dimension. The so defined trace map turns out to be the desired adjunction map. The trace map $Tr_{f,K}$ defines a duality homomorphism

$$\alpha : Hom(L, f^!(K)) \to Hom(Rf_!(L), K) ,$$

where $Hom(-, -)$ denotes the homomorphism functor in the derived categories $D(-, -)$. By definition this homomorphism maps $\phi : L \to f^!(K)$ to $Tr_{K,f} \circ Rf_!(\phi)$.

3) By a sheafification we also get a duality homomorphism

$$\beta : Rf_* R\mathcal{H}om(L, f^!(K)) \to R\mathcal{H}om(Rf_!(L), K) .$$

To be precise we demand $L \in D^-(X, \Lambda)$ and $K \in D^+(S, \Lambda)$.

4) **Claim.** These two duality homomorphisms α and β are isomorphisms. In the second case the map β is an isomorphism in the derived category.

As already explained in stating Theorem II.7.1, both statements are more or less equivalent. The proof for the absolute case – that is the case of a scheme X over the spectrum S of a field – is contained in [FK], chap. II, §1. We deduce the general case from this case. See also Verdier, A duality theorem in the etale cohomology of schemes [306]. The proof of the claim 4) is established by noetherian induction with respect to the base S and will occupy the rest of this section.

5) We start with a remark, which allows reduction to the case of constructible sheaves. This will be used for the noetherian induction later. Every sheaf is a factor sheaf of the – possibly infinite – direct sum of all its constructible subsheaves. See

[FK] I §4 for a proof. This in particular implies, that every complex $L = L^\bullet$ in $D^-(X, \Lambda)$ has a resolution $L \in D^-(X, \Lambda)$, which has components L^ν that are direct sums of constructible sheaves. Obviously one can therefore reduce the proof of claim 4) to the particular case, where L is a single sheaf, which is furthermore a direct sum of constructible sheaves. The functors $\mathrm{R}^\nu f_!$ commute with direct sums. The hyper-ext functors

$$Ext^\nu(\mathscr{F}, K)$$

transform direct sums of sheaves in the first variable to direct products. Therefore the Hom-functor

$$Hom(A, B)$$

of the derived category transforms direct sums of sheaves in the first variable into direct products of abelian groups. Hence we will assume in the following, that for the given fixed morphism

$$f : X \to S$$

the complex L – in the first variable – is a single constructible sheaf

$$L = \mathscr{F} .$$

6) Let $\mathscr{H}^\nu$ be the ν-th cohomology sheaf of the given complex $K \in D^+(S, \Lambda)$. The first isomorphism claimed is a consequence of corresponding statements where the complex is replaced by certain translates of sheaves

$$Hom(\mathscr{F}, f^!(\mathscr{H})[\nu]) \overset{\simeq}{\to} Hom(\mathrm{R}f_!(\mathscr{F}), \mathscr{H}[\nu]) \quad , \quad \forall \nu .$$

Now, since $\mathscr{F}$ and all cohomology sheaves of $\mathrm{R}f_!(\mathscr{F})$ are constructible sheaves, both sides of the identity which shall be proved commute with direct limits in the second variable $\mathscr{H}$. This is easily deduced from the corresponding fact for the hyper-ext functors $Ext^\nu(A, B)$. It is therefore enough to prove the theorem for all constructible subsheaves of the cohomology sheaf

$$\mathscr{H} = \mathscr{H}^\nu(K) .$$

We can and will furthermore assume that S is reduced.

7) If S is the spectrum of a field, then the statement of the claim is contained in [FK]. This is the start of the noetherian induction. By induction assumption the following may be assumed to be proven already: The claim is true for all proper closed subschemes S' of S and the corresponding morphisms

$$f' = f \otimes_S S' : \quad X' = X \otimes_S S' \longrightarrow S' .$$

We consider the cartesian diagram

$$\begin{array}{ccc} X' & \overset{\tilde{i}}{\hookrightarrow} & X \\ \downarrow{\scriptstyle f'} & & \downarrow{\scriptstyle f} \\ S' & \overset{i}{\hookrightarrow} & S \end{array}$$

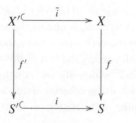

with closed embeddings $\tilde{i}, i$. By the base change theorems and elementary adjunction properties for the complexes $L \in D^-(X, \Lambda)$, $M \in D^+(S', \Lambda)$, $K = i_*(M) \in D^+(S, \Lambda)$ we get

$$Hom(L, f^!(K)) = Hom(L, \tilde{i}_*(f'^!(M))) = Hom(\tilde{i}^*(L), f'^!(M))$$

$$= Hom(Rf'_!(\tilde{i}^*(L)), M) = Hom(i^*(Rf_!(L)), M) = Hom(Rf_!(L), K) .$$

Therefore – by the reasoning above – the claim is true also for f itself and all complexes $K \in D(S, \Lambda)$, for which all constructible subsheaves of the cohomology sheaves have support in a proper subscheme of S (depending on K). Let F be the function field of one of the irreducible components of S which has a nonempty fiber over the generic point. If such a component does not exist, then the image of f is already contained in a proper closed subscheme of S and we are done by the induction hypothesis. Else consider the inclusion map

$$j : T = Spec(F) \to S$$

induced by the generic point of the component. Consider the cartesian diagram

$$\begin{array}{ccc} X_T = X \times_S T & \xrightarrow{\tilde{j}} & X \\ \Big\downarrow{\scriptstyle \tilde{f}} & & \Big\downarrow{\scriptstyle f} \\ T & \xrightarrow{j} & S \end{array} .$$

Let B be a complex in $D^+(T, \Lambda)$ and put

$$K = \mathrm{R}j_*(B) .$$

The smooth base change theorem implies $f^*\mathrm{R}j_*(B) = \mathrm{R}\tilde{j}_*(\tilde{f}^*(B))$, hence by definition of $f^! = f^*[2d](d)$

$$f^!K = f^!\mathrm{R}(j_*B) = \mathrm{R}\tilde{j}_*\tilde{f}^!(B) .$$

Using the elementary adjunction property of $\mathrm{R}\tilde{j}_*$ and $\tilde{j}^*$, we thus obtain

$$Hom(\mathscr{F}, f^!(K)) = Hom(\tilde{j}^*(\mathscr{F}), \tilde{f}^!(B)) .$$

Since T is the spectrum of a field we already know by the induction assumption the adjunction formula $Hom(\tilde{j}^*(\mathscr{F}), \tilde{f}^!(B)) = Hom(\mathrm{R}\tilde{f}_!(\tilde{j}^*\mathscr{F}), B)$. This implies

$$Hom(\mathscr{F}, f^!(K)) = Hom(\mathrm{R}\tilde{f}_!(\tilde{j}^*\mathscr{F}), B) = Hom(j^*\mathrm{R}f_!(\mathscr{F}), B) ,$$

where for the second equality we have used the proper base change theorem $\mathrm{R}\tilde{f}_!(\tilde{j}^*(\mathscr{F})) = j^*\mathrm{R}f_!(\mathscr{F})$ for the functor $\mathrm{R}f_!$. By the elementary adjunction properties of $\mathrm{R}j_*$ and j^* then $Hom(j^*\mathrm{R}f_!(\mathscr{F}), B)$ and $Hom(\mathrm{R}f_!(\mathscr{F}), \mathrm{R}j_*(B))$ can be identified. Therefore

$$Hom(\mathscr{F}, f^!(K)) = Hom(\mathrm{R}f_!(\mathscr{F}), K) .$$

We leave it as an exercise to verify the equalities stated above by checking diverse commutative diagrams!

8) Let now K be an arbitrary complex in $D^+(S, \Lambda)$. Consider the distinguished triangle

$$K \longrightarrow \mathrm{R}j_*j^*(K) \longrightarrow P \longrightarrow K[1]$$

which is defined by the adjunction map in the obvious way. By definition $j^*(P)$ is a zero object in $D^+(T, \Lambda)$. In particular, every constructible subsheaf $\mathscr{N}$ of a cohomology sheaf of the complex P has vanishing pullback $j^*(\mathscr{N})$. All such constructible subsheaves therefore have support in some proper closed subscheme of S. The isomorphism claim is therefore true in this particular case by the induction assumption. On the other hand the isomorphism claim for the complex

$$\mathrm{R}j_*j^*(K)$$

has been verified in the last step 7). Hence the claim 4) now follows for arbitrary K as above by the 5-lemma. □

II.9 Relative Duality for Closed Embeddings

In this section we continue and complete the proof of the relative duality Theorem II.7.1, which was proved for smooth morphisms in the last section. We also continue the numbering of steps in the proof, which started in the last section.

9) We consider closed embeddings

$$i : Y \to X .$$

For an etale sheaf $\mathscr{G}$ of Λ-modules on X let

$$\Gamma_Y(\mathscr{G}) = \mathscr{H}om(i_*(\Lambda_Y), \mathscr{G})$$

be the sheaf of **sections with support** in Y.

For a sheaf $\mathscr{F}$ on Y there are canonical identities

$$Hom(i_*(\mathscr{F}), \mathscr{G}) = Hom(i_*(\mathscr{F}), \Gamma_Y(\mathscr{G})) = Hom(\mathscr{F}, i^*(\Gamma_Y(\mathscr{G})) .$$

The functor

$$i^* \circ \Gamma_Y$$

is therefore right adjoint to $i_* = i_!$. So for injective sheaves $\mathscr{G}$ on X also $i^*(\Gamma_Y(\mathscr{G}))$ is injective, since the functor i_* is an exact functor. In particular, the hyper-ext functors Ext^ν for complexes $L \in D^-(Y, \Lambda)$ and $K \in D^+(X, \Lambda)$ satisfy

$$Ext^\nu(L, i^*\mathrm{R}\Gamma_Y(K)) = Ext^\nu(i_*(L), K) .$$

Therefore

$$Hom(L, i^* \mathrm{R}\Gamma_Y(K)) = Hom(i_*(L), K) .$$

Again – the homomorphism groups are understood to be the homomorphism groups in the derived category! In other words

Lemma 9.1 *The functor*

$$i^! := i^* \circ \mathrm{R}\Gamma_Y : \; D^+(X, \Lambda) \to D^+(Y, \Lambda)$$

is a (partial) right adjoint functor for the functor

$$i_! = i_* : \; D^-(Y, \Lambda) \to D^-(X, \Lambda) .$$

Remark. For smooth morphisms $f : X \to S$ and $g : Y \to S$ and closed embeddings i, such that

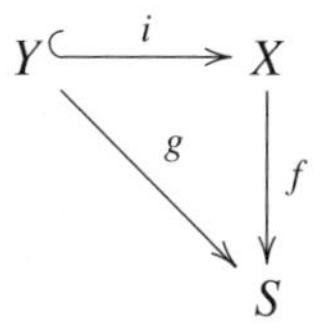

we get canonical isomorphisms

$$g^!(K) \cong i^! \circ f^!(K) \quad , \quad K \in D^+(S, \Lambda) ,$$

an immediate consequence of the relative duality theorem for the smooth case, proved in claim 4)!

The General Case

10) Let be given a smoothly embedable morphism f, i.e.

$$f = g \circ i : X \to S ,$$

can be written as the composite of a closed embedding $i : X \to T$ and a smooth (finitely generated) morphism $g : T \to S$.

For $\mathrm{R}g_!$ and also for $\mathrm{R}i_! = i_!$ we have already constructed partial right adjoint functors $g^!$ and $i^!$. The composite functor

$$f^! = i^! \circ g^! : D^+(S, \Lambda) \to D^+(X, \Lambda)$$

is the desired partial right adjoint functor for

$$\mathrm{R}f_! : D^-(X, \Lambda) \to D^-(S, \Lambda) .$$

We now restrict ourselves to the category of finitely generated schemes over a fixed base field. Then the functor f_* has finite cohomological dimension (see

Theorem II.7.6). This permits to conclude that for a closed embedding $i : Y \to X$ the functor Γ_Y, hence also the functor $i^!$, has finite cohomological dimension. Note that this is a consequence of Theorem II.7.6, using the distinguished triangle

$$\mathrm{R}\Gamma_Y(K) \longrightarrow K \longrightarrow \mathrm{R}j_*j^*(K) \longrightarrow \mathrm{R}\Gamma_Y(K)[1]\,.$$

Here $j : U = X \setminus Y \hookrightarrow X$ denotes the embedding of the open complement of Y in X.

11) The same distinguished triangle allows to deduce from Theorem II.7.7 respectively its corollaries II.7.8 and II.7.9, that for a complex

$$K \in D^b_{ctf}(X, \Lambda)$$

also

$$i^!(K) \in D_{ctf}(Y, \Lambda)$$

holds. More generally for a smoothly embedable morphisms $f : X \to S$ as in 10) (in the category of finitely generated schemes over a fixed base field) the functor $f^!$ defined above is a functor which preserves the subcategory of complexes of finite Tor-dimension

$$f^! : D^b_{ctf}(S, \Lambda) \to D^b_{ctf}(X, \Lambda)\,.$$

In other words $f^!$ is a right adjoint functor for the functor

$$\mathrm{R}f_! : D^b_{ctf}(X, \Lambda) \to D^b_{ctf}(S, \Lambda)\,.$$

12) Finally let Λ' be a different coefficient ring, subject to the assumptions made earlier, and let

$$\Lambda \to \Lambda'$$

be a ring homomorphism. Let

$$f^!_\Lambda : D^b_{ctf}(S, \Lambda) \to D^b_{ctf}(X, \Lambda)$$

$$f^!_{\Lambda'} : D^b_{ctf}(S, \Lambda') \to D^b_{ctf}(X, \Lambda')$$

be the adjoint functors, that were constructed for the underlying rings Λ and Λ'. It is easy to see from the definitions that

$$f^!_{\Lambda'}(K) = f^!_\Lambda(K) \otimes^L_\Lambda \Lambda'\,.$$

This observation now implies the existence of the functor

$$f^! : D^b_c(S, \overline{\mathbb{Q}}_l) \to D^b_c(X, \overline{\mathbb{Q}}_l)$$

under the hypothesis of Theorem II.7.1 and with all the required properties. This is shown by the limiting argument explained e.g. in Appendix A. Thus Theorem II.7.1 is proved. □

II.10 Proof of the Biduality Theorem

After having proved Theorem II.7.1 we now turn to the proof of the biduality Theorem II.7.4, again for etale sheaves of Λ-modules. The biduality theorem for $\overline{\mathbb{Q}}_l$-sheaves is then deduced from this by a limiting process.

We now restrict ourselves to consider only finitely generated schemes over a given base field k and morphisms between such schemes over k. Let

$$f : X \to S = Spec(k)$$

be a finitely generated smoothly imbedable compactifiable scheme over the base field k.

Definition 10.1 *The complex $f^!(\Lambda_S) = K_X$ will be called the Λ-dualizing complex.*

Lemma 10.2 *The Λ-dualizing complex K_X is an object of the category $D^b_{ctf}(X, \Lambda)$. It has the following finiteness property:*

There is a natural number N such that for all constructible sheaves $\mathcal{G}$ on X the following holds

$$\mathcal{H}^\nu\big(R.\mathcal{H}om(\mathcal{G}, K_X)\big) = 0 \quad , \quad \text{for } \nu > N .$$

Alternatively: K_X has a finite resolution by a complex $P^\bullet$, whose components P^ν are acyclic with respect to all functors

$$\mathcal{H}om(\mathcal{G}, -) ,$$

where $\mathcal{G}$ is a constructible sheaf on X.

Remark. If the base field k is separable closed we prove below, that the Λ-dualizing complex K_X has finite injective dimension. So there is a number r such that

$$Ext^\nu_\Lambda(\mathcal{G}, K_X) = 0 \quad \forall\ \Lambda\text{-sheaves } \mathcal{G} \quad , \quad \nu > r .$$

Equivalently K_X has a bounded injective resolution. Indeed for all Λ-sheaves $\mathcal{G}$ we have

$$Ext^\nu_\Lambda(\mathcal{G}, K_X) = 0 \quad , \quad \text{if } \nu > 0 \text{ or } \nu < -2 \cdot dim(X) .$$

Without restrictions on the base field the following holds:

$$\mathcal{H}^\nu(K_X) = 0 \quad , \quad \text{if } \nu > 0 \text{ or } \nu < -2 \cdot dim(X) .$$

Proof of the Second Assertion. The construction of the dualizing complex K_X is compatible with arbitrary base field extension. The construction of the hyperext-sheaves

$\mathscr{E}xt^{\nu}_{\Lambda}(\mathscr{G}, K_X)$ is compatible with finite separable (= etale) base field extension. For an affine projective limit X of schemes the global group $Ext^{\nu}_{\Lambda_X}(\mathscr{G}, \mathscr{F})$ is the direct limits of the corresponding group for the schemes of the projective system, provided the sheaves $\mathscr{G}$ and $\mathscr{F}$ are constructible ([FK], chap. I, proposition 4.18). The proof remains valid for more general rings Λ than the ring $\mathbb{Z}/n\mathbb{Z}$ considered in loc. cit. Hence the construction of the sheaves $\mathscr{E}xt^{\nu}_{\Lambda}(\mathscr{G}, K_X)$ is compatible with arbitrary separable base field extension. Without loss of generality we may therefore assume, that the base field k is separable closed. We now prove the claim of the remark. Since Λ is a self injective ring, for a bounded complex M of Λ-modules satisfying $H^{\nu}(M) = 0$ for all $\nu > r, \nu < s$, the hyperext-groups $Ext^{\nu}_{\Lambda}(M, \Lambda)$ vanish for all $\nu > -s, \nu < -r$. For a complex $K \in D^{-}(S, \Lambda)$ we have $Ext^{\nu}_{\Lambda_S}(K, \Lambda_S) = Ext^{\nu}_{\Lambda}(\Gamma(S, K), \Lambda)$. Since the cohomological dimension of the functors $f_!$ and $\Gamma_c(X, -)$ is bounded above by $N = 2dim(X)$, we obtain from Poincaré duality for all Λ-sheaves $\mathscr{G}$ on X

$$Ext^{\nu}_{\Lambda}(\mathscr{G}, K_X) = Ext^{\nu}_{\Lambda}(Rf_!(\mathscr{G}), \Lambda) = Ext^{\nu}_{\Lambda}(\mathrm{R}\Gamma_c(X, \mathscr{G}), \Lambda) = 0$$

for all $\nu > 0, \nu < -N$. □

Lemma II.10.2 has the following consequence: For any complex $L \in D^b_c(X, \Lambda)$ the **dual complex** $D(L) = D_X(L)$

$$D_X(L) = \mathrm{R}\mathscr{H}om(L, K_X)$$

is again contained in $D^b_c(X, \Lambda)$. If furthermore $L \in D^b_{ctf}(X, \Lambda)$, then also

$$D_X(L) \in D^b_{ctf}(X, \Lambda)$$

(Corollary II.7.9).

For a complex $L \in D^b_c(X, \Lambda)$ there exists a natural functorial homomorphism, the biduality map

$$L \to D_X(D_X(L)) .$$

This defines a natural transformation

$$Id \to D_X \circ D_X .$$

Theorem 10.3 *The biduality map is a canonical isomorphism.*

Remark. For our purposes it is enough to restrict to the case of complexes L in $D^b_{ctf}(X, \Lambda)$.

Proof. We will make use of the formula

$$(*) \qquad \mathrm{R}f_* \circ D_X = D_Y \circ \mathrm{R}f_! ,$$

for morphisms

$$f : X \to Y .$$

This was obtained in II.7.3 as a corollary of the relative duality theorem, which we proved in the last two sections.

The assertion of the theorem is local on X and S with respect to the etale topology and it is enough to consider the case where $L = \mathscr{F}$ is an etale sheaf (5-lemma). We may furthermore assume $\mathscr{F}$ to be Λ-flat (use a resolution).

The Smooth Case. Suppose X is smooth over the base field k and $\mathscr{F}$ is locally constant.

As the statement is of local nature we can assume that $\mathscr{F}$ is constant, i.e $\mathscr{F} = \mathscr{M}_X$ for some finite Λ-module $\mathscr{M}$. Furthermore we may assume that X is equidimensional. Recalling the definition of $K_X = f^!(\Lambda)$ and the definition of $f^!$ in the smooth case, we get for any geometric point $\overline{x}$ of X

$$\mathrm{R}\mathscr{H}om_\Lambda(\mathscr{M}_X, K_X)_{\overline{x}} = \mathrm{R}\mathscr{H}om_\Lambda(\mathscr{M}, (K_X)_{\overline{x}}) \cong \mathscr{H}om_\Lambda(\mathscr{M}, \Lambda)[2d](d)\,.$$

This follows from the definition of $\mathrm{R}\mathscr{H}om(-,-)$ and the Remark II.1.8. Here we used, that Λ is an injective Λ-module. We therefore get

$$\mathrm{R}\mathscr{H}om(\mathscr{M}_X, K_X) = \mathscr{H}om(\mathscr{M}, \Lambda)_X[2d](d)\,.$$

Hence the biduality theorem is an immediate consequence of the corresponding biduality theorem for finite Λ-modules

$$Hom_\Lambda(Hom_\Lambda(\mathscr{M}, \Lambda), \Lambda) = \mathscr{M}\,.$$

This identity holds for all finite Λ-modules $\mathscr{M}$, since Λ is by assumption a quasi Frobenius ring (Gorenstein of dimension 0).

The General Case. As the statement is of local nature with respect to the etale topology, we may assume that X is affine, hence a closed subscheme of some standard affine space $\mathbb{A}^m$ over k. However, because of formula $(*)$ above we can actually assume without restriction of generality, that $X = \mathbb{A}^m$. Now the plan of proof is induction on m. We want to show that for constructible sheaves $\mathscr{F}$ on $\mathbb{A}^m$ we have

$$\mathscr{F} = D_{\mathbb{A}^m}(D_{\mathbb{A}^m}(\mathscr{F}))\,.$$

The case $m = 0$ is a trivial case of the smooth case, hence already considered. So suppose $m > 0$. We can find an open dense subscheme

$$j : U \hookrightarrow X$$

of X such that $\mathscr{F}|U$ is locally constant. In this case we already know

$$D_U(D_U(\mathscr{F}|U)) = \mathscr{F}|U\,.$$

If we extend the biduality morphism on X to a distinguished triangle

$$\mathscr{F} \longrightarrow D_X(D_X(\mathscr{F})) \longrightarrow \Delta \longrightarrow \mathscr{F}[1]\,,$$

therefore $\Delta|U = 0$. Hence Δ can be considered as a complex with support on the complement $Y = X - U$, where Y is endowed with the reduced subscheme structure. We have to show $\Delta = 0$.

By assumption $dim(Y) < m = dim(X)$. The Noether normalization theorem implies the existence of a linear projection – after a linear coordinate change and possibly a base field extension

$$q : \mathbb{A}^m = \mathbb{A}^{m-1} \times \mathbb{A}^1 \longrightarrow \mathbb{A}^{m-1} ,$$

such that the restriction

$$q|Y : Y \to \mathbb{A}^{m-1}$$

is a finite morphism, i.e. proper with finite fibers. The strategy is now to show that the direct image complex of Δ with respect to q is zero using induction and then to conclude that $\Delta = 0$.

To be precise embed $\mathbb{A}^1 \hookrightarrow \mathbb{P}^1$ into projective space and consider the partial compactification $\overline{X} = \mathbb{A}^{m-1} \times \mathbb{P}^1$ of $X = \mathbb{A}^m$. Extend the sheaf $\mathscr{F}$ to a sheaf $\overline{\mathscr{F}}$ on $\overline{X}$ (extension by zero) and put $\overline{Y} = Y \cup (\mathbb{A}^{m-1} \times \{\infty\})$. The analogous projection

$$\overline{q} : \overline{X} \to \mathbb{A}^{m-1}$$

has now the advantage of being a proper map, and its restriction $\overline{q}|\overline{Y} \to \mathbb{A}^{m-1}$ is still finite. In particular $R\overline{q}_! = R\overline{q}_*$ holds and by formula $(*)$ also

$$R\overline{q}_* \circ D_{\overline{X}} = D_{\mathbb{A}^{m-1}} \circ R\overline{q}_* .$$

So, if we consider the analogous distinguished triangle

$$\overline{\mathscr{F}} \longrightarrow D_{\overline{X}}(D_{\overline{X}}(\overline{\mathscr{F}})) \longrightarrow \overline{\Delta} \longrightarrow \overline{\mathscr{F}}[1] ,$$

applying $R\overline{q}_*$ gives the distinguished triangle

$$R\overline{q}_*(\overline{\mathscr{F}}) \longrightarrow D_{\mathbb{A}^{m-1}}(D_{\mathbb{A}^{m-1}}(R\overline{q}_*(\overline{\mathscr{F}}))) \longrightarrow R\overline{q}_*(\overline{\Delta}) \longrightarrow R\overline{q}_*(\overline{\mathscr{F}})[1] .$$

One has to verify, that the first morphism of this distinguished triangle is still the biduality map, now on $\mathbb{A}^{m-1}$! We leave it to the reader. By the induction assumption the biduality theorem holds on $\mathbb{A}^{m-1}$. This implies

$$R\overline{q}_*(\overline{\Delta}) = 0 .$$

However $R\overline{q}_*(\overline{\Delta}) = 0$ forces

$$\overline{\Delta} = 0 ,$$

since the morphism $\overline{q}$ becomes a finite morphism, when it is restricted to the support $\overline{Y}$ of $\overline{\Delta}$. For the necessary facts on the direct image functor for finite morphisms see [FK], chap. I, §3. This completes the proof of the biduality theorem, since the obstruction $\Delta = \overline{\Delta}|Y$ was shown to be zero. See II.1.6. □

Remark. The functor D_X and also the biduality map

$$L \to D_X D_X(L)$$

can be defined for arbitrary complexes in the derived category $D_c(X, \Lambda)$ of arbitrary complexes with constructible cohomology sheaves, by reasons of finite cohomological dimension. The proof of the biduality theorem carries over to this more general setting.

II.11 Cycle Classes

Let $f : X \to S$ be a smooth morphism, equidimensional with fibers of dimension d. By the proof of Theorem II.7.1 (case of smooth morphisms), we may identify the two functors $f^!$ and $f^*[2d](d)$. For any sheaf complex K on S we have functorial isomorphisms

$$\theta_{X/S} : f^!(L) \cong f^*(L)[2d](d) .$$

In fact – using Poincare duality – we even had identified both sides. So, by its very definition, we see that in our special case $f^!$ is compatible with tensor products in the following sense

$$\begin{array}{ccc} f^!(K' \otimes^L K) & \xrightarrow{\cong} & f^!(K') \otimes^L f^*(K) \\ \cong \downarrow \theta_{X/S} & & \cong \downarrow \theta_{X/S} \otimes id_{f^*(K)} \\ f^*(K' \otimes^L K)[2d](d) & \xrightarrow{\cong} & f^*(K')[2d](d) \otimes^L f^*(K) \end{array}$$

which follows from the obvious fact $f^*(K' \otimes K) = f^*(K') \otimes^L f^*(K)$.

Suppose $f : X \to S$ is as above, and let $i : Y \longrightarrow X$ be a closed embedding with open complement $j : U \to X$ and the restrictions g and q of f to Y respectively U

$$\begin{array}{ccccc} Y & \xrightarrow{i} & X & \xleftarrow{j} & U \\ & {\scriptstyle g} \searrow & \downarrow {\scriptstyle f} & \swarrow {\scriptstyle q} & \\ & & S & & \end{array}$$

Then there exist distinguished triangles

$$\begin{array}{l} i_*i^!(K) \xrightarrow{adj} K \longrightarrow Rj_*j^*(K) \longrightarrow i_*i^!(K)[1] , \\ Rj_!j^*(K) \longrightarrow K \xrightarrow{adj} i_*i^*(K) \longrightarrow Rj_!j^*(K)[1] , \end{array}$$

which are dual to each other in the following sense:

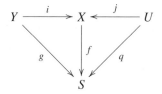

Lemma 11.1 *The dual $D(adj_{DK})$ of the adjunction morphism $adj = adj_{DK} : DK \to i_*i^*DK$ becomes the adjunction morphism $adj = adj_K : i_*i^!K \to K$.*

Proof. $D(id_{i^*(DK)}) = id_{i^!K}$ by II.7.5. Hence $D(adj_{DK}) = adj_K$, since the two adjunction morphisms correspond to the identity morphism $id_{i^!K}$ resp. id_{i^*K}. We have

$$adj \in Hom_{D(X)}(K, i_*i^*K) = Hom_{D(X)}(i^*K, i^*K) \ni id_{i^*K} \qquad \text{, and}$$
$$adj \in Hom_{D(X)}(i_*i^!K, K) = Hom_{D(X)}(i^!K, i^!K) \ni id_{i^!K} .$$

□

Theorem 11.2 *Let $X \to Spec(\kappa)$ and $Y \to Spec(\kappa)$ be smooth morphisms and suppose, that $i : Y \hookrightarrow X$ is a closed embedding of pure codimension r in X. Then for*

$$i^!(K) = i^*R\Gamma_Y(K) \quad , \quad K \in D_c^b(X)$$

there exist natural morphisms

$$\mu : i^*(K) \to i^!(K)[2r](r) .$$

Supplement. *If the complex K is (locally) the pullback*

$$K = f^*(F) \quad , \quad F \in D_c^b(S)$$

of a complex F under a smooth equidimensional morphism $f : X \to S$ (defined locally on X), such that $g = f \circ i$ is also smooth, then μ is an isomorphism

$$\mu : i^*(K) \cong i^!(K)[2r](r) .$$

Proof. *(of Theorem II.11.2)* $i^!(K) = i^*R\Gamma_Y(K)$ was obtained in the proof of Theorem II.7.1 (case of a closed embedding).

Definition of μ. To define μ we use the adjunction formula of the duality Theorem II.7.1. It implies

$$Hom_{D(Y)}(i^*(K), i^!K[2r](r)) = Hom_{D(X)}(i_*i^*(K), K[2r](r)) ,$$

via

$$\mu \mapsto adj \circ i_*(\mu) .$$

So it is enough to specify the corresponding term on the right side. By Corollary II.7.5(g) we have $i_*(i^*K) = i_*(\overline{\mathbb{Q}}_{lY}) \otimes^L K$. Hence it is enough to define $\tilde{\mu}$ corresponding to μ with respect to the following identification

$$\mu \in Hom_{D(Y)}(i^*K, i^!K[2r](r)) = Hom_{D(X)}(i_*i^*(\overline{\mathbb{Q}}_{lX}) \otimes^L K, K[2r](r)) \ni \tilde{\mu} .$$

To define $\tilde{\mu}$ put $\tilde{\mu} = \tilde{\mu}_0 \otimes^L id_K$, where $\tilde{\mu}_0$ is the morphism in the special case of the constant sheaf complex $K = \overline{\mathbb{Q}}_l$

$$\mu_0 : i^*(\overline{\mathbb{Q}}_{l X}) \to i^!(\overline{\mathbb{Q}}_{l X})[2r](r) .$$

Then it only remains to define μ_0. In the case of the constant sheaf $K = \overline{\mathbb{Q}}_l$ we are, by our assumptions, in the situation of the supplementary claim of Theorem II.11.2, if we put $S = Spec(\kappa)$ and $F = \overline{\mathbb{Q}}_l$. In particular, μ_0 should therefore be an isomorphism. Hence for the definition of μ_0 it only remains to consider the supplementary claim of the theorem.

Proof of the Supplementary Claim. Suppose, that we are in the situation of the supplementary claim. In particular, $K = f^*(F)$ and $F \in D_c^b(S)$. Then $i^*(K) = i^*(f^*(F)) = g^*(F)$. Put $d = dim(X/S)$. Since g is smooth and equidimensional, there exists an isomorphism induced by $\theta_{Y/S}^{-1}$

$$g^*(F) \cong g^!(F)[2r - 2d](r - d) .$$

By functoriality $g^! = i^! f^!$ we have $g^!(F) = i^! f^!(F)$. Thus we get another isomorphism $i^!(\theta_{X/S})$

$$i^!(f^!(F[2r - 2d](r - d))) \cong i^!(f^*(F)[2r](r)) ,$$

induced from the duality isomorphism $\theta_{X/S}$ on X for the sheaf complex $F[2r - 2d](r - d)$. If we put things together, we obtain a composed morphism

$$\begin{aligned} i^*(K) = g^*(F) &\cong g^!(F)[2r - 2d](r - d) = i^!(f^! F[2r - 2d](r - d)) \\ &\cong i^!(f^*(F)[2r](r)) , \end{aligned}$$

which defines an isomorphism $\nu = \theta_{Y/S}^{-1} \circ i^!(\theta_{X/S})$

$$\nu : i^*(K) \cong i^!(K)[2r](r) \quad , \quad K = f^*(F) ,$$

which is functorial in F.

Definition of μ_0. We put $\mu_0 := \nu_0$ in the special case $f : X \to Spec(\kappa)$ and $F = \overline{\mathbb{Q}}_l$.

Compatibility. We have now defined two such maps – μ as defined above in the general case – and ν, as defined in the situation of the supplementary claim. In fact, both maps actually coincide. For this consider the diagram

$$\begin{array}{cccc} \mu \otimes id_{i^*K} : & i^*(K') \otimes^L i^*(K) & \cong & i^!(K')[2r](r) \otimes^L i^*(K) \\ & \downarrow\cong & & \downarrow \\ \mu : & i^*(K' \otimes^L K) & \cong & i^!(K' \otimes^L K)[2r](r) \end{array} .$$

The vertical maps are defined by the cup product. The cup product has natural properties. See [SGA4$\frac{1}{2}$], page 133 ff. The diagram commutes by the definition of the map μ.

Tensor Products. The map ν defined in the special situation is also compatible with tensor products

$$\begin{array}{ccc} i^*(K') \otimes^L i^*(K) & \xrightarrow[\cong]{\nu\otimes id_{i^*(K)}} & i^!(K')[2r](r) \otimes^L i^*(K) \\ \downarrow \cong & & \downarrow \\ i^*(K' \otimes^L K) & \xrightarrow[\cong]{\nu} & i^!(K' \otimes^L K)[2r](r) \end{array}$$

where now K and K' are assumed to be of the form $K = f^*(F)$ and $K' = f^*(F')$. This holds, since ν is constructed from $\theta_{X/S}$ and $\theta_{Y/S}$. Use that for the smooth morphism f the natural isomorphisms $f^!(K' \otimes^L K) \cong f^!(K') \otimes^L f^*(K)$ are compatible with the duality identifications $\theta_{X/S}$; and similar for the smooth morphism g. See the beginning of this section.

End of the Argument. In the last two commutative diagrams put $F' = h^*(\overline{\mathbb{Q}}_l)$ for $h : S \to Spec(\kappa)$. Then for $K' = f^*(F')$ we are in the case $\nu = \nu_0 = \mu_0 = \mu$, hence the upper horizontal maps in these diagrams coincide. By the commutativity of the diagrams, then also the lower horizontal maps coincide. □

The Chern Class

After these preliminary remarks we give a short review of properties of the Chern class attached to a divisor in the etale cohomology. For further information we refer to [FK], chap. II, §2 and [SGA4$\frac{1}{2}$]. We will see, that this Chern class is related to the map μ defined above (Theorem II.11.2).

Let $\mathscr{L}$ be a line bundle on the scheme X. Using the Kummer sequence, one defines the Chern class $cl_X(\mathscr{L}) \in H^2(X, \overline{\mathbb{Q}}_l(1))$ or alternatively

$$cl_X(\mathscr{L}) \in Hom_{D(X)}\Big(\overline{\mathbb{Q}}_{lX}, \overline{\mathbb{Q}}_{lX}[2](1)\Big)\,.$$

We will not repeat the general definition here (see for instance [FK]). Assume instead, that $f : X \to Spec(\kappa)$ is smooth of dimension d. Furthermore assume, that the line bundle $\mathscr{L}$ has a section, such that the zero set Y is of pure codimension 1 in X. Let $i : Y \to X$ denote the inclusion map, let g denote the map of Y to the spectrum of the base field κ. Assume that g is a smooth morphism. In this special situation, one can also define a cohomology class $cl'^0(Y)$, as explained below. It is shown in [FK] lemma 2.6, II §2 and proposition 2.7, II §2, that the two cohomology classes $cl_X(\mathscr{L})$ and $cl'^0(Y)$ coincide.

Definition of $cl'^0(Y)$**.** Put $S = Spec(\kappa)$ and $\Lambda_S = (\overline{\mathbb{Q}}_l)_S$, then put $\Lambda_X := f^*(\Lambda_S)$ and $\Lambda_Y := i^*\Lambda_X = g^*(\Lambda_S)$. Consider the trace $Tr_g \in Hom_{D(S)}(\mathrm{R}g_!(g^!\Lambda_S), \Lambda_S)$. Then the class

$$cl'^0(Y) \in Hom_{D(X)}(\Lambda_X, i_*i^!\Lambda_X[2](1))$$

is defined from the trace map Tr_g by the left chain of identifications in the diagram

$$
\begin{array}{ccc}
Hom_{D(S)}(Rg_!g^!\Lambda_S, \Lambda_S) \ni Tr_g & & \\
\| & & \\
id \in Hom_{D(Y)}(g^!\Lambda_S, g^!\Lambda_S) & & \\
\theta_{Y/S}\uparrow\cong & & \\
Hom_{D(Y)}(\Lambda_Y[2d-2](d-1), g^!\Lambda_S) & & \\
\| & & \\
Hom_{D(Y)}(\Lambda_Y, g^!\Lambda_S[2-2d](1-d)) & \xrightarrow[\cong]{\theta_{Y/S}} & \\
\| & & \\
Hom_{D(Y)}(\Lambda_Y, i^!f^!\Lambda_S[2-2d](1-d)) & & Hom_{D(Y)}(\Lambda_Y, \Lambda_Y) \ni id \\
\cong\downarrow\theta_{X/S} & & \| \\
Hom_{D(Y)}(\Lambda_Y, i^!\Lambda_X[2](1)) & \xleftarrow[\cong]{\mu} & Hom_{D(Y)}(\Lambda_Y, i^*\Lambda_X) \\
\| & & \| \\
Hom_{D(Y)}(i^*\Lambda_X, i^!\Lambda_X[2](1)) & & Hom_{D(Y)}(i^*\Lambda_X, i^*\Lambda_X) \\
\| & & \| \\
Hom_{D(X)}(\Lambda_X, i_*i^!\Lambda_X[2](1)) & \xleftarrow[\cong]{i_*(\mu)} & Hom_{D(X)}(\Lambda_X, i_*i^*\Lambda_X) \ni adj \\
\downarrow adj[2](1) & & \\
Hom_{D(X)}(\Lambda_X, \Lambda_X[2](1)) \ni cl'^0(Y) & &
\end{array}
$$

In this commutative diagram we have identified homomorphism groups, whenever there exists a canonical identification or an adjunction formula. The elements Tr_g, id, adj and $cl'^0(Y)$ correspond to each other under the isomorphisms and identifications used in this diagram.

An Alternative Definition

We still maintain the assumptions made in the last paragraph, i.e. we assume that f and g are smooth morphisms. Let us reformulate the definition of the Chern class $cl'^0(Y) = cl_X(\mathscr{L})$ via the right side of the diagram above. In this way one obtains a new description of the Chern class

$$\eta = cl_X(\mathscr{L}) .$$

The new description gives $\eta \in Hom_{D(X)}(\Lambda_X, \Lambda_X[2](1))$ as the composite of the following three morphisms:

$$\eta : \quad \Lambda_X \xrightarrow{adj} i_*i^*\Lambda_X \xrightarrow{i_*(\mu)} i_*i^!\Lambda_X[2](1) \xrightarrow{adj[2](1)} \Lambda_X[2](1) \ .$$

The homomorphism η is a morphism in the triangulated category $D_c^b(X, \overline{\mathbb{Q}}_l)$. It depends only on the Chern class of $\mathscr{L}$, and induces corresponding homomorphisms

$$\eta \otimes^L id_K : K \longrightarrow K[2](1)$$

for every complex K on X, via the tensor product $(-) \otimes^L K$. The homomorphisms, obtained in this way, are compatible with base change, since the Chern class of $\mathscr{L}$ has this property.

Lemma 11.3 *Suppose K is a complex on X, where X is smooth over the base field κ. Suppose $\mathscr{L}$ is a line bundle on X with section s, whose zeros define a divisor $i : Y \to X$, which is smooth over κ. Let $\eta = cl_X(\mathscr{L})$ be the Chern class of $\mathscr{L}$. Then the map $\eta \otimes^L id_K$ factorizes in the form*

$$\phi_{\mathscr{L},K} : \qquad K \xrightarrow{adj} i_* i^*(K) \xrightarrow{i_*(\mu)} i_* i^!(K)[2](1) \xrightarrow{adj} K[2](1) \ .$$

Under appropriate assumptions on K – see Theorem II.11.2 – the map $i_*(\mu)$ in the middle is an isomorphism.

Proof. This follows immediately from the definition of the map μ, which was given in the proof of Theorem II.11.2, and the following diagram

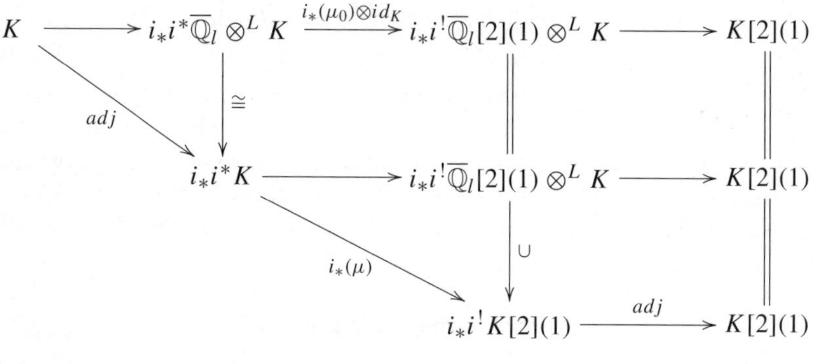

□

Let us now consider direct images with respect to a relative situation. Let f be a map

$$f : X \longrightarrow S \ ,$$

where now S is an arbitrary base scheme again. Let g denote the restriction of f to the divisor Y. The "cohomology class" η induces a homomorphism

$$\phi_{\mathscr{L},K,f} = \mathrm{R}f_*(\eta \otimes^L id_K) : \ \mathrm{R}f_*(K) \longrightarrow \mathrm{R}f_*(K)[2](1) \ .$$

From the last description of this map, one derives the following result

Theorem 11.4 *Let the assumptions be as in II.11.2 and II.11.3, and let the notations be as above except that K is denoted G: In particular $G = f^*(F)$ is a complex on*

X obtained from a smooth map $f : X \to S$, whose restriction to Y is also smooth, and some complex F on S. The cohomology class of the line bundle $\mathscr{L}$ defines a homomorphism in the derived category

$$\phi_{\mathscr{L},G,f} : Rf_*(G) \longrightarrow Rf_*(G)[2](1) ,$$

which is the composite of the natural restriction map

$$\alpha_G : Rf_*(G) \longrightarrow Rg_*i^*(G) ,$$

an isomorphism $Rg_*i^*(G) \cong Rg_*i^!(G)[2](1)$ and the natural map

$$\beta_G : Rg_*i^!(G)[2](1) \longrightarrow Rf_*(G)[2](1) .$$

Remark. If f is proper, then $\mathrm{R}f_* = \mathrm{R}f_!$. By Lemma II.11.1 and the results of Corollary II.7.5 the map β_G is the dual of the natural restriction map

$$\alpha_{\tilde{G}} : \mathrm{R}f_*(\tilde{G}) \longrightarrow \mathrm{R}g_*i^*(\tilde{G}) \qquad \tilde{G} = D(G)[-2](-1) .$$

Iterating the map $\phi_{\mathscr{L},G,f}$ and its twisted shifts r times, one obtains

Lemma 11.5 *Under the preceding assumptions and with the notations of the second part of Theorem II.11.4 the following diagram is commutative for $\phi = \phi_{\mathscr{L},G,f}$ and $\varphi = \phi_{i^*(\mathscr{L}),i^*(G),g}$:*

$$\begin{array}{ccccccc}
Rf_*(G) & \xrightarrow{\phi} \cdots \xrightarrow{\phi} & Rf_*(G)[2r-2](r-1) & \xrightarrow{\phi} & Rf_*(G)[2r](r) \\
\downarrow{\scriptstyle \alpha_G} & & \downarrow{\scriptstyle \alpha_G} & & \uparrow{\scriptstyle \beta_G} \\
Rg_*i^*(G) & \xrightarrow{\varphi} \cdots \xrightarrow{\varphi} & Rg_*i^*(G)[2r-2](r-1) & \xrightarrow{\cong} & Rg_*i^!(G)[2r](r)
\end{array}$$

Proof. For the right square use II.11.4, for the other squares use compatibility with pullback. □

II.12 Mixed Complexes

Definition 12.1 *Let X_0 be an algebraic scheme over a finite field. Then an object of the category $D^b_c(X_0, \overline{\mathbb{Q}}_l)$ is said to be τ-mixed respectively mixed, if all its cohomology sheaves are τ-mixed respectively mixed sheaves on X_0.*

Theorem 12.2 *In case of algebraic schemes over a field, the operations*

$$Rf_!, Rf_*, f^*, f^!, \otimes^L, D$$

preserve boundedness, constructibility and in case of finite base fields τ-mixedness respectively mixedness of complexes.

Proof. We restrict ourselves to the mixedness statements. τ-mixedness and constructibility statements are proved literally in the same way. In fact, since the definition of mixedness requires constructibility statements, these corresponding statements have to be proved first. Therefore it is assumed, that the corresponding proofs have been established before. So this allows to restrict ourselves to the proof of the mixedness statements.

The first two cases of the theorem involving the functors $Rf_!$ and Rf_* are the essential cases. They are covered by I.9.3 and I.9.4. The third case of the functor f^* is trivial. The fourth case involving $f^!$ is reduced to the case f^* by biduality, using the formula

$$D \circ f^! = f^* \circ D$$

and the assertion, that D preserves mixedness. The fifth case of the tensor product $\otimes^L$ follows using the Künneth formula. So it remains to prove the last case: $D(K_0)$ is mixed if K_0 is mixed.

For the proof, that $D(K_0)$ is mixed, we can assume $K_0 = \mathcal{G}_0$ to be a mixed sheaf on some algebraic reduced scheme Y_0. We will prove the assertion by induction on $dim(Y_0)$. The assertion is true, if Y_0 is smooth and if $\mathcal{G}_0$ is a smooth sheaf on Y_0. In general there exists an open dense smooth subscheme

$$j_0 : U_0 \hookrightarrow Y_0$$

of Y_0 with the complement

$$i_0 : Z_0 \hookrightarrow Y_0 ,$$

such that $j_0^*(\mathcal{G}_0)$ is smooth. We have: $dim(Z_0) < dim(Y_0)$. Then $D(j_0^*(\mathcal{G}_0))$ is mixed by our smoothness assumption, hence according to Theorem I.9.4 also

$$D(j_{0!}j_0^*(\mathcal{G}_0)) = Rj_{0*}(D(j_0^*(\mathcal{G}_0))) .$$

By the induction assumption also

$$D(i_{0*}i_0^*(\mathcal{G}_0)) = i_{0*}D(i_0^*(\mathcal{G}_0))$$

is mixed. We dualize the admissible triangle

$$j_{0!}j_0^*(\mathcal{G}_0) \longrightarrow \mathcal{G}_0 \longrightarrow i_{0*}i_0^*(\mathcal{G}_0) \longrightarrow j_{0!}j_0^*(\mathcal{G}_0)[1]$$

to obtain the admissible triangle

$$i_{0*}D(i_0^*(\mathcal{G}_0)) \longrightarrow D(\mathcal{G}_0) \longrightarrow Rj_{0*}D(j_0^*(\mathcal{G}_0)) \longrightarrow i_{0*}D(i_0^*(\mathcal{G}_0))[1] .$$

Therefore also the second complex $D(\mathcal{G}_0)$ of this triangle is mixed. This completes the proof. □

Definition 12.3 *Let X_0 be a scheme over a finite field κ. Fix an isomorphism $\tau : \overline{\mathbb{Q}}_l \cong \mathbb{C}$. For a τ-mixed sheaf B_0 on X_0 remember Definition I.2.3 of $w(B_0)$, i.e. the maximum of its punctual τ-weights for nontrivial B_0. For a τ-mixed complex $B_0 \in D_c^b(X_0, \overline{\mathbb{Q}}_l)$ let*

$$w(B_0) = max_\nu\big(w(\mathcal{H}^\nu B_0) - \nu\big)$$

denote the maximum of all the $w(\mathcal{H}^\nu B_0) - \nu$.

For a finitely generated scheme X_0 over a finite field κ let

$$D_{mixed}(X_0) = D^b_{mixed}(X_0, \overline{\mathbb{Q}}_l)$$

denote the triangulated category of mixed complexes of sheaves. If τ is fixed, this will denote the triangulated category of τ-mixed complexes of sheaves in $D_c^b(X_0, \overline{\mathbb{Q}}_l)$. Let

$$D^b_{\leq w} = D^b_{\leq w}(X_0) = \{K_0 \in D_{mixed}(X_0) \mid w(K_0) \leq w\}$$

be the - non triangulated - subcategory of (τ-)mixed complexes of (τ-)weight smaller or equal to w. Finally define

$$D^b_{\geq w} = D^b_{\geq w}(X_0) = \{K_0 \in D_{mixed}(X_0) \mid w(D(K_0)) \leq -w\}\ .$$

Here w may be any real number.

Remark. Note, that $w(K_0) \geq w$ does not imply $K_0 \in D^b_{\geq w}(X_0)$. However it is shown in II.12.7 below, that in the other direction $K_0 \in D^b_{\geq w}(X_0)$ implies $w(K_0) \geq w$. This is trivial for $X_0 = Spec(\kappa)$.

We have the following elementary properties

(1) $w(K_0[n]) = w(K_0) + n$
(2) $w(K_0(m)) = w(K_0) - 2m$ (Tate twist)

and the permanence properties

(3) $Rf_{0*}(D^b_{\geq w}(X_0)) \subset D^b_{\geq w}(Y_0)$ for a morphism $f_0 : X_0 \to Y_0$ defined over κ.
(4) $f_0^!(D^b_{\geq w}(Y_0)) \subset D^b_{\geq w}(X_0)$ for a morphism $f_0 : X_0 \to Y_0$ defined over κ

which follow by duality from the Weil conjectures I.9.3 and I.9.4:

(5) $Rf_{0!}(D^b_{\leq w}(X_0)) \subset D^b_{\leq w}(Y_0)$ for a morphism $f_0 : X_0 \to Y_0$ defined over κ.
(6) $f_0^*(D^b_{\leq w}(Y_0)) \subset D^b_{\leq w}(X_0)$ for a morphism $f_0 : X_0 \to Y_0$ defined over κ.

Homomorphisms. In particular, as a consequence

$$K_0 = R\mathcal{H}om(A_0, B_0[1]) \ \in\ D^b_{\geq w'-w+1}(X_0)$$

for $A_0 \in D^b_{\leq w}(X_0)$ and $B_0 \in D^b_{\geq w'}(X_0)$. Recall that $R\mathcal{H}om(A_0, B_0[1]) \cong D(A_0 \otimes D(B_0[1]))$.

The Galois group $Gal(k/\kappa)$ is the profinite completion of the cyclic group generated by the Frobenius element F. This implies:

Let $K_0 \in D_c^b(X_0)$. Then there is an exact sequence

$$0 \to H^{-1}(X, K)_F \to H^0(X_0, K_0) \to H^0(X, K)^F \to 0$$

in terms of coinvariants and invariants under F (Leray spectral sequence). By property (3) above we have for $K_0 \in D^b_{\geq w'-w+1}(X_0)$

$$w(H^\nu(X, K)) \geq w' - w \quad , \qquad \text{for } \nu \geq -1 \, .$$

Suppose now $w' - w > 0$. Then this forces $H^0(X_0, K_0) = 0$, because

$$H^{-1}(X, K)_F = 0 = H^0(X, K)^F \, .$$

Consider $Hom_{D_c^b(X_0)}(A_0, B_0[1])$. To compute this group, we replace B_0 by an injective resolution I_0. Then $Hom_{D_c^b(X_0)}(A_0, B_0[1])$ is the group $Hom_{K(X_0)}(A_0, I_0[1])$ of complex maps up to homotopy. See Remark II.1.8. Under our assumption on B_0 this is the same as the hypercohomology group $H^0(X_0, K_0) = H^0(X_0, \mathrm{R}\mathcal{H}om(A_0, I_0[1]))$. See Verdier [308] and also [FK], p.300. Furthermore we have

$$Hom_{D_c^b(X_0)}(A_0, B_0[1]) = H^0(X_0, \mathrm{R}\mathcal{H}om(A_0, B_0[1])) \, .$$

Proposition 12.4 *Let $A_0 \in D^b_{\leq w}(X_0)$ and $B_0 \in D^b_{\geq w'}(X_0)$. Then*

$$Hom_{D_c^b(X_0)}(A_0, B_0[1]) = 0$$

vanishes if $w' > w$.

Proposition 12.5 *Let $A_0 \in D^b_{\leq w}(X_0)$ and $B_0 \in D^b_{\geq w'}(X_0)$. Then base change to the algebraic closure k of κ*

$$Hom_{D_c^b(X_0)}(A_0, B_0[1]) \to Hom_{D_c^b(X)}(A, B[1])^F = 0$$

is the zero map, if $w' > w - 1$.

If $K = \mathrm{R}\mathcal{H}om(A_0, B_0[1])$ happens to be in $D^{\geq 0}(X_0)$, i.e has nontrivial cohomology sheaves only in degrees ≥ 0, then $H^{-1}(X, K) = 0$. We then get the stronger

Proposition 12.6 *Suppose $R\mathcal{H}om(A_0, C_0) \in D^{\geq 0}(X_0)$ and furthermore assume that $A_0 \in D^b_{\leq w}(X_0)$ and $C_0 \in D^b_{\geq w'}(X_0)$. Then*

$$Hom_{D_c^b(X_0)}(A_0, C_0) = Hom_{D_c^b(X)}(A, C)^F \, ,$$

and this group vanishes for $w' > w$.

This is a variant of the last proposition, if we put $C_0 = B_0[1]$. It holds, since the kernel of the base change restriction map of the last proposition is the cohomology group $H^{-1}(X, R\mathscr{H}om(A, B[1]))_F$ and vanishes by assumption.

Lemma-Definition 12.7 *Let $K_0 \in D_c^b(X_0, \overline{\mathbb{Q}}_l)$ be a mixed respectively τ-mixed complex with upper weight $w = w(K_0)$. If $K_0 \neq 0$ is nontrivial, then the following holds*

$$w(DK_0) \geq -w(K_0) .$$

In particular, if $K_0 \in D^b_{\geq w}(X_0)$ holds for some w, then $w \leq w(K_0)$.

The complex K_0 - suppose $K_0 \neq 0$ - is called **pure** *resp.* **τ-pure of weight** *w if equality $w(DK_0) = -w(K_0)$ holds, i.e. if*

$$K_0 \in D^b_{\leq w}(X_0) \cap D^b_{\geq w}(X_0) .$$

It is also convenient to define $K_0 = 0$ to be (τ-)pure of weight $-\infty$.

Remark. If K_0 is τ-pure of weight w, then w necessarily coincides with the upper weight $w(K_0)$

$$w = w(K_0) = max_\nu(w(\mathscr{H}^\nu(K_0)) - \nu) .$$

Warning. In case of sheaves concentrated in degree zero the notion of being τ-pure of weight w given in II.12.7 above does not coincide with the notion of (point-wise) τ-pure sheaves in the sense of Definition I I.2.1. In the following we will have to distinguish these two notions. We will therefore always add "point-wise", if the meaning of pure or τ-pure is not to be understood in the sense of the definition above.

Proof of II.12.7. Put $A_0 = K_0$ and $B_0[1] = K_0$. Then $w = w(A_0) = w(K_0)$ and $w' = -w(D(B_0)) = -1 - w(D(K_0))$. Now either $w' \leq w - 1$, i.e. $-w(D(K_0)) \leq w(K_0)$. This is the assertion of II.12.7. Or $w' > w - 1$. Then Proposition II.12.5 implies

$$id_K \in Hom(K, K)^F = 0 ,$$

hence $K = 0$. Then the stalks of the cohomology sheaves in all geometric points vanish. Therefore $K_0 = 0$ contrary to the assumption.

III. Perverse Sheaves

III.1 Perverse Sheaves

The theory of perverse sheaves historically emerged from several independent directions. One of them was the theory of intersection cohomology of Goresky-MacPherson, which originally was not defined in terms of sheaf theory but rather using explicit chain complexes. Perhaps stimulated by the Kazhdan-Lusztig conjectures it was Deligne, who gave a reformulation of the notion of intersection cohomology within the setting of sheaf theory. In this form intersection cohomology can be defined also for finitely generated schemes over a field of characteristic zero, over a finite field or over the algebraic closure of a finite field. The theory found its final form in the fundamental treatise [BBD], soon after it was realized that perverse sheaves define an abelian category inside $D_c^b(\overline{\mathbb{Q}}_l)$. This was suggested by the algebraic theory of $\mathscr{D}$-modules due to Bernstein, where it turned out that the category of holonomic $\mathscr{D}$-modules with regular singularities provide an analog via the Riemann-Hilbert correspondence. In this book we only consider the so called middle perversity. Middle perverse sheaves turn out to be most interesting perverse sheaves, since their notion is self dual with respect to Verdier duality. It is the abelian category of middle perverse sheaves, which also fits perfectly with Deligne's theory of weights. In particular, one has for it the decomposition theorem due to Gabber. The more general perversities, which are defined in [BBD], do not have this particular rigid structure. Their definitions and properties can be dealt with more or less in the same way as for the case of the middle perversity. Although they are important, we therefore do not consider them in this book. For their definition and properties we refer the reader to the original source [BBD].

Notational Remark. Throughout this chapter on perverse sheaves we often write $D(X)$ instead of $D_c^b(X, \overline{\mathbb{Q}}_l)$ (derived categories of a scheme X) and $D(K)$ or DK instead of $D_X(K)$ (Verdier dual of a complex K on X) and $\otimes$ instead of $\otimes^L$, in order to simplify notation. It seemed to us, that there is little danger of confusion.

Let X be a scheme over a base field, which is either a finite field or an algebraically closed field. Let D be now the triangulated category $D(X) = D_c^b(X, \overline{\mathbb{Q}}_l)$. The selfdual (middle) **perverse t-structure** is defined by

$$B \in {}^pD^{\leq 0}(X) \iff dim\, supp(\mathscr{H}^{-i}B) \leq i \quad , \quad \forall i \in \mathbb{Z}$$

$$B \in {}^pD^{\geq 0}(X) \iff dim\, supp(\mathscr{H}^{-i}DB) \leq i \quad , \quad \forall i \in \mathbb{Z}\,.$$

The index ^{p} refers to truncation with respect to the perverse t-structure. For the moment let us assume, that these definitions induce a t-structure on $D(X)$. The proof of this fact will be postponed to §3. From this definition is then clear, that every object of $D(X)$ is bounded with respect to the perverse t-structure. In particular almost all perverse cohomology groups of an object in $D(X)$ vanish.

Lemma-Definition 1.1 *The abelian category $Perv(X)$ of* **perverse sheaves** *on X is the core*

$$Perv(X) = {}^pD^{\leq 0}(X) \cap {}^pD^{\geq 0}(X)$$

of the category $D(X) = D^b_c(X, \overline{\mathbb{Q}}_l)$ with respect to the perverse t-structure. The corresponding cohomological functors are

$${}^pH^\nu : D(X) \longrightarrow Perv(X) .$$

For any distinguished triangle (A, B, C) in $D(X)$ this gives a long exact sequence

$$\cdots \longrightarrow {}^pH^{-1}(C) \longrightarrow {}^pH^0(A) \longrightarrow {}^pH^0(B) \longrightarrow {}^pH^0(C) \longrightarrow {}^pH^1(A) \longrightarrow \cdots$$

in $Perv(X)$.

Warning. In general a perverse sheaf on X is not a sheaf, but only represented by a complex of sheaves on X.

In the following the truncation operators ${}^p\tau$ and τ in the category $D(X)$ will be always understood as truncation operators with respect to the perverse t-structure respectively with respect to the standard t-structure, if not stated otherwise.

Remark 1.2 By definition the perverse t-structure is self dual on $D(X)$, i.e.

$$B \in {}^pD^{\geq 0}(X) \text{ iff } DB \in {}^pD^{\leq 0}(X) .$$

Let x be an arbitrary point x of X, let $i : Y \to X$ its closure in X with reduced subscheme structure. Any $B \in D(Y)$ becomes a smooth complex on a suitable smooth open neighborhood U of x. For such U the dualizing complex is $K_U \cong \overline{\mathbb{Q}}_{lU}[2d(x)](d(x))$, where $d(x) = dim(Y)$. Hence $(D_Y(B))_\eta \cong (B_\eta)^\vee[2d(x)](d(x))$ for any geometric point η over x. Here $M^\vee$ denotes the dual of M as a complex of $\overline{\mathbb{Q}}_l$-vectorspaces, where stalk complexes are considered as complexes of $\overline{\mathbb{Q}}_l$-vectorspaces.

Definition 1.3 *For L in $D(X)$ define*

$$i_x^!L = (i^!L)_\eta \quad , \quad i_x^*L = (i^*L)_\eta .$$

Since $i_x^*(D_X(L)) = (i^*(D_X(L))_\eta = (D_Y(i^!L))_\eta = i_x^!(L)[2d(x)](d(x))$, another characterization for the perverse t-structure on $D(X)$ is provided by

$$B \in {}^pD^{\leq 0}(X) \iff \mathcal{H}^\nu i_x^* B = 0 \text{ for } \nu > -d(x)$$

$$B \in {}^pD^{\geq 0}(X) \iff \mathcal{H}^\nu i_x^! B = 0 \text{ for } \nu < -d(x) ,$$

where this should hold for all points x in X.

III.2 The Smooth Case

Let X be a scheme over some arbitrary base field. Let $\tilde{X}$ denote the scheme obtained from X by base change to the algebraic closure of the base field. The scheme X is called **essentially smooth**, if the reduced scheme $\tilde{X}_{red}$ is smooth. Under the assumption that the base field is finite or algebraically closed, the scheme X is essentially smooth if and only if the reduced scheme X_{red} is smooth.

Suppose X is essentially smooth, and suppose X is equidimensional of dimension d. Then the dualizing complex on X has the form

$$\overline{\mathbb{Q}}_l[2d](d)\ .$$

It is a complex K_X, whose cohomology is concentrated in degree $-2d$ and such that its cohomology sheaf $\mathscr{H}^{-2d}(K_X)$ is isomorphic to the smooth sheaf $\overline{\mathbb{Q}}_l(d)$. For a smooth sheaf $\mathscr{G}$ on X – sheaf to be understood in the ordinary sense – the dual sheaf was denoted $\mathscr{G}^\vee$

$$\mathscr{G}^\vee = \mathscr{H}om(\mathscr{G}, \overline{\mathbb{Q}}_l)\ .$$

A sheaf complex $K \in D(X) = D^b_c(X, \overline{\mathbb{Q}}_l)$ will be called a **smooth complex**, if all its cohomology sheaves $\mathscr{H}^\nu(K)$ are smooth sheaves on X.

Proposition 2.1

(1) *Let X be an essentially smooth equidimensional scheme of dimension d and let $K \in D(X)$ be a smooth complex on X. Then*

$$\mathscr{H}^\nu(DK) \cong \mathscr{H}^{-\nu-2d}(K)^\vee(d)\ .$$

In particular $\mathscr{H}^\nu(DK)$ vanishes if and only if $\mathscr{H}^{-\nu-2d}(K)$ vanishes.

(2) *Suppose X is irreducible and let K be in $D(X)$. Then there exists an open dense essentially smooth subscheme*

$$j : U \hookrightarrow X$$

of X, such that

$$j^*(K)$$

is a sheaf complex, which is smooth on U.

Proof. Use induction on the cohomology degree in which there are non vanishing cohomology sheaves, using the "ordinary" truncation operators $\tau_{\leq s}$ for the standard t-structure on the triangulated category $D(X)$. For $K \in D(X)$ we have $\mathscr{H}^s(K)[-s] = \tau_{\geq s}\tau_{\leq s}K$. See Chap. II.6.1 for details on the standard t-structure. For a smooth complex $K \neq 0$ there exists s such that the distinguished triangle

$$\tau_{\leq s-1}K \to K \to \tau_{\geq s}K \to (\tau_{\leq s-1}K)[1]$$

defines the smooth sheaf $\mathcal{G} = \tau_{\geq s} K = \tau_{\geq s}\tau_{\leq s} K = \mathcal{H}^s(K) \neq 0$, such that $A = \tau_{\leq s-1} K$ is a smooth complex with cohomology sheaves $\mathcal{H}^\nu(A) = \mathcal{H}^\nu(K)$ for $\nu < s$ and $\mathcal{H}^\nu(A) = 0$ for $\nu \geq s$.

For smooth $\overline{\mathbb{Q}}_l$-sheaves $\mathcal{G}$ and $s \in \mathbb{Z}$ consider the complex

$$\mathcal{G}[-s] \in D_c^b(X, \overline{\mathbb{Q}}_l)$$

concentrated in degree s. Then $\mathcal{H}^\nu(\mathcal{G}[-s]) = 0$ for $\nu \neq s$ and $\mathcal{H}^s(\mathcal{G}[-s]) = \mathcal{G}$. Then, for any smooth $\overline{\mathbb{Q}}_l$-sheaves $\mathcal{F}$, $\mathcal{G}$ on X,

$$\mathcal{H}^\nu \mathrm{R}\mathcal{H}om(\mathcal{G}[0], \mathcal{F}[0]) = \begin{cases} 0 & \nu \neq 0 \\ \mathcal{H}om(\mathcal{G}, \mathcal{F}) & \nu = 0 \,. \end{cases}$$

For $\mathcal{F} = \overline{\mathbb{Q}}_l(d)$ this gives

$$D(\mathcal{G}[-s]) = \mathrm{R}\mathcal{H}om(\mathcal{G}[-s], \mathcal{F}[2d]) = \mathcal{H}om(\mathcal{G}, \mathcal{F})[2d+s] \,.$$

Therefore, if we apply the functor $\mathrm{R}\mathcal{H}om(-, \mathcal{F}[2d])$ to the triangle $(A, K, \mathcal{G}[-s])$ defined above, the long exact cohomology sequence of its cohomology sheaves

$$\mathcal{H}^\nu \mathrm{R}\mathcal{H}om(\mathcal{G}[-s], \mathcal{F}[2d]) \longrightarrow \mathcal{H}^\nu \mathrm{R}\mathcal{H}om(K, \mathcal{F}[2d]) \longrightarrow$$
$$\mathcal{H}^\nu \mathrm{R}\mathcal{H}om(A, \mathcal{F}[2d]) \longrightarrow$$

and the obvious facts

$$\mathcal{H}^\nu \mathrm{R}\mathcal{H}om(\mathcal{G}[-s], \mathcal{F}[2d]) = \begin{cases} 0 & \nu \neq -2d-s \\ \mathcal{H}om(\mathcal{G}, \mathcal{F}) & \nu = -2d-s \end{cases}$$

and $\mathcal{H}^\nu \mathrm{R}\mathcal{H}om(A, \mathcal{F}[2d]) = 0$ for $\nu \leq -2d-s$ and $\mathcal{H}^\nu \mathrm{R}\mathcal{H}om(K, \mathcal{F}[2d]) = 0$ for $\nu < -2d-s$ imply

$$\mathcal{H}^\nu \mathrm{R}\mathcal{H}om(K, \mathcal{F}[2d]) = \begin{cases} 0 & \nu < -2d-s \\ \mathcal{H}om(\mathcal{G}, \mathcal{F}) & \nu = -2d-s \\ \mathcal{H}^\nu \mathrm{R}\mathcal{H}om(A, \mathcal{F}[2d]) & \nu > -2d-s \,. \end{cases}$$

Using these identities and the induction hypothesis the proof is now complete. □

Assume that X is irreducible and essentially smooth. Recall, that this was defined at the beginning of this section. Let B be a complex in $D(X)$ such that all cohomology sheaves $\mathcal{H}^i B$ of B are smooth sheaves on X. This will be referred to as the "**smooth case**" for the rest of this section and the next section. A smooth sheaf on X has support of dimension $dim(X)$. All $\mathcal{H}^i B$ were assumed to be smooth, therefore we obtain from the last proposition.

Remark 2.2 Under the assumptions above, i.e. X essentially smooth, equidimensional and B a smooth complex, we have

$$B \in {}^pD^{\leq 0}(X) \text{ iff } \mathscr{H}^\nu B = 0 \text{ for all } \nu > -dim(X) .$$

By duality

$$B \in {}^pD^{\geq 0}(X) \text{ iff } \mathscr{H}^\nu B = 0 \text{ for all } \nu < -dim(X) .$$

□

In the smooth case the perverse t-structure therefore behaves like the standard t-structure shifted by $dim(X)$. Let X be essentially smooth and let B be a smooth sheaf complex on X. We get in this case: $B \in Perv(X)$ if and only if

$$B = \mathscr{G}[dim(X)]$$

for a smooth sheaf $\mathscr{G}$. In particular, the cohomology sheaves of B are trivial in degrees different from $-dim(X)$.

That in general the definitions preceding III.1.1 define a t-structure on $D(X)$ is not that obvious. The proof will be given in §3. In order to verify the properties (iii) and (i) of a t-structure one has to use stratifications of X and for property (iii) one has to use glueing of t-structures. These techniques allow to reduce everything to the above mentioned simple "smooth case". Details will be explained in the next section.

III.3 Glueing

Again in this section we consider finitely generated schemes over a finite field or over an algebraically closed field.

Let be given a scheme X and an open subscheme U

$$j : U \hookrightarrow X .$$

Let

$$i : Y \to X$$

be the closed complement of U in X, endowed with some subscheme structure, e.g. the reduced subscheme structure.

Let $T(U)$ be a full triangulated subcategory of $D(U)$ and let $T(Y)$ be a full triangulated subcategory of $D(Y)$. We additionally impose the condition that

$$A \in T(U) \Longrightarrow i^* \mathrm{R} j_* A \in T(Y) .$$

For our purposes it actually suffices to assume $T(Y) = D(Y)$.

Suppose we are given some t-structures on the categories $T(U)$ and $T(Y)$. In the following these t-structures will be denoted $T^{\leq 0}(U), T^{\geq 0}(U)$ respectively $T^{\leq 0}(Y), T^{\geq 0}(Y)$. Then one can glue these t-structures together to obtain a new t-structure on the full subcategory $T(X, U)$ of $D(X)$, defined by

$$T(X, U) = \{B \in D(X) \mid j^* B \in T(U), i^*(B) \in T(Y), i^! B \in T(Y)\} .$$

By our assumptions on $T(U)$, $T(Y)$, and because of $i^! \mathrm{R} j_* = 0$, we have

$$B \in T(U) \Longrightarrow \mathrm{R} j_* B \in T(X, U)\,.$$

This being said we define the glued t-structure $T^{\leq 0}(X, U)$, $T^{\geq 0}(X, U) \subset T(X, U)$ via

1) $B \in T^{\leq 0}(X, U) \iff j^* B \in T^{\leq 0}(U)$ and $i^* B \in T^{\leq 0}(Y)$
2) $B \in T^{\geq 0}(X, U) \iff j^* B \in T^{\geq 0}(U)$ and $i^! B \in T^{\geq 0}(Y)$.

Verification of the Axioms (i–iii) for Glued t-Structures (II.2.1). Axiom (ii) for the new t-structure is obvious. Axiom (i) follows from the exact sequence

$$\begin{array}{ccccccc} \longrightarrow & Hom(i_* i^* B, C) & \longrightarrow & Hom(B, C) & \longrightarrow & Hom(j_! j^* B, C) & \longrightarrow \\ & \| & & & & \| & \\ & Hom(i^* B, i^! C) & & & & Hom(j^* B, j^* C) & \end{array}$$

which is obtained using adjunction and the long exact Hom-sequence attached to the distinguished triangle $(j_! j^* B, B, i_* i^* B)$.

Axiom (iii). Let E be an object in $T(X, U)$. We consider in the category $T(U)$ – with its corresponding t-structure – the distinguished triangle

$$(\tau_{\leq 0} j^* E, j^* E, \tau_{\geq 1} j^* E)\,.$$

The corresponding morphism $j^* E \to \tau_{\geq 1} j^* E$ gives rise to an adjoint morphism $E \to \mathrm{R} j_* \tau_{\geq 1} j^* E$ in the category $T(X, U)$. Taking pull back under j^* gives back the original morphism $j^* E \to \tau_{\geq 1} j^* E$. Now we complete to a full distinguished triangle

$$F \xrightarrow{\;v\;} E \longrightarrow \mathrm{R} j_* \tau_{\geq 1} j^* E \longrightarrow F[1]\,.$$

In the same way one finds in $T(X, U)$ the following distinguished triangle

$$A \xrightarrow{\;u\;} F \longrightarrow i_* \tau_{\geq 1} i^* F \longrightarrow A[1]\,.$$

Its pull back under i^* is isomorphic to the distinguished triangle

$$(\tau_{\leq 0} i^* F, i^* F, \tau_{\geq 1} i^* F)\,.$$

The two distinguished triangles above can be viewed as part of the following diagram – with distinguished rows and columns using the TR4a octaeder axiom

$$\begin{array}{ccccccc}
 & & A[1] & = & A[1] & & \\
 & & \uparrow & & \uparrow & & \\
\mathrm{R}j_*\tau_{\geq 1}j^*E[-1] & \longrightarrow & i_*\tau_{\geq 1}i^*F & \longrightarrow & B & \longrightarrow & \mathrm{R}j_*\tau_{\geq 1}j^*E \\
\| & & \uparrow & & \uparrow & & \| \\
\mathrm{R}j_*\tau_{\geq 1}j^*E[-1] & \longrightarrow & F & \xrightarrow{v} & E & \longrightarrow & \mathrm{R}j_*\tau_{\geq 1}j^*E \\
 & & \uparrow{\scriptstyle u} & & \uparrow{\scriptstyle v\circ u} & & \\
 & & A & = & A & &
\end{array}$$

Then the octaeder axiom TR4a of triangulated categories gives for this diagram in particular the distinguished triangles $(i_*\tau_{\geq 1}i^*F, B, \mathrm{R}j_*\tau_{\geq 1}j^*E)$ and (A, E, B).

Claim. The triangle (A, E, B) is the desired truncation triangle in $T(X, U)$. In other words

$$A \in T^{\leq 0}(X, U) \quad , \quad B \in T^{\geq 1}(X, U) ,$$

with the truncation triangle

$$\tau_{\leq 0}E \longrightarrow E \longrightarrow \tau_{\geq 1}E \longrightarrow \tau_{\leq 0}E[1]$$

in $T(X, U)$ being isomorphic to the distinguished triangle

$$A \longrightarrow E \longrightarrow B \longrightarrow A[1] .$$

Proof. In order to establish the last claim, we apply the functors j^* and $i^!$ to the octaeder diagram above. Use that

$$j^*i_* = 0 \quad , \quad i^!\mathrm{R}j_* = 0 .$$

This implies the following isomorphism of distinguished triangles

$$j^*(A, E, B) \cong j^*(F, E, \mathrm{R}j_*\tau_{\geq 1}j^*E) \cong (\tau_{\leq 0}j^*E, j^*E, \tau_{\geq 1}j^*E) .$$

Then $i^*(A, F, i_*\tau_{\geq 1}i^*F) \cong (\tau_{\leq 0}i^*F, i^*F, \tau_{\geq 1}i^*F)$ and $i^!(A, F, i_*\tau_{\geq 1}i^*F) \cong i^!(A, E, B)$ by definition.

From this we get 1) $A \in T^{\leq 0}(X, U)$: In other words

$$j^*A \cong \tau_{\leq 0}j^*E \in T^{\leq 0}(U) \quad , \quad i^*A \cong \tau_{\leq 0}i^*F \in T^{\leq 0}(Y)$$

and 2) $B \in T^{\geq 1}(X, U)$: In other words

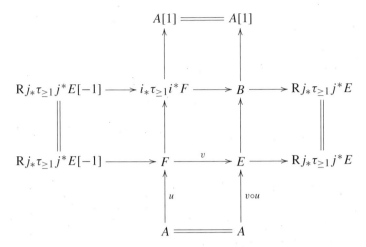

$$j^*B \cong \tau_{\geq 1} j^*E \in T^{\geq 1}(U) \quad , \quad i^!B \cong i^! i_* \tau_{\geq 1} i^*F \cong \tau_{\geq 1} i^*F \in T^{\geq 1}(Y) .$$

This finishes the proof. □

Example $\tau^Y_{\leq 0}$. One useful special case is the t-structure, which is obtained from glueing with the degenerate t-structure on $T(U)$, i.e.

$$T(U)^{\geq 1} = 0 .$$

In this special case the above construction gives

$$F \cong E \quad , \quad B \cong i_* \tau_{\geq 1} i^* F .$$

For the glued t-structure on $T(X, U)$ this determines the upper truncation functor

$$\tau^Y_{\geq 0} E \cong i_* \tau_{\geq 0} i^* E .$$

For the corresponding lower truncation operator, denoted $\tau^Y_{\leq 0}$, we therefore get a distinguished truncation triangle

$$(\tau^Y_{\leq 0} E, E, i_* \tau_{\geq 1} i^* E) .$$

Using the gluing construction and noetherian induction, we are now prepared to prove that the perverse truncation structure ${}^pD^{\leq 0}(X)$, ${}^pD^{\geq 0}(X)$

$$B \in {}^pD^{\leq 0}(X) \iff dim\ supp(\mathcal{H}^{-i}(B)) \leq i$$
$$B \in {}^pD^{\geq 0}(X) \iff dim\ supp(\mathcal{H}^{-i}(DB)) \leq i$$

imposed on $D(X)$, which was introduced in the last section §1, is indeed a t-structure.

Lemma 3.1 *Let $j : U \to X$ be an open subscheme of X, and let $i : Y \to X$ be the closed complement. Then for $B \in D(X)$ the following holds*

$$B \in {}^pD^{\leq 0}(X) \iff j^*B \in {}^pD^{\leq 0}(U) \text{ and } i^*B \in {}^pD^{\leq 0}(Y)$$
$$B \in {}^pD^{\geq 0}(X) \iff j^!B = j^*B \in {}^pD^{\geq 0}(U) \text{ and } i^!B \in {}^pD^{\geq 0}(Y).$$

Proof. First observe, that a complex E on X satisfies the semiperversity condition $E \in {}^pD^{\leq 0}(X)$ iff its restrictions to U and Y satisfy the corresponding conditions $j^*(E) \in {}^pD^{\leq 0}(U)$ and $i^*(E) \in {}^pD^{\leq 0}(Y)$. This is a trivial consequence of the definition and the exactness property of the functor f^*. By the commutation rules for the dualizing functor

$$f^! \circ D = D \circ f^* \quad , \quad f^* \circ D = D \circ f^!$$

on the category of etale sheaves, we immediately get the assertion of the second equivalence. □

The compatibility properties III.3.1 say, that the perverse truncation structure ${}^pD^{\leq 0}(X)$, ${}^pD^{\geq 0}(X)$ on X is obtained from the perverse t-structures on $D(U)$ and

$D(Y)$ by gluing (provided we already know that the Definition III.1.1 defines t-structures on $D(U)$ and $D(Y)$).

There is an obvious generalization of Lemma III.3.1, if X is a finite disjoint union of locally closed subschemes. We do not formulate this.

Proof of Lemma III.1.1. Let X be a scheme. Assume, that X is reduced without restriction of generality. Let

$$j : U \hookrightarrow X$$

be a nonempty open essentially smooth subscheme of X and let

$$i : Y \hookrightarrow X$$

its closed complement. Suppose, that all generic points of X are in U.

Then by induction on the dimension we can assume, that the perverse truncation structure already defines a t-structure on Y. Put $T(Y) = D(Y)$ together with the perverse t-structure on it. On the other hand let $T(U)$ be the full subcategory of $D(U)$, consisting of complexes with smooth cohomology sheaves. According to Remark III.2.2 the axioms for a t-structure are valid for the restriction of the perverse truncation structure to $T(U)$, by trivial reasons. Furthermore, the induced t-structure on $T(U)$ coincides with the standard t-structure up to a shift of degree. By gluing we get

$$T(X, U) = \{E \in D(X) \mid j^*E \text{ has smooth cohomology sheaves on } U\} .$$

By Lemma III.3.1 the glued t-structure on $T(X, U)$ coincides with the perverse truncation structure, which is obtained by restriction from $D(X)$ to $T(X, U)$. In other words, the perverse truncation structure satisfies the axioms (i)–(iii) of a t-structure on $T(X, U)$.

Now let E be an arbitrary complex in $D(X)$. Then there always exists an open dense essentially smooth subscheme $U \hookrightarrow X$, such that the restriction of E to U has smooth cohomology sheaves. In other words

$$D(X) = \bigcup_{U \subset X,\ \text{dense open ess. smooth}} T(X, U) .$$

Therefore axiom (iii) holds for the perverse truncation structure.

A similar argument proves axiom (i). For complexes $E \in {}^{p}D^{\leq 0}(X)$ and $E' \in {}^{p}D^{\geq 1}(X)$ there exist U and U' as above, such that $E \in T(X, U)$ and $E' \in T(X, U')$. We can replace U, U' by $U'' = U \cap U'$. Then $Hom(E, E') = 0$ can be verified in the full subcategory $T(X, U'')$. Therefore the perverse truncation structure on $D(X)$ satisfies all axioms of a t-structure. □

Truncation of Mixed Perverse Sheaves. In the remaining part of this section let the base field κ be a finite field. Let X_0 be a an algebraic variety, i.e. a finitely generated scheme over κ. A complex $K_0 \in D(X_0)$ was called τ-mixed, if all its cohomology sheaves $\mathcal{H}^\nu(K_0)$ are τ-mixed sheaves in the sense of II.2.1. The τ-mixed complexes define a full triangulated subcategory

$$D_{mixed}(X_0)$$

of $D(X_0)$. This subcategory is stable under the functors

$$Rf_*, Rf_!, f^*, f^!, \otimes^L, D .$$

Again by noetherian induction and the construction of truncated objects via gluing, one obtains

Lemma 3.2 *Let X_0 be a variety over a finite field κ. Let K_0 be a τ-mixed complex $K_0 \in D_{mixed}(X_0)$. Then the perverse truncation operations preserve the property of being τ-mixed*

$$^p\tau_{\leq 0}K_0 \in D_{mixed}(X_0) \quad \textit{and} \quad ^p\tau_{\geq 0}K_0 \in D_{mixed}(X_0) .$$

The perverse t-structure on the category $D(X_0)$ therefore induces a t-structure on $D_{mixed}(X_0)$.

Remark 3.3 Suppose given a complex $K \in D(X)$. Then there exists a "stratification" of X by finitely many locally closed essentially smooth equidimensional subschemes

$$i_\nu : Y_\nu \hookrightarrow X_\nu ,$$

such that the restrictions $i_\nu^* K$ and $i_\nu^! K$ are smooth on these subschemes, i.e have smooth cohomology sheaves. X is a disjoint union of its strata Y_ν. We can additionally assume, that the closure of each stratum is a union of strata. Fixing this stratification we consider the full triangulated subcategory of $D(X)$ of complexes $L \in D(X)$ such that

$$i_\nu^* L \quad , \quad i_\nu^! L$$

have smooth cohomology sheaves. The restriction of the perverse t-structure to this subcategory is obtained by gluing of shifted standard t-structures (depending on the dimensions; see Remark III.2.2) in the categories of smooth complexes on the strata

$$Y_\nu \hookrightarrow X .$$

This defines the t-structure in terms of iterates of the gluing method.

III.4 Open Embeddings

Let T be an additive, translation preserving functor between triangulated categories **A** and **B** with t-structures, which transforms distinguished triangles in distinguished triangles. Such a functor is called

$$\text{t-right exact iff } T(D^{\leq 0}(\mathbf{A})) \subset D^{\leq 0}(\mathbf{B})$$

and

$$\text{t-left exact iff } T(D^{\geq 0})(\mathbf{A}) \subset D^{\geq 0}(\mathbf{B}).$$

Finally T is called t-exact, if T is t-left and t-right exact.

Lemma 4.1 *Let X be a finitely generated scheme over a finite or algebraically closed field. Let $j : U \to X$ be an open embedding with closed complement $i : Y \to X$. Then the functors*

$j_!\ ,\ i^*$	*are t-right exact,*
$i_*\ ,\ j^*$	*are t-exact,*
$Rj_*\ ,\ i^!$	*are t-left exact*

for the perverse t-structures on $D(X)$ respectively $D(U)$ and $D(Y)$.

Proof. The properties are obvious for the functors $j^*, i^*, i^!$ and i_* using Lemma III.3.1. In the case of i_* recall that $j^*i_* = 0$ and $i^*i_* = i^!i_* = id$. For $\mathrm{R}j_*$ and $j_!$ the assertion follows by adjunction from the t-exactness of j^*. Namely

$$Hom\big({}^pD^{\leq -1}(X), \mathrm{R}j_*B\big) = Hom\big(j^{*p}D^{\leq -1}(X), B\big) \subset Hom\big({}^pD^{\leq -1}(U), B\big) = 0$$

for $B \in {}^pD^{\geq 0}(U)$. Therefore $\mathrm{R}j_*B \in {}^pD^{\geq 0}(X)$ by Lemma II.2.2. So $\mathrm{R}j_*$ is t-left exact. Then $j_!$ is t-right exact by duality. □

As a consequence of the t-exactness of j^* and the t-left exactness of $\mathrm{R}j_*$ in the situation of Lemma III.4.1 above we get for perverse sheaves $\overline{B} \in Perv(X)$

$${}^p\!H^{-1}\mathrm{R}j_*j^*(\overline{B}) = 0\,.$$

The distinguished triangle $(i_*i^!\overline{B}, \overline{B}, j_*j^*\overline{B})$ therefore induces the following exact sequence of perverse sheaves in the abelian category $Perv(X)$

Restriction Sequence

$$0 \longrightarrow i_*{}^p\!H^0(i^!\overline{B}) \longrightarrow \overline{B} \longrightarrow {}^p\!H^0(\mathrm{R}j_*j^*\overline{B}) \longrightarrow i_*{}^pH^1(i^!\overline{B}) \longrightarrow 0\,.$$

Now let us consider two perverse sheaves $A \in Perv(Y)$ and $\overline{B} \in Perv(X)$. We have $A = \tau_{\leq 0}A$ and $i^!\overline{B} \in D^{\geq 0}$. Furthermore i_* is fully faithful. This implies by adjunction and the long exact sequence of the Hom-functor, together with t-structure axiom (i),

$$Hom(i_*A, \overline{B}) = Hom(A, i^!\overline{B}) = Hom(A, {}^p\!H^0(i^!\overline{B})) = Hom(i_*A, i_*{}^p\!H^0(i^!\overline{B}))\,.$$

Thus $i^{*p}\!H^0(i^!\overline{B})$ is the largest perverse subobject of the perverse sheaf $\overline{B}$, which comes from $Perv(Y)$. More precisely

Lemma 4.2 *Let $i : Y \to X$ be a closed embedding and $\overline{B}$ a perverse sheaf on X. Then*

$$i_* {}^pH^0(i^! \overline{B}) \hookrightarrow \overline{B}$$

is the **largest perverse subobject** *of $\overline{B}$, which is isomorphic to an object of the subcategory $i_* Perv(Y)$ of $Perv(X)$. As the dual statement,*

$$\overline{B} \twoheadrightarrow i_* {}^pH^0(i^* \overline{B})$$

is isomorphic to the **largest perverse quotient** *of $\overline{B}$ with that property.*

Suppose X to be a finitely generated scheme over a finite or algebraically closed field. In III, §3 we obtained the truncation axiom

$$Hom_{D(X)}({}^pD^{\leq 0}(X), {}^pD^{\geq 1}(X)) = 0 .$$

This global identity has a local analog for the functor $R.\mathcal{H}om(-,-)$.

Lemma 4.3 *Let B be in ${}^pD^{\leq 0}(X)$ and let C be in ${}^pD^{\geq 0}(X)$. Then for the standard t-structure we have $R.\mathcal{H}om(B, C) \in D^{\geq 0}(X, \overline{\mathbb{Q}}_l)$, i.e.*

$$\mathcal{H}^\nu(R.\mathcal{H}om(B, C)) = 0 \quad , \quad \nu < 0 .$$

Proof. Because the proof is similar to the proof of the global case (§3), we only sketch the argument. The statement is obvious in the smooth case, i.e. if X is essentially smooth equidimensional of dimension d and B and C are smooth complexes. We then reduce to the case where $B = \mathcal{F}[n]$, $C = \mathcal{G}[m]$ are translates of sheaves. Hence $B \in D^{\leq -d}(X)$ and $C \in D^{\geq -d}(X)$ (for the standard t-structure). Then for an injective resolution I of C also $I \in D^{\geq -d}(X)$. Therefore $R.\mathcal{H}om(B, C) = \mathcal{H}om(B, I) \in D^{\geq 0}(X)$.

One reduces now the general case to the smooth case by noetherian induction and the gluing technique. Choose a suitable open dense essentially smooth subscheme $j : U \hookrightarrow X$, such that B, C become smooth on U. For the distinguished triangle $(j_! j^* B, B, i_* i^* B)$ one obtains another distinguished triangle using adjunction formulas

$$\begin{array}{ccccccc}
R.\mathcal{H}om(i_* i^* B, C) & \longrightarrow & R.\mathcal{H}om(B, C) & \longrightarrow & R.\mathcal{H}om(j_! j^* B, C) & \longrightarrow \\
\| & & \| & & \| & \\
i_* R.\mathcal{H}om(i^* B, i^! C) & \longrightarrow & R.\mathcal{H}om(B, C) & \longrightarrow & Rj_* R.\mathcal{H}om(j^* B, j^* C) & \longrightarrow
\end{array}$$

The claim then follows from the lower distinguished triangle by noetherian induction, using Lemma III.4.1. □

III.5 Intermediate Extensions

Let the situation be as in the last section. Especially let X be a finitely generated scheme over a finite or algebraically closed field. Let $j : U \hookrightarrow X$ be an embedding of an open subscheme. Let $i : Y \hookrightarrow X$ denote the closed complement of U. Let B be a perverse sheaf on U. A perverse sheaf $\overline{B}$ on X is called an extension of B, if

$$j^*\overline{B} = B\ .$$

Consider the triangle

$$i^*\overline{B} \longrightarrow i^*\mathrm{R}j_*B \longrightarrow i^!\overline{B}[1] \longrightarrow i^*\overline{B}[1]\ ,$$

which is distinguished, since the rotated adjunction triangle

$$\overline{B} \longrightarrow \mathrm{R}j_*B \longrightarrow i_*i^!\overline{B}[1] \longrightarrow \overline{B}[1]$$

is distinguished. The morphisms, which appear in these triangles, will be viewed as canonical and used in the next lemma without further explanation or notation.

Lemma 5.1 *With the preceding notations we have the following chain of equivalent conditions (1)–(6) for a perverse extension $\overline{B}$ of the perverse sheaf B:*

(1) *$\overline{B}$ has neither subobjects nor quotients from $i_*Perv(Y)$.*
(2) ${}^pH^0(i^*\overline{B}) = {}^pH^0(i^!\overline{B}) = 0$.
(3) $i^*\overline{B} \in {}^pD^{\leq -1}(Y)$ *and* $i^!\overline{B} \in {}^pD^{\geq 1}(Y)$
(3)′ $i^*\overline{B} \in {}^pD^{\leq -1}(Y)$ *and* $(i^!\overline{B})[1] \in {}^pD^{\geq 0}(Y)$.
(4) *Perverse attachment: $i^*\overline{B} \to i^*Rj_*B$ is isomorphic to the adjunction map*

$$ad : {}^p\tau_{\leq -1}(i^*Rj_*B) \to i^*Rj_*B\ .$$

(5) *Dual perverse attachment: $i^*Rj_*B \to i^!\overline{B}[1]$ is isomorphic to the adjunction map*

$$ad : i^*Rj_*B \to {}^p\tau_{\geq 0}(i^*Rj_*B)\ .$$

(6) *Perverse IC condition: $\overline{B} \to Rj_*B$ is isomorphic to the adjunction map*

$$ad : \tau^Y_{\leq -1}(Rj_*B) \to Rj_*B\ .$$

If one of these equivalent conditions holds, then there exists a distinguished triangle

(7) $(\overline{B}, Rj_*B, i_*{}^p\tau_{\geq 0}i^*Rj_*B)$.

Proof. The first equivalence is clear from III.4.2. For the second equivalence observe that

$$i^*\overline{B} \in {}^pD^{\leq 0}(Y)$$

and

$$i^!\overline{B} \in {}^pD^{\geq 0}(Y)\,,$$

because $i^*, i^!$ are t-right (resp. t-left) exact. This proves the nontrivial direction (2) $\Longrightarrow$ (3). Obviously (3) $\Longleftrightarrow$ (3)$'$.

Consider the triangle $(i^*\overline{B}, i^*\mathrm{R}j_*B, i^!\overline{B}[1])$ for the equivalence of (3)$'$, (4) and (5). The implications (3)$'$ $\Longrightarrow$ (4), (5) follow from the uniqueness (up to isomorphism) of distinguished triangles $(A, i^*\mathrm{R}j_*B, C)$ with $A \in {}^pD^{\leq -1}(Y)$ and $C \in {}^pD^{\geq 0}(Y)$. II.1.4 gives the reverse directions.

The same type of uniqueness argument, directly applied to $(\overline{B}, \mathrm{R}j_*B, i_*i^!\overline{B}[1])$, proves the next equivalence. For the definition of $\tau^Y_{\leq 0}$ see the example in Section 3.

Assertion (7) follows from the distinguished triangle $(\tau^Y_{\leq -1}E, E, i_*{}^p\tau_{\geq 0}i^*E)$, which holds by definition with $E = \mathrm{R}j_*B$, together with III.5.1(6). Lemma III.5.1 is proved. □

Lemma-Definition 5.2 (Intermediate Extension) *Under the assumptions and notations as in Lemma III.5.1 above there is (up to quasiisomorphism) a unique extension $\overline{B} \in Perv(X)$ of a perverse sheaf $B \in Perv(U)$, such that $\overline{B}$ has neither quotients nor subobjects of type i_*A for $A \in Perv(Y)$. This unique extension will be called the* **intermediate extension**

$$j_{!*}B$$

of B and defines a functor

$$j_{!*} : Perv(U) \to Perv(X)\,.$$

If U and X are algebraic schemes over a finite field, $j_{!*}$ maps mixed perverse sheaves to mixed perverse sheaves.

Remark. For an open subscheme $j' : U' \hookrightarrow U$ and a perverse sheaf $B \in Perv(U')$ we obviously have

$$(j_{!*} \circ j'_{!*})(B) = (j \circ j')_{!*}(B)\,.$$

Proof. Existence, uniqueness and functoriality of $j_{!*}B$ follows from the equivalence (1) and (6) of Lemma III.5.1, whereas III.5.1(7), III.3.2 and II.12.2 show, that $j_{!*}$ preserves mixedness. □

An important characterization of the intermediate extension $\overline{B} = j_{!*}B$, which will be frequently used later, is the one stated in III.5.1(2). It characterizes the intermediate extension, up to isomorphism, as the unique perverse sheaf $\overline{B} \in Perv(X)$ with the following properties

$$\boxed{j^*\overline{B} = B \text{ and } {}^pH^0(i^*\overline{B}) = {}^pH^0(i^!\overline{B}) = 0}$$

Suppose $\overline{B}$ satisfies these conditions. Then it is immediately clear that also $D\overline{B}$ satisfies these conditions. Hence

Corollary 5.3 *Let $j : U \to X$ be an open embedding and $B \in Perv(U)$, then*

$$\boxed{D\big(j_{!*}B\big) = j_{!*}\big(DB\big)\,.}$$

Example. Let X be smooth and equidimensional of dimension $d \geq 1$. Suppose $\overline{B} = \mathscr{G}[d]$ for a smooth sheaf $\mathscr{G}$ on X. Then $\overline{B} = j_{!*}j^*(\overline{B})$ for every open dense subscheme $j : U \to X$. For this note ${}^p\!H^0(i^*\overline{B}) = 0$ or equivalently $i^*\overline{B} = {}^p\tau_{\leq -1}i^*(\overline{B})$ holds for all closed subschemes $i : Y \to X$ of smaller dimension; see Remark III.2.2. Similarly ${}^p\!H^0(i^!\overline{B}) = 0$ by duality.

An immediate consequence of the existence of monomorphisms $i_*{}^p\!H^0 i^!\overline{B} \hookrightarrow \overline{B}$ and the epimorphisms $\overline{B} \twoheadrightarrow i_*{}^p\!H^0 i^*\overline{B}$ for arbitrary extensions $\overline{B} \in Perv(X)$ of B (Lemma III.4.2) is the next

Corollary 5.4 *Let $j : U \to X$ be an open imbedding with closed complement $i : Y \to X$. Any simple object $\overline{B}$ from $Perv(X)$ is either of the form i_*A for a simple object A in $Perv(Y)$ or of the form $j_{!*}B$ for a simple object B in $Perv(U)$.*

Using a suitable stratification with essentially smooth strata, such that the cohomology sheaves become smooth on the strata, we obtain

Corollary 5.5 *A perverse sheaf B on X is simple if and only if it is of the form $B = i_*j_{!*}\mathscr{G}[d]$, for an irreducible closed subscheme $i : Y \to X$, an open dense essentially smooth d-dimensional subscheme $j : U \to Y$ of Y and a smooth irreducible $\overline{\mathbb{Q}}_l$-sheaf $\mathscr{G}$ on U.*

Proof. One direction is clear from III.5.4. The converse is easily reduced to the case $X = Y$. For an irreducible smooth sheaf $\mathscr{G}$ on U any perverse subsheaf B of $\mathscr{G}[d]$ is again of the form $\mathscr{F}[d]$ for some smooth sheaf $\mathscr{F}$, at least on an open dense subset V of U by III.2.2. Then II.2.7 implies $\mathscr{F} = \mathscr{G}$ or $\mathscr{F} = 0$, hence $\mathscr{G}$ has either a submodule or a quotient with support in $U \setminus V$. This is impossible, as explained in the last example above. So $\mathscr{G}$ irreducible implies, that $\mathscr{G}[d]$ is simple on U. The claim therefore follows from the next lemma. □

Lemma 5.6 *Let $j : U \to X$ be an open embedding with closed complement $i : Y \to X$. If B is simple in $Perv(U)$, then $j_{!*}B$ is simple in $Perv(X)$.*

Proof. j^* is t-exact. Hence $j_{!*}B$ either has a perverse subsheaf or a perverse quotient coming from $i_*Perv(Y)$, if the lemma were not true! This is impossible because of Definition III.5.2. □

Similarly one can show

Corollary 5.7 *The category $Perv(X)$ is artinian and noetherian.*

Proof. One has to show, that a perverse sheaf $\overline{B} \in Perv(X)$ can have only finitely many perverse constituents. By noetherian induction the assertion can assumed to be true for all closed subspaces of smaller dimension. For a suitable open embedding $j : U \to X$ our $\overline{B}$ satisfies $B = j^*\overline{B} = \mathcal{G}[d]$ for smooth $\mathcal{G}$. Using the perverse cohomology sequence for the triangle III.5.1.(7) and the restriction sequence III.4.1, one obtains the following exact vertical resp. horizontal sequences for any $\overline{B} \in Perv(X)$

$$\begin{array}{ccccccccccc}
 & & & & & & 0 & & & & \\
 & & & & & & \uparrow & & & & \\
 & & & & & & i_*{}^pH^0(i^*\mathrm{R}j_*B) & & & & \\
 & & & & & & \uparrow & & & & \\
0 & \longrightarrow & i_*{}^pH^0(i^!\overline{B}) & \longrightarrow & \overline{B} & \longrightarrow & {}^pH^0(\mathrm{R}j_*B) & \longrightarrow & i_*{}^pH^1(i^!\overline{B}) & \longrightarrow & 0 \\
 & & & & & & \uparrow & & & & \\
 & & & & & & j_{!*}B & & & & \\
 & & & & & & \uparrow & & & & \\
 & & & & & & 0 & & & &
\end{array}$$

By III.5.5 the irreducible constituents of $\mathcal{G}$ correspond to the simple constituents of B, so B is of finite length in $Perv(U)$. By the induction assumption the same holds for $\overline{B}$ by the restriction sequence above, provided ${}^pH^0(\mathrm{R}j_*B)$ is of finite length. Since for a short exact sequence $0 \to B_1 \to B_2 \to B_3 \to 0$ of perverse sheaves we have an exact sequence

$${}^pH^0(\mathrm{R}j_*B_1) \to {}^pH^0(\mathrm{R}j_*B_2) \to {}^pH^0(\mathrm{R}j_*B_3)\ ,$$

induction on the length of B allows to reduce to the case of a simple perverse sheaf B. Then the induction assumptions, Lemma III.5.6 and the vertical exact sequence of the diagram finish the argument. □

We remark, that the diagram above also implies ${}^pH^0 i^*\mathrm{R}j_*B \cong {}^pH^1 i^! j_{!*}B$ and dualy ${}^pH^0 i^!\mathrm{R}j_!B \cong {}^pH^{-1} i^* j_{!*}B$.

From the diagram above it follows without difficulty

Corollary 5.8 *Let $j : U \to X$ be an open embedding and let B be a perverse sheaf on U. Then*

$$\boxed{j_{!*}B \cong image({}^pH^0(Rj_!B) \to {}^pH^0(Rj_*B))}\ ,$$

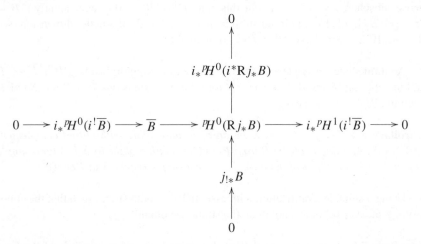

where the map on the right side is induced from the natural map $Rj_!B \to Rj_*B$.

Proof. Put $\overline{B} = image\big(({}^pH^0(\mathrm{R}j_!B) \to {}^pH^0(\mathrm{R}j_*B)\big)$. Then $\overline{B} \in Perv(X)$ is an extension of $B \in Perv(U)$. By Lemma III.5.1(1) it is therefore enough to show, that $\overline{B}$ has no nontrivial subobjects or quotients in $i_*(Perv(Y))$. This readily follows from the next Lemma III.5.9. □

Dual to the vertical monomorphism in the diagram above, used for the proof of Corollary III.5.7, and by a similar argument one gets an epimorphism from ${}^pH^0(\mathrm{R}j_!B)$ to $j_{!*}B$ with kernel $i_*{}^pH^0(i^!\mathrm{R}j_!B)$.

$$
\begin{array}{ccccccccc}
 & & & & 0 & & & & \\
 & & & & \downarrow & & & & \\
 & & & & i_*{}^pH^0(i^!\mathrm{R}j_!B) & & & & \\
 & & & & \downarrow & & & & \\
0 & \longrightarrow & i_*{}^pH^{-1}(i^*\overline{B}) & \longrightarrow & {}^pH^0(\mathrm{R}j_!B) & \longrightarrow & \overline{B} \longrightarrow i_*{}^pH^0(i^*\overline{B}) & \longrightarrow & 0 \\
 & & & & \downarrow & & & & \\
 & & & & j_{!*}B & & & & \\
 & & & & \downarrow & & & & \\
 & & & & 0 & & & &
\end{array}
$$

Lemma 5.9 *Let B be a perverse sheaf on U. Then*

- ${}^pH^0(Rj_!B)$ *has no nonzero quotient objects from $i_*(Perv(Y))$*
- ${}^pH^0(Rj_*B)$ *has no nonzero subobjects from $i_*(Perv(Y))$.*

Proof. Use the t-right exactness $\mathrm{R}j_!({}^pD^{\leq 0}(U)) \subset {}^pD^{\leq 0}(X)$ and t-exactness of i_* for the perverse t-structure. Since $Hom({}^pD^{\leq -1}, {}^pD^{\geq 0}) = 0$, the first claim is clear from

$$Hom({}^pH^0(\mathrm{R}j_!B), i_*A) = Hom(\mathrm{R}j_!B, i_*A)$$

and

$$Hom(\mathrm{R}j_!B, i_*A) = Hom(i^*\mathrm{R}j_!B, A) = 0$$

using $i^*\mathrm{R}j_! = 0$. The second statement is the dual statement and follows from $i^!\mathrm{R}j_* = 0$. □

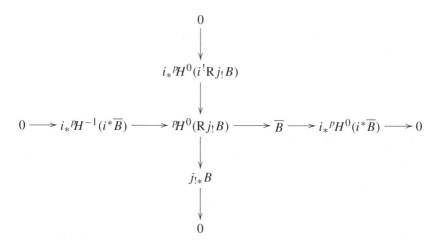

Remark 5.10 For $B \in Perv(U)$ the horizontal exact sequence of the diagram used in III.5.7 implies ${}^pH^1 i^!({}^pH^0\mathrm{R}j_*B) = 0$ (choose $\overline{B} = {}^pH^0\mathrm{R}j_*B$). By III.5.9 the maximal subobject $i_*{}^pH^0 i^!({}^pH^0\mathrm{R}j_*B) = 0$ of ${}^pH^0\mathrm{R}j_*B$ in $i_*Perv(Y)$ is zero. Finally

by Lemma III.4.2 and from considering the vertical exact sequence in the diagram of III.5.7 we get the following isomorphism ${}^pH^0 i^*({}^pH^0(\mathrm{R}j_*B)) \cong {}^pH^0(i^*\mathrm{R}j_*B)$.

Corollary 5.11 (Continuation principle) *Let the situation be as in Corollary III.5.8. Then the category $Perv(U)$ can be identified with the full subcategory $j_{!*}(Perv(U))$ of $Perv(X)$*

$$\boxed{Hom_{Perv(U)}(A,B) = Hom_{Perv(X)}(j_{!*}A, j_{!*}B) \quad , \quad A, B \in Perv(U)\ .}$$

Proof. $Hom(A,B) = Hom(j^*\mathrm{R}j_!A, B) = Hom(\mathrm{R}j_!A, \mathrm{R}j_*B)$. Since we have $\mathrm{R}j_!A \in {}^pD^{\leq 0}$ and $\mathrm{R}j_*B \in {}^pD^{\geq 0}$ by Lemma III.4.1, this homomorphism group can be identified with $Hom({}^pH^0\mathrm{R}j_!A, {}^pH^0\mathrm{R}j_*B)$. Finally III.5.9 implies

$$Hom({}^pH^0\mathrm{R}j_!A, {}^pH^0\mathrm{R}j_*B) = Hom(j_{!*}A, j_{!*}B)\ .$$

□

In view of the Definition III.5.1(1) of the intermediate extension the last corollary is a special case of the considerations made in II.3.3. Here C is the thick subcategory of $D = D^b_c(X)$ of those complexes, whose cohomology sheaves are supported in Y. Let **A** be $Perv(X)$ and **B** be $i_*Perv(Y)$. By definition the reduced objects are the perverse sheaves, which are obtained by intermediate extension from $U = X \setminus Y$.

Lemma 5.12 *If $j : U \to X$ is an open embedding with closed complement $i : Y \to X$ and $B \in Perv(U)$, then ${}^pH^\nu(i^*j_{!*}B) \cong {}^pH^\nu(i^*Rj_*B)$ for all $\nu < 0$.*

Proof. Put $\overline{B} = j_{!*}B$. Then ${}^pH^0(i^!\overline{B}) = 0$. The long exact perverse cohomology sequence deduced from the triangle in Lemma III.5.1(7) gives an exact sequence

$$0 \to j_{!*}B \to {}^pH^0(\mathrm{R}j_*B) \to i_*A \to 0$$

for some $A \in Perv(Y)$. Since i^* is t-right exact by Lemma III.4.1, one concludes

$${}^pH^\nu(i^*j_{!*}B) \cong {}^pH^\nu i^*({}^pH^0\mathrm{R}j_*B) \quad , \quad \nu \neq 0\ .$$

On the other hand the functor $\mathrm{R}j_*$ is t-left exact for the perverse t-structure. Therefore we have a distinguished triangle

$${}^pH^0\mathrm{R}j_*B \to \mathrm{R}j_*B \to {}^p\tau_{\geq 1}\mathrm{R}j_*B \to {}^pH^0\mathrm{R}j_*B[1]\ .$$

Under the restriction j^* the first morphism becomes an isomorphism. Hence $j^*(K) = 0$ for $K = {}^p\tau_{\geq 1}\mathrm{R}j_*B$. Therefore $K \cong i_*i^*(K) \in i_*{}^pD^{\geq 1}(Y, \overline{\mathbb{Q}}_l)$. This implies ${}^pH^\nu(i^*{}^pH^0\mathrm{R}j_*B) \cong {}^pH^\nu(i^*\mathrm{R}j_*B)$ for all $\nu \leq 0$. □

Lemma 5.13 *Suppose $\overline{B} \in {}^pD^{\geq 0}(X, \overline{\mathbb{Q}}_l)$ and $dim(X) \leq d$. Then*

$$\mathcal{H}^\nu(\overline{B}) = 0 \quad \textit{for } \nu < -d\ .$$

In other words ${}^pD^{\geq 0}(X, \overline{\mathbb{Q}}_l)$ *is contained in* $D^{\geq -d}(X, \overline{\mathbb{Q}}_l)$ *(with respect to the standard t-structure).*

Proof. This is proved by noetherian induction. The case $d = 0$ is clear. For the induction step assume the statement holds for smaller dimensions then d. To prove the statement, one easily reduces to the case where $\overline{B}$ is perverse and simple. Then either the statement holds by induction or $\overline{B} = j_{!*}B$ for some open subset $j : U \to X$. We can assume U to be essentially smooth of dimension d and $B = \mathcal{G}[d]$ for a smooth sheaf on U. By III.5.1(7) we have an exact sequence of cohomology sheaves

$$.. \to \mathcal{H}^{\nu-1}(i_*A) \to \mathcal{H}^\nu(\overline{B}) \to \mathcal{H}^\nu \mathrm{R}j_*B \to \mathcal{H}^\nu i_*A \to ..$$

with $A = {}^p\tau_{\geq 0} i^* \mathrm{R}j_*B$ in ${}^pD^{\geq 0}(Y, \overline{\mathbb{Q}}_l)$. Here $Y = \overline{U} \setminus U$ is of dimension $< d$. Therefore by induction assumption $\mathcal{H}^\nu(\overline{B}) \cong \mathcal{H}^\nu \mathrm{R}j_*B = \mathrm{R}^{d+\nu} j_*\mathcal{G}$ for all $\nu \leq -d$. This completes the proof. □

Corollary 5.14 *In the situation of Corollary III.5.5 we have*

$$\mathcal{H}^{-d}(i_* j_{!*}\mathcal{G}[d]) = i_* j_* \mathcal{G}\ .$$

Proof. This follows from the last exact sequence for $\nu = -d$. □

Example. Let X be a smooth curve over a field, $j : U \hookrightarrow X$ an open dense subset and $\mathcal{G}$ a smooth, constructible $\overline{\mathbb{Q}}_l$-sheaf on U. Then $B = \mathcal{G}[1]$ is a perverse sheaf on U and

$$j_{!*}B = j_{!*}(\mathcal{G}[1]) = j_*(\mathcal{G})[1]\ .$$

This follows from III.5.1.(6). Namely $j_{!*}B = \tau^Y_{\leq -1}(\mathrm{R}j_*B)$. Therefore $j_{!*}B$ is equal to $\tau^Y_{\leq 0}(\mathrm{R}j_*\mathcal{G})[1] = j_*(\mathcal{G})[1]$.

In particular Lemma I.6.2 is a special case of Corollary III.5.3, since $D(j_*(\mathcal{G})[1])$ coincides with $\mathrm{R}\mathcal{H}om(j_*(\mathcal{G}), \overline{\mathbb{Q}}_l)[1]$ up to the twist (1) and $D(\mathcal{G}[1])$ is $\mathcal{G}^\vee[1]$ up to the same twist (1).

III.6 Affine Maps

From Artin's estimate of the cohomological dimension of an affine scheme we get

Theorem 6.1 (M. Artin) *Let* $f : X \to Y$ *be an affine map. Then* Rf_* *is t-right exact and* $Rf_!$ *is t-left exact (dual statement) with respect to the perverse t-structures.*

Proof. The t-right exactness

$$\mathrm{R}f_*({}^pD^{\leq 0}(X)) \subset {}^pD^{\leq 0}(Y)$$

follows from Milne, p. 253, theorem VI.7.3(d) or Artin [SGA4]. In fact, one has to use the local-global spectral sequence $R^\nu f_*(\mathcal{H}^\mu A) \Longrightarrow \mathcal{H}^{\nu+\mu}(Rf_*A)$, in order to able to apply the results of loc. cit. □

Corollary 6.2 *For affine open embeddings $j : U \to X$ (more generally affine quasifinite maps) the functors Rj_* and $Rj_! = j_!$ are t-exact for the perverse t-structures, hence*

$$Rj_!, Rj_* : Perv(U) \to Perv(X)$$

preserve perversity.

Let $j : U \to X$ be an open affine embedding with complement $i : Y \to X$. The distinguished triangle $(Rj_!j^*\overline{B}, \overline{B}, i_*i^*\overline{B})$ and the corresponding perverse cohomology sequence for any $\overline{B}$ in $Perv(X)$ therefore gives the following exact sequence

$$0 \longrightarrow i_*{}^pH^{-1}(i^*\overline{B}) \longrightarrow Rj_!B \longrightarrow \overline{B} \longrightarrow i_*{}^pH^0(i^*\overline{B}) \longrightarrow 0$$

in $Perv(X)$, where

$$B = j^*\overline{B}$$

and

$${}^pH^\nu(i^*\overline{B}) = 0$$

for all $\nu \neq -1, 0$. More generally for $\overline{B} \in {}^pD^{[a,b]}(X) = {}^pD^{\geq a}(X) \cap {}^pD^{\leq b}(X)$ this implies $i^*\overline{B} \in {}^pD^{[a-1,b]}(Y)$.

Let us come back to the exact sequence stated above. Its right term $i_*{}^pH^0(i^*\overline{B})$ is the maximal quotient of $\overline{B}$ isomorphic to an object in $i_*(Perv(Y))$. Stated in other terms this means, that for any perverse sheaf $\overline{B}$ on X without quotients in $i_*Perv(Y)$ the complex $i^*\overline{B}[-1]$ is a perverse sheaf on Y, provided $U = X \setminus Y \to X$ is an affine embedding. Because a perverse sheaf has only finitely many simple perverse constituents by III.5.7, this gives

Lemma 6.3 *Let $j_t : U_t \to X$ be an infinite family of affine open embeddings. Let $i_t : X_t = X \setminus U_t \hookrightarrow X$ be the closed complements. Assume $X_t \cap X_{t'} = \emptyset$ for all $t \neq t'$. Then for any $\overline{B}$ in $Perv(X)$*

$$i_t^*\overline{B}[-1] \in Perv(X_t)$$

holds for almost all t.

Remark 6.4 Let $j : U \hookrightarrow X$ be affine open emdebbing with the complement $i : Y \hookrightarrow X$. Suppose Y is of dimension $\leq d-1$. Suppose $B \in Perv(X)$ and $B \in D^{\leq -d}(X, \overline{\mathbb{Q}}_l)$ (for the standard t-structure). Then

$$i^*B[-1] \in Perv(Y) .$$

Proof. Since $i^*(B) \in {}^pD^{[-1,0]}(Y, \overline{\mathbb{Q}}_l)$, it is enough to show $Hom(i^*(B), L) = 0$ for all $L \in Perv(Y)$. However this follows from the truncation property II.2.1(i), since $i^*(B) \in D^{\leq -d}(Y, \overline{\mathbb{Q}}_l)$ and $L \in D^{\geq 1-d}(Y, \overline{\mathbb{Q}}_l)$ (Lemma III.5.13).

In particular

Lemma 6.5 *Suppose* Y *is a local complete intersection of dimension* d*. Then* $\overline{\mathbb{Q}}_{lY}[d]$ *is a perverse sheaf on* Y.

Proof. This statement is local in the etale topology. So assume Y is a complete intersection in a smooth scheme. Then the claim follows by repeated application of the statement of the last remark. □

Universal Extensions. Let $j : U \to X$ be an affine open embedding with closed complement $i : Y \to X$ as above. In the abelian quotient category $Perv(X)/i_*(Perv(Y))$, every perverse sheaf $\overline{B}$ on X becomes isomorphic to $j_{!*}B$ for $B = j^*\overline{B}$. The maximal quotient $\overline{B}_q$ of $\overline{B}$ in $i_*(Perv(Y))$ is isomorphic to $i_*{}^pH^0(i^*\overline{B})$. Note that for $A \in Perv(Y)$

$$Ext^1_{Perv(X)}(Rj_!B, i_*A) = Hom_{Perv(X)}(Rj_!B[-1], i_*A) = 0$$

by II.3.2 and the adjunction formula, since $i^*Rj_! = 0$. Furthermore $Rj_!B$ is a right reduced perverse sheaf on X, i.e. without nontrivial quotients in $i_*Perv(Y)$. See Lemma III.5.9. Suppose $\overline{B} \in Perv(X)$ is also right reduced, i.e $\overline{B}_q = 0$. Then we have the exact sequence

$$0 \longrightarrow i_*{}^pH^{-1}(i^*\overline{B}) \longrightarrow Rj_!B \longrightarrow \overline{B} \longrightarrow 0\ .$$

From the long exact $Hom(-, -)$-sequence we conclude

$$Hom_{Perv(X)}(i_*{}^pH^{-1}(i^*\overline{B}), i_*A) \cong Ext^1_{Perv(X)}(\overline{B}, i_*A)$$

for $A \in Perv(Y)$. Hence the exact sequence is the universal extension of $\overline{B}$ with kernel supported in Y. This means, that any surjection $E \to \overline{B}$ for $E \in Perv(X)$ with kernel $i_*A \in i_*(Perv(Y))$ gives rise to an extension

$$0 \longrightarrow i_*A \longrightarrow E \longrightarrow \overline{B} \longrightarrow 0$$

obtained as pushout with respect to a morphism in $Hom_{Perv(Y)}(A_{univ}, A)$ for $A_{univ} = {}^pH^{-1}(i^*\overline{B})$. In fact, a morphism $\rho : A_{univ} \to A$ induces a commutative diagram

$$\begin{array}{ccc} Hom_{Perv(X)}(i_*A_{univ}, i_*A_{univ}) & \longrightarrow & Hom_{Perv(X)}(i_*A_{univ}, i_*A) \\ \cong\downarrow\delta & & \cong\downarrow\delta \\ Ext^1_{Perv(X)}(\overline{B}, i_*A_{univ}) & \longrightarrow & Ext^1_{Perv(X)}(\overline{B}, i_*A) \end{array}$$

The identity $id_{i_*(A_{univ})}$ maps to $i_*(\rho)$ horizontally and to the extension class defined by $Rj_!B$ vertically. The lower horizontal map corresponds to the pushout construction. From the assumption ${}^pH^0(i^*\overline{B}) = 0$ we get the vanishing ${}^pH^\nu i^*\overline{B} = 0$ for $\nu \neq -1$, hence

$$i^*\overline{B}[-1] = {}^pH^{-1}(i^*\overline{B}) = A_{univ} \in Perv(Y) .$$

III.7 Equidimensional Maps

Lemma 7.1 *Let $f : X \to Y$ be a morphism with equidimensional fibers of dimension d. Then the functors $f^*[d], Rf_![d]$ are t-right exact and the functors $f^![-d], Rf_*[-d]$ are t-left exact with respect to the perverse t-structures.*

Proof. By duality (Remark III.1.2) and adjunction (Lemma II.2.2) one reduces this to the almost trivial case of the pull back f^*. Use $f^*\mathcal{H}^{-\nu} = \mathcal{H}^{-\nu} f^*$ and that pullback increases support dimension by d

$$dim\ supp(f^*\mathcal{H}^{-\nu}(B)) \leq dim\ supp(\mathcal{H}^{-\nu}(B)) + d .$$

Together with the definition this immediately proves the assertion. □

Relative Poincare duality (in particular Definition II.8.1) now implies

Theorem 7.2 *If f is equidimensional of relative dimension d and smooth, then the functor*

$$f^*[d] = f^![-d](-d) : D(Y) \to D(X)$$

is t-exact for the perverse t-structures.

An argument similar to III.7.1 can also be applied under some weaker assumptions: Suppose $f : X \to Y$ is a morphism between finitely generated schemes over k, such that there exists a finite partition $\mathcal{P}$ of Y into locally closed subsets $Y = \coprod_{\alpha=0}^r Y_\alpha$ with $U = Y_0$ open and dense in Y. Furthermore let $j : U \hookrightarrow Y$ be the inclusion map. The complement of U in Y is denoted Z. Then $Z = \coprod_{\alpha>0} Y_\alpha$. Put $n = dim(X)$.

Definition 7.3 *With respect to the partition $\mathcal{P}$ the morphism f will be called*

(1) **semi-small**, *if $2 \cdot dim(f^{-1}(y)) \leq n - dim(Y_\alpha)$ holds for all α and all $y \in Y_\alpha$.*
(2) **small**, *if f is semi-small and $2 \cdot dim(f^{-1}(y)) < n - dim(Y_\alpha)$ holds for all $\alpha > 0$ and all $y \in Y_\alpha$.*

The estimate for the cohomological dimension of the fiber F of f by $2 \cdot dim(F)$ implies

Lemma 7.4 *Let $\mathcal{L}$ be an etale $\overline{\mathbb{Q}}_l$-sheaf on X in the ordinary sense and let $K^\bullet = Rf_!\mathcal{L}[n]$. Then*

(1) For semi-small maps f we have $K^\bullet \in {}^pD^{\leq 0}(Y)$.
(2) For small maps f we have $K^\bullet|Z \in {}^pD^{\leq -1}(Z)$.

Using duality and the characterization of $j_{!*}$ given in III.5.1(3), one concludes

Lemma 7.5 *Suppose X is smooth and equidimensional of dimension n over the base field k. Suppose $f : X \to Y$ is proper and suppose $A = \mathcal{G}[n]$ is a smooth perverse sheaf on X, i.e $\mathcal{G}$ is a smooth etale sheaf on X. Then*

(1) If f is semi-small, then $Rf_*A \in Perv(Y)$.
(2) If f is small, then $Rf_*A = j_{!*}(Rf_*A|U)$.

For later applications in the Chaps. V and VI we now consider pullbacks in a more general situation. This will not be needed in this chapter before §11. Hence the reader may first skip the details.

Pullback and Homomorphisms. Let us return to the more general situation, where $f : X \to Y$ is a compactifiable morphism with equidimensional fibers of dimension d. Let A, B be two complexes in $D(Y)$.

Then by Poincare duality we get

$$Hom_{D(X)}(f^*[d]A, f^![-d](-d)B) = Hom_{D(Y)}(\mathrm{R}f_!\overline{\mathbb{Q}}_{l\,X}[2d](d) \otimes^L A, B)\ .$$

The right side can be identified with

$$H^0(Y, \mathrm{R}.\mathcal{H}om(\mathrm{R}f_!\overline{\mathbb{Q}}_{l\,X}[2d](d) \otimes^L A, B))\ ,$$

which, by the usual Hom- and $\otimes$-formulas, equals

$$H^0\big(Y, \mathrm{R}.\mathcal{H}om(\mathrm{R}f_!\overline{\mathbb{Q}}_l[2d](d), \mathrm{R}.\mathcal{H}om(A, B))\big)$$

$$\cong\ Hom_{D(Y)}(\mathrm{R}f_!\overline{\mathbb{Q}}_l[2d](d), \mathrm{R}.\mathcal{H}om(A, B))\ .$$

Observe, that

$$\mathrm{R}f_!\overline{\mathbb{Q}}_l[2d](d)\ \in\ D^{\leq 0}(Y, \overline{\mathbb{Q}}_l)$$

(standard t-structure). For $A \in {}^pD^{\leq 0}(Y)$ and $B \in {}^pD^{\geq 0}(Y)$ it was proved in Lemma III.4.3 that

$$\mathrm{R}.\mathcal{H}om(A, B)\ \in\ D^{\geq 0}(Y, \overline{\mathbb{Q}}_l)$$

(standard t-structure). Therefore, by truncation for the standard t-structure we obtain

Lemma 7.6 *Let $f : X \to Y$ be a compactifiable morphism with equidimensional fibers of dimension d. Then*

$$Hom_{D(X)}\big(f^*[d]A, f^![-d](-d)B\big) = Hom_{D(Y)}\big(R^{2d} f_!(\overline{\mathbb{Q}}_l)(d), R.\mathcal{H}om(A, B))\big)\ .$$

The Trace Map. Assume now that the morphism $f : X \to Y$ is also flat. Then we have a trace morphism

$$Tr_f : \mathrm{R}^{2d} f_! \overline{\mathbb{Q}}_l \to \overline{\mathbb{Q}}_l(-d) ,$$

which satisfies certain natural compatibility properties. E.g. it commutes with base change, and if all geometric fibers are nonempty and irreducible of dimension d, then the trace morphism Tr_f is an isomorphism. (Actually for the existence of Tr_f it is enough to assume that the fiber dimensions are $\leq d$. See [SGA4], expose XVIII, p. 553).

Remark. By passing to a suitable open subset the construction of Tr_f in loc. cit is reduced to the case, where the morphism

$$f = f_2 \circ f_1 : X \xrightarrow{f_1} Z \xrightarrow{f_2} Y$$

is the composite of a flat quasifinite morphism f_1 and a smooth morphism f_2; one may furthermore assume that $f_2 : \mathbb{A}_S^N \to S$ is the canonical projection. Then Tr_f is obtained as the composite of Tr_{f_1} and Tr_{f_2}. The smooth case was considered in Chap. II, §7 of this book. For the quasifinite flat case see also [FK], chap. II, §1, lemma 1.1. The statement, that the trace map is an isomorphism, follows from Deligne's remarque 2.10.1 in loc. cit, if the fibers are geometrically irreducible. This is a $\overline{\mathbb{Q}}_l$-statement, and in general is not true on the level of torsion sheaves!

Corollary 7.7 *Let $f : X \to Y$ be a compactifiable flat morphism, with nonempty geometrically irreducible equidimensional fibers of dimension d. Let $A \in {}^p D^{\leq 0}(Y)$ and $B \in {}^p D^{\geq 0}(Y)$ be complexes on Y. Then there exists a canonical isomorphism*

$$Hom_{D(X)}\big(f^*[d]A, f^![-d](-d)B\big) \cong Hom_{D(Y)}\big(A, B\big) .$$

Proof. This is an obvious consequence of the isomorphism $\mathrm{R}^{2d} f_!(\overline{\mathbb{Q}}_l)(d) \to \overline{\mathbb{Q}}_l$ induced by the trace map and the formulas following from duality theory discussed above. □

In particular, if A and B in Corollary III.7.7 are perverse sheaves on Y, then

$$Hom_{D(X)}(f^*[d]A, f^![-d](-d)B) \cong Hom_{Perv(Y)}(A, B) .$$

By Lemma III.7.1 we also have $f^*[d]A \in {}^p D^{\leq 0}(X)$ and $f^![-d](-d)B \in {}^p D^{\geq 0}(X)$, whence

$$Hom_{Perv(X)}\big({}^pH^0(f^*[d]A), {}^pH^0(f^![-d](-d)B)\big) \cong Hom_{Perv(Y)}(A, B) .$$

Hence for $B = A$ the identity map induces a natural morphism

$${}^pH^0(f^*[d]A) \longrightarrow {}^pH^0(f^![-d](-d)A) .$$

Lemma-Definition 7.8 *Let $f : X \to Y$ be a compactifiable flat morphism, with nonempty geometrically irreducible equidimensional fibers, of dimension d. In analogy with the intermediate extension we may then define the intermediate pullback of a perverse sheaf $A \in Perv(Y)$ as the perverse sheaf*

$$f^{*!}A = image\Big({}^pH^0(f^*[d]A) \longrightarrow {}^pH^0(f^![-d](-d)A) \Big) .$$

This defines a functor $f^{*!} : Perv(Y) \to Perv(X)$ such that

$$Hom_{Perv(X)}(f^{*!}A, f^{*!}B) = Hom_{Perv(Y)}(A, B) .$$

Proof. To give a homomorphism $\phi : A \to B$ between perverse sheaves is to give a homomorphism $\tilde{\phi} : {}^pH^0 f^*[d]A \to {}^pH^0 f^![-d](-d)B$ by Corollary III.7.7 and the remarks following it. This gives a commutative diagram

$$\begin{array}{ccc} {}^pH^0 f^*[d]A & \longrightarrow & {}^pH^0 f^*[d]B \\ \downarrow{\scriptstyle \tilde{id}_A} & \searrow{\scriptstyle \tilde{\phi}} & \downarrow{\scriptstyle \tilde{id}_B} \\ {}^pH^0 f^![-d](-d)A & \longrightarrow & {}^pH^0 f^![-d](-d)B \end{array} .$$

This clearly defines a functorial map $f^{*!}(\phi)$ between the perverse images $f^{*!}A$ and $f^{*!}B$ of the vertical maps. □

III.8 Fourier Transform Revisited

Consider integers n, r, s such that $n = r + s$. Consider the n-dimensional affine space $\mathbb{A}_0^n = \mathbb{A}_0^r \times \mathbb{A}_0^s$ and the morphism $m : \mathbb{A}_0^r \times \mathbb{A}_0^r \to \mathbb{A}_0^1$, defined by the scalar product $m(x, y) = \sum_{i=1}^r x_i y_i$. On the one dimensional affine space $\mathbb{A}_0^1$ we have the Artin-Schreier sheaf $\mathscr{L}_0(\psi)$. Consider the diagram

$$\begin{array}{ccccc} & & \mathbb{A}_0^r \times \mathbb{A}_0^r \times \mathbb{A}_0^s & \xrightarrow{m_{12}} & \mathbb{A}_0^1, \ \mathscr{L}_0(\psi) \\ & \swarrow{\scriptstyle p_{13}} & & {\scriptstyle p_{23}}\searrow & \\ \mathbb{A}_0^r \times \mathbb{A}_0^s & & & & \mathbb{A}_0^r \times \mathbb{A}_0^s \end{array}$$

Here p_{13}, p_{23} are the obvious projections and m_{12} is defined by m, using projection onto $\mathbb{A}_0^r \times \mathbb{A}_0^r$ first. With morphism as defined in the diagram, the **(partial) Fourier transform** $T_\psi : D_c^b(\mathbb{A}_0^n, \overline{\mathbb{Q}}_l) \to D_c^b(\mathbb{A}_0^n, \overline{\mathbb{Q}}_l)$ with respect to the affine subspace $\mathbb{A}_0^r$

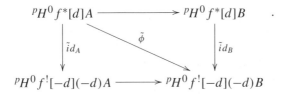

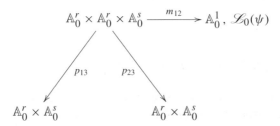

is defined by

$$T_\psi B = \mathrm{R}(p_{13})_!(p_{23}^* B \otimes m_{12}^* \mathscr{L}_0(\psi))[r] .$$

In a similar way one defines the Fourier transform over a base scheme Y_0, such that

$$T_\psi : D_c^b(\mathbb{A}_0^n \times Y_0) \to D_c^b(\mathbb{A}_0^n \times Y_0) .$$

It follows from Theorem II.12.2, that the Fourier transform T_ψ preserves τ-mixedness. Recall that for a τ-mixed complex $B \in D(X_0) = D_c^b(X, \overline{\mathbb{Q}}_l)$ we defined $w(B)$ to be the maximum of all $w(\mathscr{H}^\nu(B)) - \nu$. See also II.2.3.

Theorem 8.1 *For $B \in D(\mathbb{A}_0^n \times Y_0)$ the following holds*

1) Fourier Inversion: $T_{\psi^{-1}} T_\psi B = B(-r)$
2) If B is mixed, then $w(B) \le w \iff w(T_\psi B) \le w + r$
3) $T_\psi : Perv(\mathbb{A}_0^n \times Y_0) \to Perv(\mathbb{A}_0^n \times Y_0)$ is an equivalence of categories.

Proof. 1) By the proper base change theorem this is easily reduced to the case $r = 1$ ($s = 0$) and Y_0 the spectrum of an algebraically closed field.

2) The implication $\Rightarrow$ follows from the Weil conjectures. Its converse $\Leftarrow$ then by 1).

3) p_{23} is smooth equidimensional, therefore $p_{23}^*[r]$ is t-exact. Also $- \otimes m_{12}^* \mathscr{L}_0(\psi)$ is t-exact. Finally p_{13} is affine, hence $\mathrm{R}(p_{13})_!$ is t-left exact. This shows t-left exactness

$$T_\psi({}^p D^{\ge 0}(\mathbb{A}_0^n \times Y_0)) \subset {}^p D^{\ge 0}(\mathbb{A}_0^n \times Y_0) .$$

For B in $Perv(\mathbb{A}_0^n \times Y_0)$ we get a distinguished triangle $({}^p H^0 T_\psi B, T_\psi B, {}^p\tau_{\ge 1} T_\psi B)$ and from that a distinguished triangle

$$(T_{\psi^{-1}}({}^p H^0 T_\psi B), B(-r), T_{\psi^{-1}}({}^p\tau_{\ge 1} T_\psi B)) .$$

The map from $B(-r) \in {}^p D^{\le 0}(\mathbb{A}_0^n \times Y_0)$ to $T_{\psi^{-1}}({}^p\tau_{\ge 1} T_\psi B)$, which lies in $T_{\psi^{-1}}({}^p D^{\ge 1}(\mathbb{A}_0^n \times Y_0)) \subset {}^p D^{\ge 1}(\mathbb{A}_0^n \times Y_0)$, is the zero map by $Hom({}^p D^{\le 0}, {}^p D^{\ge 1}) = 0$. The same holds by Fourier inversion for the original map $T_\psi B \to {}^p\tau_{\ge 1} T_\psi B$. This implies

$${}^p H^0 T_\psi B[1] \cong cone(T_\psi B \xrightarrow{0} {}^p\tau_{\ge 1} T_\psi B) = T_\psi B[1] \oplus {}^p\tau_{\ge 1} T_\psi B .$$

A shift by [-1] and further truncation by ${}^p\tau_{\ge 1}$ yields ${}^p\tau_{\ge 1} T_\psi B = 0$, hence $T_\psi B \in Perv(\mathbb{A}_0^n \times Y_0)$ for any $B \in Perv(\mathbb{A}_0^n \times Y_0)$.

For a stronger result see Katz-Laumon [177], p. 157–159.

Exercise. Give a proof of III.8.1(3) using III.11.3 instead of using the Theorem III.6.1 of Artin-Grothendieck.

Exercise. Show that the abelian category $Perv(\mathbb{A}^n)$ has no nontrivial injective (nor projective) objects I for $n \geq 1$. Hint: Use Beilinson's result [Be1], that the groups $Ext^i_{Perv(\mathbb{A}^n)}(X, Y)$ coincide with the groups $Hom_{D(\mathbb{A}^n)}(X, Y[i])$. Thus evaluating against $X = \mathscr{L}(\psi_y)[n]$ for $Y = I$ implies, that $T_\psi(I)$ is a skyscraper sheaf. Next consider X to be a skyscraper sheaf. □

III.9 Key Lemmas on Weights

Let X_0 denote a scheme over a finite field κ. We fix an isomorphism $\tau : \overline{\mathbb{Q}}_l \to \mathbb{C}$. Let B_0 be a τ-mixed complex in $D(X_0) = D^b_c(X_0)$. For any τ-mixed complex B_0 we define $w(B_0)$ as in II.12.3.

In the following we study perverse sheaves, whose complexes are τ-mixed. The τ-mixed perverse sheaves in $D_{mixed}(X_0)$ define an abelian subcategory of $Perv(X_0)$, which is the core of the triangulated category $D_{mixed}(X_0)$ with respect to perverse truncation.

Recall that a τ-mixed complex $B_0 \in D(X_0)$ is called τ-pure of weight w if $B_0 \in D_{\leq w}(X_0) \cap D_{\geq w}(X_0)$. By definition this is equivalent to $w(B_0) \leq w$ and $w(DB_0) \leq -w$. From Proposition II.12.6 and III.4.3 we get

Proposition 9.1 *Let A_0, B_0 be τ-mixed perverse sheaves on X_0. Then*

$$Hom_{Perv(X_0)}(A_0, B_0) = Hom_{Perv(X)}(A, B)^F .$$

This group $Hom_{Perv(X)}(A_0, B_0)^F$ vanishes if B_0 is τ-pure of weight w and $w(A_0) < w$

$$Hom_{Perv(X_0)}(A_0, B_0) = 0 .$$

Lemma I (Semicontinuity of Weights) *Let $j : U_0 \to X_0$ be an open embedding with closed complement $i : Y_0 \to X_0$. Let $\overline{B}_0 \in Perv(X_0)$ be a τ-mixed perverse sheaf on X_0, such that*

$$j^*(\overline{B}_0) = B_0 \quad , \quad {}^pH^0(i^*(\overline{B}_0)) = 0 .$$

Then

$$w(\overline{B}_0) \leq w(B_0) .$$

In particular

$$w(j_{!*}(B_0)) \leq w(B_0) .$$

Let us discuss an immediate consequence of Lemma I. Suppose B_0 as in Lemma I is τ-pure of weight w. Then Lemma I and the formula $Dj_{!*} = j_{!*}D$ implies, that $j_{!*}(B_0)$ is also τ-pure of weight w. Therefore the classification of simple perverse sheaf (III.5.5) together with Theorem I.2.8 (3) permits to conclude from Lemma I the

Corollary 9.2 *Any τ-mixed simple perverse sheaf B_0 on X_0 is τ-pure of weight $w = w(B_0)$.*

Lemma II (Subquotients) *Let $B_0 \in Perv(X_0)$ be a τ-mixed perverse sheaf. Then $w(A_0) \le w(B_0)$ holds for any perverse subquotient A_0 of B_0 in $Perv(X_0)$.*

Lemma III (Weight Filtration) *In the abelian category $Perv(X_0)$ any τ-mixed perverse sheaf E_0 on X_0 has a – canonical – finite increasing τ-weight filtration $W = (E_0^{(w_i)})$*

$$0 = E_0^{(-\infty)} \subset E_0^{(w_1)} \subset ... \subset E_0^{(w_r)} = E_0 ,$$

such that the graded components $Gr^W_{w_i}(E_0) = E_0^{(w_i)}/E_0^{(w_{i-1})}$ are either zero or τ-pure perverse sheaves of weight w_i such that

$$w_i < w_j \quad \text{for} \quad i < j .$$

In the way stated above such weight filtrations are of course not unique. One can always "fill in" additional redundant filtration steps. The new graded pieces, which arise in this way, are zero.

The weight filtrations, whose existence is stated above, is therefore **canonical** only in the following sense:

(a) If we demand all graded pieces $Gr^{(w_i)}(E_0)$ to be nontrivial, the filtration is uniquely determined.
(b) Let $\phi_0 : E_0 \to F_0$ be a homomorphism of a τ-mixed perverse sheaf E_0 into a τ-mixed perverse sheaf F_0, then there exist weight filtrations $E_0^{(w_i)}$, $F_0^{(w_i)}$ on E_0, F_0, whose graded pieces are zero or τ-pure of weight w_i such that ϕ_0 maps $E_0^{(w_i)}$ to $F_0^{(w_i)}$.

The proofs of Lemma I and II are easily deduced from properties of the Deligne-Fourier transform. (For a proof not using the Fourier transform see analyse et topologie sur les espaces singuliers, asterisque 100).

Lemma III is obtained from Lemma I, II by standard methods. All three lemmas are proved simultaneously by induction on $dim(X_0)$ in consecutive order. For $dim(X_0) = 0$ all statements are trivial.

Proof of Lemma I. We start with some auxiliary result

Proposition 9.3 *Let $i : W_0 \to Z_0$ be a closed embedding and $A_0 \in Perv(Z_0)$ be a perverse sheaf on Z_0. Then*

$$i^*(A_0) \in {}^pD^{\le 0}(W_0)$$

is equivalent to any of the following assertions

(1) ${}^pH^0(i^*A_0) = 0$
(2) $i^*(A_0) \in {}^pD^{\le -1}(W_0)$

(3) $^P\tau_{\leq -1}(A_0) = A_0$
(4) $dim\ supp\ \mathscr{H}^{-v}(i^*A_0) \leq v - 1$ for all $v \in \mathbb{Z}$.

Proof. A_0 being perverse implies $i^*(A_0) \in {}^PD^{\leq 0}(W_0)$, therefore $i^*(A_0) = {}^P\tau_{\leq 0}i^*(A_0)$. Via the distinguished triangle $({}^P\tau_{\leq -1}i^*A_0, i^*A_0, {}^P\tau_{\geq 0}i^*A_0)$ one obtains

$$^PH^0(i^*A_0) = {}^P\tau_{\geq 0}{}^P\tau_{\leq 0}(i^*A_0) = {}^P\tau_{\geq 0}(i^*A_0) = 0 \Longleftrightarrow {}^P\tau_{\leq -1}(i^*A_0) = i^*A_0$$

$$\Longleftrightarrow i^*A_0 \in {}^PD^{\leq -1}(W_0)\ .$$

$\square$

We now return to the proof of Lemma I. Recall $\overline{B}_0$ was an extension of a perverse sheaf B_0 on $U_0 \subset X_0$ over the closed complement Y_0.

First Step (Reduction to $X_0 = \mathbb{A}_0^n$). The statement of Lemma I is local on X_0. So it is enough to consider affine schemes X_0. We can replace X_0 by $\mathbb{A}_0^n$, using some closed embedding $i_X : X_0 \to \mathbb{A}_0^n$ into affine space.

$$
\begin{array}{ccccc}
Y_0 & \overset{i}{\hookrightarrow} & X_0 & \overset{i_X}{\hookrightarrow} & \mathbb{A}_0^n \\
& & \uparrow{\scriptstyle j} & & \uparrow{\scriptstyle j} \\
& & U_0 & \overset{i_X}{\hookrightarrow} & \mathbb{A}_0^n \setminus Y_0
\end{array}
$$

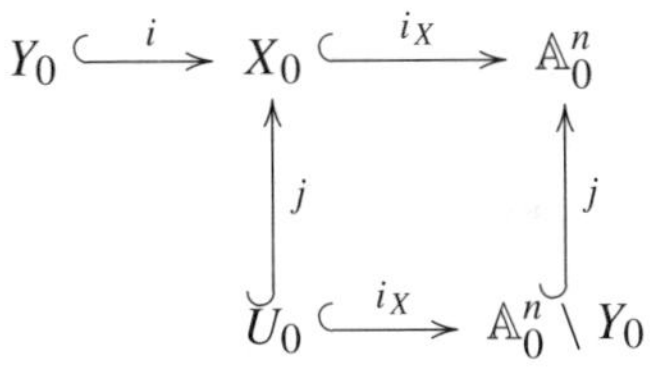

Now $j^*(i_X)_* \overline{B}_0 = (i_X)_* j^* \overline{B}_0$ and $^PH^0((i_Xi)^*(i_X)_*\overline{B}_0) = 0$, as $(i_X)^*(i_X)_* = id$. Therefore the assumptions carry over. As $(i_X)_*$ is weight preserving, we can therefore replace X_0 by $\mathbb{A}_0^n$ for the proof of Lemma I. So we will assume $X_0 = \mathbb{A}_0^n$ from now on. In this case the proof will be given by induction on n.

Second Step (Reduction to $dim(Y_0) = 0$). Choose some linear projection $X_0 = \mathbb{A}_0^n \to \mathbb{A}_0^1$. It has disjoint affine fibers X_{0t}, t closed in $T = \mathbb{A}_0^1$. The fibers are divisors, hence the complementary open embedding $X_0 - X_{0t} \hookrightarrow X_0$ is an affine morphism. Consider

$$
\begin{array}{ccccc}
U_0 & \overset{j}{\hookrightarrow} & X_0 & \overset{i}{\longleftarrow} & Y_0 \\
\uparrow{\scriptstyle i_t} & & \uparrow{\scriptstyle i_t} & & \uparrow{\scriptstyle i_t} \\
U_0 \cap X_{0t} = U_{0t} & \overset{j}{\hookrightarrow} & X_{0t} & \overset{i}{\longleftarrow} & Y_{0t} = Y_0 \cap X_{0t}
\end{array}
$$

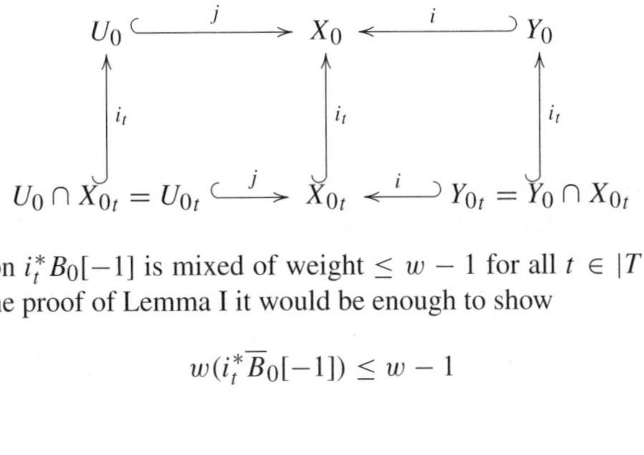

By assumption $i_t^* B_0[-1]$ is mixed of weight $\leq w - 1$ for all $t \in |T|$, where $w = w(B_0)$. For the proof of Lemma I it would be enough to show

$$w(i_t^* \overline{B}_0[-1]) \leq w - 1$$

for all $t \in T$. We show slightly less, namely that this equality holds for almost all t in $|T|$.

First of all $i_t^*\overline{B}_0[-1] \in Perv(X_{0t})$ and $i_t^* B_0[-1] \in Perv(U_{0t})$ holds for almost all t, by Lemma III.6.3. Next observe ${}^pH^0 i^*\overline{B}_0 = 0$, hence $dim\ supp\,\mathcal{H}^{-\nu} i^*\overline{B}_0 \leq \nu - 1$ by Prop. III.9.3. For general t in $|T|$ therefore $dim\,(X_{0t} \cap supp\,\mathcal{H}^{-\nu} i^*\overline{B}_0) \leq \nu - 2$ holds for all ν or equivalently

$$ {}^pH^0(i^*(i_t^*\overline{B}_0[-1])) = 0 \,, $$

because of Proposition III.9.3.

For the moment fix such a t. By assumption $w(j^*i_t^*\overline{B}_0[-1]) = w(i_t^* B_0[-1]) \leq w - 1$. Therefore by induction ($dim\ X_{0t} < dim\ X_0$) the mixed sheaf $i_t^*(\overline{B}_0)[-1]$ satisfies the same inequality.

This shows $w(i_t^*\overline{B}_0[-1]) \leq w - 1$ for almost all t in $|T|$. Repeating the above procedure for other projection maps $X_0 = \mathbb{A}_0^n \to \mathbb{A}_0^1$ gives $w(\overline{B}_0) \leq w$ on X_0 outside a subscheme $Y_0' \subset Y_0$ of dimension zero. So we can and will therefore assume $dim(Y_0) = 0$ from now on. (Replace Y_0 by Y_0' and use Proposition III.9.3 for ${}^pH^0(\overline{B}_0|Y_0') = 0$).

Last Step (Proof in Case $dim(Y_0) = 0$ and $X_0 = \mathbb{A}_0^n$). By assumption we now have $H^0 i^*\overline{B}_0 = {}^pH^0 i^*\overline{B}_0 = 0$, for the inclusion $i : Y_0 \to \mathbb{A}_0^n$ and the perverse sheaf $\overline{B}_0$ on $\mathbb{A}_0^n$. Let us show, that this implies $w(T_\psi \overline{B}_0) \leq w + n$ for the total Fourier transform of $\overline{B}_0$. $T_\psi \overline{B}_0$ is perverse on $\mathbb{A}_0^n$, so all cohomology stalks are concentrated in degrees $\nu = -n, -n+1, .., -1, 0$ by III.5.13 and III.1.3. The stalks $\mathcal{H}^\nu(T_\psi \overline{B}_0)_x$ of $T_\psi \overline{B}_0 = \mathrm{R}(p_1)_!(p_2^*\overline{B}_0 \otimes m^*\mathcal{L}_\psi)[n]$ are given by

$$ H_c^{\nu+n}(\mathbb{A}^n, \overline{B} \otimes \mathcal{L}_{\psi_x}) \qquad (0 \leq m = \nu + n \leq n)\,. $$

On the other hand we have $i^*\overline{B}_0 \in {}^pD^{\geq -n}(Y_0) \cap {}^pD^{\leq -1}(Y_0)$ from ${}^pH^0(i^*\overline{B}_0) = 0$. From $dim(Y_0) = 0$ we then conclude

$$ H_c^m(Y, i^*\overline{B} \otimes \mathcal{L}_{\psi_x}) = 0 \text{ for all } m \geq 0\,. $$

The long exact sequence

$$ \to H_c^m\big(\mathbb{A}^n \setminus Y, B \otimes \mathcal{L}_{\psi_x}\big) \to H_c^m\big(\mathbb{A}^n, \overline{B} \otimes \mathcal{L}_{\psi_x}\big) \to H_c^m\big(Y, i^*\overline{B} \otimes \mathcal{L}_{\psi_x}\big) \to $$

shows, that for all $m \geq 0$ the fiber $\mathcal{H}^{m-n} T_\psi(\overline{B})_x$ is a quotient of the cohomology group $H_c^m(\mathbb{A}^n \setminus Y, B \otimes \mathcal{L}_{\psi_x})$. The latter has weight $\leq w + m$ for all x by the Weil conjecture I.7.1. This shows $w(T_\psi \overline{B}_0) \leq w + n$ and Fourier inversion finally implies $w(\overline{B}_0) \leq w = w(B_0)$. Lemma I is proved. □

Proof of Lemma II. We can assume X_0 to be affine and choose some closed embedding $i_X : X_0 \to \mathbb{A}_0^n$. Since $(i_X)_*$ is t-exact the perverse subquotients A_0 of $B_0 \in Perv(X_0)$ define perverse subquotients $(i_X)_*(A_0)$ of $(i_X)_*(B_0)$ in $Perv(\mathbb{A}_0^n)$.

Also $w((i_X)_*(A_0)) \leq w$ is equivalent to $w(A_0) \leq w$. Therefore we can assume $X_0 = \mathbb{A}^n_0$.

We will show, that every simple perverse constituent A_0 of B_0, where A_0 is a perverse sheaf on $\mathbb{A}^n_0$, has weight $w(A_0) \leq w(B_0)$. This immediately implies the statement for arbitrary constituents of B_0.

The simple perverse sheaf A_0 is of the form $A_0 = i_* j_{!*} C_0$, for some smooth sheaf $C_0 \neq 0$ on a locally closed irreducible smooth subscheme U_0 placed in degree $-dim(U_0)$. The closure Y_0 of U_0 is an irreducible subscheme Y_0 of affine space, of dimension say $s = n - r$.

$$U_0 \xrightarrow{j} Y_0 \xrightarrow{i} \mathbb{A}^n_0 \quad , \qquad dim(U_0) = dim(Y_0) = s \ .$$

For pedagogical reasons, we will distinguish three cases:

Case 1 (Generic Support $Y_0 = X_0 = \mathbb{A}^n_0$). We may replace U_0 be a sufficiently small open dense subscheme, and may then assume

$$j^*(B_0) = \mathscr{F}_0[n]$$

$$C_0 = \mathscr{G}_0[n] \ .$$

Here $\mathscr{F}_0$ and $\mathscr{G}_0$ are smooth sheaves – in the ordinary, not in the perverse sense – and $\mathscr{G}_0$ is a subquotient of $\mathscr{F}_0$ in the category of smooth sheaves. The assertion of Lemma II is obvious in this case.

Case 2 (Zero Dimensional Support Y_0). We use a Fourier transform $T_\psi : D(\mathbb{A}^n_0) \to D(\mathbb{A}^n_0)$ on $\mathbb{A}^n_0$ to reduce to case 1) already considered. By Theorem III.8.1 the Fourier transform $T_\psi(B_0)$ is a perverse sheaf with the property $w(T_\psi(B_0)) = w(B_0) + n$. $T_\psi(A_0)$ is a simple perverse subquotient of $T_\psi(B_0)$. Since the support of $A_0 \neq 0$ is zero dimensional by assumption, it consists of finitely many closed points. Hence the Fourier transform is easily calculated, such that

$$supp\ T_\psi(A_0) = \mathbb{A}^n_0 \ .$$

Then according to case 1)

$$w(T_\psi(A_0)) \leq w(T_\psi(B_0)) \ .$$

But as desired this implies

$$w(A_0) \leq w(B_0) \ ,$$

by Theorem III.8.1 (2).

Case 3) The General Case $0 \leq s \leq n$. We use the same trick as in case 2) – a thickening of the support Y_0 of A_0 – however now by a properly chosen partial Fourier transform (see Chap. III, §8).

For this we use the Noether normalization theorem. After a change of coordinates and – if necessary – a change of base field we find a projection

$$q_0 : \mathbb{A}_0^n = \mathbb{A}_0^r \times \mathbb{A}_0^s \to \mathbb{A}_0^s ,$$

such that the restriction

$$q_0 \,|Y_0 : \quad Y_0 \to \mathbb{A}_0^s$$

is a finite surjective, possibly ramified covering. We now apply the partial Fourier transform $T_\psi : D(\mathbb{A}_0^n) \to D(\mathbb{A}_0^n)$ with respect to the subspace $\mathbb{A}_0^r$ (see §8). Then $T_\psi(A_0)$ is a simple perverse subquotient of the perverse sheaf $T_\psi(B_0)$ and we have $w(T_\psi(B_0)) = w(B_0) + r$.

For some open dense subscheme

$$V_0 \subset \mathbb{A}_0^s ,$$

the perverse sheaf B_0 becomes smooth on the intersection of $q_0^{-1}(V_0)$ with Y_0, i.e.

$$q_0^{-1}(V_0) \cap Y_0 \subset U_0 .$$

The stalks of the Fourier transform $T_\psi(A_0)$ in a geometric point $x = (\rho, \sigma)$ of $\mathbb{A}_0^r \times \mathbb{A}_0^s$ are

$$T_\psi(A_0)_x = R\Gamma_c\big(q_0^{-1}(\sigma), A_0|q_0^{-1}(\sigma) \otimes \mathscr{L}_0(\psi_\rho)\big)[r] .$$

So we obtain

$$supp\ T_\psi(A_0) \supseteq q_0^{-1}(V_0) = \mathbb{A}_0^r \times V_0 .$$

So for the perverse subquotient $T_\psi(A_0)$ of $T_\psi(B_0)$ we are again in the situation of the first case and it follows that

$$w(T_\psi(A_0)) \leq w(T_\psi(B_0)) ,$$

hence $w(A_0) \leq w(B_0)$. □

Proof of Lemma III. The proof uses induction on the length $l(E_0)$ of the perverse sheaf E_0. E_0 is a simple perverse sheaf and therefore τ-pure by Corollary III.9.2 if $l(E_0) = 1$.

Assume $l(E_0) > 1$ and let $B_0 \neq 0$ be a simple perverse subsheaf of E_0, necessarily τ-pure say of weight w'. We consider in the abelian category $Perv(X_0)$ the short exact sequence

$$0 \longrightarrow B_0 \longrightarrow E_0 \longrightarrow A_0 \longrightarrow 0 .$$

By induction A_0 already has a weight filtration. Pick the largest weight w in this filtration for which

$$w < w' .$$

First assume $A_0 = A_0^{(w)}$. In order to prove Lemma III we have to show that the extension defined by the exact sequence of perverse sheaves

$$0 \longrightarrow B_0 \longrightarrow E_0 \longrightarrow A_0 \longrightarrow 0$$

splits. The existence of such a splitting follows from the two properties $A_0 \in D_{\leq w}(X_0)$ (obvious) and $B_0 \in D_{\geq w'}(X_0)$ (Corollary III.9.2). We thus invoke Lemma II.3.2 and Proposition II.12.4, to derive the desired vanishing of

$$Ext^1_{Perv(X_0)}(A_0, B_0) = Hom_{D(X_0)}(A_0, B_0[1]) = 0 .$$

If $A_0 \neq A_0^{(w)}$, we may replace A_0 by $A_0^{(w)}$ and E_0 by the preimage of $A_0^{(w)}$ in E_0, and thereby reduce to the case already considered.

The filtration is canonical; this follows immediately from Proposition III.9.1. □

III.10 Gabber's Theorem

Theorem 10.1 *Let B_0 be a τ-mixed perverse sheaf in $Perv(X_0)$. Then*

$$w(B_0) \leq w$$

holds iff for every irreducible subscheme Y_0 of X_0 (of dimension say d) there is an open dense subscheme $U_0 \xrightarrow{j} Y_0$, such that

$$w(\mathscr{H}^{-d} B_0 | U_0) \leq w - d .$$

Proof. Consider the nontrivial direction: Choose a short exact sequence $0 \to A_0 \to B_0 \to Q_0 \to 0$ in $Perv(X_0)$ with Q_0 simple of maximal weight $w(Q_0) = w(B_0)$. Existence of such a sequence follows from the weight filtration (§9, Lemma III). Then $Q_0 = i_* j_{!*} C_0$ for a smooth sheaf C_0 on U_0 placed in degree $-d = -dim(U_0)$ with open resp. closed embeddings $U_0 \xrightarrow{j} Y_0 \xrightarrow{i} X_0$ and Y_0 is irreducible. We have to estimate $w(Q_0)$. Since $A_0 \in {}^pD^{\leq 0}(X_0)$ we have $dim\ supp\, \mathscr{H}^{-d+1}(A_0) \leqq d - 1$. On the open dense subset $V_0 = U_0 \setminus supp\, \mathscr{H}^{-d+1} A_0$ of U_0 the map $\mathscr{H}^{-d} B_0 \to \mathscr{H}^{-d} Q_0$ is surjective. This surjectivity together with the assumptions implies $w(\mathscr{H}^{-d} Q_0 | V_0) \leq w - d$, possibly after a further shrinking of V_0. Hence $w(B_0) = w(Q_0) \leq w$, using perverse semicontinuity of weights (Lemma I of §9). □

Corollary 10.2 *Let B_0 be τ-mixed in $D_c^b(X_0, \overline{\mathbb{Q}}_l)$, then*

$$w(B_0) \leq w \iff w({}^pH^\nu B_0) \leq w + \nu \qquad \forall \nu \in \mathbb{Z} .$$

Proof. The direction $\Longleftarrow$ is easily derived from the long exact cohomology sequences attached to the various distinguished perverse truncation triangles. So consider the converse direction $\Longrightarrow$:

Assume $w(B_0) \leq w$ and $w({}^pH^\nu B_0) \leq w + \nu$ for all $\nu > l$. Then $w({}^p\tau_{>l} B_0) \leq w$, using the $\Longleftarrow$-direction. Let us show $w({}^pH^l B_0) \leq w + l$, which inductively will imply $w({}^pH^\nu B_0) \leq w + \nu$ for all ν.

For simplicity we may assume $l = 0$. Fix some integer $0 \leq d \leq dim(X_0)$. Using the long exact cohomology sequence for the triangle $({}^p\tau_{\leq 0}B_0, B_0, {}^p\tau_{>0}B_0)$, the weight estimates $w(B_0), w({}^p\tau_{>0}B_0) \leq w$ imply $w({}^p\tau_{\leq 0}B_0) \leq w$. Therefore $w(\mathcal{H}^{-d}\ {}^p\tau_{\leq 0}B_0) \leq w - d$. Since $dim\ supp\ \mathcal{H}^{-d+1}({}^p\tau_{<0}B_0) < d$ (perversity condition), this weight estimate and the exact sequence

$$\mathcal{H}^{-d}({}^p\tau_{\leq 0}B_0) \longrightarrow \mathcal{H}^{-d}({}^pH^0 B_0) \longrightarrow \mathcal{H}^{-d+1}({}^p\tau_{<0}B_0)$$

imply the conditions of the weight criterion III.10.1 for the perverse sheaf ${}^pH^0 B_0$. Therefore Theorem III.10.1 shows $w({}^pH^0 B_0) \leq w$. This proves the inductive step. Thus the nontrivial direction of the corollary follows. □

Exercise 10.3 Let $j : U \hookrightarrow X$ be an open embedding, let $B \in Perv(U)$ be τ-pure of weight w. Then $\overline{B} = j_{!*}B$ is the graded component of weight w in the weight filtration of ${}^pH^0 Rj_! B$. (Use the diagram of III.5.8). Similar for ${}^pH^0 Rj_* B$.

The last corollary and its dual imply

Corollary 10.4 *A τ-mixed complex $B_0 \in D_c^b(X_0, \overline{\mathbb{Q}}_l)$ is τ-pure of weight w iff all ${}^pH^\nu B_0$ are τ-pure of weight $w + \nu$.*

Corollary 10.5 *In the situation of III.10.4 the τ-mixed perverse sheaves ${}^pH^\nu B_0$ are τ-pure of weight $w + \nu$ iff all their simple constituents are τ-pure of weight $w + \nu$.*

Proof. §9, Lemma II and its dual for the nontrivial direction $\Longrightarrow$. □

This together with an argument similar to the one used in the proof of §9, Lemma III implies one of the most striking results, namely Gabber's decomposition theorem

Theorem 10.6 (Semisimplicity) *Let $B_0 \in D_c^b(X_0, \overline{\mathbb{Q}}_l)$ be τ-pure of weight w. Consider the base change B of B_0 to the algebraic closure $B \in D_c^b(X, \overline{\mathbb{Q}}_l)$, $X = X_0 \times_\kappa k$. Then B is isomorphic to a finite direct sum of translates $A[i]$, $i \in \mathbb{Z}$ of simple perverse sheaves A on X*

$$B \cong \bigoplus A[i] \quad , \quad A \in Perv(X)\ .$$

Proof. Using induction on the (finite) number of nonvanishing perverse cohomology sheaves ${}^pH^\nu(B_0) \neq 0$ we first prove $B \cong \bigoplus_\nu {}^pH^\nu(B)[-\nu]$. For $B_0 \neq 0$ as in the theorem consider the distinguished triangle

$$({}^p\tau_{<n}B_0, B_0, {}^pH^n(B_0)[-n])\ ,$$

with respect to the last nonvanishing perverse cohomology sheaf ${}^pH^n(B_0)$. By Corollary III.10.4 and the induction assumption it is immediately clear that ${}^p\tau_{<n}B_0$ and

${}^pH^n(B_0)[-n]$ are again τ-pure of weight w. This implies that the triangle splits into a direct sum. For this it is enough to show the vanishing of

$$Hom_{D(X)}\Big({}^pH^n(B)[-n], {}^p\tau_{<n}B[1]\Big)$$

$$= \bigoplus_{\nu<n} Hom_{D(X)}\Big({}^pH^n(B)[-n], {}^pH^\nu(B)[-\nu][1]\Big) = 0,$$

because of Corollary II.1.7. The necessary vanishing statement for the homomorphism group follows from Proposition II.12.5. The assumptions are satisfied, since $w' = w({}^pH^\nu(B_0)[-\nu]) = w > w-1$ and $w = w({}^pH^n(B_0)[-n])$. So our first claim has been established by induction.

Now we can assume B_0 to be a translate $C_0[i]$ of a τ-pure perverse sheaf C_0. Without restriction of generality we can assume $i = 0$ and then use induction on the length of C_0. The argument is similar to the one above, now using Corollary III.10.5 instead of Corollary III.10.4. The details will be skipped. □

Exercise 10.7 Let X be smooth of dimension $d \geq 2$ over an algebraically closed field k. Let Y be a Weil divisor with r irreducible components and open complement $j : U \hookrightarrow X$. Let $\tilde{j} : Y^0 \hookrightarrow Y$ be an essentially smooth dense open subset, and define $A = \tilde{j}_{!*}(\overline{\mathbb{Q}}_{l Y^0}[d-1])(-1)$. Show

(a) $\overline{B} = Rj_*\overline{\mathbb{Q}}_{l U}[d]$ is in $Perv(X)$ (use III.6.2).
(b) $\overline{B}$ has a weight filtration, with $Gr_d^W = \overline{\mathbb{Q}}_{l X}[d]$ (III.10.3).
(c) $A \hookrightarrow Gr_{d+1}^W$ (use III.10.6 and purity).
(d) Gr_{d+1}^W/A and all Gr_w^W, $w \geq d+2$ are perverse sheaves with support in $Y \setminus Y^0$, which has dimension $\leq d-2$.
(e) Show $H^{1-d}(Y, A) \cong H^0(Y^0, \mathbb{Q}_l)(-1)$ (Lemma III.5.13 and its corollary).
(f) Deduce the long exact sequence

$$0 \to H^1(X, \overline{\mathbb{Q}}_l) \to H^1(U, \overline{\mathbb{Q}}_l) \to \overline{\mathbb{Q}}_l(-1)^r \to H^2(X, \overline{\mathbb{Q}}_l) \to H^2(U, \overline{\mathbb{Q}}_l).$$

(g) Describe the connecting morphism $\delta : \overline{\mathbb{Q}}_l(-1)^r \to H^2(X, \overline{\mathbb{Q}}_l)$ in terms of the divisor class group of X.
(h) Show injectivity of δ, if there exists a proper map $\pi : X \to \tilde{X}$, which is an isomorphism on U and satisfies $dim(\pi(Y)) \leq d-2$ (reduce to normal $\tilde{X}$ and use (g)).

III.11 Adjunction Properties

Let $u^* : \mathbf{A} \to \mathbf{B}$ be an exact fully faithful functor between abelian categories $\mathbf{A}$ and $\mathbf{B}$. Suppose that the functor u^* admits a right adjoint functor u_* and a left adjoint functor $u_!$. Then the following conditions are equivalent:

(a) For B in **B** the adjunction map

$$u^* u_* B \to B$$

is a monomorphism.

(b) For B in **B** the adjunction map

$$B \to u^* u_! B$$

is an epimorphism.

(c) u^* identifies **A** with a subcategory of **B**, which is stable under taking subquotients in **B**.

This is left as an exercise for the reader.

If property (a) is fulfilled,

$$u^* u_* B \hookrightarrow B$$

is the largest subobject of B, which is image of an object of **A** under u^*. Namely

$$Hom(u^* A, B) = Hom(A, u_* B) = Hom(u^* A, u^* u_* B)$$

holds by our assumptions.

Similarly $u^* u_! B$ is the largest quotient of B, which is image of an object from **A** under u^*.

Lemma 11.1 *If every object of* **A** *has finite length and if for every simple object A of* **A** *the object $u^* A$ remains simple in* **B**, *then the equivalent conditions (a)-(c) are fulfilled.*

We restrict ourselves to sketch a proof of (a):

Suppose the adjunction morphism $u^* u_* B \to B$ is not a monomorphism for some B in **B**. Let $K \neq 0$ be the kernel. By the assumptions there exists a finite composition series of $u^* u_* B$ by nontrivial simple objects of the form $u^* S_\nu$ for $S_\nu \in$ **A**. Then the same is true for the kernel K. Hence $K \hookrightarrow u^* u_* B$ contains a simple subobject $S = u^*(S_0)$ with $S_0 \neq 0$ simple in **A**. However $Hom(u^*(S_0), u^* u_* B) = Hom(S_0, u_* B)$. The corresponding nontrivial morphism $S_0 \to u_* B$ in $Hom(S_0, u_* B) = Hom(u^* S_0, B)$ is necessarily a monomorphism, S_0 being simple. It corresponds to the zero map in $Hom(u^* S_0, B) = Hom(S, B)$. This is a contradiction and proves Lemma III.11.1. □

Let now $f : X \to Y$ be a smooth morphism between finitely generated schemes of relative dimension d. Then $f^*[d] : D(Y) \to D(X)$ maps perverse sheaves to perverse sheaves, by Theorem III.7.2. Suppose that in addition all fibers of f are geometrically connected and nonempty. Then

Lemma 11.2 *The functor*

$$f^*[d] : Perv(Y) \to Perv(X)$$

is fully faithful under the assumptions above.

Proof. f is smooth, thus $f^{*!}K \cong f^*K[d] \cong f^!K[-d](-d)$. Hence this is a special case of Lemma III.7.8. □

So under the assumption on $f : X \to Y$ made above, the functor

$$f^*[d] : Perv(Y) \to Perv(X)$$

is fully faithful and exact. Let us show, that it has the left adjoint functor

$${}^p\!H^d\big(\mathrm{R}f_!(d)\big) : Perv(X) \to Perv(Y)\ ,$$

and the right adjoint functor

$${}^p\!H^{-d}\big(\mathrm{R}f_*\big) : Perv(X) \to Perv(Y)\ .$$

Proof of Assertions. Since $\mathrm{R}f_*[-d]B \in {}^pD^{\geq 0}(Y)$ holds for $B \in {}^pD^{\geq 0}(X)$, we get $Hom(f^*[d]A, B) = Hom(A, \mathrm{R}f_*[-d]B) = Hom(A, {}^p\!H^{-d}(\mathrm{R}f_*B))$. Recall $A \in {}^pD^{\leq 0}(Y)$ for perverse A. The proof of the second assertion is similar. □

So far all assumption made in the beginning of this section are fulfilled in the present situation with $u^* = f^*[d]$, $\mathbf{A} = Perv(Y)$ and $\mathbf{B} = Perv(X)$. It will be shown in the proof of the next theorem, that also the assumptions of Lemma III.11.1 are satisfied.

Therefore $f^*[d]$ identifies the abelian category $Perv(Y)$ with a full subcategory of $Perv(X)$, which is closed under taking subquotients. However, in general, this subcategory is not a thick subcategory (Serre category) of $Perv(X)$, i.e it is not closed under extensions.

Theorem 11.3 *Let $f : X \to Y$ be a smooth morphism with geometrically connected nonempty fibers of pure dimension d and let B be a perverse sheaf on X. Then*

$$f^*[d]\, {}^p\!H^{-d}\big(Rf_*B\big) \hookrightarrow B$$

is the **largest perverse subsheaf** *of B of the form*

$$f^*[d]A \qquad A \in Perv(Y)\ .$$

Similarly

$$B \twoheadrightarrow f^*[d]{}^p\!H^d\big(Rf_!B(d)\big)\ .$$

defines the **largest perverse factor sheaf** *of B in $f^*[d](Perv(Y))$.*

Proof. We have to show, that the assumptions of Lemma III.11.1 are fulfilled. The first assertion follows from Corollary III.5.7. It remains to show that $f^*[d]$ preserves simple objects. For this we use the description of simple objects in III.5.4. We are free to replace Y by a closed subscheme of Y, the support of A. The verification of the second assertion is then easily reduced to the case, where Y is irreducible and the perverse sheaf A given on Y is of the form $A = \mathcal{G}[dimY]$ for some smooth irreducible $\overline{\mathbb{Q}}_l$-sheaf $\mathcal{G}$ on Y.

The Smooth Case. We show, that the pullback $f^*(\mathcal{G})$ to X of an irreducible smooth sheaf $\mathcal{G}$ on Y remains irreducible. This follows from the fact, that the fundamental group $\pi_1(X, a)$ maps surjectively onto the corresponding fundamental group $\pi_1(Y, f(a))$; irreducible etale coverings of Y remain irreducible (= connected) – if Y is replaced by one of its connected components. Remember, that the fibers of f were supposed to be connected.

Reduction Step. $f^*[d]$ commutes with intermediate extensions. To show this, let $j : U \to Y$ be an open embedding with closed complement $i : S \to Y$. Let $\tilde{j}$ and $\tilde{i}$ denote the corresponding pullbacks under the map f. Let $\overline{A} \in Perv(Y)$ be the intermediate extension of some perverse sheaf A on U. Observe

$$f^*[d]j^* = \tilde{j}^* f^*[d]$$

and

$$f^*[d]i^* = \tilde{i}^* f^*[d] \text{ and } f^*[d]i^! = \tilde{i}^! f^*[d] .$$

For the last assertion use $f^*[d] = f^![-d](-d)$. This implies, that condition III.5.1(3) in paragraph 5 characterizing intermediate extensions is preserved by the t-exact functor $f^*[d]$.

This being said, it is clear that $f^*[d]\overline{A} \in Perv(X)$ is the intermediate extension of the sheaf $f^*[d]A \in Perv(f^{-1}(U))$, obtained via pullback from the restricted map $f : f^{-1}(U) \to U$. □

Additional Assumptions. For the rest of this section let us maintain all previous assumptions on f. Add the further assumptions that $R^\nu f_!(\overline{\mathbb{Q}}_l) = 0$ for $\nu = 2d - 1, 2d - 2, \ldots, 2d - m$ and some $m \geq 1$.

Then for $A, B \in Perv(Y)$ the same argument, which was used for the proof of III.11.2, shows that

$$Ext^\nu_{Perv(X)}(f^*[d]A, f^*[d]B) = Ext^\nu_{Perv(Y)}(A, B) \quad \nu = 0, .., m .$$

Therefore we get for $m = 1$

Corollary 11.4 *Under the assumptions on f in III.11.3 suppose, that $R^{2d-1} f_! \overline{\mathbb{Q}}_l = 0$. Then*

$$f^*[d]Perv(Y) \subset Perv(X)$$

is a thick abelian subcategory of the abelian category $Perv(X)$ (a Serre subcategory).

An object $B \in Perv(X)$ is called left reduced, if it has no nontrivial subobject isomorphic to an object of the thick subcategory $f^*[d]Perv(X)$. By Theorem III.11.3 this is equivalent to the assumption that ${}^pH^{-d}Rf_*B = 0$.

Any B in $Perv(X)$ can be reduced

$$0 \longrightarrow f^*[d]B_s \longrightarrow B \longrightarrow B_r \longrightarrow 0\,,$$

if one divides by the maximal subobject $f^*[d]B_s$ in the essential image of $Perv(Y)$. So B_r is left reduced. The long exact Ext-sequence for the abelian category $Perv(X)$ yields

$$0 \to Ext^1(f^*[d]A, f^*[d]B_s) \longrightarrow Ext^1(f^*[d](A), B) \longrightarrow Ext^1(f^*[d]A, B_r) \to Ext^2(f^*[d]A, f^*[d]B_s) \to \dots\,.$$

Lemma 11.5 *Under the assumptions of III.11.4 the following holds: For* $A \in Perv(Y)$ *and left reduced* $B \in Perv(X)$ *one has a canonical isomorphism*

$$Ext^1_{Perv(X)}(f^*[d]A, B) = Hom_{Perv(Y)}(A, {}^pH^{1-d}Rf_*B)\,.$$

Proof. Use again $Ext^1_{Perv(X)}(U, V) = Hom_{D(X)}(U, V[1])$. By adjunction $Ext^1_{Perv(X)}(f^*[d]A, B)$ is equal to

$$Hom_{D(X)}(A, Rf_*B[1-d]) = Hom_{D(X)}(A, {}^p\tau^{\leq 0}Rf_*B[1-d])\,.$$

Note that Lemma III.7.1 implies $Rf_*B[1-d] \in {}^pD^{\geq -1}(Y)$. By assumption B is left reduced, hence ${}^pH^{-1}(Rf_*B[1-d]) = {}^pH^{-d}Rf_*B = 0$. Therefore we have ${}^p\tau^{\leq 0}Rf_*B[1-d] = {}^pH^{1-d}Rf_*B)$. □

By III.11.4 and III.11.5 we get with notations as above, B_s being the maximal subsheaf of B coming from $Perv(Y)$,

$$0 \to Ext^1_Y(A, B_s) \longrightarrow Ext^1_X(f^*[d](A), B) \longrightarrow Hom^0_X(A, {}^pH^{1-d}Rf_*B_r) \to Ext^2_X(f^*[d]A, f^*[d]B_s)\,.$$

Ext-groups are with respect to the abelian categories $Perv(Y)$ resp. $Perv(X)$.

III.12 The Dictionary

Let X_0 be an algebraic scheme over a finite field κ with q elements. Let K_0 be a complex in $D^b_c(X_0, \overline{\mathbb{Q}}_l)$. Fix an isomorphism $\tau : \overline{\mathbb{Q}}_l \cong \mathbb{C}$. Remember the functions

$$f^{K_0} : X(\kappa) \to \mathbb{C}$$

defined by

$$f^{K_0}(x) = \tau \sum_{\nu} (-1)^\nu Tr\big(F; \mathscr{H}^\nu(K)_{\overline{x}}\big) .$$

They were already considered in the particular case of sheaves, in I.2.12 and in Lemma I.5.6. F is the geometric Frobenius of the field κ. For any field extension κ_m of κ with $q_m = q^m$, $m \geq 1$ elements we consider the extension of (X_0, K_0) from κ to κ_m. The correspondingly defined function of the extension with respect to the new base field κ_m is denoted $f_m^{K_0}$. In particular $f_1^{K_0} = f^{K_0}$.

Theorem 12.1 *Let X_0 be an algebraic scheme over the finite field κ. Then the functions f^{K_0} for $K_0 \in D_c^b(X_0)$ have the following properties:*

(1) $f^{M_0}(x) = f^{K_0}(x) + f^{L_0}(x)$ for $K_0, M_0, L_0 \in D_c^b(X_0)$, if there exists a distinguished triangle (K_0, M_0, L_0).
(2) $f^{K_0 \otimes L_0}(x) = f^{K_0}(x) \cdot f^{L_0}(x)$ for $K_0, L_0 \in D_c^b(X_0)$.
(3) Let K_0 and L_0 be semisimple perverse sheaves on X_0. Then the equality

$$f_m^{K_0}(x) = f_m^{L_0}(x) \quad x, \quad \in X(\kappa_m)$$

for all finite fields κ_m holds if and only if $K_0 \cong L_0$.
(4) Let $g : X_0 \to Y_0$ be a morphism defined over κ and let $K_0 \in D_c^b(X_0, \overline{\mathbb{Q}}_l)$. Then

$$f^{Rg_!(K_0)}(y) = \sum_{x \in X_0(\kappa_m), g(x) = y} f^{K_0}(x) .$$

(5) In the situation of (4) we have for $L_0 \in D_c^b(Y_0, \overline{\mathbb{Q}}_l)$

$$f^{g^*(L_0)}(x) = f^{L_0}(g(x)) .$$

(6) Let K_0 be a τ-mixed perverse sheaf on X_0, or more generally a τ-mixed complex in $D_c^b(X_0)$, which is τ-pure of weight w. Then

$$f^{DK_0}(x) = q^{-w} \cdot \overline{f^{K_0}(x)} .$$

(7) Let K_0 be as in (6). Then $w(K_0[d]) = w(K_0) + d$ and

$$f^{K_0[d]}(x) = (-1)^d f^{K_0}(x) .$$

(8) Let K_0 be as in (6). Then $w(K_0(n)) = w(K_0) - 2n$ and

$$f^{K_0(n)}(x) = q^{-n} f^{K_0}(x) .$$

Proof. Properties (1) and (2) are trivial. By noetherian induction and III.5.5 the proof of property (3) is easily reduced to the smooth case, where it follows from the Chebotarev density theorem. Property (4) is an immediate consequence of the Grothendieck trace formula I.1.1. (5), (7) and (8) are trivial. The remaining property (6) is shown as follows:

For simplicity assume $w = 0$. Let $x \in X_0(\kappa_m)$. We have to show

$$f^{DK_0}(x) = \overline{f^{K_0}(x)} \,.$$

We can reduce to the case $\kappa = \kappa_m$ by a base field extension. The question is local at x with respect to the etale topology. So we can assume X_0 to be affine. By a closed embedding and then intermediate extension with respect to an open embedding we can assume $X_0 = (\mathbb{P}^1_0)^N$ and x to be the zero point of the affine space $(\mathbb{A}^1_0)^N$. There always exists an elliptic curve $\mathbb{E}_0$ over the finite field κ with a κ-rational closed point $e \in \mathbb{E}_0(\kappa)$ and a finite map $\pi : \mathbb{E}_0 \to \mathbb{P}^1_0$, which is etale in a neighborhood of e in $\mathbb{E}_0$. By a translation e can be assumed to be the zero point of the group law on $\mathbb{E}_0$ and $\pi(e)$ can be assumed to be the origin in $\mathbb{A}^1_0$. We can therefore replace $x \in (\mathbb{A}^1_0)^N$ by $x = (e, e, .., e)$ in $\mathbb{E}_0^N$ (etale neighborhood).

It is therefore enough to prove the statement for spaces $X_0 = \mathbb{E}_0^N$ at the origin x. Then the proof goes by induction on N. By a simple reduction we can furthermore assume K_0 to be a simple perverse sheaf on $\mathbb{E}_0^N$.

Let $g : \mathbb{E}_0^N \to \mathbb{E}_0^{N-1}$ be the projection onto the last $N-1$ factors. This map is proper and smooth with connected fibers of dimension $d = 1$. Hence by Lemma III.7.1

$$ {}^p\!H^\nu(Rg_!(K_0)) \;=\; {}^p\!H^\nu(Rg_*(K_0)) = 0$$

holds for $\nu \neq -1, 0, 1$. Especially $Rg_*(K_0)$ is a perverse sheaf if and only if

$$ {}^p\!H^{-1}(Rg_*(K_0)) = {}^p\!H^1(Rg_*(K_0)) = 0 \,.$$

According to III.11.3 these two cohomology groups vanish if and only if our simple mixed perverse sheaf K_0 does not satisfy

$$K_0 \in g^*[1]Perv(\mathbb{E}_0^{N-1}) \,.$$

So we distinguish two cases

Case 1 $K_0 \in g^*[1]Perv(\mathbb{E}_0^{N-1})$:

In this case $K_0 = g^*[1](L_0)$ holds for some simple perverse sheaf L_0. As K_0 is pure of weight 0, L_0 is pure of weight -1. Then by II.7.5 and III.7.2 and III.12.1 we get

$$f^{DK_0}(x) = f^{Dg^*[1]L_0}(x) = f^{g^![-1]DL_0}(x) = f^{g^*[1](1)(DL_0)}(x) = -q^{-1} f^{DL_0}(g(x))$$

whereas

$$f^{K_0}(x) = f^{g^*[1]L_0}(x) = -f^{L_0}(g(x)) \,.$$

From the induction assumption one has

$$\overline{f^{L_0}(g(x))} = q^{-1} f^{DL_0}(g(x)) \,,$$

because L_0 is perverse and pure of weight -1. This completes the proof in case 1.

Case 2 $Rg_*(K_0)$ is again perverse:

Consider the morphism

$$\wp : \mathbb{E}_0 \to \mathbb{E}_0$$

defined by $\wp(x) = F(x) - x$ with kernel $\mathbb{E}_0(\kappa)$. This is an etale morphism. See e.g. [SGA $4\frac{1}{2}$], p. 171 for further information on Lang-torsors in general. We have

$$\wp_* \overline{\mathbb{Q}}_l = \bigoplus_{\psi} \mathscr{L}_\psi .$$

The sum ranges over all characters $\psi : \mathbb{E}_0(\kappa) \to \overline{\mathbb{Q}}_l^{\,*}$ of the finite abelian covering group $\mathbb{E}_0(\kappa)$ of the map $\wp$. Now the fiber of the map $g : \mathbb{E}_0^N(\kappa) \to \mathbb{E}_0^{N-1}(\kappa)$ over the origin $g(x)$ can be identified with the set $\{(y, e, ..., e) | y \in \mathbb{E}_0(\kappa)\}$. By III.12.1(4) from above we have

$$\sum_{x', g(x')=g(x)} f^{DK_0}(x') = f^{Rg_!(DK_0)}(g(x)) = f^{D(Rg_*K_0)}(g(x)) .$$

But Rg_*K_0 was supposed to be a perverse sheaf. By the Weil conjectures it is again pure of weight 0. Therefore by the induction assumption the last term is equal to the complex conjugate of

$$f^{Rg_*K_0}(g(x)) = f^{Rg_!K_0}(g(x)) = \sum_{x', g(x')=g(x)} f^{K_0}(x') .$$

This shows

$$\sum_{y \in E(\kappa)} \big(f^{DK_0} - \overline{f^{K_0}}\big)((y, e, e, .., e)) = 0 .$$

We repeat the same conclusion for the complex K_0 replaced by $K_0 \otimes \mathscr{L}_\psi$, where by abuse of notation we assume $\mathscr{L}_\psi$ to be a sheaf on $\mathbb{E}_0^N$ by pullback from projection onto the first factor. Then $D(K_0 \otimes \mathscr{L}_\psi) = D(K_0) \otimes \mathscr{L}_\psi^{-1}$. As $\mathscr{L}_\psi$ is smooth of rank 1 and weight 0, we have $f^{\mathscr{L}_\psi^{-1}}(x) = \overline{f^{\mathscr{L}_\psi}(x)}$. So our claim for the complex K_0 on $\mathbb{E}_0^N$ would follow from the corresponding assertion for the complex $K_0 \otimes \mathscr{L}_\psi$. This allows us to assume all twists $K_0 \otimes \mathscr{L}_\psi$ to be not in $g^*[1]Perv(\mathbb{E}_0^{N-1})$. It shows

$$\sum_{y \in E(\kappa)} \overline{f^{\mathscr{L}_\psi}}(y)\big(f^{DK_0} - \overline{f^{K_0}}\big)((y, e, e, .., e)) = 0 .$$

The geometric Frobenius element F_x for a point $x \in \mathbb{E}_0(\kappa)$ in the covering group $\mathbb{E}_0(\kappa)$ is given by $-x$. This uses the same argument as in the proof of I.5.6, and it is an explicit form of the reciprocity law. The function $\overline{f^{\mathscr{L}_\psi}}(y)$ on $\mathbb{E}_0(\kappa)$ is therefore equal to the character $\psi(y)$. Hence these functions separate the points $y \in \mathbb{E}_0(\kappa)$. This proves $f^{DK_0}(x) = \overline{f^{K_0}(x)}$ at the origin $x \in \mathbb{E}_0(\kappa)^N$. □

Example 12.2 If K_0 is an irreducible perverse sheaf on $X_0 = \mathbb{A}_0^r$, let w be its weight and let $T_\psi(K_0)$ be the Fourier transform of K_0, which is of weight $w + r$. Part (3) of the last theorem implies $D(T_\psi(K_0)) \cong T_{\psi^{-1}}(D(K_0)(r))$, since the corresponding functions satisfy

$$f_m^{D(T_\psi(K_0))} = q^{-w-r}\overline{f_m^{T_\psi(K_0)}} = q^{-r} f_m^{T_{\psi^{-1}} D(K_0)} .$$

This uses the formula (6) of the same theorem. A stronger result is the following.

In the definition of the Fourier transform T_ψ, the direct image with proper support $\mathrm{R}p_{2!}$ can be replaced by the direct image $\mathrm{R}p_{2*}$, and the result is a second Fourier transform T'_ψ. It is easy to show $T'_\psi(K) = D\big(T_{\psi^{-1}}(D(K)(r))\big)$ from the definitions. There is a natural map $T_\psi \to T'_\psi$ induced by $\mathrm{R}p_{2!} \to \mathrm{R}p_{2*}$. Katz and Laumon have shown ([Ka-L], theorem 2.13 and corollaire 2.1.5) the more difficult statement, that this natural map

$$T_\psi \to T'_\psi$$

is an isomorphism.

Theorem 12.3 *Let D be the dualizing functor. Then there exist functorial isomorphisms*

$$D \circ T_\psi(K^\bullet) \cong T_{\psi^{-1}} \circ D(K^\bullet)(r) \quad , \quad K \in D_c^b(\mathbb{A}_0^r, \overline{\mathbb{Q}}_l)$$

induced by the natural transformation $Rp_{2!} \to Rp_{2*}$.

III.13 Complements on Fourier Transform

Let $\mathbb{F}_q$ be a finite field of characteristic p. Let k be an algebraically closed field extension of $\mathbb{F}_q$. Let S be a finitely generated scheme over k. Furthermore choose an auxiliary prime $l \neq p$ and a nontrivial character

$$\psi : \mathbb{F}_q \to \overline{\mathbb{Q}}_l^{\,*} .$$

Let $\mathscr{L}_0(\psi)$ be the **Artin-Schreier** sheaf on the affine line over $\mathbb{F}_q$ and let $\mathscr{L}(\psi)$ be its pullback to the affine line $\mathbb{A}_S$ over S.

For a vector bundle E of rank r over S the dual vector bundle will be denoted E'. One has the natural evaluation pairing

$$\mu : E \times_S E' \to \mathbb{A}_S$$

and projection maps

$$\begin{array}{ccccc} & & E \times_S E' & \xrightarrow{\quad \mu \quad} & \mathbb{A}_S \,. \\ & \swarrow^{p_1} & & \searrow^{p_2} & \\ E & & & & E' \end{array}$$

Similar to Chap. I, §5 we now define the relative Fourier transform $T_\psi^{E/S}$ for the vector bundle E/S

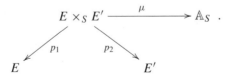

$$T_\psi^{E/S}: D_c^b(E,\overline{\mathbb{Q}}_l) \longrightarrow D_c^b(E^{'},\overline{\mathbb{Q}}_l)$$

by

$$T_\psi^{E/S}(K^\bullet) = \mathrm{R}p_{2!}\Big(p_1^*(K^\bullet)\otimes \mu^*(\mathscr{L}(\psi))\Big)[r]\,.$$

We often write $T_\psi^{E/S} = T_\psi^E$ or $T_\psi^{E/S} = T_\psi$, if the underlying vector bundle is understood from the context.

This generalized Fourier transform for vector bundles E/S enjoys a number of properties, whose proofs are easy or similar to the special cases already considered in Chap. I, §5 and Chap. III, §8. Hence no proofs are given for the following theorems.

Theorem 13.1 (Inversion formula) *Let E/S be a vector bundle of rank r over S. For complexes $K^\bullet \in D_c^b(E,\overline{\mathbb{Q}}_l)$, the equation*

$$(T_{\psi^{-1}}\circ T_\psi)(K^\bullet) = K^\bullet(-r)$$

canonically holds.

Theorem 13.2 (Functoriality) *For the Fourier transform the following holds:*

1) *Compatibility $T_\psi^{E\times_S T/T}(f^*K^\bullet) = f^*(T_\psi^{E/S}(K^\bullet))$ for arbitrary base change $f: T\to S$.*
2) *Suppose given a homomorphism u of vector bundles over S and its dual*

$$u: E\to F \quad , \quad u': F^{'} \to E^{'}\,.$$

Put $d = \mathrm{rank}(F) - \mathrm{rank}(E)$. Then for arbitrary complexes $K^\bullet \in D_c^b(E,\overline{\mathbb{Q}}_l)$ and $L^\bullet \in D_c^b(F,\overline{\mathbb{Q}}_l)$ the following holds:

$$(a) \qquad T_\psi^{F/S}(Ru_!K^\bullet) = u'^*(T_\psi^{E'/S}(K^\bullet))[d]\,,$$

$$(b) \qquad T_\psi^{E/S}(u^*(L^\bullet)) = Ru'_!\,T_\psi^{F'/S}(L^\bullet)[d](d)\,.$$

The arguments for (1) and (2a) are obvious. (2b) follows from (2a) by the inversion formula.

Theorem 13.3 (Direct images) *Let E be a vector bundle over S, let $B\to S$ be a base change map and suppose $K^\bullet \in D_c^b(B\times_S E,\overline{\mathbb{Q}}_l)$. Consider the projections*

$$pr: B\times_S E\to E, \qquad pr': B\times_S E^{'} \to E^{'}\,.$$

Then there is a natural isomorphism $Rpr'_!(T_\psi^{B\times_S E'/B}(K^\bullet)) = T_\psi^{E/S}(Rpr_!(K^\bullet))$.

Again this is obvious from the definitions and the proper base change theorem.

Remark. The corresponding relations hold, if $Rpr_!$ and $Rpr'_!$ are replaced by Rpr_* and Rpr'_*. This can be deduced from Theorem III.12.3. Since we do not need this, it is left as an easy exercise to the reader.

Theorem III.13.2, (2a) and (2b) above gives the useful

Corollary 13.4 *Let V be a (locally split) subbundle of rank μ of the vector bundle E of rank ν over S*

$$i : V \hookrightarrow E\ .$$

Consider the dual map i' for the dual bundles. For $V^\perp = ker(i')$

$$0 \longleftarrow V' \xleftarrow{\ i'\ } E' \xleftarrow{\ \tilde{i}\ } V^\perp \longleftarrow 0$$

Put $\delta_V = i_*(\overline{\mathbb{Q}}_{l,V})$ and $\delta_{V^\perp} = \tilde{i}_*(\overline{\mathbb{Q}}_l)_{V^\perp}$. Then there is a canonical isomorphism

$$T_\psi^{E/S}(\delta_V) \ \cong\ \delta_{V^\perp}[\nu - 2\mu](-\mu)\ .$$

Other Formulas. Let $e : S \to E$ be the zero section of the bundle E of rank ν over S and let $e' : S \to E'$ the zero section of the dual bundle $E' \overset{\pi}{\to} S$ of E. We define

$$\delta_0 = e_*(\overline{\mathbb{Q}}_{l,S}), \qquad \delta'_0 = e'_*(\overline{\mathbb{Q}}_{l,S})\ .$$

Put $\delta_E = \overline{\mathbb{Q}}_{l\,E}$ and $\delta_{E'} = \overline{\mathbb{Q}}_{l\,E'}$ and similar put $\delta_S = \overline{\mathbb{Q}}_{l\,S}$. Then special cases of the last Corollary III.13.4 are

$$T_\psi^{E/S}(\delta_E[\nu]) = \delta'_0(-\nu) \quad , \quad T_\psi^{E/S}(\delta_0) = \delta_{E'}[\nu]\ .$$

From $(e')^!\pi^!(\delta_S) = \mathrm{id}^!(\delta_S) = \delta_S$ and $\pi^!(\delta_S) = \delta_{E'}(\nu)[2\nu]$ we therefore obtain $(e')^!(\delta_{E'}[\nu]) = \delta_S[-\nu](-\nu)$. Hence

$$e'_!(e')^!(\delta_{E'}[\nu]) = \delta'_0[-\nu](-\nu)\ .$$

Restriction and Corestriction Maps. Let E/S be a vector bundle of rank ν and let E'/S be its dual bundle. Let $W \hookrightarrow E'$ be a locally split subbundle of E' of rank μ_W. Let $V = W^\perp$ be its associated orthogonal bundle in E, which is of rank $\mu = \nu - \mu_W$. We write $\langle 2m\rangle$ instead of $[2m](m)$, and also write $\delta_E = \overline{\mathbb{Q}}_{l\,E}$ and $\delta_W = i_{W*}\overline{\mathbb{Q}}_{l\,W}$ and similarly for V. The adjunction map $\overline{\mathbb{Q}}_{l\,E} \to i_{V*}i_V^*\overline{\mathbb{Q}}_{l\,E}$ for the inclusion $i_V : V \to E$ is called the restriction map

$$res_{E,V} : \delta_E \to \delta_V\ .$$

Similarly we have the restriction map $res_{W,0} : \delta_W \to \delta'_0$. Consider its Fourier transform $T_\psi^{E'/S}(res_{W,0}) : T_\psi^{E'/S}(\delta_W) \to T_\psi^{E'/S}(\delta'_0)$. The identities $T_\psi^{E'/S}(\delta_W) =$

$\delta_V[\nu]\langle -2\mu_W\rangle$ and $T_\psi^{E'/S}(\delta_0') = \delta_E[\nu]$ give a corestriction map. Twist by $[-\nu]\langle 2\nu\rangle$, then $cores_{V,E} = T_\psi^{E'/S}(res_{W,0})[\nu]$ such that

$$cores_{V,E} : \delta_V\langle 2\mu\rangle \to \delta_E\langle 2\nu\rangle \ .$$

Note $i_{V!}i_V^!\delta_E\langle 2\nu\rangle = Di_{V*}i_V^* D\delta_E\langle 2\nu\rangle = \delta_V\langle 2\mu\rangle$, if the base S is smooth over k.

Lemma 13.5 *With the notations above either assume $V = E$ or V is of rank 0 or assume S to be smooth over the base field k. Then the corestriction map*

$$cores_{V,E} : \delta_V\langle 2\mu\rangle \to \delta_E\langle 2\nu\rangle$$

admits a unique factorization $\delta_V\langle 2\mu\rangle \xrightarrow[\cong]{\rho} i_{V!}i_V^!\delta_E\langle 2\nu\rangle \xrightarrow{adj} \delta_E\langle 2\nu\rangle$. *The map on the right is the adjunction map for the inclusion $i_V : V \hookrightarrow E$, and the map ρ on the left is an isomorphism.*

Proof. Since $\delta_V\langle 2\mu\rangle$ is supported in V, the corestriction map $cores_{V,E}$ uniquely factorizes over the functor $i_{V!}i_V^!$. It only remains to show, that the induced map ρ is an isomorphism. For this it is enough to consider the case, where the base S is connected. Then $\rho \in Hom(\delta_V\langle 2\mu\rangle, \delta_V\langle 2\mu\rangle) \cong \overline{\mathbb{Q}}_l$, since $i_{V!}i_V^!\delta_E\langle 2\nu\rangle = \delta_V\langle 2\mu\rangle$. Hence ρ is either zero or an isomorphism. But ρ can not be zero, since this implies $cores_{V,E} = 0$. By Fourier inversion this implies $res_{W,0} = 0$, which obviously is a contradiction. □

Remark 13.6 If the base S is smooth, then $D(res_{E,V}) : D(\delta_V) \to D(\delta_E)$ gives another map

$$D_{E/S}(res_{E,V}) : \delta_V\langle 2\mu\rangle \to \delta_E\langle 2\nu\rangle \ ,$$

where $D_{E/S}(.) = D(.)\langle -2dim(S)\rangle$. The last lemma can now be restated in the following form: There exists a unique isomorphism ρ, which makes the following diagram commutative

$$\begin{CD}
\delta_E\langle 2\nu\rangle @<{D_{E/S}(res_{E,V})}<< \delta_V\langle 2\mu\rangle \\
@| @AA{\cong\ \rho}A \\
\delta_E\langle 2\nu\rangle @<{cores_{V,E}}<< \delta_V\langle 2\mu\rangle
\end{CD}$$

Remark 13.7 Similar to the last lemma one has a more general result for a pair of two locally split subvectorbundles $V \hookrightarrow W \hookrightarrow E$ of the vector bundle E.

III.14 Sections

Consider the following situation. Let $E \stackrel{\pi}{\to} S$ be a vector bundle of rank r over S, let $E' \stackrel{\pi'}{\to} S$ be the dual bundle. Let $e : S \to E$ and $e' : S \to E'$ denote the zero sections of these bundles. Let $\delta_0 = \delta_{0,E} = e_*(\overline{\mathbb{Q}}_{lS})$ respectively $\delta'_0 = \delta_{0,E'} = e'_*(\overline{\mathbb{Q}}_{lS})$ be the constant skyscraper sheaves on the zero sections of E resp. E'.

Suppose $s : S \to E$ is another section of the bundle $\pi : E \to S$, possibly different from the zero section e. Put

$$\delta_\Delta = s_*(\overline{\mathbb{Q}}_{lS}) \quad , \qquad \Delta = s(S) \subset E \ .$$

Then δ_Δ is a skyscraper sheaf on E concentrated on Δ

$$\begin{array}{ccc} \Delta & \hookrightarrow & E \\ & {}^{s}\nwarrow & \downarrow \pi \\ & & S \end{array} \ .$$

The section s defines a morphism ν_s from the dual vector bundle to the affine space, which we call the characteristic morphism of the section

$$\nu_s : E' \to \mathbb{A}_S \ .$$

It is defined by $\nu_s(y) = y(s(\pi'(y)))$, where $y(x) = \langle y, x \rangle$ denotes the evaluation map for the linear function y on E.

Lemma 14.1 *Let $K \in D^b_c(E, \overline{\mathbb{Q}}_l)$ be given. Let T_{-s} denote the translation by $-s$ in the fibers of $E \to S$. Then*

$$T_\psi\big(T^*_{-s}(K)\big) \cong T_\psi(K) \otimes \nu_s^*(\mathscr{L}\big(\psi)\big) \ .$$

Before we give a proof for the lemma, we give an application.

Recall from Theorem III.13.5, that the Fourier transform $T_\psi(res_{E,0})$ of the natural restriction map $\overline{\mathbb{Q}}_{lE} \to \delta_{0,E}$ essentially coincides with the adjunction map $(*)$

$$(*) \qquad adj : \quad e'_! e'^{!}(\overline{\mathbb{Q}}_{lE'}[r]) = \delta'_0(-r)[-r] \longrightarrow \overline{\mathbb{Q}}_{lE'}[r] \ .$$

Note that $T^*_{-s}(\delta_0) = \delta_\Delta$. Using the last lemma, one obtains a canonical isomorphism

$$T_\psi(\delta_\Delta) = T_\psi(\delta_{0,E}) \otimes \nu_s^*\big(\mathscr{L}(\psi)\big) = \nu_s^*\big(\mathscr{L}(\psi)\big)[r] \ .$$

The Fourier transform $T_\psi(h)$ of the natural restriction map $h : \overline{\mathbb{Q}}_{lE} \longrightarrow \delta_\Delta$ is obtained from the adjunction map denoted $(*)$ above, by tensoring with $\nu_s^*\big(\mathscr{L}(\psi)\big)$

$$\begin{array}{ccc}
T_\psi(\overline{\mathbb{Q}}_{l\,E}) & \xrightarrow{\quad T_\psi(h)\quad} & T_\psi(\delta_\Delta) \\
\Vert & & \Vert \\
\delta'_0(-r)[-r] & & \Vert \\
\Vert & & \Vert \\
\delta'_0(-r)[-r]\otimes \nu_s^*(\mathscr{L}(\psi)) & \xrightarrow{adj\otimes \nu_s^*\mathscr{L}(\psi)} & T_\psi(\delta_{0,E})\otimes \nu_s^*\mathscr{L}(\psi)\,.
\end{array}$$

Remark. We used, that the restriction of ν_s to the zero section is the zero map. Hence

$$\nu_s^*(\mathscr{L}(\psi))|\text{zero section} \;=\; \mathbb{Q}_{l,E'}|\text{zero section}\,.$$

Remark. For a complex $\mathscr{G}^\bullet$ on E' the cup product defines an isomorphism

$$e'^!(\mathscr{G}^\bullet)\otimes e^*(\nu_s^*\mathscr{L}(\psi)) \xrightarrow{\;\cong\;} e'^!(\mathscr{G}^\bullet\otimes \nu_s^*\mathscr{L}(\psi))\;\cdot$$

Proof. For the proof of Lemma III.14.1 recall

$$\begin{array}{ccccc}
 & & E\times_S E' & \xrightarrow{\quad\mu\quad} & \mathbb{A}_S\,. \\
 & \swarrow^{p_1} & & \searrow^{p_2} & \\
E & & & & E'
\end{array}$$

Consider the two morphisms

$$\gamma,\tilde{\gamma} : E\times_S E' \longrightarrow E\times_S E'\,,$$

defined by $\gamma(x,y) = (x - s(\pi(x)), y)$ and $\tilde{\gamma}(x,y) = (x + s(\pi(x)), y)$. These morphisms are inverse to each other $\gamma\circ\tilde{\gamma} = \tilde{\gamma}\circ\gamma = \text{id}$. From the next Lemma III.14.2 we obtain

$$\begin{aligned}
\gamma^*(\mu^*(\mathscr{L}(\psi)) &= (\mu\circ\gamma)^*(\mathscr{L}(\psi))) = (\mu - \nu_s\circ p_2)^*(\mathscr{L}(\psi)) \\
&= \mu^*(\mathscr{L}(\psi)) \;\otimes\; p_2^*(\nu_s^*\mathscr{L}(\psi))^{-1}\,.
\end{aligned}$$

The pullback of K in E under the morphism $T_{-s} : E \ni x \mapsto x - s(\pi(x)) \in E$ is $T_{-s}^*(K)$. Hence $p_1^*(T_{-s}^*(K)) = \gamma^*(p_1^*(K))$. This implies

$$p_1^*(T_{-s}^*(K))\otimes\mu^*(\mathscr{L}(\psi)) = \gamma^*(p_1^*(K))\otimes\gamma^*(\mu^*(\mathscr{L}(\psi)))\otimes p_2^*(\nu_s^*(\mathscr{L}(\psi)))\,.$$

We obtain the Fourier transform $T_\psi(T_{-s}^*(K))$ as

$$\mathrm{R}p_{2!}(p_1^*(T_{-s}^*(K))\otimes\mu^*(\mathscr{L}(\psi))) \;=\; \mathrm{R}p_{2!}\big(\gamma^*\big(p_1^*(K)\otimes\mu^*(\mathscr{L}(\psi))\big)\big)\otimes\nu_s^*(\mathscr{L}(\psi))$$

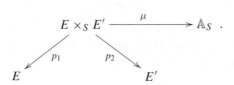

$$= \mathrm{R}p_{2!}\big(p_1^*(K) \otimes \mu^*(\mathscr{L}(\psi))\big) \otimes \nu_s^*(\mathscr{L}(\psi))\,,$$

since $\mathrm{R}p_{2!}\gamma^* = \mathrm{R}p_{2!}\tilde{\gamma}_!\gamma_!\gamma^* = \mathrm{R}p_{2!}$. In fact $\gamma_! = \gamma_*$ and $\gamma_*\gamma^* \cong id$ and $p_2\tilde{\gamma} = p_2$. Therefore $T_\psi\big(T_{-s}^*(K)\big) = T_\psi\big(K\big) \otimes \nu_s^*(\mathscr{L}(\psi))$. This was the claim. □

Lemma 14.2 *Let π_1, π_2 be the two projections $\pi_i : \mathbb{A}_S \times_S \mathbb{A}_S \to \mathbb{A}_S$ and let $\mathscr{F} = \mathscr{L}(\psi)$ be the Artin-Schreier sheaf on $\mathbb{A}_S$. Then, for $a : \mathbb{A}_S \times_S \mathbb{A}_S \to \mathbb{A}_S$ defined by $a(x, y) = x \pm y$, one has a canonical isomorphism*

$$a^*\big(\mathscr{F}\big) = \pi_1^*\big(\mathscr{F}\big) \otimes \pi_2^*\big(\mathscr{F}^{\pm 1}\big)\,.$$

Proof. Obvious. □

III.15 Equivariant Perverse Sheaves

In the following, consider schemes X of finite type over the field k together with an action $a : G \times X \to X$ of an algebraic group G over k. We will consider objects K of the derived category of etale $\overline{\mathbb{Q}}_l$-sheaves on that scheme, together with a G-action, which is compatible with the G-action on the scheme. We will consider both left and right actions of G on X, and for finite groups we will often consider right actions on X instead of left actions (sic). Of course, this is only a matter of convention.

Finite Groups. First let $G = W$ be a finite group. W can be considered as an algebraic group. We assume, that it acts on a scheme X from the right. We also write $\sigma(x) = a(\sigma, x)$ for $\sigma \in W$. So we have morphisms $\sigma : X \to X$ for $\sigma \in W$, such that

$$(\sigma\tau)(x) = \tau(\sigma(x)) \quad x \in X,\ \sigma, \tau \in W\,.$$

Definition 15.1 *Let $K^\bullet \in D_c^b(X, \overline{\mathbb{Q}}_l)$ be a complex. By a compatible right action of the group W on the complex $K^\bullet$, i.e an action which is compatible with the given right action of W on X, we understand a family $(\varphi_\sigma)_{\sigma \in W}$ of isomorphisms (in the derived category)*

$$\varphi_\sigma : \sigma^*(K^\bullet) \overset{\cong}{\to} K^\bullet$$

satisfying the following cocycle conditions

a) $\varphi_{\sigma\tau} = \varphi_\sigma \circ \sigma^*(\varphi_\tau)$ holds for all $\sigma, \tau \in W$.
b) For the unit element $e \in W$ we have $\varphi_e = \mathrm{Id} : K^\bullet \to K^\bullet$.

If W acts on X trivially, then a compatible action of W on $K^\bullet$ is simply a homomorphism

$$W \to \mathrm{Aut}_{D_c^b(X,\overline{\mathbb{Q}}_l)}(K^\bullet) \subset \mathrm{End}_{D_c^b(X,\overline{\mathbb{Q}}_l)}(K^\bullet)\,.$$

Remark 15.2 The definition above has an obvious analog if W is assumed to act on X from the left. The corresponding definition of a compatible left action on a complex $K^\bullet$ is the same as above, except that the first cocycle condition has to be changed into the condition $\phi_{\sigma\tau} = \phi_\tau \circ \tau^*(\phi_\sigma)$. But right actions have the "advantage", that they induce left actions on cohomology groups, i.e. ordinary representations of W. Sometimes it is useful to convert a right action on $(X, K^\bullet)$ into a left action (or conversely). This is done by setting $l_\sigma = \sigma^{-1} : X \to X$, which defines a left action; and by setting $\phi_\sigma = \varphi_{\sigma^{-1}} = (\sigma^{-1})^*(\varphi_\sigma^{-1})$, which defines isomorphisms $\phi_\sigma : l_\sigma^*(K^\bullet) \to K^\bullet$ such that $\phi_{\sigma\tau} = \phi_\tau \circ l_\tau^*(\phi_\sigma)$ holds.

Remark 15.3 There are certain obvious permanence properties. For instance the following facts will be relevant:

a) Let W act on a second scheme Y and let

$$f : Y \to X$$

be a W-equivariant morphism. If W acts on the complex $K^\bullet$ compatible with the action on X, then W acts in a natural way on the pullback $f^*(K^\bullet)$ compatible with the action on Y. For example let $f : X \to S$ be the structure morphism to the base scheme, which we assume to be equivariant with respect to the trivial action of W on S. This induces a "standard" W-structure on the constant sheaf $\overline{\mathbb{Q}}_{l\,X}$. If f is a morphism over a base scheme S, then the "standard" action on the constant sheaf $\overline{\mathbb{Q}}_{l\,X}$ obviously base pulls back to the "standard" action on the constant sheaf $\overline{\mathbb{Q}}_{l\,Y}$. Furthermore, let $K \in D_c^b(Y, \overline{\mathbb{Q}}_l)$, $L \in D_c^b(X, \overline{\mathbb{Q}}_l)$ be complexes with W-actions, that are compatible with the underlying W-action on Y resp. X. Then the complexes

$$\mathrm{R}f_!(K) \in D_c^b(X, \overline{\mathbb{Q}}_l), \quad f^!(L) \in D_c^b(Y, \overline{\mathbb{Q}}_l)$$

have natural W-group actions consistent with the action of W on X resp. Y. The adjunction mapping

$$\mathrm{R}f_! f^!(L) \to L$$

is W-equivariant.

b) Let $j : U \hookrightarrow X$ be an open W-equivariant embedding. Let $K^\bullet$ be a W-equivariant perverse sheaf on U. Then $j_{!*}K^\bullet$ becomes a W-equivariant sheaf, using the canonical isomorphism

$$End_{Perv(U)}(K^\bullet) = End_{Perv(X)}(j_{!*}K^\bullet)\,.$$

See III.5.11.

c) Suppose $K^\bullet \in D_c^b(X, \overline{\mathbb{Q}}_l)$ and assume W acts on X from the right. A compatible action of W on $K^\bullet$ from the right induces a W-action on the Verdier dual $D(K^\bullet)$: If the W-action on $K^\bullet$ is given by the cocycle $\varphi_w : w^*(K^\bullet) \cong K^\bullet$, then $D(\varphi_w) : D(K^\bullet) \to D(w^*K^\bullet) = w^! D(K^\bullet) = w^* D(K^\bullet)$. Therefore $\tilde{\varphi}_w = w^* D(\varphi_{w^{-1}}) : w^* D(K^\bullet) \to D(K^\bullet)$. An easy computation shows $\tilde{\varphi}_{\sigma\tau} = \tilde{\varphi}_\sigma \circ \sigma^*(\tilde{\varphi}_\tau)$. Using this for a given W-right equivariant complex $K^\bullet$, we can canonically equip the dual complex $D(K^\bullet)$ with a compatible W-right action in a canonical way – again compatible with the underlying W-right action.

d) Suppose X is essentially smooth. Suppose $\pi : Y \to X$ is an etale Galois covering with Galois group W acting from the right. Then $\pi_*\overline{\mathbb{Q}}_{lY}$ is a smooth W-equivariant sheaf on X. The trivial case $f = id_X, X = S, K = \pi_*\overline{\mathbb{Q}}_{lX}$ of the next Lemma III.15.4 implies that $\pi_*\overline{\mathbb{Q}}_{lY}$ can be decomposed into ϕ-isotypic components, where ϕ runs over the irreducible characters of W

$$\pi_*\overline{\mathbb{Q}}_{lY} = \bigoplus_{\phi\in\hat{W}} \mathscr{F}_\phi \otimes_{\overline{\mathbb{Q}}_l} V_\phi \ .$$

Here $\mathscr{F}_\phi$ are smooth $\overline{\mathbb{Q}}_l$-sheaves on X and V_ϕ are $\overline{\mathbb{Q}}_l$-vectorspaces, such that W acts on $\pi_*\overline{\mathbb{Q}}_{lY}$ via irreducible representations $\phi : W \to Gl(V_\phi)$. Obviously in all stalks $\pi_*(\overline{\mathbb{Q}}_l)_x$ the group W acts through the regular representation. In particular $\mathscr{F}_\phi \neq 0$ for all ϕ. Furthermore

$$\pi^*(\mathscr{F}_\phi) \cong (\overline{\mathbb{Q}}_{lY})^{r(\phi)}$$

for the ranks $r(\phi)$ of the smooth sheaves $\mathscr{F}_\phi$. In particular $\#W = \sum_{\phi\in\hat{W}} r(\phi) \cdot deg(\phi)$. Since $R\pi_! = \pi_! = \pi_*$ and $\pi^! = \pi^*$ the adjunction formula gives

$$Hom_{D_c^b(Y)}(\overline{\mathbb{Q}}_{lY}, \pi^*(C)) = Hom_{D_c^b(X)}(\pi_*(\overline{\mathbb{Q}}_{lY}), C) \ .$$

For $C = \pi_*\overline{\mathbb{Q}}_{lY}$ this implies $\#W = dim_{\overline{\mathbb{Q}}_l} End_{D_c^b(X)}(\pi_*\overline{\mathbb{Q}}_{lY})$, hence

$$\#W = \sum_{\phi_1,\phi_2} deg(\phi_1)\cdot deg(\phi_2)\cdot dim_{\overline{\mathbb{Q}}_l} Hom_{D_c^b(X)}(\mathscr{F}_{\phi_1}, \mathscr{F}_{\phi_2}) \ .$$

Similar for $C = \mathscr{F}_\phi$ the rank $r(\phi) = dim_{\overline{\mathbb{Q}}_l} Hom_{D_c^b(Y)}(\overline{\mathbb{Q}}_{lY}, \pi^*(\mathscr{F}_\phi))$ is equal to $dim_{\overline{\mathbb{Q}}_l} Hom_{D_c^b(X)}(\bigoplus_\chi \mathscr{F}_\chi \otimes_{\overline{\mathbb{Q}}_l} V_\chi, \mathscr{F}_\phi)$ by the adjunction formula. Hence

$$r(\phi) = \sum_\chi deg(\chi)\cdot dim_{\overline{\mathbb{Q}}_l} Hom_{D_c^b(X)}(\mathscr{F}_\chi, \mathscr{F}_\phi) \geq deg(\phi) \ .$$

Now it is easy to see, that the character formula $\#W = \sum_{\phi\in\hat{W}} deg(\phi)^2$ and the inequality $r(\phi) \geq deg(\phi)$ imply $r(\phi) = deg(\phi)$ for all $\phi \in \hat{W}$. Furthermore we obtain, that $\mathscr{F}_{\phi_1} \cong \mathscr{F}_{\phi_2}$ are isomorphic as $\overline{\mathbb{Q}}_l$-sheaves iff $\phi_1 \cong \phi_2$ are isomorphic as representations of W. More precisely $dim_{\overline{\mathbb{Q}}_l} Hom_{D_c^b(X)}(\mathscr{F}_{\phi_1}, \mathscr{F}_{\phi_2}) = 1$, if $\phi_1 \cong \phi_2$, and is zero else. In particular the sheaves $\mathscr{F}_\phi$ are irreducible smooth sheaves.

e) Let $E \overset{\pi}{\to} S$ be a vector bundle. Assume, that the finite group W acts on E and S so that π is equivariant

$$\begin{array}{ccc} E & \xrightarrow{\sigma} & E \\ \downarrow{\scriptstyle\pi} & & \downarrow{\scriptstyle\pi} \\ S & \xrightarrow{\sigma} & S \end{array} \qquad \sigma \in W \ .$$

Then $E \to S$ is called an *equivariant vector bundle* iff the fiber mappings $\sigma : E_s \longrightarrow E_{\sigma(s)}$ for $s \in S$ and $\sigma \in W$ are linear. If $E \to S$ is equivariant, then also the dual bundle $E' \to S$. The action $\sigma : E'_s \longrightarrow E'_{\sigma(s)}$ on E', for $s \in S, \sigma \in W$, is given in terms of the fibers by $\langle \sigma(y), x \rangle = \langle y, \sigma^{-1}(x) \rangle$, where $y \in E'_s$ and $x \in E_{\sigma(s)}$. By the permanence properties of the Fourier transform (Theorem III.13.2, III.13.3) there are natural isomorphisms

$$(*) \quad T_\psi(\sigma^*(K)) \cong \sigma^*(T_\psi(K)), \qquad \sigma \in W,\ K \in D_c^b(E, \overline{\mathbb{Q}}_l).$$

With their help one constructs a canonical W-action on

$$T_\psi(K) \in D_c^b(E', \overline{\mathbb{Q}}_l)$$

for each complex K with W-action. This construction is consistent with the Fourier inversion formula.

f) Suppose that $(K^\bullet, (\varphi_\sigma)_{\sigma \in W})$ is a W-equivariant complex on X. Furthermore let $\gamma : L^\bullet \cong K^\bullet$ be an isomorphism in the derived category $D_c^b(X, \overline{\mathbb{Q}}_l)$. Then $(L^\bullet, (\gamma^{-1}\varphi_\sigma \sigma^*(\gamma))_{\sigma \in W})$ is a W-equivariant complex on X.

Lemma 15.4 *Let $f : X \to S$ be an equidimensional smooth morphism with geometrically irreducible fibers. Especially, all fibers are nonempty. Suppose a finite group W acts on X over S, i.e. such that f is W-equivariant and such that W acts trivially on S. Let $K^\bullet \in D_c^b(S, \overline{\mathbb{Q}}_l)$ be a complex, such that there is an integer n with $K^\bullet[n] \in Perv(S)$. Assume W acts on the pullback $f^*(K^\bullet)$ compatible with the action of W on the scheme X. Then there is a representation of W on $K^\bullet$, i.e. there exists a group homomorphism*

$$W \longrightarrow \mathrm{Aut}_{D_c^b(S, \overline{\mathbb{Q}}_l)}(K^\bullet),$$

such that the action of W on $f^*(K^\bullet)$ is obtained from that representation of W by pullback.

Proof. We abbreviate $L^\bullet = f^*(K^\bullet) \in D_c^b(X, \overline{\mathbb{Q}}_l)$. Then according to III.7.2 and III.7.8 of Chap. III we have $\mathrm{End}_{D_c^b(S, \overline{\mathbb{Q}}_l)}(K^\bullet) = \mathrm{End}_{D_c^b(X, \overline{\mathbb{Q}}_l)}(L^\bullet)$, hence

$$\mathrm{Aut}_{D_c^b(S, \overline{\mathbb{Q}}_l)}(K^\bullet) = \mathrm{Aut}_{D_c^b(X, \overline{\mathbb{Q}}_l)}(L^\bullet).$$

For every group element $\sigma \in W$ the complex $\sigma^*(L^\bullet)$ is canonically isomorphic to the complex $L^\bullet$

$$\alpha(\sigma) : L^\bullet \xrightarrow{\cong} \sigma^*(L^\bullet) \quad , \quad \sigma \in W.$$

These isomorphisms $\alpha(\sigma)$ satisfy cocycle conditions $\sigma^*(\alpha(\tau))\alpha(\sigma) = \alpha(\sigma\tau)$ for the obvious reasons. Furthermore, for any homomorphism $h : K^\bullet \to K^\bullet$ in $D_c^b(S, \overline{\mathbb{Q}}_l)$ the pullback $f^*(h) = \tilde{h} : L^\bullet \to L^\bullet$ of this homomorphism satisfies

$$(*) \qquad \sigma^*(\tilde{h}) = \alpha(\sigma) \circ \tilde{h} \circ \alpha(\sigma)^{-1}.$$

Since the given group action of W on $L^\bullet$ is compatible with the action on the scheme X, this gives rise to a family of isomorphisms $\varphi_\sigma : \sigma^*(L^\bullet) \to L^\bullet$ for $\sigma \in W$ as defined in Definition III.15.1. Then $c(\sigma) = \varphi_\sigma \circ \alpha(\sigma)$ is an automorphism of $L^\bullet$

$$c(\sigma) : L^\bullet \xrightarrow{\cong} L^\bullet \quad , \quad \sigma \in W .$$

The cocycle conditions and the identity $(*)$ imply $\varphi_\sigma \circ \alpha(\sigma) \circ \varphi_\tau \circ \alpha(\tau) = \varphi_\sigma \circ \sigma^*(\varphi_\tau \circ \alpha(\tau)) \circ \alpha(\sigma) = \varphi_{\sigma\tau} \circ \alpha_{\sigma\tau}$. Hence $c(\sigma\tau) = c(\sigma) \circ c(\tau)$ for all $\sigma, \tau \in W$. Since $\mathrm{Aut}(K^\bullet) = \mathrm{Aut}(L^\bullet)$ there is a unique automorphism

$$c_0(\sigma) : K^\bullet \to K^\bullet$$

such that $f^*(c_0(\sigma)) = c(\sigma)$. Furthermore $c(e) = \mathrm{Id}_{L^\bullet}$ and $c_0(e) = \mathrm{Id}_{K^\bullet}$ for the unit element e of W. It follows, that

$$c_0(\sigma\tau) = c_0(\sigma) \circ c_0(\tau) \quad , \quad \sigma, \tau \in W$$

defines a representation of W, i.e. a homomorphism $c_0 : W \to \mathrm{Aut}(K^\bullet)$. The pullback of this action on $K^\bullet$ is the action of W on $L^\bullet$. This proves the lemma. □

Now let G be a general group again. Also assume now, that G acts on X from the left. Let $m : G \times G \to G$ be the multiplication and let $e : X \to G \times X$ be the zero section. Let $a : G \times X \to X$ denote the left action on X and let p_2 denote the projection $p_2 : G \times X \to X$. A complex $K^\bullet \in D_c^b(X, \overline{\mathbb{Q}}_l)$ is called G-equivariant, if there exists an isomorphism φ in $D_c^b(G \times X, \overline{\mathbb{Q}}_l)$

$$\varphi : a^*(K^\bullet) \cong p_2^*(K^\bullet)$$

which is rigidified along the zero section $e^*(\varphi) = id_{K^\bullet}$, such that furthermore the cocycle condition $(m \times id_X)^*(\varphi) = p_{23}^*(\varphi) \circ (id_G \times a)^*(\varphi)$ holds over $G \times G \times X$. To be precise, we demand commutativity of the diagram

$$\begin{array}{ccc}
(p_2 \circ (id_G \times a))^*(K) & =\!=\!=\!=\!= & (a \circ p_{23})^*(K) \\
\Big\uparrow {\scriptstyle (id_G \times a)^*(\varphi)} & & \Big\downarrow {\scriptstyle p_{23}^*(\varphi)} \\
(a \circ (id_G \times a))^*(K) & & (p_2 \circ p_{23})^*(K) \\
\Big\| & & \Big\| \\
(a \circ (m \times id_X))^*(K) & \xrightarrow{(m \times id_X)^*(\varphi)} & (p_2 \circ (m \times id_X))^*(K)
\end{array}$$

for the complex $K = K^\bullet$.

Connected Groups. From now on, we consider the case of a smooth geometrically connected algebraic group G of dimension d over the base field. Suppose, that the complex $K = K^\bullet$ is a perverse sheaf $K \in Perv(X)$. If there exists an isomorphism $\varphi : a^*(K) \cong p_2^*(K)$, it can be easily modified to become rigidified. Simply replace φ by $p_2^*(i^*(\varphi)^{-1})\circ\varphi$, where $i : X \to G \times X$ is the zero section. Note that $i^*a^*(K) = K$ and $i^*p_2^*(K) = K$, thus $i^*(\varphi)$ is an isomorphism $i^*(\varphi) : K \to K$ of K.

So suppose φ is rigidified along the zero section. For a rigidified isomorphism φ the cocycle condition is automatically satisfied, since K was supposed to be a perverse sheaf. In fact, a deviation from the cocycle condition induces an automorphism of the perverse sheaf $(p_2[d]\circ p_{23}[d])^*(K) = p_3^*[2d](K)$ on $G\times G\times X$. This automorphism is the identity over the zero section $X \to G \times G \times X$, hence it is the identity. Since the projection $p_3 : G \times G \times X \to X$ is smooth with connected fibers, this follows from lemma III.11.2. This implies the cocycle condition.

Suppose G is a linear connected algebraic group over the base field k. Suppose H is a geometrically connected closed algebraic subgroup of G. Then the quotient $X = G/H$ exists. If k is a separable closed field, then any G-equivariant perverse sheaf K on X is constant as a perverse sheaf, i.e. $K \cong \overline{\mathbb{Q}}_l^{\,r}[dim(X)]$ holds for some integer r. To show this it is enough to consider the case $H = 1$ by lemma III.11.2, since the quotient map $G \to X = G/H$ is smooth and equidimensional with connected fibers. But for $H = 1$ the action a is the multiplication $m : G \times G \to G$. Let $i : G \to G \times G$ be the inclusion $i(g) = (g, e)$. Then $K = id_G^*(K) = i^*a^*(K) \cong i^*p_2^*(K) = \overline{\mathbb{Q}}_{lG} \otimes_{\overline{\mathbb{Q}}_l} K_0$, where K_0 is the stalk at the unit element of G. Hence K is a constant perverse sheaf.

Suppose G is a linear connected algebraic group over the field k. Suppose K is a G equivariant perverse sheaf on $G \times X$, where G acts trivially on X and acts by left multiplication on G. Let $p_2 : G \times X \to X$ be the projection. Then $K \cong p_2^*[dim(G)](L)$ for some $L \in Perv(X)$. This is a variant of the argument used above: $L = i^*[-dim(G)](K)$ for the zero section $i : X \to G \times X$. Using III.11.3 one can generalize this to obtain

Lemma 15.5 *Let $f : X \to Y$ be a locally trivial principal homogeneous G-bundle, then every G-equivariant perverse sheaf on X is of the form $f^*(L)[dim(G)]$ for some perverse sheaf L on Y.*

The following properties of a G-equivariant perverse sheaves for smooth connected algebraic groups G are left as exercises:

Exercise 15.6 If $f : X \to Y$ is a G-equivariant morphism for G-varieties X and Y, then ${}^pH^i(Rf_!K)$ and ${}^pH^i(f^*L)$ are G-equivariant for G-equivariant perverse sheaves K, L on X respectively Y. Any subquotient of a G-equivariant perverse sheaf is G-equivariant.

Hint for the Last Assertion. Apply Lemma III.11.2 for the morphisms a and p_2 and use Lemma III.11.1 to study the Jordan-Hölder series. See also the proof of Theorem III.11.3. Finally use pullback by the zero section.

III.16 Kazhdan-Lusztig Polynomials

This section contains a brief outline of the most basic properties of the Kazhdan-Lusztig polynomials attached to a finite Coxeter group W. Where possible, we tried to give self contained proofs. The results of this section will not be applied elsewhere in this book.

Let κ be a finite field with q elements. We fix an isomorphism $\tau : \overline{\mathbb{Q}}_l \cong \mathbb{C}$. For a finitely generated scheme X_0 over κ consider the abelian category of τ-mixed $\overline{\mathbb{Q}}_l$-sheaves on X_0 of integral weight, i.e. the standard core of $D^b_{mixed}(X_0, \overline{\mathbb{Q}}_l)$. Its objects will be simply called $\overline{\mathbb{Q}}_l$-sheaves in this section. Alternatively consider the abelian category of τ-mixed Weil sheaves on X_0 of integral weight, called Weil sheaves in this section. Fix $q^{1/2} \in \overline{\mathbb{Q}}_l$ via the isomorphism $\tau : \overline{\mathbb{Q}}_l \cong \mathbb{C}$. Then half integral Tate twists $\mathscr{F}_0(\frac{1}{2})$ of Weil sheaves $\mathscr{F}_0$ are defined.

$\mathbb{Z}[F]$. For $X_0 = Spec(\kappa)$ let $K_0(X, F_X)$ denote the Grothendieck group of the abelian category of Weil sheaves in this sense. $K_0(X, F_X)$ can be identified with the group $\mathbb{Z}[F]$ of all finite sums $Q = \sum_\alpha n_\alpha \alpha$, where $n_\alpha \in \mathbb{Z}$ and α runs over elements in $\overline{\mathbb{Q}}_l^*$, such that $|\tau(\alpha)|^2 = q^m$ for some $m \in \mathbb{Z}$. Upper weights can be defined by $w(0) = -\infty$ and otherwise

$$w(\sum_\alpha n_\alpha \alpha) \;=\; max_\alpha\big(log_q|\tau(\alpha)|^2\big) ,$$

where the maximum is over all α such that $n_\alpha \neq 0$. The tensor product induces a ring structure on $\mathbb{Z}[F]$ with unit element 1 given by $\alpha = 1$. We also consider the subring R generated by $\alpha = q^{n/2}$ for $n \in \mathbb{Z}$

$$R = \mathbb{Z}[q^{1/2}, q^{-1/2}] \subset \mathbb{Z}[F] .$$

Let $p^*(\sum_\alpha n_\alpha \alpha) = \sum_\alpha n_\alpha$ be the augmentation.

Define $D : \mathbb{Z}[F] \to \mathbb{Z}[F]$ by $D(\sum n_\alpha \alpha) = \sum n_\alpha \alpha^{-1}$. An element Q in $Z[F]$ is called pure of weight w, if $w(Q) \leq w$ and $w(D(Q)) \leq -w$. Any Q with $w(Q) \leq w$ can be uniquely written in the form

$$Q = \nu + q^{-1/2} \cdot Q' \quad , \quad w(Q') \leq w ,$$

where $\nu \in \mathbb{Z}[F]$ is pure of weight w (integrality of weights).

Let G be a semisimple connected algebraic group over the algebraic closure k of the finite field κ. We assume G to be of Chevalley type. Suppose the corresponding group G_0 acts on X_0, such that the induced action of G on X (over the algebraic closure k of κ) satisfies

16.1 Assumptions.

a) The stabilizers are connected subgroups of G.
b) There are only finitely many G-orbits, each containing a κ-rational point.

Then by Lang's theorem each orbit $\mathscr{O} = G/G_x$ (for a fixed κ-rational point x) satisfies $\mathscr{O}(\kappa) = G(\kappa)/G_x(\kappa)$. For the G-orbit $\mathscr{O}$ in X let $d(\mathscr{O})$ denote the dimension of $\mathscr{O}$. For a G_0-equivariant Weil sheaf $\mathscr{F}_0$ on X_0 the extension $\mathscr{F}$ on X is both a G- and F_X-equivariant $\overline{\mathbb{Q}}_l$-sheaf on X. Both actions are compatible (with notations as in Chap. III §15)

$$\begin{array}{ccc} a^*(F_X^*(\mathscr{F})) & \xrightarrow[\sim]{F_X^*(\varphi)} & p_2^*(F_X^*(\mathscr{F})) \\ \Big\downarrow{\scriptstyle a^*(F^*)} & & \Big\downarrow{\scriptstyle p^*(F^*)} \\ a^*(\mathscr{F}) & \xrightarrow[\sim]{\varphi} & p_2^*(\mathscr{F}) \end{array} .$$

A G- and F_X-equivariant $\overline{\mathbb{Q}}_l$-sheaf on X is called a G-equivariant Weil sheaf, if the actions are compatible in this sense. If the Weil sheaf structure $F^* : F_X^*(\mathscr{F}) \xrightarrow{\sim} \mathscr{F}$ on $\mathscr{F}$ is understood from the context, by abuse of notation we also write $w(\mathscr{F})$ for the weight $w(\mathscr{F}_0)$.

The Grothendieck Group. In the situation above let $K_0(G, X, F_X)$ denote the Grothendieck group of the abelian category of all G-equivariant Weil sheaves. In the following we will define two standard bases of this Grothendieck group, the **T** and the **I** basis. These two bases are related by "triangular" coordinate transformations. In the remaining part of this section this will be applied in two related situations, where the coefficients of the coordinate transformation define the KAZDHAN-LUSZTIG polynomials up to some normalization.

The T-Basis. For each orbit define a G_0-equivariant Weil sheaf on X_0 by

$$T_{\mathscr{O}_0} = \overline{\mathbb{Q}}_{l\,\mathscr{O}_0}\Big(\frac{d(\mathscr{O})}{2}\Big)$$

extended from $\mathscr{O}_0$ to X_0 by zero. The corresponding Weil sheaf has upper weight $w(T_{\mathscr{O}}) = -d(\mathscr{O})$. The class of $T_{\mathscr{O}}$ in the Grothendieck group is denoted $\mathbf{T}_{\mathscr{O}}$.

Under the assumptions a) and b) above $K_0(G, \mathscr{O}, F_X) \cong \mathbb{Z}[F]$. The isomorphism is induced by restriction to the fixed κ-rational base point x of the orbit $\mathscr{O}$. Similarly

Lemma 16.2 *Under the assumptions above the classes* $\mathbf{T}_{\mathscr{O}}$ *form a* $\mathbb{Z}[F]$ *basis of the Grothendieck group of* G_0*-equivariant Weil sheaves on* X_0

$$K_0(G, X, F_X) \cong \bigoplus_{\mathscr{O}} \mathbb{Z}[F] \cdot \mathbf{T}_{\mathscr{O}} .$$

The classes of $\overline{\mathbb{Q}}_{l\overline{\mathcal{O}}_0}(\frac{d(\mathcal{O})}{2})$, corresponding to the closures $\overline{O}$ of the orbits $\mathcal{O}$, define another natural basis of the Grothendieck group. However, there is a better choice:

Consider the category whose objects are complexes in $D_c^b(X, \overline{\mathbb{Q}}_l)$, together with a compatible G-equivariant and F_X-equivariant structure. Morphisms are assumed to be G and F_X equivariant. This need not define a triangulated subcategory, but this category is stable under Verdier duality and the other functors. The subcategory, where the complexes in addition are pure complexes of weight zero, is denoted $H(G, X, F_X)$. If we forget the Weil sheaf structure, by Gabber's theorem each object I in $H(G, X, F_X)$ is a finite sum of translates of G-equivariant perverse sheaves. Since the cohomology sheaves of I are G-equivariant, we can define the class of I

$$cl(I) = \sum_\nu (-1)^\nu cl(\mathcal{H}^\nu(I))$$

in the Grothendieck group $K_0(G, X, F_X)$.

The I-Basis. Consider the inclusions $\mathcal{O} \overset{j}{\hookrightarrow} \overline{\mathcal{O}} \overset{i}{\hookrightarrow} X$. Let

$$I_{\mathcal{O}} = i_* j_{!*} \overline{\mathbb{Q}}_{l\mathcal{O}} \langle d(\mathcal{O}) \rangle \quad , \quad w(I_{\mathcal{O}}) = 0$$

be the intermediate extension of $\overline{\mathbb{Q}}_{l\mathcal{O}}\langle d(\mathcal{O})\rangle$ to X, where $\langle m \rangle = [m](\frac{m}{2})$. In the obvious way this defines a G-equivariant Weil complex on X_0, which is pure of weight zero. Let $\mathbf{I}_{\mathcal{O}} = (-1)^{d(\mathcal{O})} cl(I_{\mathcal{O}})$ be its class in the Grothendieck group. Then

$$(*) \qquad \mathbf{I}_{\mathcal{O}} = \sum_{\mathcal{O}' \subset \overline{\mathcal{O}}} Q_{\mathcal{O}'\mathcal{O}}(F) \cdot \mathbf{T}_{\mathcal{O}'} \ ,$$

for some $Q_{\mathcal{O}'\mathcal{O}} \in \mathbb{Z}[F]$ such that $Q_{\mathcal{O}\mathcal{O}}(F) = 1$ and

Lemma 16.3 *The weights of the coefficients $Q_{\mathcal{O}'\mathcal{O}}$ in $\mathbb{Z}[F]$ satisfy $w(Q_{\mathcal{O}'\mathcal{O}}) < 0$ for $\mathcal{O}' \subsetneqq \overline{\mathcal{O}}$, if $\mathcal{O}' \neq \mathcal{O}$.*

Proof. $I_{\mathcal{O}}$ is pure of weight zero. In particular $w(\mathcal{H}^{-\nu}(I_{\mathcal{O}})) \leq -\nu$. Furthermore $-\nu < -d(\mathcal{O}')$ if $\mathcal{H}^{-\nu}(I_{\mathcal{O}})|\mathcal{O}' \neq 0$. Since the cohomology sheaves $\mathcal{H}^{-\nu}(I_{\mathcal{O}})|\mathcal{O}'$ are G-equivariant, this follows from the support condition III.9.3 satisfied by a intersection cohomology sheaf. Since $w(T_{\mathcal{O}'}) = -d(\mathcal{O}')$ and $w(\mathcal{H}^{-\nu}(I_{\mathcal{O}})|\mathcal{O}') \leq -\nu < -d(\mathcal{O}')$, the lemma immediately results from $cl(\mathcal{H}^{-\nu}(I_{\mathcal{O}}))|\mathcal{O}') \in \mathbb{Z}[F] \cdot \mathbf{T}_{\mathcal{O}'}$. □

The $\mathbf{I}_{\mathcal{O}}$ form a $\mathbb{Z}[F]$-basis of the Grothendieck group $K_0(G, X, F_X)$. Verdier duality induces a map D on $K_0(G, X, F_X)$, for which this basis is self dual

$$D(\mathbf{I}_{\mathcal{O}}) = \mathbf{I}_{\mathcal{O}} \quad , \quad D(\sum n_\alpha \alpha) = \sum n_\alpha \alpha^{-1} \ .$$

Inductively we deduce from $(*)$ – by inverting a triangular matrix – the similar relations

$$(**) \qquad \mathbf{T}_{\mathcal{O}} = \mathbf{I}_{\mathcal{O}} + \sum_{\mathcal{O}' \subsetneq \overline{\mathcal{O}}} S_{\mathcal{O}'\mathcal{O}}(F) \cdot \mathbf{I}_{\mathcal{O}'} ,$$

for some $S_{\mathcal{O}'\mathcal{O}}(F) \in \mathbb{Z}[F]$ with upper weight $w(S_{\mathcal{O}'\mathcal{O}}) < 0$.

Lemma 16.4 *The elements $\mathbf{I}_{\mathcal{O}}$ in the Grothendieck group are uniquely determined by self duality and the equations $(*)$ together with the weight conditions of Lemma III.16.3.*

Proof. For another choice $\mathbf{I}'_{\mathcal{O}}$ we get $\mathbf{I}'_{\mathcal{O}} = \mathbf{I}_{\mathcal{O}} + \sum_{\mathcal{O}' \subsetneq \overline{\mathcal{O}}} R_{\mathcal{O}'\mathcal{O}}(F)\mathbf{I}_{\mathcal{O}'}$ from $(**)$, such that weight $w(R_{\mathcal{O}'\mathcal{O}}) < 0$ holds. Self duality forces $R_{\mathcal{O}'\mathcal{O}} = 0$. □

Correspondences on $\mathscr{B}$

Let B_0 be a Borel group of G_0 defined over κ and let $T_0 \subset B_0$ be a maximal torus in B_0 defined over κ. By assumption this is a split torus of dimension say r. Let $\mathscr{B}_0 = G_0/B_0$ denote the flag variety. Let k denote the algebraic closure of κ and let G, B, T denote the corresponding groups over k. Let $W = N(T)/T$ denote the Weyl group. For $\sigma \in W$ let $l(\sigma)$ be the length of σ (with respect to B). Let σ_0 denote the element of maximal length $l(\sigma_0) = N$ where $N = dim(\mathscr{B})$, or $N = dim(B) - r$.

G-Orbits and Bruhat Decomposition. Consider $X = \mathscr{B}$ and $Y = X \times X$. The G-orbits Y_σ on Y are in one to one correspondence with the elements $\sigma \in W$ of the Weyl group. We write $\sigma' < \sigma$ if $\sigma' \neq \sigma$ and $Y_{\sigma'} \subset \overline{Y}_\sigma$ (**Bruhat ordering**). Consider the G-equivariant map

$$Y \to B \setminus G/B \cong W ,$$

with trivial action on W, defined by $(g_1B, g_2B) \mapsto Bg_1^{-1}g_2B$. Then $Y_\sigma = \{(gB, g\sigma B) \,|g \in G\}$ is the inverse image of the double coset $B\sigma B$ in $B \setminus G/B$. Note

$$Y = \bigsqcup_{\sigma \in W} Y_\sigma \quad , \quad Y_\sigma \cong G/(B^\sigma \cap B) ,$$

where $B^\sigma = \sigma B \sigma^{-1}$. In fact $B_\sigma = B^\sigma \cap B$ is the stabilizer of the point $(1B, \sigma B) \in Y_\sigma)$. Hence $dim(Y_\sigma) = dim(G) - r - l(\sigma\sigma_0) = N + l(\sigma)$. All assumptions of III.16.1 are satisfied for the action of G on Y. Hence we have the complexes T_{Y_σ} and I_{Y_σ}, and the two bases

$$\mathbf{T}_{Y_\sigma} \quad , \quad \mathbf{I}_{Y_\sigma}$$

of the Grothendieck group at our disposal.

Both X and Y have canonical G-actions from the left, such that the two projections $pr_i : Y \to X$ are G-equivariant. Then $Y \times_X Y \cong X \times X \times X$. Put $s = pr_{12}, t = pr_{23}$ and $m = pr_{13} : Y \times_X Y \to Y$ for the third projection. These maps are G-equivariant.

Convolution. For sheaf complexes $K_1, K_2 \in D_c^b(Y, \overline{\mathbb{Q}}_l)$ on Y define the convolution product

$$K_1 * K_2 = \mathrm{R}m_!\Big(s^*(K_1) \otimes^L t^*(K_2)\Big)\langle -N\rangle \ .$$

By the proper base change theorem $(K_1 * K_2) * K_3 \cong K_1 * (K_2 * K_3)$.

If $K_1, K_2 \in D_c^b(Y, \overline{\mathbb{Q}}_l)$ are G-equivariant, so is $K_1 * K_2$ and the associativity isomorphism is an isomorphism of G-equivariant sheaves.

16.5 Some Special Cases

1.Example. $I_{Y_{\sigma_0}} = \overline{\mathbb{Q}}_{l\,Y}\langle dim(Y)\rangle$. Then $I_{Y_{\sigma_0}} * I_{Y_{\sigma_0}} = I_{Y_{\sigma_0}} \otimes_{\overline{\mathbb{Q}}_l} H^*(\mathscr{B}, \overline{\mathbb{Q}}_l)\langle dim(\mathscr{B})\rangle$ with class $(-1)^N q^{-N/2}(\sum_{\sigma\in W} q^{l(\sigma)}) \cdot cl(I_{Y_{\sigma_0}})$. Since $dim(Y) = 2N$ is even $\mathbf{I}_{Y_{\sigma_0}} = (-1)^{dim(Y)} cl(I_{Y_{\sigma_0}}) = cl(I_{Y_{\sigma_0}})$. Hence for the longest element $\sigma_0 \in W$

$$\mathbf{I}_{Y_{\sigma_0}} * \mathbf{I}_{Y_{\sigma_0}} = (-1)^N q^{-N/2}(\sum_{\sigma\in W} q^{l(\sigma)}) \cdot \mathbf{I}_{Y_{\sigma_0}} \ .$$

2.Example. $I_{Y_1} = \overline{\mathbb{Q}}_{l\,Y_1}\langle N\rangle$ is a unit element for the convolution product; note $Y_1 \cong X \hookrightarrow X \times X$ is the diagonal, hence is closed.

3.Example. $G = Sl(2)$ and $W = \{1, s\}$. Then $X = \mathbb{P}^1$ and $Y = \mathbb{P}^1 \times \mathbb{P}^1$. Furthermore $Y_1 = \mathbb{P}^1$ is the diagonal in $Y = \mathbb{P}^1 \times \mathbb{P}^1$ and $Y_s = Y \setminus Y_1$ for $s \neq 1$ is the open complement. Since $I_{Y_s} = \overline{\mathbb{Q}}_{l\,Y}[2](1)$ and $T_{Y_s} = \overline{\mathbb{Q}}_{l\,Y_s}(1)$ and $T_{Y_1} = \overline{\mathbb{Q}}_{l\,Y_1}(\frac{1}{2})$ we have

$$\mathbf{I}_{Y_s} = \mathbf{T}_{Y_s} + q^{-1/2} \cdot \mathbf{T}_{Y_1} \ .$$

In particular the coefficient defined in Lemma III.16.3 is $Q_{Y_1,Y_s}(F) = q^{-1/2}$. By example 1 therefore $\mathbf{I}_{Y_s} * \mathbf{I}_{Y_s} = -(q^{1/2} + q^{-1/2}) \cdot \mathbf{I}_{Y_s}$. The same computation carries over for any $\mathbf{T}_{Y_s}$ and $\mathbf{T}_{Y_s}$ defined by a simple reflection s in the case of an arbitrary semisimple group G

$$\mathbf{I}_{Y_s} * \mathbf{I}_{Y_s} = (-1)^N (q^{1/2} + q^{-1/2}) \cdot \mathbf{I}_{Y_s} \ .$$

Lemma 16.6 *Suppose that both K_1, K_2 are translates of G-equivariant perverse Weil sheaves. Then convolution commutes with Verdier duality $D(K_1 * K_2) \cong D(K_1) * D(K_2)$.*

For the proof we need

Another Description

Consider

$$\begin{array}{ccccc} G \times \mathscr{B} & \xrightarrow{\ \mu\ } & G \times^B \mathscr{B} & \xrightarrow{\ \pi\ } & \mathscr{B} \\ \downarrow{\scriptstyle \nu} & & & & \\ \mathscr{B} \times \mathscr{B} & & & & \end{array} \qquad \begin{array}{ccc} \nu^*(K\langle -N\rangle) = (\pi\mu)^*(T) & \longleftarrow & T \ . \\ \uparrow & & \\ K\langle -N\rangle & & \end{array}$$

Here $\nu(g, hB) = (gB, hB)$ and $\mu(g, hB) = g^{-1} \times^B hB$ and $\pi(g^{-1} \times^B hB) = g^{-1}hB$. By III.15.5 a G-left equivariant perverse sheaf K on $Y = \mathscr{B} \times \mathscr{B}$ corresponds to a G-left, B-right equivariant perverse sheaf $\nu^*(K)$ on $G \times \mathscr{B}$ (up to twists and shift):

$$\nu^*(Perv_G(Y))\langle N + r\rangle = Perv_{G\times B}(G \times \mathscr{B}) .$$

Similarly

$$(\pi\mu)^*(Perv_B(X)\langle 2N + r\rangle = Perv_{G\times B}(G \times \mathscr{B}) .$$

Hence the

Matching Condition. $\nu^*(K)\langle N + r\rangle = (\pi\mu)^*(T)\langle 2N + r\rangle$
defines a correspondence

$$Perv_G(Y) \ni K \mapsto T = T_K \in Perv_B(X) .$$

This correspondence matches G- respectively B-equivariant perverse sheaves on Y respectively X and respects Verdier duality. In the case of Weil sheaf complexes, it matches Weil sheaf complexes which are pure of weight zero. We write $T = T_1 * T_2$, if T corresponds to $K_1 * K_2$ and the T_i correspond to the K_i.

This new convolution can be phrased directly in terms of the associated B-equivariant sheaves T_i on $X = \mathscr{B}$. Since $T_1 \boxtimes T_2$ on $X \times X$ is $B \times B$-equivariant, the sheaf $\nu^*(T_1 \boxtimes T_2) \cong \mu^*(T)$ descends to a B-equivariant sheaf on $G \times^B \mathscr{B}$. We claim

$$T_1 * T_2 \ = \ R\pi_!(T) .$$

This claim immediately implies Lemma III.16.6, since π is proper $D\, R\pi_! = R\pi_!\, D$, and since the maps μ and ν are smooth of the same relative dimension $dim(B)$. Use III.7.2. The proof of the claim itself is left as an exercise. We only give the following indication

The Dictionary. Suppose K and T are matching. Look at the functions $f^K(xB, yB)$ for $x, y \in G^F$. (See Chap.III §12). By G-equivariance it is enough to consider $x = 1, y = \sigma \in W^F$ since $G^F = B^F W^F B^F$. By the Grothendieck trace formula

$$f^{K_1 * K_2}(xB, zB) = (-1)^N q^{N/2} \sum_{y\in G^F/B^F} f^{K_1}(xB, yB) f^{K_2}(yB, zB) .$$

In terms of B^F biinvariant functions $f^{T_i}(g)$ on G^F we write $f^{T_i}(Bx^{-1}yB) = (-1)^N q^{N/2} \cdot f^{K_i}(xB, yB)$. Then the right side corresponds to convolution

$$f^{T_1 * T_2}(g) = \sum_{h\in G^F/B^F} f^{T_1}(h) f^{T_2}(h^{-1}g) .$$

The Hecke Ring $\mathscr{H}$

Consider B-equivariant complexes on the flag variety $X = \mathscr{B}$. The B-orbits are the **Bruhat cells** X_σ in X. The Bruhat cells X_σ or $\mathscr{B}_\sigma$ are defined by $B\sigma B/B \cong B/(B^\sigma \cap B)$ in $\mathscr{B}$. They are of dimension $l(\sigma)$

$$X = \bigsqcup_{\sigma \in W} X_\sigma \quad , \quad X_\sigma = B\sigma B/B \ .$$

The B action on X satisfies the assumption of III.16.1. The corresponding complexes will be denotes T_σ and I_σ, the elements in the Grothendieck group will be denoted $\mathbf{T}_\sigma, \mathbf{I}_\sigma$.

Suppose T is an object of the category $H(B, X, F_X)$, represented by a B- and F_X-equivariant complex on X. Then $T \mapsto cl(T) = \sum_\nu (-1)^\nu cl(\mathscr{H}^\nu(T)) \in K_0(B, X, F_X)$ is well defined in the Grothendieck group $K_0(B, X, F_X)$ of B-equivariant Weil sheaves. Convolution respects distinguished $(B,\ F_X)$-equivariant triangles in $D_c^b(X, \overline{\mathbb{Q}}_l)$, hence defines a ring structure on the Grothendieck group. We call this ring the **Hecke ring** $\mathscr{H}$.

The category $H(B, X, F_X)$ of B-equivariant pure weight zero complexes in the category $D_c^b(B; X, \overline{\mathbb{Q}}_l)$ may be called the Hecke category. It contains all translates $I_\sigma\langle n\rangle$ for $n \in \mathbb{Z}$ and $\sigma \in W$.

Fact (Weil Conjectures). $w(T_1 * T_2) \leq w(T_1) + w(T_2)$ holds for (B, F_X)-equivariant complexes T_i in $D_c^b(X, \overline{\mathbb{Q}}_l)$.

From the last Lemma III.16.6 and the Weil conjectures we obtain

Lemma 16.7 *The category $H(B, X, F_X)$ is stable under convolution. If T_1, T_2 are in $H(B, X, F_X)$, then $T_1 * T_2$ is in $H(B, X, F_X)$.*

For $\sigma \in W$ consider the basic Weil sheaves $T_\sigma = \overline{\mathbb{Q}}_{l X_\sigma}(\frac{d(X_\sigma)}{2})$ (extended to X by zero). Similarly $I_\sigma = i_* j_{!*}(\overline{\mathbb{Q}}_{l X_\sigma})(\frac{d(X_\sigma)}{2})$. Let the $\mathbf{T}_\sigma = cl(T_\sigma)$ and $\mathbf{I}_\sigma = (-1)^{d(X_\sigma)} cl(I_\sigma)$ be the corresponding elements in the Hecke ring $\mathscr{H} = K_0(B, X, F_X)$. The passage from Y to X gives for all $\sigma \in W$ the following matching conditions for the T and I-bases:

Lemma 16.8

1) *The perverse sheaves I_{Y_σ} on Y match with the perverse sheaves I_σ on X.*
2) *The complex $T_{Y_\sigma}[N]$ matches with the complex T_σ on X.*

This follows from the definitions and the formula $d(Y_\sigma) = N + d(X_\sigma)$. As a consequence, we get

Lemma 16.9 *The assignment $\mathbf{T}_{Y_\sigma} \mapsto (-1)^N \mathbf{T}_\sigma$, $\mathbf{I}_{Y_\sigma} \mapsto (-1)^N \mathbf{I}_\sigma$ respects the convolution products on $K_0(G, Y, F_Y)$ and $K_0(B, X, F_X)$.*

In particular from III.16.5 we deduce

$$\mathbf{I_s} = \mathbf{T}_s + q^{-1/2}\mathbf{T}_1$$

and

$$(10) \qquad \mathbf{I}_s * \mathbf{I}_s = (q^{1/2} + q^{-1/2}) \cdot \mathbf{I}_s \,.$$

BN-Pair Properties. We note the following facts

1) If T_1 is supported in X_{σ_1} and T_2 in X_{σ_2}, then $T_1 * T_2$ is supported in the union of the X_σ, for which $\sigma \in B\sigma_1 B\sigma_2 B$. Use that $g_1 B\sigma B \ni g_2 B$ and $g_2 B\sigma' B \ni g_3 B$ implies $g_1 B\sigma B\sigma' B \ni g_3 B$.

2) Suppose $l(\sigma\sigma') = l(\sigma) + l(\sigma')$. Then $a_i \in B(k)\sigma B(k), b_i \in B(k)\sigma' B(k)$ and $a_1 b_1 = a_2 b_2$ implies $a_2 = a_1 b, b_2 = b^{-1} b_1$ for some $b \in B(k)$, the converse being trivial. In particular $(B \cdot B^{\sigma^{-1}}) \cap (B \cdot B^{\sigma'}) = B$, by the special case $a_1 = \sigma, b_1 = \sigma'$. These facts are general consequences of the BN-pair properties. (In our case these statements are related to counting formulas over finite fields ; in fact for $|(B\sigma B)^F| = |B^F \sigma B^F| = q_\sigma |B^F|$ with $q_\sigma = |X_\sigma^F| = |B^F|/|(B \cap B^\sigma)^F|$ the first statement means $q_\sigma q_{\sigma'} = q_{\sigma\sigma'}$ – or since we are in the split case – $q_\sigma = q^{l(\sigma)}$. The counting formula conversely implies the statement above).

3) Abbreviate $A * B = s^{-1}(A) \cap t^{-1}(B)$ for constructible subsets A, B in Y. Then for $l(\sigma\sigma') = l(\sigma) + l(\sigma')$ we have a well defined morphism

$$m : \; Y_\sigma * Y_{\sigma'} \;\to\; Y_{\sigma\sigma'}$$

by 1). Consider its fiber over a point of $Y_{\sigma\sigma'}$. By equivariance it is enough to consider the base point $(\sigma^{-1}B, \sigma' B) \in Y_{\sigma\sigma'}$. A closed point in the fiber has the form $(\sigma^{-1}B, gB) \times (gB, \sigma' B)$ such that $g \in G(k)$ satisfies $\sigma g \in (B\sigma B)(k) = B(k)\sigma B(k)$ and similar $g^{-1}\sigma' \in B(k)\sigma' B(k)$. By 2) above with $a_1 = \sigma, b_1 = \sigma'$ we get $g \in B(k)$.

This implies that the morphism m is injective on the level of closed points. Both spaces are smooth of the same dimension over k and by G-equivariance the map m is surjective on closed points. Hence m is a purely inseparable morphism, which in fact would suffice for our purposes. However – using that the maps $G(R) \to (G/B_\sigma)(R)$ are all surjective for R-valued points of an Artin ring R – the same argument can also be applied for R-valued points. Thus the differential of m is injective. Then by Zariski's main theorem

$$m : \; Y_\sigma * Y_{\sigma'} \;\cong\; Y_{\sigma\sigma'} \,.$$

Recall that $T_{Y_\sigma} = \overline{\mathbb{Q}}_{l Y_\sigma}(\frac{dim(Y_\sigma)}{2})$ extended to Y by zero. From the isomorphism m discussed above we get

$$T_{Y_\sigma}[N] * T_{Y_{\sigma'}}[N] = T_{Y_{\sigma\sigma'}}[N] \qquad \text{if } l(\sigma\sigma') = l(\sigma) + l(\sigma') \,,$$

since $d(Y_\sigma) + d(Y_{\sigma'}) = 2N + l(\sigma) + l(\sigma') = N + d(Y_{\sigma\sigma'})$. Hence by Lemma III.16.9 we get the statement $(H1)$ of the next lemma

Lemma 16.10 *Multiplication in the Hecke algebra $\mathcal{H}$ is given by*

$$(H1) \qquad \mathbf{T}_\sigma * \mathbf{T}_{\sigma'} = \mathbf{T}_{\sigma\sigma'} \quad , \quad l(\sigma\sigma') = l(\sigma) + l(\sigma') \,.$$

$$(H2) \qquad \mathbf{T}_s * \mathbf{T}_\sigma = \mathbf{T}_{s\sigma} + (q^{1/2} - q^{-1/2}) \cdot \mathbf{T}_\sigma \quad , \quad l(s\sigma) = l(\sigma) - 1 \,.$$

Proof. For $(H2)$ write $s\sigma = ss\sigma'$ for some σ' of length $l(\sigma) - 1$. Then $(H2)$ is a consequence of $(H1)$ and the special case $\mathbf{T}_s^2 = \mathbf{T}_1 + (q^{1/2} - q^{-1/2})\mathbf{T}_s$ of (HS), where $\sigma = s$. Theses special relations are equivalent to $\mathbf{I}_s * (\mathbf{T}_s - q^{-1/2}) = (\mathbf{T}_s + q^{-1/2}) * (\mathbf{T}_s - q^{-1/2}) = 0$ and follow from $(I0)$. □

Verdier Duality

For $X = \mathscr{B}$ the B-orbits are the Bruhat cells X_σ. Consider the intermediate extensions I_σ and the basis $\mathbf{I}_\sigma = (-1)^{l(\sigma)} cl(I_\sigma)$ of the Grothendieck group (Hecke algebra). In the case of a simple reflection s recall

$$\mathbf{I_s} = \mathbf{T}_s + q^{-1/2}\mathbf{T}_1$$

and

$$\mathbf{I}_s * \mathbf{I}_s = (q^{1/2} + q^{-1/2}) \cdot \mathbf{I}_s \,.$$

The sheaf complex I_σ on Y is G-equivariant and pure of weight 0.

Verdier duality acts on the Grothendieck group $\mathbb{Z}[F]$ by $D(\sum n_\alpha \alpha) = \sum n_\alpha \alpha^{-1}$, hence $D(q^{1/2}) = q^{-1/2}$. It induces a ring homomorphism on the Hecke algebra (Lemma III.16.6). Furthermore $D(\mathbf{T}_s) = D(\mathbf{I}_s - q^{-1/2}\mathbf{I}_1) = \mathbf{I}_s - q^{1/2}\mathbf{I}_1 = \mathbf{T}_s - (q^{1/2} - q^{-1/2})\mathbf{T}_1 = \mathbf{T}_s^{-1}$. Hence by $(H1)$ for all $\sigma \in W$

$$D(\mathbf{T}_\sigma) = (\mathbf{T}_{\sigma^{-1}})^{-1} \quad , \quad D(q^{\nu/2}) = q^{-\nu/2} \,.$$

This completely describes the involutive ring automorphism D of the Hecke algebra $\mathscr{H}$. Note

$$\mathbf{T}_{\sigma_0} * D(\mathbf{T}_\sigma) = \mathbf{T}_{\sigma_0\sigma}$$

for the longest element $\sigma_0 \in W$. This follows from Lemma III.16.10 $(H1)$ by reduction to the case of simple reflections.

From Lemma III.16.3 recall

$$(*) \qquad \mathbf{I}_\sigma = \mathbf{T}_\sigma + \sum_{\sigma' < \sigma} Q_{\sigma'\sigma}(F) \cdot \mathbf{T}_{\sigma'}$$

where the $Q_{\sigma',\sigma} \in \mathbb{Z}[F]$ are coefficients of weight $w(Q_{\sigma'\sigma}(F)) \leq -1$. Multiplying by $q^{1/2} \in \mathbb{Z}[F]$, the coefficients $\nu(\sigma', \sigma) \in \mathbb{Z}[F]$ are obtained from

$$q^{1/2} \cdot Q_{\sigma',\sigma}(F) = \nu(\sigma', \sigma) + \text{terms of weight} < 0 \quad (\text{in } \mathbb{Z}[F])$$

as the pure parts of highest weight 0.

Lemma 16.11 *Suppose* $l(s) = 1$ *and* $l(s\sigma') = l(\sigma') + 1$. *Then for* $\sigma = s\sigma'$

$$(I1) \qquad \mathbf{I}_s * \mathbf{I}_{\sigma'} = \mathbf{I}_{s\sigma'} + \sum_{\sigma'' \prec \sigma'} \nu(\sigma'', \sigma') \cdot \mathbf{I}_{\sigma''} .$$

The sum is over all $\sigma'' \prec \sigma'$, which means $\sigma'' < \sigma'$ such that $l(s\sigma'') = l(\sigma'') - 1$. (Over the algebraic closure k) the intersection cohomology complexes satisfy

$$I_s * I_{\sigma'} \cong I_{s\sigma'} \oplus \bigoplus_{\sigma'' \prec \sigma'} \mu(\sigma'', \sigma') \cdot I_{\sigma''} \qquad \in D_c^b(X, \overline{\mathbb{Q}}_l) ,$$

with $\mu(\sigma'', \sigma') = (-1)^{l(\sigma)-l(\sigma'')} p^*(\nu(\sigma'', \sigma')) \in \mathbb{N}$, where $p^* : \mathbb{Z}[F] \to \mathbb{Z}$ is the augmentation.

Corollary 16.12 *Suppose $l(s) = 1$ and $l(s\sigma) = l(\sigma) - 1$. Then*

$$(I2) \qquad \mathbf{I}_s * \mathbf{I}_\sigma = (q^{1/2} + q^{-1/2}) \cdot \mathbf{I}_\sigma .$$

Proof. (Use induction on the lenght of σ). Put $\sigma' = s\sigma$. Then $l(s\sigma') = l(\sigma') + 1$ and $\sigma = s\sigma'$. Multiply the formula $(I1)$ of the last lemma by $\mathbf{I}_s$ from the left to obtain $\mathbf{I}_s^2 * \mathbf{I}_{\sigma'} = \mathbf{I}_s * \mathbf{I}_\sigma + \sum_{\sigma'' \prec \sigma'} \nu(\sigma'', \sigma') \cdot \mathbf{I}_s * \mathbf{I}_{\sigma''}$. Put $c = q^{1/2} + q^{-1/2}$. Since $l(s\sigma'') < l(\sigma'')$, and also $l(\sigma'') < l(\sigma)$ one obtains from $\mathbf{I}_s^2 = c \cdot \mathbf{I}_s$ and $\mathbf{I}_s * \mathbf{I}_{\sigma''} = c \cdot \mathbf{I}_{\sigma''}$ the statement by induction. □

We now come to the proof of Lemma III.16.11:

Proof. Put $\mathbf{I} := \mathbf{I}_s * \mathbf{I}_{\sigma'}$. Then $D(\mathbf{I}) = D(\mathbf{I}_s) * D(\mathbf{I}_{\sigma'}) = \mathbf{I}$. From (*) and $\mathbf{I}_s = \mathbf{T}_s + q^{-1/2}\mathbf{T}_1$ we obtain

$$\mathbf{I}_s * \mathbf{I}_{\sigma'} = \mathbf{T}_s * \mathbf{T}_{\sigma'} + \sum_{\sigma'' < \sigma'} Q_{\sigma'', \sigma'}(F) \cdot \mathbf{T}_s * \mathbf{T}_{\sigma''} + ...$$

plus terms with coefficients of weight < 0 in the $\mathbf{I}$-basis. Now $\mathbf{T}_s * \mathbf{T}_{\sigma'} = \mathbf{T}_\sigma$ by $(H1)$. By $(H2)$ the products $\mathbf{T}_s * \mathbf{T}_{\sigma''} = q^{1/2}\mathbf{T}_{\sigma''} + ...$ increase weight by 1, if $l(s\sigma'') = l(\sigma'') - 1$. They do not increase weights otherwise. From the weight increasing terms – with $\nu(\sigma'', \sigma')$ denoting the highest coefficient of $Q_{\sigma'', \sigma'}$, i.e the weight 0 term of $q^{1/2} Q_{\sigma'', \sigma'}$ – we get

$$\mathbf{I} = \mathbf{I}_\sigma + \sum_{\sigma'' < \sigma', s\sigma'' < \sigma''} \nu(\sigma'', \sigma') \cdot \mathbf{I}_{\sigma''} + ...$$

plus terms with coefficients in $\mathbb{Z}[F]$ of weight < 0 for the basis $\mathbf{T}_\tau$, $\tau \in W$, or alternatively the basis $\mathbf{I}_\tau$, $\tau \in W$ using (**). Self duality of $\mathbf{I}$ implies – using the latter basis – that all the additional terms are zero. This proves the statement of the lemma in the Grothendieck group. But $w(I) \leq 0$ and I is self dual, hence I_0 is pure of weight zero.

To prove the second assertion, one first verifies that $I_s * I_{\sigma'}$ is in ${}^pD^{\leq 0}(Y, \mathbb{Q}_l)$. This uses a direct inspection of the fibers of the map $m : \overline{Y}_s * \overline{Y}_{\sigma'} \to \overline{Y}_\sigma$ as in the proof of the next lemma. We skip the argument. We only mention that the IC-property of $I_{\sigma'}$ is used (Proposition III.9.3(4)), and the even stronger but trivial fact, that the cohomology sheaves of I_s are concentrated in one single degree.

Once we know $I_s * I_{\sigma'} \in {}^pD^{\leq 0}(Y, \mathbb{Q}_l)$, we conclude $I = I_s * I_{\sigma'} \in Perv(Y)$. This follows, since I is self dual (Lemma III.16.6).

Next we bring all term in $(I1)$ with negative coefficients $\mu(\sigma'', \sigma')$ to the left side. We then apply the Theorem III.12.1(3). It gives an isomorphism between perverse sheaves over k. Since after our rearrangement of $(I1)$ all irreducible constituent $I_{\sigma''}$ on the left side are not isomorphic to any remaining constituent on the right side, all $I_{\sigma''}$ on the left necessarily occur with multiplicity $\mu(\sigma'', \sigma') = 0$. This proves the second assertion of Lemma III.16.11. □

Purity Properties

A Weil complex K is said to have the purity property, if on all stalks of the semisimplifications $\mathscr{H}^\nu(K)^{ss}_x(\frac{\nu}{2})$ of the cohomology sheaves the Frobenius F_x acts trivially. In addition we say that K has even or odd parity, if all these stalks vanish unless ν has the given fixed parity mod 2.

Lemma 16.13 (Purity) *For the complexes I_σ the Frobenius F_x acts trivially on the stalks of the semisimplifications $\mathscr{H}^\nu(I_\sigma)^{ss}_x(\frac{\nu}{2})$. These stalks vanish unless ν has a parity equal to $l(\sigma)$ mod 2.*

Comment. In particular $cl(\mathscr{H}^\nu(I_\sigma)|X_{\sigma'}) = q^{\nu/2} \cdot rank\big(\mathscr{H}^\nu(I_\sigma)|X_{\sigma'}\big) \cdot cl(\overline{\mathbb{Q}}_{l\,X_{\sigma'}})$.

The class of Weil complexes K with the purity (and parity) property is closed under direct factors, tensor products, twists by $\overline{\mathbb{Q}}_l\langle n\rangle, n \in \mathbb{Z}$ and pullbacks. Concerning direct images: Suppose K_0 is a Weil complex on X_0 with this purity property. Suppose X_0 has a finite stratification, for which the restrictions of the cohomology sheaves of K_0 to the strata are geometrically smooth constant $\overline{\mathbb{Q}}_l$-sheaves. Suppose $f_0 : X_0 \to Y_0$ has geometric fibers of dimension ≤ 1 and for any closed point y of Y_0 suppose, that the reduced inverse image Z_{y0} of y (over $Spec(\kappa(y))$) contains a finite union of $\kappa(y)$-rational closed points, whose complement is an open subset $U_0 \subset Z_{0y}$, which is contained in one of the strata and which over $\kappa(y)$ is isomorphic to affine space $\mathbb{A}_0$ or empty. Under these assumptions $Rf_!K_0$ again has the purity property. Here $\kappa(y) \supset \kappa$ as usual denotes the residue field of y.

Since the reduced fibers are of the form $Z_0 = \mathbb{A}_0 \bigcup Z_0'$, with $dim(Z_0') = 0$, by the proper base change theorem and the Leray spectral sequence it is enough to show the following: Suppose $\mathscr{F}_0$ is a sheaf on Z_0 and Frobenius F_x acts trivially on its stalks $\mathscr{F}_x$, and such that $\mathscr{F}|\mathbb{A} \cong \overline{\mathbb{Q}}_{l\,\mathbb{A}}$. Then F acts on $H^\nu_c(Z, \mathscr{F})$ by $q^{\nu/2}$. However this follows from the exact sequence

$$... \to H^\nu_c(\mathbb{A}, \overline{\mathbb{Q}}_l) \to H^\nu_c(Z, \mathscr{F}) \to H^\nu_c(Z', \overline{\mathbb{Q}}_l) \to ... \quad .$$

This being said, we now come to the

Proof. The Purity Lemma is shown again by induction. Choose σ' such that $\sigma = s\sigma'$ and $l(\sigma) = l(\sigma') + 1$. According to lemma III.16.11 and induction it is enough to prove the similar statement for the pure complex $I_s * I_{\sigma'}$, since – as a sheaf complex in $D_c^b(X, \overline{\mathbb{Q}}_l)$ – I_σ is a direct factor of this complex (the factor belonging to some dense open stratum). For the proof we may replace the I_σ by the I_{Y_σ} (Lemma III.16.8). The purity statement in this case follows from Gabber's theorem by considering the natural map (a birational isomorphism)

$$m : \overline{Y}_s * \overline{Y}_{\sigma'} \to \overline{Y}_\sigma .$$

I_{Y_s} is a smooth perverse sheaf on $\overline{Y}_s$. The fibers Z of the map m are either points or of dimension one. We want determine the fibers. For this consider the closed cells $\overline{Y}_{\sigma'} = \coprod_{\sigma'' \leq \sigma'} Y_{\sigma''}$ and $\overline{Y}_s = Y_s \cup Y_1$. The map m induces an isomorphism $m : Y_1 * Y_{\sigma''} \cong Y_{\sigma''}$ and induces isomorphisms $m : Y_s * Y_{\sigma''} \cong Y_{s\sigma''}$ if $l(s\sigma'') > l(\sigma'')$. If $l(s\sigma'') < l(\sigma'')$, then

$$Y_s * Y_{\sigma''} \cong Y_s * (Y_s * Y_{s\sigma''}) = (Y_s * Y_s) * Y_{s\sigma''} \longrightarrow \Big(Y_s \cup Y_1\Big) * Y_{s\sigma''} \cong Y_{\sigma''} \cup Y_{s\sigma''}$$

and similarly

$$\overline{Y}_s * Y_{\sigma''} \cong \overline{Y}_s * Y_s * Y_{s\sigma''} \longrightarrow \Big(Y_s \cup Y_1\Big) * Y_{s\sigma''} \cong Y_{\sigma''} \cup Y_{s\sigma''} .$$

This reduces to look at the fibers of

$$\overline{Y}_s * Y_s \longrightarrow Y_s \cup Y_1 .$$

These fibers are affine lines isomorphic to $\mathbb{A}$ and this isomorphism is defined over the residue field of the base point of the fiber. Since the cohomology sheaves of I_s respectively I_σ are constant on $\overline{Y}_s$ respectively $Y_s * Y_{s\sigma''} \cong Y_{\sigma''}$, we can apply the remarks above. □

Remark 16.14 The coefficients $\mu(\sigma'', \sigma')$ and $\nu(\sigma'', \sigma')$ in $\mathbb{Z}[F]$ defined in Lemma III.16.11 are nonnegative integers

$$\nu(\sigma'', \sigma') = \mu(\sigma'', \sigma') \in \mathbb{N} .$$

They are integers, because they are D-selfdual Laurent polynomials in R of weight zero; they are nonnegative, since $l(\sigma) - l(\sigma'') \in 2\mathbb{Z}$ holds by Lemma III.16.13. In fact I_σ and I_σ occur as constituents of $I_s * I_{\sigma'}$, and therefore have the same parity.

If we take into account $cl(I_\sigma) = (-1)^{l(\sigma)} \cdot \mathbf{I}_\sigma$ and $cl(\overline{\mathbb{Q}}_{l\,X_{\sigma'}}) = q^{l(\sigma')/2} \cdot \mathbf{T}_{\sigma'}$, we obtain from a variable shift $\nu = \mu - l(\sigma)$ and the comment after the Lemma III.16.13

Corollary 16.15 *We have* $q^{l(\sigma)/2} \cdot \mathbf{I}_\sigma = \sum_{\sigma'} P_{\sigma\sigma'} \cdot q^{l(\sigma')/2} \cdot \mathbf{T}_{\sigma'}$, *where* $P_{\sigma\sigma'} = q^{(l(\sigma)-l(\sigma'))/2} \cdot Q_{\sigma\sigma'}$ *is given by*

$$P_{\sigma\sigma'} = \sum_{\mu=0}^{l(\sigma)-l(\sigma')} (-1)^\mu \cdot q^{\mu/2} \cdot rank\Big(\mathcal{H}^{\mu-l(\sigma)}\big(j_{!*}\overline{\mathbb{Q}}_{l\,X_\sigma}[l(\sigma)]\big) \mid X_{\sigma'}\Big) .$$

Proposition 16.16 $q^{l(\sigma)/2} \cdot \mathbf{I}_\sigma \in \sum_{\sigma'} \mathbb{Z}[q] \cdot q^{l(\sigma')/2} \cdot \mathbf{T}_{\sigma'}$.

Proof. This follows from the parity property (Lemma III.16.13) and the last proposition. □

Hence the $P_{\sigma'\sigma}(q)$ are polynomials in q. They are called the **Kazhdan-Lusztig polynomials**. Since $w(Q_{\sigma'\sigma}) < 0$ for $\sigma' \neq \sigma$, they are polynomials $P_{\sigma'\sigma}(q)$ of q-degree $< (l(\sigma)-l(\sigma'))/2$ unless $\sigma = \sigma'$. Their coefficient of degree $(l(\sigma)-l(\sigma')-1)/2$ is either zero or one of the coefficients $\mu(\sigma', \sigma) > 0$. In this case we must have $l(\sigma)-l(\sigma')-1 \in 2\mathbb{Z}$. Corollary III.16.15 and the last Proposition furthermore imply

Corollary 16.17 *Since the ranks in the odd degrees must vanish, the coefficients of the polynomial* $P_{\sigma'\sigma}(t)$ *are the ranks of the smooth sheaves* $\mathcal{H}^{2\mu-l(\sigma)}\big(j_{!*}\overline{\mathbb{Q}}_{l\,X_\sigma}[l(\sigma)]\big)| X_{\sigma'}$. *Hence they are nonnegative integers.*

Corollary 16.18 *There exist symmetric polynomials*

$$b(\sigma_1, \sigma_2, \sigma)(q^{1/2}) = b(\sigma_1, \sigma_2, \sigma)(q^{-1/2})$$

with nonnegative integral coefficients in $q^{1/2}$, *such that*

$$(I3) \qquad \mathbf{I}_{\sigma_1} * \mathbf{I}_{\sigma_2} = \sum_\sigma b(\sigma_1, \sigma_2, \sigma) \cdot \mathbf{I}_\sigma .$$

Proof. Any $\mathbf{I}$ with $D(\mathbf{I}) = \mathbf{I}$ is a linear combination of the $\mathbf{I}_\sigma$ with symmetric coefficients. That the coefficients are Laurent polynomials in $q^{1/2}$ with nonnegative coefficients, follows from $(I1)$ and $(I2)$ by induction using $\nu(\sigma'', \sigma') \geq 0$ from the last remark. □

Cells

1.Example. $\mathbf{I}_{\sigma_0} = \sum_\sigma q^{-l(\sigma\sigma_0)/2}\mathbf{T}_\sigma = q^{-N/2} \sum_\sigma q^{l(\sigma)/2}\mathbf{T}_\sigma$ holds for the longest element σ_0. In particular $Q_{\sigma\sigma_0}(q) = q^{(l(\sigma)-l(\sigma_0))/2}$ or respectively $P_{\sigma\sigma_0}(q) = 1$. Furthermore we have

$$f * \mathbf{I}_{\sigma_0} = 1_{\mathcal{H}}(f) \cdot \mathbf{I}_{\sigma_0}$$

for all f in the Hecke algebra $\mathscr{H}$. Here $1_{\mathscr{H}}(\mathbf{T}_\sigma) = q^{l(\sigma)/2}$ is the deformation of the trivial character of the Weyl group W. In other words for $q \mapsto 1$ the Hecke algebra becomes the group algebra of the Weyl group W and the character $1_{\mathscr{H}}$ becomes the trivial character of this algebra.

2.Example. The deformation $\varepsilon_{\mathscr{H}}(\mathbf{T}_\sigma) = (-1)^{l(\sigma)}q^{-l(\sigma)/2}$ of the ε_W-character of the Coxeter group W annihilates $\mathbf{I}_\sigma$ for all $\sigma \neq 1$ and is 1 on $\mathbf{I}_1$. This follows from $(I1)$ by induction using $1 \not< \sigma'$. Hence $\mathbf{I}_s * \sum_{\sigma\neq 1} R\cdot\mathbf{I}_\sigma \subset \sum_{\sigma\neq 1} R\cdot\mathbf{I}_\sigma$.

In fact $\{\sigma_0\}$ and $\{1\}$ are the extremal two-sided cells in the Weyl group. These two-sided cells are defined from an ascending chain of two sided ideals in the Hecke algebra $\mathscr{H}$ refining

$$0 \subset R\cdot\mathbf{I}_{\sigma_0} \subset \ldots \subset \sum_{\sigma\neq 1} R\cdot\mathbf{I}_\sigma \subset \mathscr{H} .$$

There are also the left sided cells. Define left ideals

$$\sum_{x\leq_L \sigma} R\cdot\mathbf{I}_x .$$

Here

$$x \leq_L \sigma$$

holds by definition iff a translate of the perverse sheaf I_x is a direct summand of $I_{\sigma'} * I_\sigma$ for some $\sigma' \in W$ (as perverse sheaf over k). The equivalence classes in W generated by this ordering $\leq_L$ define the left cells in W. Note $\sigma \simeq_L \sigma'$ for this equivalence relation iff $\sigma \leq_L \sigma'$ and $\sigma' \leq_L \sigma$. Each left cell defines a left module of the Hecke algebra.

The order $\leq_L$ can also be build up recursively from the elementary relations $x \leq_{KL} y$ meaning, that either a translate of I_x is a direct summand of $I_s * I_y$ for some simple reflection s or $x = y$.

Exercise (see [Cu], 5.3(ii)). $x \leq_{KL} y$ (for $x \neq y$) is equivalent to $\mu(x,y) \neq 0$ or $\mu(y,x) \neq 0$ together with $sx < x$ and $sy > y$ (in the Bruhat ordering) for some simple reflection s.

The further study of cells and their properties has deep implications, for instance for the classification of the representations of finite groups of Lie type [Lu1]. We also remark, that the positivity property of Corollary III.16.18 has some remarkable consequences for the structure of the Hecke ring in this context. We refer to [Cu] for a detailed exposition of this.

IV. Lefschetz Theory and the Brylinski–Radon Transform

IV.1 The Radon Transform

In the following assume $d \geq 1$. Let κ be a finite or an algebraically closed field. Let $\mathbb{P}^d$ be the d-dimensional projective space defined over the base field κ. Let $\check{\mathbb{P}}^d$ be the dual projective space over κ, which parameterizes the hyperplanes in $\mathbb{P}^d$. Let Y be a finitely generated scheme over κ, which will play the role of a base scheme in the following. Consider the diagram

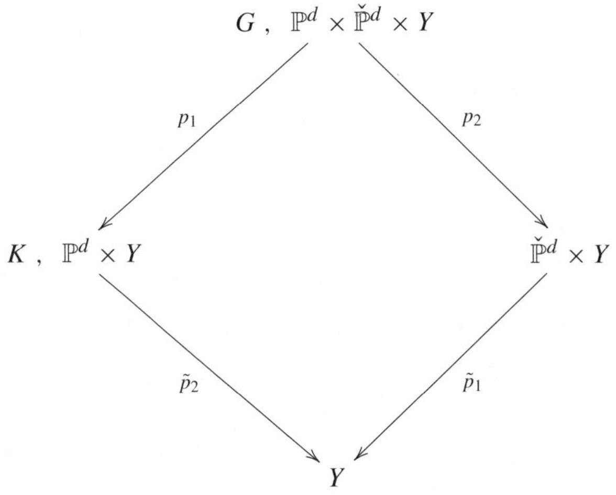

Products are fiber products over $Spec(\kappa)$, if not stated otherwise. The maps $p_1, p_2, \tilde{p}_1, \tilde{p}_2$ are the obvious projections. For a complex K on $\mathbb{P}^d \times Y$ we often write G for $p_1^*[d]K$.

For a sheaf complex K on $\mathbb{P}^d \times Y$ there are two ways to go to $\check{\mathbb{P}}^d \times Y$, which coincide by the proper base change theorem

$$Rp_{2*}(p_1^*[d]K) \cong \tilde{p}_1^*[d](R\tilde{p}_{2*}K) .$$

Put $\mathscr{K} = R\tilde{p}_{2*}(K)$. In fact we are interested in the perverse direct image sheaves

$$\mathscr{K}^n = {}^pH^n(R\tilde{p}_{2*}K)$$

(in general these are not the components of a complex representing $\mathscr{K}$ in the derived category!) and their pullbacks

$$\tilde{p}_1^*[d]\,\mathscr{K}^n \;:=\; \tilde{p}_1^*[d]\,{}^pH^n(\mathrm{R}\tilde{p}_{2*}K)\,,$$

which are perverse sheaves on $\check{\mathbb{P}}^d \times Y$; of special interest is the case, where K itself is a perverse sheaf on $\mathbb{P}^d \times Y$.

Now we consider the **universal incidence relation** $H \times Y$ over Y, where $(x,h) \in \mathbb{P}^d \times \check{\mathbb{P}}^d$ is in H if and only if the point x is contained in the hyperplane h

$$i : H \times Y \hookrightarrow \mathbb{P}^d \times \check{\mathbb{P}}^d \times Y\,.$$

The map i is the obvious inclusion. The compositions

$$\pi_1 = p_1 \circ i \quad , \quad \pi_2 = p_2 \circ i$$

are smooth maps defining $\mathbb{P}^{d-1}$-fiber bundles.

$$(*)\qquad \begin{array}{ccccc} & & \tilde{G}\,,\ H \times Y & & \\ & \swarrow^{\pi_1} & & \searrow^{\pi_2} & \\ K\,,\ \mathbb{P}^d \times Y & & & & \check{\mathbb{P}}^d \times Y \\ & \searrow_{\tilde{p}_2} & & \swarrow_{\tilde{p}_1} & \\ & & Y & & \end{array}$$

We often write $\tilde{G}$ for $\pi_1^*[d-1]K$.

Definition 1.1 *The* **Brylinski-Radon transform**

$$Rad : D_c^b(\mathbb{P}^d \times Y, \overline{\mathbb{Q}}_l) \to D_c^b(\check{\mathbb{P}}^d \times Y, \overline{\mathbb{Q}}_l)$$

is defined by

$$Rad(K) \;=\; R\pi_{2*}\pi_1^*(K)[d-1]\,.$$

For $K \in D_c^b(\mathbb{P}^d \times Y, \overline{\mathbb{Q}}_l)$ we also define the perverse sheaves

$$Rad^n(K) = {}^pH^n\,Rad(K)\,.$$

Weights. If K is mixed or τ-mixed we obviously have

$$w(Rad(K)) \le w(K) + d - 1\,.$$

Similar, if K is pure resp. τ-pure of weight w, then $Rad(K)$ is pure resp. τ-pure of weight $w + d - 1$.

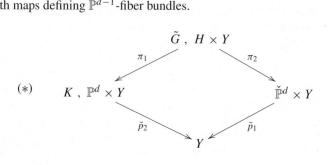

Duality. By III.7.2 the Radon transform commutes with duality up to a $(1-d)$-fold Tate twist

$$Rad(K) \xrightarrow{\cong} D\big(Rad(DK))\big)(1-d)\ .$$

Remark. The case $d = 1$ is exceptional and in a certain sense trivial. In this case $Rad = Rad^0$ is an exact functor. It defines an equivalence of categories for trivial reasons, since then π_1 and π_2 are isomorphisms.

Constant Sheaves

We will see later that – also for $d \geq 2$ – the Radon transform can almost be inverted. In fact, this is true only up to error terms which come from the base Y. A sheaf complex on $\mathbb{P}^d \times Y$, which comes from the base Y, should therefore be ignored to some extent. For this purpose let us use the following convenient terminology

Definition. A complex in $D_c^b(\mathbb{P}^d \times Y, \overline{\mathbb{Q}}_l)$ or $D_c^b(\check{\mathbb{P}}^d \times Y, \overline{\mathbb{Q}}_l)$ will be called **constant**, if it comes from the base Y by pull back via $\tilde{p}_2$ respectively $\tilde{p}_1$. A perverse sheaf on $\mathbb{P}^d \times Y$ or $\check{\mathbb{P}}^d \times Y$ will be called constant if it is isomorphic to a perverse sheaf contained in the thick abelian subcategory $\tilde{p}_2^*[d]Perv(Y)$ of the abelian category $Perv(\mathbb{P}^d \times Y)$ respectively the thick abelian subcategory $\tilde{p}_1^*[d]Perv(Y)$ of the abelian category $Perv(\check{\mathbb{P}}^d \times Y)$. For thickness use III.11.4.

In the following sections perverse sheaves on Y or perverse sheaves on $\mathbb{P}^d \times Y$ or $\check{\mathbb{P}}^d \times Y$, which are constant in this sense, will usually be written with script letters. As an exercise let us compute the Radon transform of a constant sheaf complex K. Let $K = \tilde{p}_2^*[d](\tilde{K})$ – with $\tilde{K} \in D_c^b(Y)$ – be a constant complex in the sense above. Then

$$Rad(K) = \tilde{p}_1^*[d](\tilde{K}) \otimes^L Rad(\overline{\mathbb{Q}}_l)\ .$$

More precisely, by the Lemma IV.1.3 below

$$Rad(\overline{\mathbb{Q}}_l) \ = \ \bigoplus_{i=0}^{d-1} \overline{\mathbb{Q}}_l[d-1-2i](-i)\ .$$

This follows from the well known tensor identities related to the Künneth theorem. See II.7.5. So the last formula implies, that the perverse sheaves $Rad^n(K) \cong \tilde{p}_1^*[d](\bigoplus_{i=0}^{d-1}\ {}^pH^{n+d-1-2i}(\tilde{K}(-i))\ )$ are constant for all n. Hence the Radon transform maps constant sheaf complexes to constant sheaf complexes. It is interesting to see, that properties of the Radon transform depend on the parity of the dimension d. In particular, for a constant perverse sheaf K we obtain the following

Lemma 1.2 (Parity Law) *Let $K = \tilde{p}_2^*[d](\tilde{K})$ be a constant perverse sheaf on $\mathbb{P}^d \times Y$, then*

(i) $Rad^n(K) \cong \tilde{p}_1^*[d]\tilde{K}(\frac{1-d-n}{2})$ for $n \equiv d-1 \mod 2$ and $|n| \leq d-1$.
(ii) $Rad^n(K) = 0$ else.

Lemma 1.3 *Let $p : Z \longrightarrow S$ be a $\mathbb{P}^n$-fibration of relative dimension n, i.e. a locally trivial projective fiber bundle whose fibers are projective spaces of dimension n. Let η be the cohomology class of the corresponding relative ample line bundle. For a complex L in $D_c^b(S, \overline{\mathbb{Q}}_l)$ the following map is an isomorphism*

$$\sum_i \eta^i : \bigoplus_{i=0}^n L[-2i](-i) \longrightarrow Rp_* p^*(L) .$$

Proof. The Künneth formula II.7.5 allows to reduce to the case $L = \overline{\mathbb{Q}}_l$. The classes η^i define morphisms $\overline{\mathbb{Q}}_l[-2i](-i) \to \overline{\mathbb{Q}}_l$ in the derived category, hence induced morphisms $\mathrm{R}p_*\overline{\mathbb{Q}}_l[-2i](-i) \to \mathrm{R}p_*\overline{\mathbb{Q}}_l$. Since $\overline{\mathbb{Q}}_l[-2i](-i) \cong \tau^{\leq 2i}(\mathrm{R}p_*\overline{\mathbb{Q}}_l[-2i](-i))$, we have a natural morphism $\sum \eta^i$ in the derived category, as stated in the lemma. That it actually defines an isomorphism, can be checked on geometric points. This allows to reduce to the case $S = Spec(k)$ and $L = \overline{\mathbb{Q}}_l$, where the statement is well known. □

Radon Inversion

We now introduce an analogue of the Radon transform, with the roles of $\mathbb{P}^d$ and $\check{\mathbb{P}}^d$ interchanged: Define $Rad^\vee = \mathrm{R}\pi_{1*}\pi_2^*[d-1]$ to be the **dual Radon transform**

$$Rad^\vee : D_c^b(\check{\mathbb{P}}^d \times Y, \overline{\mathbb{Q}}_l) \to D_c^b(\mathbb{P}^d \times Y, \overline{\mathbb{Q}}_l) .$$

We then get the following inversion formula

Lemma 1.4 (Radon Inversion Formula) *For every $K \in D_c^b(\mathbb{P}^d \times Y, \overline{\mathbb{Q}}_l)$ define a constant complex $\tilde{p}_2^*[d]\Phi(K) \in \tilde{p}_2^*(D_c^b(Y, \overline{\mathbb{Q}}_l))$ by*

$$\Phi(K) = \bigoplus_{i=0}^{d-2} \mathcal{K}[d-2-2i](-i) ,$$

where $\mathcal{K} = R\tilde{p}_{2*}K$. Then the following formula holds

$$(Rad^\vee \circ Rad)(K) \cong K(1-d) \oplus \tilde{p}_2^*[d]\Phi(K) .$$

Proof. We may assume $d \geq 2$. For simplicity of notation first assume that $Y = Spec(\kappa)$. By the proper base change theorem

$$(Rad^\vee \circ Rad)(K) = \mathrm{R}u_*(\Omega \otimes^L v^*(K))$$

where $u, v : \mathbb{P}^d \times \mathbb{P}^d \to \mathbb{P}^d$ are the two projections and Ω is the "operator kernel"

$$\Omega = (\mathrm{R}\pi_!(\overline{\mathbb{Q}}_l)_X)[2d-2]$$

such that $\pi : X = H \times_{\check{\mathbb{P}}^d} H \to \mathbb{P}^d \times \mathbb{P}^d$ is the inclusion of the incidence variety defined by all (x, h, x') in $\mathbb{P}^d \times \check{\mathbb{P}}^d \times \mathbb{P}^d$ for which x, x' are both in h. The variety X is smooth over $Spec(\kappa)$ (use the projection to $\check{\mathbb{P}}^d$), hence $(\overline{\mathbb{Q}}_l)_X$ is a pure complex on X of weight 0. Although the morphism π is not a smooth morphism, it is proper. Hence Ω is again a pure complex. By Gabber's Theorem III.10.6 it has to be a direct sum of irreducible perverse sheaves. These can be computed by Lemma IV.1.3

$$\Omega[2-2d] \cong \Delta_*((\overline{\mathbb{Q}}_l)_{\mathbb{P}^d})[2-2d](1-d) \oplus \bigoplus_{i=0}^{d-2} (\overline{\mathbb{Q}}_l)_{\mathbb{P}^d \times \mathbb{P}^d}[-2i](-i) ,$$

where $\Delta : \mathbb{P}^d \hookrightarrow \mathbb{P}^d \times \mathbb{P}^d$ is the diagonal embedding. Note that π is a $\mathbb{P}^{d-1}$-fibration over the image of Δ and a $\mathbb{P}^{d-2}$-fibration outside this image.

The claim of the lemma now follows from the last formula for the operator kernel Ω.

The proof for general Y is the same, since now

$$Rad^{\vee} \circ Rad(K) = \mathrm{R}u_*(w^*(\Omega) \otimes^L v^*(K)) ,$$

where $u : \mathbb{P}^d \times \mathbb{P}^d \times Y \to \mathbb{P}^d \times Y$, $v : \mathbb{P}^d \times \mathbb{P}^d \times Y \to \mathbb{P}^d \times Y$ and $w : \mathbb{P}^d \times \check{\mathbb{P}}^d \times Y \to \mathbb{P}^d \times \check{\mathbb{P}}^d$ are defined by $u(x, x', y) = (x, y)$, $v(x, x', y) = (x', y)$ and $w(x, x', y) = (x, x')$. $\square$

Exercise. Give a direct proof for the formula for Ω, which does not use the decomposition theorem.

IV.2 Modified Radon Transforms

The Brylinski-Radon transform does not preserve the category of perverse sheaves. To phrase it in a different way: For $n \neq 0$ the perverse sheaves $Rad^n(K)$ attached to a perverse sheaf $K \in Perv(\mathbb{P}^d \times Y)$ are nontrivial in general. This is already evident in the case of constant perverse sheaves K.

To analyze this in more detail, we define a modified Radon transform. For this we introduce some further notations. Consider the open complement U of the closed subscheme $H \times Y$ of $\mathbb{P}^d \times \check{\mathbb{P}}^d \times Y$

$$\begin{array}{ccccc} U & \overset{j}{\hookrightarrow} & \mathbb{P}^d \times \check{\mathbb{P}}^d \times Y & \overset{i}{\hookleftarrow} & H \times Y \\ & {\scriptstyle q} \searrow & \downarrow {\scriptstyle p_2} & \swarrow {\scriptstyle \pi_2} & \\ & & \check{\mathbb{P}}^d \times Y & & \end{array}$$

The restriction

$$q : U \longrightarrow \check{\mathbb{P}}^d \times Y$$

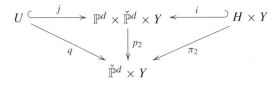

of the projection p_2 to U is an affine map; its fibers are affine spaces of dimension d. In particular, $Rq_!$ is t-left exact for the perverse t-structures by III.6.1. This implies

Lemma-Definition 2.1 *Define the* **modified Radon transform** *by*

$$Rad_!(K) = Rq_!\big(p_1^*[d](K)|U\big) \quad , \quad Rad_!^n(K) \; = \; {}^pH^n\big(Rad_!(K)\big) \; .$$

It preserves upper semi-perversity $Rad_! : {}^pD^{\geq 0}(\mathbb{P}^d \times Y) \to {}^pD^{\geq 0}(\check{\mathbb{P}}^d \times Y)$. *There is a sheaf complex homomorphism*

$$Rad(K) \to Rad_!(K)$$

with constant mapping cone

$$\tilde{p}_1^*[d]\,\mathscr{K}$$

defined by $\mathscr{K} = R\tilde{p}_{2*}(K)$.

In particular there is a long exact sequence of perverse sheaves on $\check{\mathbb{P}}^d \times Y$

$$\cdots \to Rad^\nu(K) \to Rad_!^\nu(K) \to \tilde{p}_1^*[d]\,\mathscr{K}^\nu \to Rad^{\nu+1}(K) \to \cdots \; .$$

Remark. Similar one defines the modified Radon transform $Rad_*(K) = Rq_*(p_1^*[d](K)|U)$ and $Rad_*^n(K)$. Their properties are obviously dual to those of $Rad_!(K)$.

Proof. The adjunction map $adj : id \to i_*i^*$ induces a natural distinguished triangle

$$(Rp_{2*}p_1^*(K), R\pi_{2*}\pi_1^*(K), Rq_!(p_1^*(K)|U)[1]) \; ,$$

hence the distinguished triangle $(Rad(K), Rad_!(K), Rp_{2*}p_1^*[d](K))$. Now use $Rp_{2*}p_1^*[d]K \cong \tilde{p}_1^*[d]R\tilde{p}_{2*}K$. □

We get as an immediate consequence the following

Lemma 2.2 *For upper semiperverse complex* $K \in {}^pD^{\geq 0}(\mathbb{P}^d \times Y)$ *we have the following:*

(i) For $n < 0$ *the perverse sheaves* $Rad^n(K)$ *fit into a commutative diagram*

$$\begin{array}{ccc} \tilde{p}_1^*[d]\,\mathscr{K}^{n-1} & \xrightarrow{\;\simeq\;} & Rad^n(K) \\ \downarrow \cong & & \downarrow \cong \\ {}^pH^{n-1}(Rp_{2*}(p_1^*[d]K)) & \xrightarrow{\;\simeq\;} & {}^pH^n(Rp_{2*}(i_*i^*p_1^*[d-1]K)) \end{array} .$$

In particular

$$Rad^n(K) \cong \tilde{p}_1^*[d]\,\mathscr{K}^{n-1} \quad , \quad (n < 0)$$

and for perverse K also

$$Rad^n(K) \cong \tilde{p}_1^*[d]\, \mathcal{K}^{n+1}(1) \quad , \quad (n > 0)$$

are constant perverse sheaves.

(ii) For $n = 0$ one has the following exact sequence of perverse sheaves, called **Lefschetz sequence***:*

$$0 \to {}^pH^{d-1} Rp_{2*}p_1^*(K) \to Rad^0(K) \to Rad_!^0(K) \to {}^pH^d Rp_{2*}p_1^*(K) \to Rad^1(K)$$

Note, that all the perverse sheaves

$${}^pH^\nu \mathrm{R}p_{2*}p_1^*(K) \cong \tilde{p}_1^*[d]\, \mathcal{K}^{\nu-d} \quad , \quad (\nu \in \mathbb{Z})$$

are constant perverse sheaves.

Proof. The statements for $n < 0$ and (ii) for $n = 0$ follow from IV.2.1. The statement (i) for $n > 0$ follows by duality using III.7.2. □

Definition. For perverse sheaves K on $\mathbb{P}^d \times Y$ and $G = p_1^*[d]\, K$ we define the constant sheaves $\mathcal{P}rim^n(K)$ to be the kernels of the restriction morphisms

$$\mathcal{P}rim^n(K) = Kernel\Big({}^pH^n(\mathrm{R}p_{2*}G) \to {}^pH^n(\mathrm{R}p_{2*}i_*i^*G)\Big) .$$

Evidently $\mathcal{P}rim^n(K) = 0$ for all $n < 0$ by Lemma IV.2.2.

Remark. Lemma IV.2.2 expresses this basic phenomenon that arises in *Lefschetz* theory: Statements on ${}^pH^n\mathrm{R}p_{2*}$ can be reduced to statements on ${}^pH^n\mathrm{R}\pi_{2*}$ by using restriction – i.e. the map induced by the adjunction morphism $id \to i_*i^*$. This is possible except for the two exceptional degrees $n = d - 1, d$. In the formulation used above above it is then the functor $Rad_!^0$, which controls the "critical" cokernel for $n = d - 1$ and the "critical" kernel for $n = d$.

Corollary 2.3 *For perverse sheaves K and $n \neq 0$ the perverse sheaves $Rad^n(K)$ are constant on $\check{\mathbb{P}}^d \times Y$. Also all the perverse cohomology sheaves $Rad_!^n(K)$ are constant perverse sheaves for $n \neq 0$.*

Proof. By duality, it is enough to show the first statement for $n < 0$. This case is covered by IV.2.2(i). The second assertion then follows from the long exact sequence stated in IV.2.1. □

Quotient Categories

The fact, that for perverse K and $n \neq 0$ the complex $Rad^n(K)$ is a constant perverse sheaf, obviously emphasizes the special role played by the functor Rad^0. It implies that, although Rad^0 is not an exact functor itself, it nevertheless induces an exact functor on certain abelian quotient categories.

To be more precise, consider the abelian quotient categories which are obtained by dividing the abelian categories $Perv(\mathbb{P}^d \times Y)$ resp. $Perv(\check{\mathbb{P}}^d \times Y)$ by the Serre subcategories of constant perverse sheaves $\tilde{p}_2^*[d]Perv(Y)$ resp. $\tilde{p}_1^*[d]Perv(Y)$. Then Lemma IV.1.2 and IV.2.3 and the long perverse cohomology sequences imply, that the functor Rad^0 induces an exact functor

$$Perv(\mathbb{P}^d \times Y)/\tilde{p}_2^*[d]Perv(Y) \ \to \ Perv(\check{\mathbb{P}}^d \times Y)/\tilde{p}_1^*[d]Perv(Y) \ ,$$

$$K \to K^\vee \ .$$

In fact, both functors Rad^0 and $Rad^0_!$ induce the same functor on the level of these quotient categories.

Remark. The abelian categories of perverse sheaves are noetherian and artinian abelian categories. Let **A** be such an abelian category and let **B** be a Serre subcategory. For the obvious reasons the subcategory **B** will be called the full subcategory of constant objects. The exact quotient functor

$$\Pi : \mathbf{A} \ \to \ \mathbf{A}/\mathbf{B}$$

admits an additive right inverse functor

$$\mathbf{A}/\mathbf{B} \ \to \ \mathbf{A} \ ,$$

which attaches to each object in the quotient category a reduced representative in **A**.

Let K be an object of **A**. Then let

$$K_s, \ K_q, \ {}_rK, \ K_r$$

denote the maximal constant subsheaf resp. quotient sheaf, the left reduced quotient K/K_s and the right reduced kernel of $K \to K_q$. K is called reduced, if it has no nontrivial constant subobject or quotient object. Then for arbitrary K in **A**

$$K_{red} := {}_r(K_r) \cong ({}_rK)_r$$

is **reduced**. The subquotient K_{red} is called the reduced representative of K. In the quotient categories defined above, every object K becomes isomorphic to its subquotient K_{red}. Two objects K, L in **A** become isomorphic in the quotient category **A**/**B** if and only if their reduced subquotients K_{red}, L_{red} are isomorphic in **A**. For more details see also II.3.3.

The Radon inversion formula IV.1.3 can be restated in the following way:

Corollary 2.4 *The exact functor Rad^0 induces an equivalence*

$$K \mapsto K^\vee$$

of the two abelian categories

$$Perv(\mathbb{P}^d \times Y)/\tilde{p}_2^*[d]Perv(Y) \approx Perv(\check{\mathbb{P}}^d \times Y)/\tilde{p}_1^*[d]Perv(Y) .$$

Convention. Suppose $K \in Perv(\mathbb{P}^d \times Y)$. Then we will consider $K^\vee$ also as an object of $Perv(\check{\mathbb{P}}^d \times Y)$ - represented by a reduced representative in $Perv(\check{\mathbb{P}}^d \times Y)$ (unique up to isomorphism). In this sense $K^\vee$ considered as a reduced perverse sheaf depends only on the reduced perverse sheaf of K. In this sense $K \mapsto K^\vee$ gives rise to a 1-1 correspondence between the reduced perverse sheaves on $\mathbb{P}^d \times Y$ and the reduced perverse sheaves on $\check{\mathbb{P}}^d \times Y$.

Extensions

Now we consider the Lefschetz sequence

$$0 \longrightarrow \tilde{p}_1^*[d]\,\mathscr{K}^{-1} \longrightarrow Rad^0(K) \longrightarrow Rad_!^0(K) ,$$

which was defined in Lemma IV.2.2, in more detail. We give another interpretation of this sequence in terms of the functor

$$K \mapsto K^\vee .$$

It will turn out that (the image of)

$\tilde{p}_1^*[d]\,\mathscr{K}^{-1}$ is the maximal constant perverse subsheaf $Rad^0(K)_s$ of $Rad^0(K)$.

In fact, this follows if we know that $Rad_!^0(K)$ is left reduced, i.e. has no constant nontrivial perverse subsheaf. This will be shown in Corollary IV.2.7 below. Let us assume this for the moment. Then the left monomorphism in the Lefschetz sequence can be identified with the inclusion monomorphism

$$Rad^0(K)_s \hookrightarrow Rad^0(K) .$$

So the Lefschetz sequence gives rise to a monomorphism

$$_rRad^0(K) \hookrightarrow Rad_!^0(K)$$

between left reduced perverse sheaves, whose cokernel is the constant perverse sheaf $\mathscr{P}rim^0(K)$. The reduced Radon transform $K^\vee = Rad^0(K)_{red}$ is a perverse subsheaf of $_rRad^0(K)$, so we can also consider the induced exact sequence

$$0 \longrightarrow K^\vee \longrightarrow Rad_!^0(K) \longrightarrow \mathscr{V}(K) \longrightarrow 0 ,$$

where $\mathscr{V}(K)$ is defined to be the cokernel, which is an interesting constant perverse sheaf on $\check{\mathbb{P}}^d \times Y$ attached to K. In fact for $d \geq 2$, this extension of the constant perverse sheaf $\mathscr{V}(K)$ by the reduced perverse sheaf $K^\vee$ turns out to be the universal left reduced constant perverse sheaf extensions of $K^\vee$ (see IV.2.8). By this we mean the following: Every short exact sequence of perverse sheaves

$$0 \longrightarrow K^\vee \longrightarrow E \longrightarrow \mathscr{C} \longrightarrow 0$$

where E is a left reduced perverse sheaf on $\check{\mathbb{P}}^d \times Y$ and where $\mathscr{C}$ is a constant perverse sheaf on $\check{\mathbb{P}}^d \times Y$, is the pullback

$$\begin{array}{ccccccccc} 0 & \longrightarrow & K^\vee & \longrightarrow & Rad_!^0(K) & \longrightarrow & \mathscr{V}^\cdot(K) & \longrightarrow & 0 \\ & & \| & & \uparrow & & \uparrow & & \\ 0 & \longrightarrow & K^\vee & \longrightarrow & E & \longrightarrow & \mathscr{C} & \longrightarrow & 0 \end{array}$$

of the extension defined by $Rad_!^0(K)$, with respect to a homomorphism of constant perverse sheaves $\mathscr{C} \to \mathscr{V}^\cdot(K)$. In particular, we get from III.11.5

Corollary 2.5 *For $d \geq 2$ and perverse K we have*

$$\mathscr{V}^\cdot(K) \cong \tilde{p}_1^*[d]\big({}^pH^{1-d}R\tilde{p}_{1*}(K^\vee)\big) =: \tilde{p}_1^*[d](\mathscr{K}^\vee)^{1-d} .$$

Of course we may then also consider $Rad_!^0(K)$ as the universal left reduced constant extension of the left reduced perverse sheaf ${}_rRad^0(K)$. In fact

$$0 \longrightarrow K^\vee \longrightarrow {}_rRad^0(K) \longrightarrow \mathscr{C} \longrightarrow 0 ,$$

where $\mathscr{C}$ is the cokernel of the natural map from the maximal constant perverse subsheaf to the maximal constant perverse quotient

$$\mathscr{C}o\mathscr{P}rim^{-1}(K) := cokernel\big(Rad^0(K)_s \to Rad^0(K)_q\big) .$$

The perverse sheaf $\mathscr{C}o\mathscr{P}rim^{-1}(K)$ is a perverse subsheaf of the constant perverse sheaf $\mathscr{V}^\cdot(K)$.

For perverse sheaves K on $\mathbb{P}^d \times Y$ we can therefore rephrase the Lefschetz sequence in terms of the exact sequences

$$0 \longrightarrow \mathscr{C}o\mathscr{P}rim^{-1}(K) \longrightarrow \mathscr{V}^\cdot(K) \longrightarrow \mathscr{P}rim^0(K) \longrightarrow 0 .$$

The Proofs

Essentially all statements made in the following will be a consequence of the next

Lemma 2.6 (Seesaw-Lemma) *For $K \in {}^pD^{\geq 0}(\mathbb{P}^d \times Y)$ we have*

(i) $R\tilde{p}_{1*}(Rad_!(K)) \cong R\tilde{p}_{2*}(K)[-d](-d)$.
(ii) $R\tilde{p}_{1*}(Rad_!(K)) \in {}^pD^{\geq 0}(Y)$.

Proof. The implication (i) $\Rightarrow$ (ii) is the fact that $R\tilde{p}_{2*}[-d]$ maps ${}^pD^{\geq 0}(\mathbb{P}^d \times Y)$ to ${}^pD^{\geq 0}(Y)$. See III.7.1. For (i) observe $\tilde{p}_1 p_2 = \tilde{p}_2 p_1$ and $\tilde{p}_1 \pi_2 = \tilde{p}_2 \pi_1$. Hence

$$\begin{array}{ccc}
(R\tilde{p}_{1*}Rp_{2*})p_1^*(K) & = & (R\tilde{p}_{2*}Rp_{1*})p_1^*(K) \\
\downarrow \text{“}adj\text{”} & & \downarrow \text{“}adj\text{”} \\
(R\tilde{p}_{1*}Rp_{2*})i_*i^*p_1^*(K) & = & (R\tilde{p}_{2*}Rp_{1*})i_*i^*p_1^*(K) \\
\| & & \| \\
R\tilde{p}_{1*}R\pi_{2*}\pi_1^*(K) & = & R\tilde{p}_{2*}R\pi_{1*}\pi_1^*(K)
\end{array}$$

The left side appears in the distinguished triangle, that is obtained by applying $R\tilde{p}_{1*}$ to the first morphism of the distinguished triangle

$$Rp_{2*}p_1^* \xrightarrow{adj} R\pi_{2*}\pi_1^* \longrightarrow Rad_!(1-d] \longrightarrow$$

– see also the proof of IV.2.1. We will see, that this direct image triangle splits.

To understand the relevant morphism “adj” look at the right side of the upper diagram. In fact, the vertical map is a projection onto a direct summand. To be more precise, by Lemma IV.1.3 the right vertical arrow induces an isomorphism – even before applying $R\tilde{p}_{2*}$ – if we restrict to the direct summand

$$\begin{array}{ccc}
\bigoplus_{i=0}^{d-1}\eta^i : \bigoplus_{i=0}^{d-1} K[-2i](-i) & \hookrightarrow & Rp_{1*}p_1^*K \\
\downarrow & & \downarrow adj \\
\bigoplus_{i=0}^{d-1}\tilde{\eta}^i : \bigoplus_{i=0}^{d-1} K[-2i](-i) & \xrightarrow{\simeq} & R\pi_{1*}\pi_1^*K
\end{array}$$

and it is zero on the image of η^d. To show this vanishing property, use truncation ${}^p\tau_{\leq 2d-2}$ and reduce to $K = \overline{\mathbb{Q}}_l$ and $S = Spec(k)$.

Therefore we conclude $R\tilde{p}_{1*}\ Rad_!(K)[-d] = R\tilde{p}_{2*}\ \eta^d\big(K[-2d](-d)\big) \cong R\tilde{p}_{2*}\ K[-2d](-d)$. This completes the proof. □

Corollary 2.7 *For $K \in {}^pD^{\geq 0}(\mathbb{P}^d \times Y)$ the perverse sheaf $Rad_!^0(K) \in Perv(\check{\mathbb{P}}^d \times Y)$ is left reduced, i.e. does not have a nontrivial constant perverse subsheaf coming from $Perv(Y)$.*

Proof. We have a distinguished triangle $(Rad_!^0(K), Rad_!(K), {}^p\tau^{\geq 1}Rad_!(K))$. Apply $R\tilde{p}_{1*}$ to it and recall $d \geq 1$. Now ${}^pH^\nu R\tilde{p}_{1*}{}^pD^{\geq 1}(\check{\mathbb{P}}^d \times Y) = 0$ for $\nu \leq -d$ by III.7.1, and ${}^pH^\nu R\tilde{p}_{1*}Rad_!(K) = 0$ for all $\nu < 0$ by Lemma IV.2.6. Therefore the perverse exact sequence

$${}^pH^{-d-1}R\tilde{p}_{1*}{}^p\tau^{\geq 1}Rad_!(K) \to {}^pH^{-d}R\tilde{p}_{1*}Rad_!^0(K) \to {}^pH^{-d}R\tilde{p}_{1*}(Rad_!(K))$$

gives ${}^{p}H^{-d}\mathrm{R}\tilde{p}_{1*}Rad_!^0(K) = 0$. So we can use III.11.3 to show that $Rad_!^0(K)$ is left reduced. □

Variation of the Theme. For $d \geq 2$ one obtains under the same assumptions as in IV.2.7 and by the same argument as in the proof of IV.2.7, that

$$ {}^{p}H^{1-d}\left(\mathrm{R}\tilde{p}_{1*}Rad_!^0(K)\right) = 0 $$

and for $d \geq 3$ also ${}^{p}H^{2-d}\mathrm{R}\tilde{p}_{1*}Rad_!^0(K) = {}^{p}H^{1-d}\mathrm{R}\tilde{p}_{1*}Rad_!^1(K)$. Since $Rad_!^0(K)$ is left reduced we get from III.11.5 for $d \geq 2$

$$ Ext^1_{Perv(\mathbb{P}^d \times Y)}(\tilde{p}_1^*[d]L, Rad_!^0(K)) = 0 $$

for all perverse sheaves $L \in Perv(Y)$.

Corollary 2.8 *Let K be a perverse sheaf on $\mathbb{P}^d \times Y$. The perverse sheaf $Rad_!^0(K)$ is a constant extension of the reduced Radon transform $K^\vee = Rad^0(K)_{red}$, i.e. there exists a constant perverse sheaf $\mathscr{V}(K)$ such that*

$$ 0 \longrightarrow K^\vee \longrightarrow Rad_!^0(K) \longrightarrow \mathscr{V}(K) \longrightarrow 0\,. $$

For $d \geq 2$ this extension defines the universal left reduced constant extension of the reduced perverse sheaf $K^\vee$, i.e. for all $L \in \tilde{p}_1^*[d]Perv(Y)$ we have

$$ Ext^1_{Perv(\mathbb{P}^d \times Y)}\left(L, K^\vee\right) = Hom_{Perv(\mathbb{P}^d \times Y)}\left(L, \mathscr{V}(K)\right). $$

Corollary 2.9

(i) The functor $Rad_!^0$ is a left exact functor from the abelian category $Perv(\mathbb{P}^d \times Y)$ to the abelian category $Perv(\check{\mathbb{P}}^d \times Y)$. All objects $Rad_!^0(K)$ are left reduced.

(ii) $Rad_!^n(L) = 0$ for $n \neq d$, if $L \in \tilde{p}_2^*[d]Perv(Y)$.

(iii) For $K \in Perv(\mathbb{P}^d \times Y)$ we have $Rad_!^0(K) = Rad_!^0(K_r)$. For $d \geq 2$ also $Rad_!^0(K) = Rad_!^0({}_rK) = Rad_!^0(K_{red})$.

(iv) For $d \geq 2$ the left exact functor

$$ \Psi = Rad_!^{\vee 0} \circ Rad_!^0(d-1) $$

is the functor, which assigns to $K \in Perv(\mathbb{P}^d \times Y)$ the universal left reduced constant extension of K_{red}.

Proof. $Rad_!^0(K)$ is always left reduced for a perverse sheaf K by IV.2.7. The statement (ii) follows from vanishing of the cohomology of affine space $\mathbb{A}^d$. (iii) follows now from (i) and (ii). Finally (iv) follows from the analog of IV.2.8 for $Rad_!^{\vee 0}$, since $\left(Rad^{\vee 0} \circ Rad^0(K)\right)_{red} = K_{red}(1-d)$ (follows now from the Radon inversion formula). □

IV.3 The Universal Chern Class

3.1 Cup Product. Let θ be the cohomology class of the relative divisor $H \times Y$ of the morphism $\mathbb{P}^d \times \check{\mathbb{P}}^d \times Y \to \check{\mathbb{P}}^d \times Y$. The class θ defines a homomorphism $\overline{\mathbb{Q}}_l \to \overline{\mathbb{Q}}_l[2](1)$ in the derived category $D_c^b(\mathbb{P}^d \times \check{\mathbb{P}}^d \times Y, \overline{\mathbb{Q}}_l)$, as explained in II §11. Via tensor product $- \otimes^L G$ this defines new morphisms

$$\theta_G : G \to G[2](1)$$

in the derived category. In the following we are interested in the induced morphisms

$$\cup\theta : {}^pH^n\mathrm{R}p_{2*}G \to {}^pH^{n+2}\mathrm{R}p_{2*}G(1) ,$$

i.e. the cup product with θ. Let us give a more explicit description of this morphism.

3.2 A Factorization. Let us make the following assumption. Suppose

$$G = p_1^*[d]K \quad , \quad K \in Perv(\mathbb{P}^d \times Y) .$$

Then for the inclusion $i : H \times Y \hookrightarrow \mathbb{P}^d \times \check{\mathbb{P}}^d \times Y$ we have a situation as in Theorem II.11.2. All assumptions made for the supplementary statement of this theorem are satisfied in the "universal" situation, which is considered above. In particular, there exists an isomorphism

$$\mu : i^*(G) \cong i^![2]G(1) .$$

Furthermore we know that $i^*[-1](G) = \pi_1^*[d-1](K) \in Perv(H)$. In loc. cit. we finally got a factorization of the morphism $G \to G[2](1)$ introduced above, namely

$$\theta_G : \quad G \xrightarrow{adj} i_*i^*G \xrightarrow{i_*(\mu)} i_*i^!G[2](1) \xrightarrow{adj} G[2](1) .$$

If we apply the functor ${}^pH^n\mathrm{R}p_{2*}$, we obtain the following commutative diagram $(*)$ for the cup product morphism

$$(*) \qquad \begin{array}{ccccc} {}^pH^n\mathrm{R}p_{2*}G & & \xrightarrow{\quad\cup\theta\quad} & & {}^pH^{n+2}\mathrm{R}p_{2*}G(1) \\ \downarrow {\scriptstyle {}^pH^n\mathrm{R}p_{2*}(adj)} & & & & \uparrow {\scriptstyle {}^pH^{n+2}\mathrm{R}p_{2*}(adj)} \\ {}^pH^n\mathrm{R}p_{2*}i_*i^*G & \xrightarrow{\cong} & {}^pH^n\mathrm{R}p_{2*}i_*i^![2](1)G & = & {}^pH^{n+2}\mathrm{R}p_{2*}i_*i^!G(1) \end{array}$$

where the first lower horizontal map ${}^pH^n Rp_{2*}i_*(\mu)$ is induced by the isomorphism μ.

(a) The left vertical morphism is the same as the n-th perverse cohomology of the morphism $\alpha_G = \mathrm{R}p_{2*}\big(adj\big)$ – where adj denotes the adjunction map $G \to i_*i^*(G)$ – that appeared in the study of the cycle class. See II.11.4.

(b) The left vertical morphism also appears in Lemma IV.2.1 as the n-th perverse cohomology of the third morphism of the distinguished triangle $\big(Rad(K),\ Rad_!(K),\ \mathrm{R}p_{2*}p_1^*[d](K)\big)$. In fact the adjunction map $id \to i_*i^*$ induces the third morphism

$$\mathrm{R}p_{2*}p_1^*[d](K) \to \mathrm{R}p_{2*}i_*i^*p_1^*[d](K) = \mathrm{R}\pi_{2*}\pi_1^*[d](K) = Rad(K)[1]\,.$$

For ${}^pH^n$ of this map we know from IV.2.2 the following

3.3 Fact. The left vertical morphism ${}^pH^n\mathrm{R}p_{2*}(adj)$ is a monomorphism for all $n \leq -1$ and an isomorphism for $n < -1$.

There are two cohomology degrees of particular interest, namely the degrees $n = -1$ and $n = 0$! We now discuss these cases in greater detail.

First Assume $n = 0$. For $n = 0$ the right vertical morphism ${}^pH^2\mathrm{R}p_{2*}(adj)$ in the diagram $(*)$ above is an isomorphism. This follows from fact IV.3.3 in the case $n = -2$ using duality. Hence

$$kernel\Big(\cup\,\theta : {}^pH^0\mathrm{R}p_{2*}G \to {}^pH^2\mathrm{R}p_{2*}G(1)\Big)$$

$$\cong kernel\Big({}^pH^0\mathrm{R}p_{2*}(adj) : {}^pH^0\mathrm{R}p_{2*}G \to {}^pH^0\mathrm{R}p_{2*}i_*i^*G\Big)$$

Corollary 3.4 *For $n = 0$ in the situation IV.3.2 and perverse sheaves K on $\mathbb{P}^d \times Y$ and $G = p_1^*[d]K$ we have*

$$kernel\Big(\cup\,\theta : {}^pH^0Rp_{2*}G \to {}^pH^2Rp_{2*}G(1)\Big) = \mathscr{P}rim^0(K)\,.$$

Now Assume $n = -1$. The description IV.3.2 (b) of the left vertical morphism in IV.3.2 gives the interpretation as the inclusion of the maximal constant perverse subsheaf; this follows from Corollary IV.2.7 and IV.2.2. By duality the right vertical morphism can then be viewed as the map to the maximal constant perverse factor sheaf.

Since the lower horizontal isomorphism in the diagram IV.3.2 necessarily maps the maximal constant subsheaf isomorphically into itself – and similar factor maximal constant perverse factor sheaf – we get

Corollary 3.5 *For $n = -1$ there exist isomorphisms, such that the following diagram – for $G = p_1^*[d]K$ – commutes*

$$\begin{array}{ccc}
{}^pH^{-1}Rp_{2*}G & \xrightarrow{\cup\theta} & {}^pH^{1}Rp_{2*}G(1) \\
\exists\downarrow\cong & & \exists\uparrow\cong \\
Rad^0(K)_s & & Rad^0(K)_q \\
\downarrow & & \uparrow \\
Rad^0(K) & = & Rad^0(K)
\end{array} \quad .$$

The lower vertical arrows are – on the left – the inclusion of the maximal constant perverse subsheaf $Rad^0(K)_s$ into $Rad^0(K)$ – respectively on the right – the quotient map onto a maximal constant quotient sheaf $Rad^0(K)_q$ of the perverse sheaf $Rad^0(K)$

$$Rad^0(K)_s \hookrightarrow Rad^0(K) \twoheadrightarrow Rad^0(K)_q \ .$$

In particular we get

$$cokernel\Big({}^pH^{-1}Rp_{2*}G \overset{\cup\theta}{\longrightarrow} {}^pH^{1}Rp_{2*}G(1)\Big) \cong \mathscr{C}o\mathscr{P}rim^{-1}(K)\ .$$

IV.4 Hard Lefschetz Theorem

In the following we use the conventions of Chap. I, §1. This means, that an index $_0$ indicates schemes, morphism etc. defined over a finite field κ. If we drop the index, we get the corresponding scheme, morphism etc. obtained by base change to k, where k is the algebraic closure of κ.

Let

$$f_0 : X_0 \longrightarrow Y_0$$

be a projective morphism over the base field κ and let $\eta_0 = \eta_0(X_0/Y_0)$ be the Chern class of the corresponding relative ample line bundle in $H^2(X_0, \overline{\mathbb{Q}}_l(1)) = Hom_{D(X_0)}\big(\overline{\mathbb{Q}}_l, \overline{\mathbb{Q}}_l[2](1)\big)$. Thus η_0 induces a morphism

$$\eta_0 : \overline{\mathbb{Q}}_l \to \overline{\mathbb{Q}}_l[2](1)$$

in the derived category. Let K_0 be a complex in $D(X_0) = D^b_c(X_0, \overline{\mathbb{Q}}_l)$. The tensor product $-\otimes^L K_0$ gives morphisms

$$\eta_0 : K_0 \longrightarrow K_0[2](1)\ ,$$

and

$$\mathrm{R}f_{0*}(K_0) \longrightarrow \mathrm{R}f_{0*}(K_0)[2](1)$$

in the derived categories. By an n-fold iteration one obtains a morphism

$$\mathrm{R}f_{0*}(\eta_0^n) : \mathrm{R}f_{0*}(K_0) \longrightarrow \mathrm{R}f_{0*}(K_0)[2n](n).$$

Theorem 4.1 (Hard Lefschetz Theorem) *Let $f_0 : X_0 \to Y_0$ be a projective morphism defined over the finite field κ. Let K_0 be a τ-pure perverse sheaf on X_0. Then for every positive integer r the homomorphisms induced by η_0^r between the perverse cohomology groups*

$$^p\!H^{-r}Rf_{0*}(K_0) \xrightarrow{\simeq} {}^p\!H^{r}Rf_{0*}(K_0)(r)$$

are isomorphisms.

Corollary 4.2 *The morphisms $L_n : {}^p\!H^n Rf_{0*}(K_0) \longrightarrow {}^p\!H^{n+2}Rf_{0*}(K_0)(1)$, which are induced by η_0 are epimorphisms for $n \geq -1$ and monomorphisms for $n \leq -1$. Furthermore*

$$^p\!H^0 Rf_{0*}(K_0) = Image(L_{-2}) \oplus Ker(L_0)\ .$$

Proof. The first two statements of the corollary are obvious consequences of the Theorem IV.4.1. Also L_0 has to be a monomorphism on $Image(L_{-2})$, hence $Image(L_{-2}) \cap Ker(L_0) = 0$. Furthermore ${}^p\!H^0\mathrm{R}f_{0*}(K_0)$ is a pure perverse sheaf, hence a direct sum of simple perverse constituents. Each such constituent M is either in $Ker(L_0)$ or – consider the simple perverse sheaf $L_0(M)$ in the image of $L_0 \circ L_{-2}$ – in $Image(L_{-2}) \oplus Ker(L_0)$. □

An important particular case of the last theorem is the following: Let $Y_0 = Spec(\kappa)$ and let $f_0 : X_0 \to Y_0$ be a smooth morphism. Furthermore let X_0 be pure of dimension d. Then $K_0 = \overline{\mathbb{Q}}_l[d]$ is a pure perverse sheaf. We get from the last theorem the isomorphisms

$$\cup\eta^r : H^{d-r}(X, \overline{\mathbb{Q}}_l) \longrightarrow H^{d+r}(X, \overline{\mathbb{Q}}_l)(r)\ ,\ r \geq 0$$

induced by the cup product with the class η^r. Here η is the pull back of the class η_0. By well known reduction methods one concludes

Corollary 4.3 *Let $f : X \to S$ be a smooth projective morphism of arbitrary schemes, not necessarily finitely generated schemes over a field. Let l be invertible on S and assume the fibers of f to be equidimensional of dimension d. Then the cohomology class η of the corresponding relatively ample line bundle induces isomorphisms*

$$\eta^r : R^{d-r} f_*\overline{\mathbb{Q}}_l \cong R^{d+r} f_*\overline{\mathbb{Q}}_l(r)$$

for all $r \geq 0$.

Proof of Theorem IV.4.1

1. Reduction to the Universal Situation. The assertion is local on Y_0 and also η_0 can be replaced by a positive multiple. Shrinking Y_0 allows to embed X_0 and Y_0 into smooth varieties over the base field. Therefore we can assume without restriction of generality, that X_0 and Y_0 are smooth over κ and that f_0 decomposes into

$$X_0 \quad \hookrightarrow \quad \mathbb{P}_0^d \times Y_0 \quad \longrightarrow \quad Y_0 \,,$$

where $\mathbb{P}_0^d$ denotes the projective space of dimension d over κ. We can then furthermore assume

$$X_0 = \mathbb{P}_0^d \times Y_0 \,,$$

and that $f_0 = (\tilde{p}_2)_0$ is the projection

$$f_0 : \mathbb{P}_0^d \times Y_0 \overset{(\tilde{p}_2)_0}{\longrightarrow} Y_0 \,,$$

thereby replacing K_0 by its direct image complex in $D_c^b(\mathbb{P}_0^d \times Y_0, \overline{\mathbb{Q}}_l)$, which remains to be a pure perverse sheaf on $\mathbb{P}_0^d \times Y_0$. So we are in a preferable universal situation.

2. A Further Pullback. We now want to replace the class η_0 by a universal Chern class. In order to achieve this, we further replace: $f_0 = (\tilde{p}_2)_0$ by its pullback $(p_2)_0$, K_0 by the pure perverse sheaf $G_0 = (p_1)_0^*(K_0)[d]$ and X_0 by $\mathbb{P}_0^d \times \check{\mathbb{P}}_0^d \times Y_0$. See the diagram below for the notations.

The pull back of the cohomology class η_0 to $\mathbb{P}_0^d \times \check{\mathbb{P}}_0^d \times Y_0$ is the cohomology class of the relative ample line bundle of the map $(p_2)_0$. Let θ_0 be the Chern class of the divisor $H_0 \times Y_0$. Although this class is not the pullback of the class η_0^r with respect to $(p_1)_0^*[d]$, it is enough to prove that

$$(**) \qquad \theta_0^r : {}^pH^{-r}\mathrm{R}p_{2*}(G_0) \longrightarrow {}^pH^r\mathrm{R}p_{2*}(G_0)(r) \,, \; r \geq 0$$

is an isomorphism. The original claim follows from this by a specialization $s_0^*[d]$ with respect to a section $s_0 : Spec(\kappa) \to \check{\mathbb{P}}_0^d$. Note, that the sheaf complexes specialize to the original complexes and the morphism also specializes to the original one. Since an isomorphism remains an isomorphism after this specialization, the original claim obviously follows.

For the proof we may finally pass to the algebraic closure k of κ

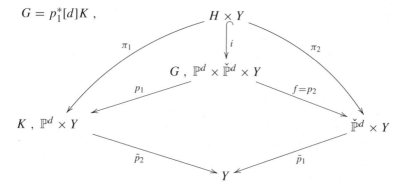

$G = p_1^*[d]K \,,$

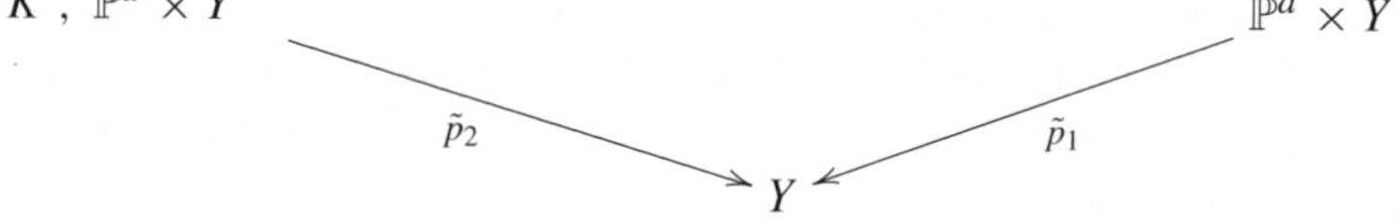

3. An Induction Argument. (Assume $r \geq 2$)
If $r \geq 2$ we use the hard Lefschetz theorem for the smooth projective map

$$(\pi_2)_0 : H_0 \times Y_0 \longrightarrow \check{\mathbb{P}}^d_0 \times Y_0 \,,$$

and the pure perverse sheaf

$$\tilde{G}_0 = i_0^*(G_0)[-1] = (\pi_1)_0^*(K_0)[d-1]$$

Recall that π_1 is equidimensional smooth of relative dimension $d-1$. By induction with respect to the relative dimension, we can assume that this case has already been established. To be more precise:

Let θ_0 be the class introduced in step 2). The description of cycle classes proved in Chap. II shows, that the map θ^r can be decomposed into the three maps

$$\begin{array}{ccc} {}^p\!H^{-r}\mathrm{R}p_{2*}G & \xrightarrow{\quad\theta^r\quad} & {}^p\!H^{r}\mathrm{R}p_{2*}G(r) \\ {\scriptstyle adj}\downarrow\cong & & {\scriptstyle adj}\uparrow\cong \\ {}^p\!H^{-(r-1)}\mathrm{R}\pi_{2*}\tilde{G} & \xrightarrow{\quad\tilde{\theta}^{r-1}\quad} & {}^p\!H^{r-1}\mathrm{R}\pi_{2*}\tilde{G}(r-1) \end{array} .$$

The first vertical map is an isomorphism by IV.2.2(i). Similar for the last map (Poincare dual of IV.2.2(i)). The lower horizontal morphism is an isomorphism by induction, since $\tilde{\theta}_0$ is the class of a relative ample line bundle for the projective morphism $H_0 \times Y_0 \to \check{\mathbb{P}}^d_0 \times Y_0$ and $\tilde{G}_0$ is a pure perverse sheaf.

4. The Essential Case. $(r = 1)$
It is enough to show that the map θ (pull back of θ_0 to the algebraic closure k of κ)

$$\theta : {}^p\!H^{-1}\mathrm{R}p_{2*}(G) \longrightarrow {}^p\!H^{1}\mathrm{R}p_{2*}(G)(1)$$

is an isomorphism. We now use properties of the Radon transform. This gives another interpretation for this map:

We have seen in IV.3.5, that it can be identified with the natural map from the maximal constant perverse subsheaf of $Rad^0(K)$ to the maximal constant perverse quotient sheaf. By duality – replace K by DK – it is enough to show, that this natural morphism is an epimorphism, i.e. that $\mathscr{C}o\mathscr{P}rim^{-1}(K) = 0$. We will give two different proofs for this fact:

First Argument. Since by assumption K_0 is a pure perverse sheaf, the same is true for $Rad^0(K_0)$. Hence $Rad^0(K)$ is semisimple by Gabber's Theorem III.10.6. This implies, that the map under consideration is an isomorphism.

Second Argument (For Simplicity $d \geq 2$). Suppose K is pure of weight say w. Then $Rad^0(K)$, hence also its perverse quotient sheaf $\mathscr{C}o\mathscr{P}rim^{-1}(K)$, is pure of weight $w + d - 1$, if it is nontrivial. On the other hand, the sheaf $\mathscr{C}o\mathscr{P}rim^{-1}(K)$ is a subsheaf of $\mathscr{V}(K) \cong \tilde{p}_1^*[d]\big({}^p\!H^{1-d}\mathrm{R}\tilde{p}_{1*}(K^\vee)\big)$, which is a pure perverse sheaf of weight $w + d$. Hence $\mathscr{C}o\mathscr{P}rim^{-1}(K) = 0$. This proves the claim. Furthermore $\mathscr{V}(K) = \mathscr{P}rim^0(K)$.

Hence the assertion is proved in case $r = 1$. This completes the proof of Theorem IV.4.1. □

IV.5 Supplement: A Spectral Sequence

Let D be a triangulated category with t-structure and the corresponding cohomology functors H^i. Suppose that every object of D is bounded, i.e. contained in $D^{\geq a}$ and $D^{\leq b}$ for some integers a, b. Let $F : D \to \mathbf{A}$ be a cohomology functor on D, i.e. an additive functor into an abelian category $\mathbf{A}$, such that for $F^i(K) = F(K[i])$ distinguished triangles in D give rise to a long exact cohomology sequence for the functors F^i. Then

Theorem 5.1 *There exists a spectral sequence*

$$E_2^{pq} := F^p H^q(X) \implies F^{p+q}(X)$$

for every object X of D.

Proof. The differentials $d_r : E_r^{pq} \to E_r^{p+r,q-r+1}$ for $r \geq 2$ arise from the more general construction of a "spectral object". See Deligne, Decompositions dans la categorie derivee [76]. □

If $H^q(X) = 0$ for $q \neq 0, 1$, the spectral sequence defined in IV.5.1 is nothing but the long exact sequence of the functors F^i attached to the distinguished triangle $(H^0(X), X, H^1(X)[-1])$.

Theorem IV.5.1 allows to obtain Leray spectral sequences in the context of perverse sheaves. In particular for $X = Rad(K)$ and $F = Rad^{\vee 0}$ we obtain

Corollary 5.2 *There exists a spectral sequence*

$$Rad^{\vee p} \circ Rad^q(K) \implies (Rad^\vee \circ Rad)^{p+q}(K)\,.$$

If we take into account the Radon inversion formula IV.1.4 we get

Corollary 5.3 *There exist spectral sequences*

$$\begin{aligned} E_2^{pq} &:= Rad^{\vee p} \circ Rad^q(K) \\ &\implies {}^p\!H^{p+q}(K)(1-d) \oplus \tilde{p}_2^*[d] \bigoplus_{\substack{n \equiv d(2) \\ |n| \leq d-2}} \mathscr{K}^{p+q+n}\left(\tfrac{n+2-d}{2}\right). \end{aligned}$$

Here we used the abbreviation $\mathscr{K}^\nu = {}^p\!H^\nu(R\tilde{p}_{2*}K)$.

Suppose that K is a perverse sheaf on $\mathbb{P}^d \times Y$. Recall that $Rad^q(K)$ is a constant perverse sheaf for all $q \neq 0$

$$Rad^q(K) =: \tilde{p}_1^*[d]\,\mathscr{R}^q \quad , \quad q \neq 0\,.$$

Therefore the left side of the spectral sequence considerably simplifies by IV.1.2. We leave it as an exercise for the reader to work out the details. Let us assume, that the base field is a finite field and that K is a pure perverse sheaf. Then the spectral sequence degenerates at the level E_2^{pq} by weight reasons. Therefore we get

5.4 Formula. Let K_0 be a pure perverse sheaf on $\mathbb{P}_0^d \times Y_0$. Then there following statements holds

(i) The perverse sheaves

$$\tilde{p}_2^*[d]\Big(\bigoplus_{\substack{q\neq 0\\ q\equiv d-1(2)}} \mathscr{R}^q(\tfrac{1-d+q}{2})\Big) \oplus \big(Rad^{\vee 0}\circ Rad^0\big)(K)$$

and

$$\tilde{p}_2^*[d]\Big(\bigoplus_{\substack{|n|\leq d-2\\ n\equiv d(2)}} \mathscr{K}^n(\tfrac{2-d+n}{2})\Big) \oplus K(1-d)$$

are isomorphic.

(ii) For all $m \neq 0$ the perverse sheaves

$$\tilde{p}_2^*[d]\Big(\bigoplus_{\substack{q\neq 0\ \&\ |m-q|\leq d-1\\ m-q\equiv d-1(2)}} \mathscr{R}^q(\tfrac{1-d-m+q}{2})\Big) \oplus Rad^{\vee m}\big(Rad^0(K)\big)$$

and

$$\tilde{p}_2^*[d]\bigoplus_{\substack{|n|\leq d-2\\ n\equiv d(2)}} \mathscr{K}^{n+m}(\tfrac{2-d+n}{2})$$

are isomorphic.

Recall the isomorphisms $\mathscr{R}^n \cong \mathscr{K}^{n-1}$ for $n < 0$ and $\mathscr{R}^n \cong \mathscr{K}^{n+1}(1)$ for $n > 0$. Using these formulas we can compare both sides of the formulas 5.4. By the semisimplicity of the category of pure perverse sheaves of fixed weight we can cancel terms, which coincide. Furthermore the pullback $\tilde{p}_2^*[d].\mathscr{K}^{-d}$ is the maximal constant perverse subsheaf of the perverse sheaf K. It therefore vanishes, if K is reduced. Furthermore $\tilde{p}_1^*[d].\mathscr{K}^{-1}$ is the maximal constant perverse subsheaf of $Rad^0(K)$. See IV.3.5. Now 5.4(i) gives the next theorem by a straight forward calculation using these facts together with the parity law IV.1.2.

Notation. Let $K^\vee$ denote the reduced Radon transform $Rad^0(K)_{red}$ of K. Define the groups

$$\big(\mathscr{K}^\vee\big)^n = {}^pH^n R\tilde{p}_{1*}(K^\vee)\,.$$

$$Rad^{\vee m}(K^\vee) = \tilde{p}_2^*[d]\big(\mathscr{R}^\vee\big)^m$$

for arbitrary n and for $m \neq 0$. Again $(\mathscr{R}^\vee)^m \cong (\mathscr{K}^\vee)^{m-1}$ for $m < 0$ etc.

Theorem 5.5 *Suppose $d \geq 2$. Let K_0 be a reduced pure perverse sheaf on $\mathbb{P}_0^d \times Y_0$.*

(a) *We get*

$$(\mathscr{K}^\vee)^{-1} \cong \begin{cases} \mathscr{K}^0(\frac{2-d}{2}) & d \equiv 0(2) \\ \mathscr{K}^1(\frac{3-d}{2}) & d \not\equiv 0(2) \end{cases}$$

(b) *Furthermore for* $0 < m \le d-2$, *hence* $2-d \le m+1-d < 0$, *we have*

$$(\mathscr{R}^\vee)^m \oplus \mathscr{R}^{m+1-d}(1-d) \cong \begin{cases} \mathscr{K}^0(\frac{2-d-m}{2}) & d \equiv m(2) \\ \mathscr{K}^1(\frac{3-d-m}{2}) & d \not\equiv m(2) \end{cases}$$

(c) *Finally for* $2-d \le m < 0$, *hence* $0 < m+d-1 \le d-2$, *we have*

$$(\mathscr{R}^\vee)^m \oplus \mathscr{R}^{m+d-1} \cong \begin{cases} \mathscr{K}^0(\frac{2-d-m}{2}) & d \equiv m(2) \\ \mathscr{K}^1(\frac{3-d-m}{2}) & d \not\equiv m(2) \end{cases}$$

These formulas express the perverse cohomology sheaves $(\mathscr{K}^\vee)^n$ of $K^\vee$ in terms of the perverse cohomology sheaves of K for all $n \neq 0$.

In the remaining part of this section let us recollect some of the information gathered in the proof of the Hard Lefschetz theorem, and formulate this in terms of the Radon transform.

We would like to emphasize that – for pure perverse sheaves K_0 as above – this gives a rather satisfactory way to express the perverse cohomology sheaves ${}^pH^n(R\tilde{p}_2K)$ in terms of the "simpler" perverse cohomology sheaves $Rad^{n+1}(K)$. Of course it is enough to consider the cases $n \le d$ using duality; and as already remarked in Lemma IV.2.2 it is also enough to restrict to the two cases $n = d-1$ and $n = d$, which are linked by the Lefschetz sequence. In fact, for pure perverse sheaves K_0 the Lefschetz sequence may be completely expressed in terms of the perverse sheaves $Rad^0(K)$ and $Rad^1(K)$ as follows

Corollary 5.6 *Suppose* $d \ge 2$ *and let* K_0 *be a pure perverse sheaf on* $\mathbb{P}_0^d \times Y_0$ *of weight* w. *Then the Lefschetz sequence is obtained by composing the following short exact sequences:*

(1) *The inclusion of the maximal constant perverse subsheaf of* $Rad^0(K)$ *into* $Rad^0(K)$

$$0 \to \tilde{p}_1^*[d]({}^pH^{d-1}R\tilde{p}_{2*}K) \to Rad^0(K) \to K^\vee \to 0\,.$$

Since $Rad^0(K_0)$ *is a pure perverse sheaf of weight* $w+d-1$, *this sequence splits by Gabber's theorem. The quotient is the reduced perverse sheaf* $Rad^0(K)_{red} = K^\vee$.

(2) *The second short exact sequence is the universal left reduced constant perverse extension*

$$0 \to K^\vee \to Rad_!^0(K) \to \mathscr{V}(K) \to 0\,,$$

where now $K_0^\vee$ *is pure of weight* $w+d-1$ *and* $\mathscr{V}(K_0)$ *is pure of weight* $w+d$.

(3) The last short exact sequence is

$$0 \to \mathscr{V}(K) \to \tilde{p}_1^*[d]({}^pH^d R\tilde{p}_{2*}K) \to Rad^1(K) \to 0 .$$

It splits, since all these perverse sheaves arise from pure perverse sheaves of weight $w + d$. *Furthermore the sheaf* $Rad_!^1(K_0)$ *is pure of weight* $w + d + 1$.

Proof. The assertions (1) respectively (2) were already obtained in step 4) of the proof of the Hard Lefschetz theorem, in the first respectively second argument given. For the sequence (3) surjectivity on the right follows from Corollary IV.4.2 and IV.3.4. Injectivity on the left follows from $\mathscr{C}o\mathscr{P}rim^{-1}(K) = 0$, which was also obtained in step 4) of the proof of the Hard Lefschetz theorem. □

For pure perverse sheaves K_0 the corollary therefore implies, that $Rad^0(K)$ uniquely determines the maximal constant subsheaf $\tilde{p}_1^*[d]({}^pH^{d-1}\mathrm{R}\tilde{p}_{2*}K)$, uniquely determines $K^\vee$ and therefore uniquely determines $\mathscr{V}(K)$. Hence $\tilde{p}_1^*[d]({}^pH^d\mathrm{R}\tilde{p}_{2*}K) \cong \mathscr{V}(K) \oplus Rad^1(K)$ is uniquely determined by $Rad^0(K)$ and $Rad^1(K)$. Therefore all the perverse sheaves

$$\mathscr{K}^\nu = {}^pH^\nu(\mathrm{R}\tilde{p}_{2*}K) \quad , \quad \nu \in \mathbb{Z}$$

are uniquely determined by the perverse sheaves $Rad^n(K)$ for $n \in \mathbb{Z}$.

V. Trigonometric Sums

V.1 Introduction

This chapter has its own bibliography.

One of the most beautiful applications of Grothendieck's trace formula and Deligne's theory of weights (La conjecture de Weil II [Del]) are certain non trivial estimates for trigonometric sums. These trigonometric sums are exponential sums, which for instance generalize the well known KLOOSTERMAN sums. It is the DELIGNE Fourier transform, which provides the link between the trigonometric sums and the etale cohomology theory. Hence, one of the essential tools is the ARTIN-SCHREIER sheaf.

Let l be a fixed prime. For every prime $p \neq l$ let us choose a non-trivial character

$$\psi_p : \mathbb{F}_p = \mathbb{Z}/p\mathbb{Z} \longrightarrow \overline{\mathbb{Q}}_l^* .$$

The Fourier transform defined by the character ψ_p on the affine line $\mathbb{A}_{\mathbb{F}_p}$ over the finite field $\mathbb{F}_p$ is denoted

$$T_{\psi_p} : \mathrm{D}_c^b(\mathbb{A}_{\mathbb{F}_p}, \overline{\mathbb{Q}}_l) \longrightarrow D_c^b(\mathbb{A}_{\mathbb{F}_p}, \overline{\mathbb{Q}}_l) .$$

KATZ and LAUMON have shown, that these Fourier transforms for the various primes have the remarkable property of being uniform: Let

$$K \in D_c^b(\mathbb{A}_{\mathbb{Z}}, \overline{\mathbb{Q}}_l)$$

be a sheaf complex on the affine space $\mathbb{A}_{\mathbb{Z}}$ over $\mathbb{Z}$ and let $K_p \in D_c^b(\mathbb{A}_{\mathbb{F}_p}, \overline{\mathbb{Q}}_l)$ be the pull back of K to the affine spaces $\mathbb{A}_{\mathbb{F}_p}$ over $\mathbb{F}_p$. Then, for almost all primes p, each of the Fourier transforms $T_{\psi_p}(K_p)$ has similar properties, from a topological point of view. As a collection they behave, as if there actually exists a global Fourier transform

$$T(K) \in D_c^b(\mathbb{A}_{\mathbb{Z}}, \overline{\mathbb{Q}}_l)$$

– a Fourier transform over $\mathbb{Z}$ – with the property

$$T(K) \times \mathrm{Spec}(\mathbb{F}_p) = T_{\psi_p}(K_p)$$

for almost all primes p ([Ka-L], theorem 4.1, corollary 4.2).

This property of the Deligne-Fourier transform of being uniform is truly remarkable. KATZ and LAUMON deduced from this fact uniform estimates for trigonometric sums, which therefore hold for almost all primes p. In this book, our intention is merely to present some of the most striking results in this area mainly due to DELIGNE, KATZ, and LAUMON. Our emphasis will be to outline the underlying ideas of methods and proofs. We do not intend to give the strongest results obtainable. In particular we concentrate on some of the simple cases, in order to give an idea of the arguments involved.

In the following, we choose a prime l and an isomorphism

$$\tau : \overline{\mathbb{Q}}_l \xrightarrow{\cong} \mathbb{C}$$

fixed from now on. We consider characters

$$\psi : \mathbb{F}_q \to \overline{\mathbb{Q}}_l^*,$$

of the additive groups of the underlying finite fields. If necessary, we will view them via τ as complex valued characters

$$\mathbb{F}_q \to \mathbb{C}^*.$$

We therefore often write $|\psi(x)|$ instead of $|\tau\psi(x)|$. For a point x of a scheme, $\overline{x}$ will denote a geometric point above x. Let R be a ring and S be a scheme. We then let $\mathbb{A}_R$ resp. $\mathbb{A}_S$ denote the affine line over $\mathrm{Spec}(R)$ respectively S.

V.2 Generalized Kloosterman Sums

Let q be a power of a prime p. Let ψ be a nontrivial character of the additive group of the finite field $\mathbb{F}_q$ with q elements. Let $a \in \mathbb{F}_q^*$ be an element of the multiplicative group $\mathbb{F}_q^*$.

Fix a natural number $m \geq 1$. Then **Generalized Kloosterman Sums** are defined as the trigonometric sums

$$\mathrm{Kloos}_m(q, \psi, a) = \sum_{\substack{x_1,\ldots,x_m \in \mathbb{F}_q^* \\ x_1 \ldots x_m = a}} \psi(x_1 + \ldots + x_m) \in \mathbb{C}.$$

Example.

$$\sum_{\substack{x_1,\ldots,x_m \in \mathbb{F}_p \\ x_1 \ldots x_m = a}} e^{\frac{2\pi i}{p}(x_1+\ldots+x_m)}, \qquad a \in \mathbb{F}_p^*.$$

For this sum a trivial estimate is

$$|\mathrm{Kloos}_m(q, \psi, a)| \leq (q-1)^{m-1}.$$

If we vary a, the Kloosterman sums $\mathrm{Kloos}_m(q, \psi, a)$ define a function on the multiplicative group $\mathbb{F}_q^*$, briefly called $\mathrm{Kloos}_m(a)$, in the following. We may then calculate the Mellin transform $\widehat{\mathrm{Kloos}}_m(\chi)$ of the Kloosterman sum, i.e. the Fourier transform with respect to the multiplicative group $\mathbb{F}_q^*$ of the field $\mathbb{F}_q$. Then $\widehat{\mathrm{Kloos}}_m(\chi)$ is a function on the set of all characters χ of $\mathbb{F}_q^*$. To determine it we may use the classical Gauss sums

$$g(\chi) = \sum_{x \in F_q^*} \psi(x)\chi(x).$$

Then the following formulas are easily verified

$$g(\chi)^m = \sum_{a \in \mathbb{F}_q^*} \chi(a) \cdot \mathrm{Kloos}_m(a) = \widehat{\mathrm{Kloos}}_m(\chi)$$

and

$$\mathrm{Kloos}_m(a) = \frac{1}{q-1} \sum_{\chi} \overline{\chi(a)} g(\chi)^m.$$

On the other hand the PLANCHEREL formula applied for the character χ gives

$$|g(\chi)|^2 = g(\chi) \cdot \overline{g(\chi)} = \begin{cases} 1 & \chi \equiv 1 \\ q & \text{else} . \end{cases}$$

This in turn implies

$$\left|\widehat{\mathrm{Kloos}}_m(\chi)\right|^2 = \begin{cases} 1 & \chi \equiv 1 \\ q^m & \text{else} . \end{cases}$$

To obtain an estimate for Kloos_m, we can again apply the PLANCHEREL formula

$$\sum_{a \in \mathbb{F}_q^*} |\mathrm{Kloos}_m(a)|^2 = \frac{1}{q-1} \sum_{\chi} |\widehat{\mathrm{Kloos}}_m(\chi)|^2 = \frac{1}{q-1}(1 + (q-2)q^m) =$$

$$= q^m - \sum_{\nu=0}^{m-1} q^{\nu} .$$

Hence

$$|\mathrm{Kloos}_m(q, \psi, a)|| \leq q^{\frac{m}{2}} .$$

For $m > 2$ this estimate is better than the trivial estimate. A much better bound – in fact the best possible one – was found by DELIGNE using cohomological methods. We will give a proof for this estimate later at the end of this chapter, and now merely state the result

Theorem 2.1 (Deligne ([De])

$$|\mathrm{Kloos}_m(q, \psi, a)| \leq m \cdot q^{\frac{m-1}{2}} = m \cdot \#(\mathbb{F}_q)^{\frac{m-1}{2}}$$

always holds.

For any natural number $a \neq 0$ and fixed m

$$\mathrm{Kloos}_m(p, \psi, a \bmod p)$$

can be viewed as a function depending on the primes p (not dividing a) and the corresponding character ψ of $\mathbf{F}_p$. The estimate of Deligne stated above is uniform in p, i.e. it is of the form

$$|\mathrm{Kloos}_m(q, \psi, a \bmod p)| \leq A \cdot q^w \qquad (q = p)\,,$$

such that the constants A and w do not depend on the chosen prime p and the chosen character ψ.

In [Ka1], Katz examined a more general class of trigonometric sums, for which he was also able to give similar uniform estimate. To explain this result, let

$$h : X \longrightarrow \mathbb{A}_{\mathbb{Z}}$$

be a finitely generated scheme over the affine line $\mathbb{A}_{\mathbb{Z}}$ above $\mathbb{Z}$. For every prime p, every power q of it, and every non-trivial character ψ of F_q, one defines the following generalized **Trigonometric Sum**

$$\sum_{x \in X(\mathbb{F}_q)} \psi(h(x))\,.$$

Let N be the supremum of the dimensions of the fibers of the complexified morphism

$$h \otimes \mathbb{C} : X \otimes \mathbb{C} \longrightarrow \mathbb{A}_{\mathbb{C}}\,.$$

Theorem 2.2 (Katz) *([Ka1], Theorem 2.3.1) Under the hypotheses above, there exists a constant A not depending on q and ψ, such that for almost all primes p, any power $q = p^r$ ($r \in \mathbb{Z}$, $r \geq 1$) of it, and for all nontrivial characters ψ of $\mathbb{F}_q$ the following estimate holds*

$$(i) \qquad \left| \sum_{x \in X(\mathbb{F}_q)} \psi(h(x)) \right| \leq A \cdot q^{N+\frac{1}{2}}\,.$$

If moreover the generic fiber $X_{\overline{\eta}}$ of the morphism h is irreducible, or if $\dim X_{\overline{\eta}} < N$ *holds, then the following stronger estimate holds*

$$(ii) \qquad \left| \sum_{x \in X(\mathbb{F}_q)} \psi(h(x)) \right| \leq A \cdot q^{N}\,.$$

Proof. We give a proof of this theorem in Section nine. □

Corollary 2.3 *There exists a constant A' such that for* any *prime p*

$$\left| \sum_{x \in X(\mathbb{F}_p)} \psi(h(x)) \right| \le A' \cdot p^{N+\frac{1}{2}}$$

respectively under the stronger assumption

$$\left| \sum_{x \in X(\mathbb{F}_p)} \psi(h(x)) \right| \le A' \cdot p^{N}$$

holds.

Remark. Let $\mathbb{A}^m_{\mathbb{Z}}$ be the m-dimensional affine space over $\mathbb{Z}$, $a \neq 0$ an integer, and $X \subseteq \mathbb{A}^m_{\mathbb{Z}}$ the closed subscheme of $\mathbb{A}^m_{\mathbb{Z}}$ defined by the equation $x_1 \cdot \ldots \cdot x_m = a$. For the choice of

$$h : X \longrightarrow \mathbb{A}_{\mathbb{Z}}$$

$$x = (x_1, \ldots, x_m) \mapsto x_1 + \ldots + x_m$$

the corresponding trigonometric sums give back the generalized Kloosterman sums

$$\mathrm{Kloos}_m(p, \psi, a \bmod p).$$

Suppose $m \ge 2$. Then in this case the fiber dimensions of h are $\le m-2$. For $m \ge 3$, the generic fiber $X_{\overline{\eta}}$ is irreducible. Corollary V.2.3 therefore implies the following estimate

$$|\mathrm{Kloos}_2(p, \psi, a \bmod p)| \le A' \cdot p^{\frac{1}{2}},$$
$$|\mathrm{Kloos}_m(p, \psi, a \bmod p)| \le A' \cdot p^{m-2}.$$

in the case $m = 2$ respectively $m \ge 3$, which is valid for all primes p. However, this estimate coincides only in the two cases $m = 2, 3$ with the strong estimate of DELIGNE, at least up to a proper choice of the constant A'!

In the next sections we want to present the proof of Theorem V.2.2, essentially following KATZ. This requires some preparations.

V.3 Links with l-adic Cohomology

In the following fix a prime $p \neq l$ and a power q of p. We also fix a finitely generated morphism

$$h : X \longrightarrow \mathbb{A}_{\mathbb{F}_q} = \mathbb{A}_0 .$$

For a nontrivial character $\psi : \mathbb{F}_q \longrightarrow \overline{\mathbb{Q}}_l^* \xrightarrow[\cong]{\tau} \mathbb{C}^*$ we use the trace $\mathrm{Tr}_{\mathbb{F}_{q^n}/\mathbb{F}_q} = \mathrm{Tr}_n : \mathbb{F}_{q^n} \to \mathbb{F}_q$ of the corresponding field extension of degree n to define

$$\psi_n = \psi \circ \mathrm{Tr}_{\mathbb{F}_{q^n}/\mathbb{F}_q} : \mathbb{F}_{q^n} \to \overline{\mathbb{Q}}_l ,$$

which is a non-trivial character of $\mathbb{F}_{q^n}$. Let $\mathscr{L}(\psi^{-1})$ be the *Artin-Schreier* sheaf on $\mathbb{A}_0$ attached to the character ψ^{-1}, as explained in Chap. I, §5. Let y be a point of $\mathbb{A}_0(\mathbb{F}_q) = \mathbb{F}_q$. The geometric Frobenius element $F_{\overline{y}}$ acts on the stalk $\mathscr{L}(\psi^{-1})_{\overline{y}}$ by multiplication with the character value $\psi(y)$. More generally, for $y \in \mathbb{A}_0(\mathbb{F}_{q^n})$, the Frobenius element $F_{\overline{y}}$ acts on the stalk $\mathscr{L}(\psi^{-1})_y \cong \overline{\mathbb{Q}}_l$ by multiplication with $\psi(\mathrm{Tr}_{\mathbb{F}_{q^n}/\mathbb{F}}(y))$. See Chap. I, §5. Let x be a point in $X(\mathbb{F}_{q^n})$. Then the action of Frobenius $F_{\overline{x}} : h^*(\mathscr{L}(\psi^{-1}))_{\overline{x}} \to h^*(\mathscr{L}(\psi^{-1}))_{\overline{x}}$ on the stalk of the pullback of $\mathscr{L}(\psi^{-1})$ on X at the point $\overline{x}$ is given by

$$F_{\overline{x}} : a \quad \mapsto \quad \psi(\mathrm{Tr}_{\mathbb{F}_{q^n}/\mathbb{F}_q}(h(x))) \cdot a .$$

For every integer $n \geq 1$, we now define the trigonometric sum

$$f_n(X, h, \psi) = \sum_{x \in X(\mathbb{F}_{q^n})} \psi(\mathrm{Tr}_{\mathbb{F}_{q^n}/\mathbb{F}_q}(h(x))) .$$

Let F be the endomorphism of $\mathrm{R}\Gamma_c(X \otimes \overline{\mathbb{F}}_q, h^*(\mathscr{L}(\psi^{-1})))$ induced by the Frobenius morphism of X over $\mathbb{F}_q$. Then we obtain from *Grothendieck's trace formula*

$$f_n(X, h, \psi) = \mathrm{Tr}(F^n) = \sum_\nu (-1)^\nu \mathrm{Tr}(F^n | H_c^\nu(X \otimes \overline{\mathbb{F}}_q, h^*(\mathscr{L}(\psi^{-1}))) .$$

Let the elements $\alpha_{\nu\mu}$ denote the eigenvalues of F on $H_c^\nu(X \otimes \overline{\mathbb{F}}_q, h^*(\mathscr{L}(\psi^{-1})))$. Then

$$f_n(X, h, \psi) = \sum_{\nu,\mu} (-1)^\nu \alpha_{\nu\mu}^n .$$

For an upper bound w_ν of the weights of $H_c^\nu(X \otimes \overline{\mathbb{F}}_q,\ h^*(\mathscr{L}(\psi^{-1})))$, we obtain $|\alpha_{\nu\mu}| \leq q^{\frac{1}{2}w_\nu}$. This implies the following estimate for the trigonometric sums f_n

Theorem 3.1 $|f_n(X, h, \psi)| \ \leq \ \sum_\nu q^{\frac{1}{2}nw_\nu} \cdot \dim H_c^\nu(X \otimes \overline{\mathbb{F}}_q, h^*(\mathscr{L}(\psi^{-1})))$

Remark 3.2 Let $K = \mathrm{R}h_!(\overline{\mathbb{Q}}_l)$. Then

$$\mathrm{R}\Gamma_c(X \otimes \overline{\mathbb{F}}_q, h^*(\mathscr{L}(\psi^{-1}))) \ = \ \mathrm{R}\Gamma_c(\mathbb{A}_0 \otimes \overline{\mathbb{F}}_q, K \otimes \mathscr{L}(\psi^{-1})) .$$

V.4 Deligne's Estimate

Let $\mathbb{A}_0^m$ be the m-dimensional affine space over $\mathbb{F}_q$. For $a \in \mathbb{F}_q^*$ and

$$X := \{x = (x_1, \ldots, x_m) \in \mathbb{A}_0^m : x_1 \cdot \ldots \cdot x_m = a\}$$

and the morphism $h : X \longrightarrow \mathbb{A}_0$ defined by

$$h : x = (x_1, \dots, x_m) \mapsto x_1 + \dots + x_m$$

we get $\mathrm{Kloos}_m(q^n, \psi \circ \mathrm{Tr}_{\mathbb{F}_{q^n}/\mathbb{F}_q}, a) = f_n(X, h, \psi)$.

Deligne proves in [De] theorem 7.4

Theorem 4.1 *Suppose l does not divide q. Then for all nontrivial characters ψ : $\mathbb{F}_q \to \overline{\mathbb{Q}}_l^*$ the following holds:*

$$H_c^\nu(X \otimes \overline{\mathbb{F}}_q, h^*(\mathscr{L}(\psi)^{-1})) = 0 \quad \textit{for} \quad \nu \neq m - 1$$

and

$$\dim_{\overline{\mathbb{Q}}_l} H_c^{m-1}(X \otimes \overline{\mathbb{F}}_q, h^*(\mathscr{L}(\psi)^{-1})) = m \, .$$

The sheaf $h^*(\mathscr{L}(\psi^{-1}))$ is pointwise of weight zero for all its stalks. Therefore the generalized *Weil conjectures* imply, that $m - 1$ is an upper bound for the weights of $H_c^{m-1}(X \otimes \overline{\mathbb{F}}_q, h^*(\mathscr{L}(\psi^{-1})))$. This implies *Deligne's estimate*

$$|\mathrm{Kloos}_m(q^n, \psi \circ \mathrm{Tr}_{\mathbb{F}_{q^n}/\mathbb{F}^q}, a)| \leq m \cdot q^{\frac{1}{2}(m-1)\cdot n} \, .$$

Remark. Deligne even proved, that

$$H_c^{m-1}(X \otimes \overline{\mathbb{F}}_q, h^*(\mathscr{L}(\psi^{-1})))$$

is pure of weight $m - 1$. This implies, that *Deligne's estimate* is best possible.

V.5 The Swan Conductor

This section is a report on [Se1], chap. IV, V, [Se2], §10, or [Lau2].

Let $\mathfrak{o}$ be a strict Henselian discrete valuation ring with algebraically closed residue class field of characteristic $p > 0$ and let K be the quotient field of $\mathfrak{o}$. Let I be the Galois group of the separable closure $\overline{K}$ of K. Then the group I is a pure ramification group. It has the wild ramification group P as normal subgroup. P is closed in I; P is the profinite p-Sylow group of I. The quotient group $I^t = I^{tame} = I/P$ is called the group of *tame ramification*. There is a natural isomorphism

$$I^{tame} = \prod_{(l,p)=1} \mathbb{Z}_l(1)$$

$$\mathbf{Z}_l(1) = \varprojlim_n \mu_{l^n} \, ,$$

where $\mu_{l^n} \subseteq \mathfrak{o}^*$ is the group of l^n-th roots of unity.

Let $l \neq p$ be a fixed prime and $\mathbb{F}$ be a finite field of characteristic $l \neq p$. Consider continuous representations of I on finite dimensional vector spaces M over $\mathbb{F}$. Then M is an I-module. The term "continuous" means, that an open normal subgroup H of I acts trivially on M. H is an open and closed subgroup of finite index in I. The action of I on M factorizes over the finite quotient group $G = I/H$. M is a G-module.

Definition 5.1 *We say that I acts* tamely *on M if P acts trivially on M. Then M is called a tame or tamely ramified I-module. M then is a module over the tame ramification group I^t.*

Let H be an arbitrary closed normal subgroup of I which is of finite index in I such that $H \cap P$ acts trivially on M. Then the factor group $PH/H = P/P \cap H$ acts on M. Let $L = \overline{K}^H$ be the finite Galois extension of K corresponding to the Galois group $G = I/H$. Let $\mathfrak{o}'$ be the valuation ring of L, $\pi \in \mathfrak{o}'$ a primitive element of the maximal ideal of $\mathfrak{o}'$. Let $v(\cdot)$ be the additive valuation of L normalized such that $v(\pi) = 1$. Now consider the higher ramification groups of $L|K$:

$$\begin{aligned} G_i &= \{\sigma \in G : v(\sigma(\pi) - \pi) \geq i + 1\} \\ &= \{\sigma \in G : v(\sigma(x) - x) \geq i + 1 \ \forall x \in \mathfrak{o}'\}, \\ G_0 &= G \supseteq G_1 \supseteq G_2 \supseteq \cdots \supseteq G_r = \{1\}. \end{aligned}$$

G_1 is the wild ramification group of $L|K$. In other words

$$G_1 = PH/H = P/H \cap P \ .$$

G_1 acts on M. M is tamely ramified if and only if the finite group G_1 acts trivially on M.

Let P_i be the preimage of G_i in P. P_i is a normal subgroup of I. The decreasing sequence of groups

$$P = P_1 \supseteq P_2 \supseteq P_3 \supseteq \ldots.$$

depends on the choice of H. P being a pro-p-group, the action of P and its subgroups on M is semisimple, because the characteristic of the field $\mathbb{F}$ is different from p by assumption.

Theorem 5.2 *(see [Se1], IV §2 and VI §2, [Se2], §19)*

(1) The rational number

$$\sum_{i \geq 1} \frac{1}{[I : H \cdot P_i]} \ \dim_{\mathbb{F}}(M/M^{P_i})$$

$$= \sum_{i \geq 1} \frac{1}{[G_0 : G_i]} \ \dim_{\mathbb{F}}(M/M^{G_i})$$

is independent of the choice of H.

(2) The number

$$\sum_{i\geq 1} \frac{1}{[I : H \cdot P_i]} \dim_{\mathbb{F}}(M/M^{P_i})$$

is an integer.

Remark on (1). For the proof of independence one needs another numbering of the higher ramification groups, the so called upper numbering. This upper numbering defines a filtration of G_0 by the family of ramification groups G^x where $x > 0$ is in $\mathbb{R}$. See [Se1], chap. IV, §3. This allows to rewrite the formula

$$\sum_{i\geq 1} \frac{1}{[G_0 : G_i]} \dim_{\mathbb{F}}(M/M^{G_i})$$

into a formula involving the upper indices x and the groups G^x. See also [Ka2], Chap. IV 1. For a normal subgroup $\Gamma \subset G$ the image of G^x in G/Γ is $(G/\Gamma)^x$ by the theorem of Herbrand. See [Se1], chap. IV, §3, proposition 14 respectively lemma 5. The claimed independence is a consequence of this fact.

Concerning (2). If H is chosen small enough we can suppose M to be a G-module. If G is abelian, the integrality property follows from the Hasse-Arf Theorem. See [Se1], chap. IV, §1, p.84. For representations of G on a finite dimensional vector space V over a field of characteristic 0, the integrality property for V is deduced from the abelian case via Brauer's Theorem ([Se2], theorem 24). For details see [Se1], chap. VI, §2. Let $\mathfrak{o}_E$ be a complete discrete valuation ring with quotient field E of characteristic 0 and residue field $\mathbb{F}$. Then there exists a virtual $\mathfrak{o}_E$-module V_0 which is free as a virtual $\mathfrak{o}_E$-module, such that $\mathbb{F} \otimes_{\mathfrak{o}_E} V_0$ and M have the same image in the Grothendieck group of finite $\mathbb{F}[G]$-modules. This is also a consequence of Brauer's Theorem. See [Se2], theorem 33 resp. theorem 42. Put $V = E \otimes_{\mathfrak{o}_E} V_0$. Since the characteristic l of E does not divide the order of the wild ramification group G_1, the functor of Γ-invariants is exact in the category of $\mathbb{F}[G]$-modules for any subgroup Γ of G_1. Hence the following holds true

$$dim_{\mathbb{F}}(M/M^{G_i}) = rank(V_0/V_0^{G_i}) = dim_E(V/V^{G_i}) \quad , \quad i \geq 1 .$$

Hence the desired integrality property for M follows from the corresponding statement for V in the characteristic 0 case.

For a related topic – the construction of the ARTIN and the SWAN representations – see [Se2] 19.1 and 19.2.

Definition 5.3 *The integer*

$$\mathrm{Swan}(M) := \sum_{i\geq 1} \frac{1}{[G_0 : G_i]} \dim(M/M^{G_i})$$

$$= \sum_{i\geq 1} \frac{1}{[I : H \cdot P_i]} \dim(M/M^{P_i})$$

is called the Swan conductor *of* M.

Remark. M is tamely ramified if and only if Swan(M) is zero. Due to the semisimple action of P and its subgroups, the invariant Swan(M) is additive for short exact sequences of I modules.

Now consider a continuous representation of I on a finite dimensional vector space V over $\overline{\mathbb{Q}}_l$ for $l \neq p$. Continuous means, that there exists a continuous representation of I on a finite dimensional vector space V_0 over a finite extension field $E \subset \overline{\mathbb{Q}}_l$ of $\mathbb{Q}_l$ such that

$$V = \overline{\mathbb{Q}}_l \otimes_E V_0$$

(as representation modules). An open and closed normal subgroup of finite index in P acts trivially on V, because P is a pro-p-group and because $l \neq p$. Using the fact that I is compact, one knows that there exists a finite free $\mathfrak{o}_E$-module

$$F_0 \subseteq V_0$$

with

$$F_0 \otimes_{\mathfrak{o}_E} E = V_0$$

and $IF_0 \subseteq F_0$. Here $\mathfrak{o}_E$ is the valuation ring of E.

Since P is a pro-p-group and the action of P on F_0 factorizes over a finite quotient group, P and all its subgroups act semisimply on F_0. We choose an open normal subgroup H of I with finite index, such that $I \cap P$ acts trivially on V. Consider the higher ramification subgroups of $G = I/H$

$$G = G_0 \supseteq G_1 = P/P \cap H \supseteq G_2 \supseteq ... \supseteq G_\nu = \{1\}$$

and the preimages in P

$$P = P_1 \supseteq P_2 \supseteq P_3 \supseteq \ldots .$$

Let $\mathbb{F}$ be the residue field of $\mathfrak{o}_E$. Then put

$$M = F_0 \otimes_{\mathfrak{o}_E} \mathbb{F} .$$

Then $F_0^{P_i}$ is a direct summand of F_0 and we have

$$F_0^{P_i} \otimes_{\mathfrak{o}_E} \mathbb{F} = M^{P_i} \quad , \quad (F_0/F_0^{P_i}) \otimes_{\mathfrak{o}_E} \mathbb{F} = M/M^{P_i},$$

$$F_0^{P_i} \otimes_{\mathfrak{o}_E} \overline{\mathbb{Q}}_l = V^{P_i} \quad , \quad (F_0/F_0^{P_i}) \otimes_{\mathfrak{o}_E} \overline{\mathbb{Q}}_l = V/V^{P_i} .$$

I acts continuously on the finite vector space M over the finite field $\mathbb{F}$. Therefore

$$\sum_{i\geq 1} \frac{1}{[I : H \cdot P_i]} \dim_{\overline{\mathbb{Q}}_l}(V/V^{P_i}) = \text{Swan}(M) .$$

Theorem 5.4 *(see [Se1], chap. VI, §2 and [Se2], §19)*

(1) $\sum_{i\geq 1} \frac{1}{[I:H\cdot P_i]} \dim_{\overline{\mathbb{Q}}_l}(V/V^{P_i})$ *is independent of the choice of* H.
(2) The rational number

$$\sum_{i\geq 1} \frac{1}{[I:H\cdot P_i]} \dim_{\overline{\mathbb{Q}}_l}(V/V^{P_i})$$

is an integer.

Definition 5.5 $\mathrm{Swan}(M) := \sum_{i\geq 1} \frac{1}{[I:H\cdot P_i]} \dim_{\overline{\mathbb{Q}}_l}(V/V^{P_i})$ *is called the* **Swan conductor** *of* V.

Remark. The Swan conductor $\mathrm{Swan}(V)$ vanishes if and only if P acts trivially on V, i.e. if and only if V is tamely ramified. $\mathrm{Swan}(V)$ is additive with respect to short exact sequences of I-modules.

Let k be a perfect field of characteristic $p > 0$. Let X be a projective smooth irreducible curve over k. Let $U \subseteq X$ be an open nonempty subscheme of X, and let $\mathscr{G}$ be a constructible sheaf of $\overline{\mathbb{Q}}_l$-vector spaces on U. Here again $l \neq p$.

Then there is an open and non-empty subscheme $V \subseteq U$ so that $\mathscr{G}|V$ is a smooth sheaf. Let η be a generic point of X, $\eta \in V \subseteq U \subseteq X$. Let s be a closed point of X, $\mathfrak{o}$ the strict Henselization of the local ring $O_{X,s}$ of X in s in a geometric point $\overline{s}$ of X above s and $I_{\overline{s}}$ the Galois group of the separable closure of the quotient field K of $\mathfrak{o}$. Then $I_{\overline{s}}$ depends up to isomorphism only on s and is called the *ramification group* of X in $\overline{s}$ resp. in s. We simply write $I_{\overline{s}} = I_s$.

The wild ramification group $P_{\overline{s}} \subseteq I_{\overline{s}}$ is referred to as the wild ramification group of X in $\overline{s}$ respectively in s. The group $I_{\overline{s}}$ acts continuously on the generic stalk $\mathscr{G}_{\overline{\eta}} = M$. This action is uniquely defined up to conjugation by a linear isomorphism

$$\sigma : M \xrightarrow{\sim} M\ .$$

σ is an element of the Galois group of the separable closure of the function field of $O_{X\overline{\eta}}$ over $O_{X\eta}$ which operates continuously on M. $\mathscr{G}$ is called *tamely ramified in* s iff the wild ramification group P_s operates trivially on $\mathscr{G}_{\overline{\eta}}$, i.e. iff M is tamely ramified.

Definition 5.6 *The Swan conductor of* $\mathscr{G}$ *in* $\overline{s}$ *respectively in* s *is defined as the Swan conductor of the* $I_{\overline{s}}$*-module* $\mathscr{G}_{\overline{\eta}} = M$. *Hence*

$$\mathrm{Swan}_{\overline{s}}(\mathscr{G}) = \mathrm{Swan}_s(\mathscr{G}) = \mathrm{Swan}(M).$$

Remark 5.7 The Swan conductor of $\mathscr{G}$ has the following properties:

(1) $\mathscr{G}$ is tamely ramified in s iff $\mathrm{Swan}_s(\mathscr{G}) = 0$.
(2) If $s \in V$ holds then $\mathrm{Swan}_s(\mathscr{G}) = 0$.

(3) If k' is an algebraic field extension of k, $\mathscr{G}'$ the preimage of $\mathscr{G}$ on $X \otimes k' = X'$ and $\bar{s}'$ a geometric point of X' over $\bar{s}$, then

$$\mathrm{Swan}_{\bar{s}}(\mathscr{G}) = \mathrm{Swan}_{\bar{s}'}(\mathscr{G}') .$$

(4) Let $\mathscr{F}$ be a sheaf on U that is tamely ramified in s. Then

$$\mathrm{Swan}_s(\mathscr{F} \otimes \mathscr{G}) = \mathrm{Swan}_s(\mathscr{G}) \cdot \dim \mathscr{F}_{\bar{\eta}} .$$

V.6 The Ogg–Shafarevich–Grothendieck Theorem

Let k be a separably closed field, X a projective smooth curve over k, and $\emptyset \neq U \subseteq X$ an open subscheme, and $\mathscr{G}$ a sheaf on U.

Definition 6.1 *The* **Euler-Poincaré** *characteristic is defined as*

(1) $\chi_c(U, \mathscr{G}) = \sum_\nu (-1)^\nu \dim H_c^\nu(U, \mathscr{G}) = \sum_{\nu=0}^{2} (-1)^\nu \dim H_c^\nu(U, \mathscr{G})$,
(2) $\chi(U, \mathscr{G}) = \sum_\nu (-1)^\nu \dim H^\nu(U, \mathscr{G}) = \sum_{\nu=0}^{2} (-1)^\nu \dim H^\nu(U, \mathscr{G})$.

The following lemma can be proved (see [Ka1], 4.6):

Lemma 6.2 $\chi_c(U, \mathscr{G}) = \chi(U, \mathscr{G})$.

It is easy to see that the following lemma also holds:

Lemma 6.3 *Let $V \subseteq U$ be an open and dense subscheme of U. Then*

$$\chi(U, \mathscr{G}) = \chi(V, \mathscr{G}) + \sum_{x \in U \setminus V} \dim \mathscr{G}_{\bar{x}} .$$

For the following, let X be irreducible and k be algebraically closed. Let g_X be the genus of the curve X. If k has characteristic zero and $\mathscr{G}|V$ is a smooth sheaf, then one knows by topological methods

$$\chi(V, \mathscr{G}) = \mathrm{Rank}(\mathscr{G}) \cdot \chi(V) ,$$

where $\chi(V)$ is given by $2 - 2g_X - \#(X \setminus V)$.

If the base field has positive characteristic $p > 0$, then the above equation is false in general. For this reason, the **tame Euler-Poincaré characteristic** is introduced.

Definition 6.4 *Let $V \subseteq U$ be an open and dense subscheme so that $\mathscr{G}|V$ is a smooth sheaf. Then the tame Euler-Poincare characteristic is defined to be*

$$\chi_c^{tame}(U, \mathscr{G}) = \chi^{tame}(U, \mathscr{G}) =$$

$$= \dim \mathcal{G}_{\overline{\eta}} \cdot \chi(V) + \sum_{x \in U \setminus V} \dim \mathcal{G}_{\overline{x}}.$$

Here, $\chi(V) = 2 - 2g_X - \#(X \setminus V)$, and η is a generic point of V. So the above equation can be abbreviated:

$$\chi^{tame} = \chi^t = \chi_c^{tame}.$$

Theorem 6.5 (Ogg–Shafarevich–Grothendieck) (see [Gr]) *Let the characteristic of the basic field be $p > 0$ and let $\mathcal{G}$ be a constructible $\overline{\mathbb{Q}}_l$ sheaf. Assume $l \neq p$. Then*

$$\chi(U, \mathcal{G}) = \chi^{tame}(U, \mathcal{G}) - \sum_{\substack{s \in X \\ s \text{ closed}}} \mathrm{Swan}_s(\mathcal{G}).$$

An Example. Let $q = p^r$. Let $\psi : \mathbb{F}_q \to \overline{\mathbb{Q}}_l^*$ be a non-trivial character. Let $\mathcal{L}_0(\psi)$ be the *Artin-Schreier* sheaf on the affine line $\mathbb{A}_0$ over $\mathbb{F}_q$ (see Lemma I.5.1) and let $\mathcal{L}(\psi)$ be its pullback to $\mathbb{A}$, the affine line over the algebraic closure $\overline{\mathbb{F}}_q$ of $\mathbb{F}_q$. Let $\mathbb{P}^1$ be the projective line over $\overline{\mathbb{F}}_q$. Then

$$\mathbb{A} \subseteq \mathbb{P}^1 = \mathbb{A} \cup \{\infty\}$$

and we have the following

Lemma 6.6 $\mathrm{Swan}_\infty(\mathcal{L}(\psi)) = 1.$

Proof. From Lemma I.5.1 and I.5.2 we get

$$H^\nu(\mathbb{A}, \mathcal{L}(\psi)) = 0 \quad \text{for } \nu \geq 1,$$

and

$$H^0(\mathbb{A}, \mathcal{L}(\psi)) = \{\lambda \in \overline{\mathbb{Q}}_l : \psi(\sigma) \cdot \lambda = \lambda \ \forall \sigma \in \mathbb{F}_q\} = 0.$$

Hence $\chi(\mathbb{A}, \mathcal{L}(\psi)) = 0$. The rank of $\mathcal{L}(\psi)$ is one and the genus of the projective line is zero. Therefore we obtain $\chi^{tame}(\mathbb{A}, \mathcal{L}(\psi)) = 1 \cdot (2 - 2 \cdot 0 - 1) = 1$. The theorem of **Ogg–Shafarevich–Grothendieck** implies

$$0 = \chi(\mathbb{A}, \mathcal{L}(\psi)) = 1 - \mathrm{Swan}_\infty(\mathcal{L}(\psi)),$$

so that $\mathrm{Swan}_\infty(\mathcal{L}(\psi)) = 1$. □

V.7 The Main Lemma

Again, let $\mathbb{A}_0$ be the affine line over the field $\mathbb{F}_q$ with q elements. Let $\mathbb{A} = \mathbb{A}_0 \otimes \overline{\mathbb{F}}_q$ be the affine line over the algebraic closure $\overline{\mathbb{F}}_q$. Let $\psi : \mathbb{F}_q \to \overline{\mathbb{Q}}_l^*$ be a nontrivial character, and assume that l does not divide q. Let $\mathscr{L}_0(\psi)$ be the *Artin-Schreier* sheaf on $\mathbb{A}_0$, and let $\mathscr{L}(\psi)$ be its pullback to $\mathbb{A}$. Let $\mathbb{P}^1$ be the projective line over $\overline{\mathbb{F}}_q$. Consider the inclusion

$$\mathbb{A} \overset{j}{\hookrightarrow} \mathbb{P}^1 = \mathbb{A} \cup \{\infty\} .$$

Lemma 7.1 (Main Lemma) *Let $\mathscr{G}$ be a sheaf on the affine line $\mathbb{A}$, which is tamely ramified at the point ∞. Then:*

(1) The canonical map

$$H_c^*(\mathbb{A}, \mathscr{G} \otimes \mathscr{L}(\psi)) \to H^*(\mathbb{A}, \mathscr{G} \otimes \mathscr{L}(\psi))$$

is an isomorphism.

(2) $H_c^\nu(\mathbb{A}, \mathscr{G} \otimes \mathscr{L}(\psi)) = 0$ holds for $\nu \geq 2$.

(3) Let η be a generic point of $\mathbb{A}$. Then

$$\mathrm{Swan}_\infty(\mathscr{G} \otimes \mathscr{L}(\psi)) = \dim \mathscr{G}_{\overline{\eta}} .$$

(4) If $\mathscr{G}$ is tamely ramified at all points of $\mathbb{P}^1$ then

$$\chi_c(\mathbb{A}, \mathscr{G} \otimes \mathscr{L}(\psi)) = \chi_c(\mathbb{A}, \mathscr{G}) - \dim \mathscr{G}_{\overline{\eta}} .$$

(5) If $\mathscr{G}$ is tamely ramified at all points, then

$$\dim H_c^0(\mathbb{A}, \mathscr{G} \otimes \mathscr{L}(\psi)) = \dim H_c^0(\mathbb{A}, \mathscr{G})$$

and

$$\dim H_c^1(\mathbb{A}, \mathscr{G} \otimes \mathscr{L}(\psi)) = \dim H_c^1(\mathbb{A}, \mathscr{G}) - \dim H_c^2(\mathbb{A}, \mathscr{G}) + \dim \mathscr{G}_{\overline{\eta}} .$$

Corollary 7.2 *Let $K \in D_c^b(\mathbb{A}, \overline{\mathbb{Q}}_l)$ be a tamely ramified complex, i.e. all cohomology sheaves $\mathscr{H}^\nu(K)$ of the complex are tamely ramified in every point of the projective line. For a sheaf $\mathscr{G}$, we use the abbreviated notation*

$$h_c^\nu(\mathscr{G}) = \dim_{\overline{\mathbb{Q}}_l} H_c^\nu(\mathbb{A}, \mathscr{G}).$$

$rg(\mathscr{G})$ denotes the dimension of the stalk $\mathscr{G}_{\overline{\eta}}$ in a generic point. Then for all integers m

$$\dim_{\overline{\mathbb{Q}}_l} H_c^m(\mathbb{A}, K \otimes \mathscr{L}(\psi)) =$$

$$= h_c^0(\mathscr{H}^m(K)) + h_c^1(\mathscr{H}^{m-1}(K)) - h_c^2(\mathscr{H}^{m-1}(K)) + \mathrm{rg}(\mathscr{H}^{m-1}(K))$$

holds.

Proof of Corollary V.7.2. First of all $\mathscr{H}^\nu(K \otimes \mathscr{L}(\psi)) = \mathscr{H}^\nu(K) \otimes \mathscr{L}(\psi)$. From the main lemma, equation (2) we get for $\mu \geq 2$

$$H_c^{\mu}(\mathbb{A}, \mathscr{H}^{\nu}(K) \otimes \mathscr{L}(\psi)) = 0$$

Hence the spectral sequence

$$H_c^p(\mathbb{A}, \mathscr{H}^q(K \otimes \mathscr{L}(\psi))) \Rightarrow H_c^{p+q}(\mathbb{A}, K \otimes \mathscr{L}(\psi))$$

gives the short exact sequences

$$0 \longrightarrow H_c^1(\mathbb{A}, \mathscr{H}^{m-1}(K) \otimes \mathscr{L}(\psi)) \longrightarrow H_c^m(\mathbb{A}, K \otimes \mathscr{L}(\psi)) \longrightarrow$$

$$\longrightarrow H_c^0(\mathbb{A}, \mathscr{H}^m(K) \otimes \mathscr{L}(\psi)) \longrightarrow 0\,.$$

From this we obtain the assertion of Corollary V.7.2 from the assertion (5) of the main lemma. □

Proof of the Main Lemma. (2) follows directly from (1). (3) is a consequence of Lemma V.6.6 and Remark V.5.7. This implies $\mathrm{Swan}_\infty(\mathscr{G} \otimes \mathscr{L}(\psi)) = \mathrm{Swan}_\infty(\mathscr{L}(\psi)) \cdot \dim \mathscr{G}_{\overline{\eta}} = \dim \mathscr{G}_{\overline{\eta}}$. (4) is a direct consequence of the **Ogg–Shafarevich–Grothendieck** theorem.

Proof of (5). Let $k : U \to \mathbb{A}$ be an open dense embedding, such that $\mathscr{G}|U$ is smooth. The sheaf $\mathscr{F} = \ker(\mathscr{G} \to k_*k^*(\mathscr{G}))$ and also the kernel sheaf $\mathscr{F}_\psi = \ker\big(\mathscr{G} \otimes \mathscr{L}(\psi) \to k_*k^*(\mathscr{G} \otimes \mathscr{L}(\psi))\big) = \mathscr{F} \otimes \mathscr{L}(\psi)$ have finite supports. Furthermore

$$H_c^0(\mathbb{A}, \mathscr{G}) = H_c^0(\mathbb{A}, \mathscr{F}), \quad H_c^0(\mathbb{A}, \mathscr{G} \otimes \mathscr{L}(\psi)) = H_c^0(\mathbb{A}, \mathscr{F}_\psi),$$

$$\dim_{\overline{\mathbb{Q}}_l} H_c^0(\mathbb{A}, \mathscr{F}) = \dim_{\overline{\mathbb{Q}}_l} H_c^0(\mathbb{A}, \mathscr{F} \otimes \mathscr{L}(\psi)).$$

This implies $\dim_{\overline{\mathbb{Q}}_l} H_c^0(\mathbb{A}, \mathscr{G} \otimes \mathscr{L}(\psi)) = \dim_{\overline{\mathbb{Q}}_l} H_c^0(\mathbb{A}, \mathscr{G})$. The second part of (5) finally follows from (4) and (2).

It remains to show, that the natural map

$$H_c^{\nu}(\mathbb{A}, \mathscr{G} \otimes \mathscr{L}(\psi)) \to H^{\nu}(\mathbb{A}, \mathscr{G} \otimes \mathscr{L}(\psi))$$

is an isomorphism. For this, it suffices to prove that for all integers $\nu \geq 0$ the following holds

$$\mathrm{R}^{\nu} j_*(\mathscr{G} \otimes \mathscr{L}(\psi))_{\overline{\infty}} = 0\,.$$

For this we use the ramification group I_∞ respectively the wild ramification group $P_\infty \subseteq I_\infty$ of the projective line at the point ∞. I_∞ acts on $\mathscr{G}_{\overline{\eta}} \otimes \mathscr{L}(\psi)_{\overline{\eta}}$. Hence the equations

$$\mathrm{R}^{\nu} j_*(\mathscr{G} \otimes \mathscr{L}(\psi))_{\overline{\infty}} = H^{\nu}(I_\infty, \mathscr{G}_{\overline{\eta}} \otimes \mathscr{L}(\psi)_{\overline{\eta}}) =$$

$$= H^{\nu}(I_\infty/P_\infty, (\mathscr{G}_{\overline{\eta}} \otimes \mathscr{L}(\psi)_{\overline{\eta}})^{P_\infty})$$

hold true. Since by assumption P_∞ acts trivially on $\mathscr{G}_{\overline{\eta}}$, we get

$$(\mathscr{G}_{\overline{\eta}} \otimes \mathscr{L}(\psi)_{\overline{\eta}})^{P_\infty} \cong \{u \in \mathscr{G}_{\overline{\eta}} : \psi(\sigma) \cdot u = u\,,\ \forall \sigma \in \mathbb{F}_q\} = 0.$$

□

V.8 The Relative Abhyankar Lemma

See also [SGA 1], exposé XIII, remarques 2.3 and proposition 5.5 for a more detailed account. Let $R \subseteq \mathbb{C}$ be a subring of $\mathbb{C}$ finitely generated over $\mathbb{Z}$. Put $S = \mathrm{Spec}(R)$. Let $X \to S$ be a projective, smooth, irreducible relative curve above S, $U \subseteq X$ be an open and non-empty subscheme of X, and let $\mathcal{G}$ be a constructible $\overline{\mathbb{Q}}_l$-sheaf on U. Then there exists an open dense subscheme $V \subseteq U$ of U, such that $\mathcal{G}|U$ is a smooth sheaf.

If we replace S by a sufficiently small open affine dense subscheme we can achieve that $X \setminus V$, with the reduced scheme structure, is finite etale over S. Then we have

Theorem 8.1 *For every field k of positive characteristic $p \neq l$ and every morphism* $\mathrm{Spec}(k) \overset{s}{\to} S$ *the preimage of $\mathcal{G}$ on $U_s = U \times_S \mathrm{Spec}(k)$ is tamely ramified in all points of $X \times_S \mathrm{Spec}(k)$.*

Now consider the affine and the projective line over $S = \mathrm{Spec}(R)$.

$$\begin{array}{ccc} \mathbb{A}_S & \longrightarrow & \mathbb{P}_S \\ & \overset{h}{\searrow} & \downarrow \\ & & S \end{array}$$

Let l be a prime, invertible in R, and let $\mathcal{G}$ be a constructible $\overline{\mathbb{Q}}_l$-sheaf on $\mathbb{A}_S$.

Let $V \subseteq \mathbb{A}_S$ be an open and non-empty subscheme of $\mathbb{A}_S$ so that $\mathcal{G}|V$ is smooth of rank d. Then there is an open and non-empty subscheme S' of S with the following properties:

a) The assumptions of Theorem V.8.1 are fulfilled.
b) All fibers of V over S' are non-empty.
c) All direct images $\mathrm{R}^\nu h_! \mathcal{G}$ are smooth on S'.

Let d_ν be the rank of $\mathrm{R}^\nu h_! \mathcal{G}|S'$. Then, obviously, almost all numbers d_ν are zero.

Under these assumptions, for every morphism $t : \mathrm{Spec}(\overline{\mathbb{F}}_q) \to S'$ and the pull back $\mathcal{G}_t$ of the sheaf $\mathcal{G}$ to the affine line $\mathbb{A}_t$ over $\overline{\mathbb{F}}_q$ we have:

1) $\mathcal{G}_t$ is tamely ramified on the projective line over $\overline{\mathbb{F}}_q$, therefore satisfies the assumptions of the main lemma. In the generic point of $\mathbb{A}_t$ the rank of $\mathcal{G}_t$ is d. Furthermore $\dim H_c^\nu(\mathbb{A}_t, \mathcal{G}_t) = d_\nu$ holds.

2) More generally, let K be a complex in $D_c^b(\mathbb{A}_S, \overline{\mathbb{Q}}_l)$. Then V and S' may be chosen simultaneously for all (the finitely many) cohomology sheaves of K. We use Corollary V.7.2 to deduce

Theorem 8.2 *Let $R \subseteq \mathbb{C}$ be a subring finitely generated over $\mathbb{Z}$. Let l be a prime invertible in R, $S = \mathrm{Spec}(R)$, and $K \in D_c^b(\mathbb{A}_S, \overline{\mathbb{Q}}_l)$. Then there is an open dense subscheme S' of S and there are integers $d_\nu \geq 0$, such that for all morphisms $\mathrm{Spec}(\mathbb{F}_q) \to S'$ and all nontrivial characters $\psi : \mathbb{F}_q \to \overline{\mathbb{Q}}_l^*$*

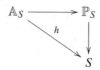

$$\dim H_c^\nu(\mathbb{A}_{\overline{\mathbb{F}}_q}, K_{\overline{\mathbb{F}}_q} \otimes \mathscr{L}(\psi)) = d_\nu$$

holds. Here, $K_{\overline{\mathbb{F}}_q}$ *is the preimage of* K *on the affine line* $\mathbb{A}_{\overline{\mathbb{F}}_q}$*, and* $K_{\overline{\mathbb{F}}_q}$ *is tamely ramified on the whole projective line over* $\overline{\mathbb{F}}_q$*.*

Remark 8.3 The statement of the last Theorem V.8.2 is a uniform statement, concerning all Fourier transforms $T_\psi(K_{\mathbb{F}_q})$ for the different pullbacks of K to the various affine lines over $\mathbb{F}_q$. By [Lau2], theorem 2.1.1, corollary 2.1.2, even the following stronger assertion can be proved:

a) All cohomology sheaves of $T_\psi(K_{\mathbb{F}_q})$ are smooth on $\mathbb{A}_{\mathbb{F}_q} \setminus \{0\}$.
b) The rank of $\mathscr{H}^\nu(T_\psi(K_{\mathbb{F}_q}))$ on $A_{\mathbb{F}_q} \setminus \{0\}$ does not depend on the choice of $\mathrm{Spec}(\mathbb{F}_q) \to S'$ nor the choice of ψ.
c) The same is true for $\mathscr{H}^\nu(T_\psi(K_{\mathbb{F}_q}))|\{0\}$. In other words, the Fourier transforms $T_\psi(K_{\mathbb{F}_q})$ are uniformly constructible. .

See also [Ka-L], theorem 4.1 and corollary 4.2.

V.9 Proof of the Theorem of Katz

In this section we prove Theorem V.2.2. To estimate trigonometric sums in a uniform way as stated in this theorem, we need uniform estimates both for the dimension and for the weights of the cohomology groups involved. Uniform estimates for the dimension were obtained in Theorem V.8.2.

As before, let $R \subseteq \mathbb{C}$ be a finitely generated ring over $\mathbb{Z}$, and let be $S = \mathrm{Spec}(R)$. We assume that a finitely generated scheme

$$h : X \longrightarrow \mathbb{A}_S$$

over the one dimensional affine space $\mathbb{A}_S$ over S is given. Let N be the supremum of the dimensions of all fibers of the morphism

$$X \otimes_R \mathbb{C} \longrightarrow \mathbb{A}_S \otimes_R \mathbb{C}\,.$$

By shrinking S, we may furthermore assume:

(a) The prime l is invertible in R.
(b) Theorem V.8.2 holds for the complex $K = Rh_!\overline{\mathbb{Q}}_l \in D_c^b(\mathbb{A}_S, \overline{\mathbb{Q}}_l)$ and $S' = S$.
(c) For every morphism

$$t : \mathrm{Spec}(\mathbb{F}_q) \to S$$

the dimensions of all fibers of the induced morphisms

$$h_t = h \otimes \overline{F}_q : \quad X_t = X \otimes \overline{\mathbb{F}}_q \to \mathbb{A}_t = \mathbb{A}_S \otimes \overline{\mathbb{F}}_q\,,$$

are $\leq N$.

Let $t, h_t, \ldots$ be as in c), and let $\psi : \mathbb{F}_q \to \overline{\mathbb{Q}}_l^*$ be an arbitrary nontrivial character. Let K_t be the pullback of K to $\mathbb{A}_t$. In particular, b) implies that for the complex $K_t = \mathrm{R}h_{t!}(\overline{\mathbb{Q}}_l)$ the dimension

$$\dim_{\overline{\mathbb{Q}}_l} H_c^\nu(\mathbb{A}_t, K_t \otimes \mathscr{L}(\psi)) = \dim_{\overline{\mathbb{Q}}_l} H_c^\nu(X_t, h_t^*(\mathscr{L}(\psi))),$$

can be estimated independently of t and ψ. K_t is tamely ramified on the projective line, i.e. all cohomology sheaves $\mathscr{H}^\nu(K_t)$ are tamely ramified in all points – they satisfy the assumptions of the main lemma.

We want to estimate the weights of all cohomology groups $H_c^\nu(\mathbb{A}_t, K_t \otimes \mathscr{L}(\psi))$. To do this, we look at the spectral sequence

$$E_2^{ab} = H_c^a(\mathbb{A}_t, \mathrm{R}^b h_{t!}\overline{\mathbb{Q}}_l \otimes \mathscr{L}(\psi))$$

which converges to

$$H_c^{a+b}(X_t, h_t^*(\mathscr{L}(\psi))) = H_c^{a+b}(\mathbb{A}_t, K_t \otimes \mathscr{L}(\psi)).$$

For $b > 2N$ or $a > 1$ we have $E_2^{ab} = 0$ using V.7.1(2). The weight of $\mathrm{R}^b h_{t!}(\overline{\mathbb{Q}}_l) \otimes \mathscr{L}(\psi)$ is $\leq b$. Hence the weight of E_2^{ab} is $\leq a + b \leq 2N + 1$. This proves part (i) of the theorem.

Now we assume the generic fiber of this mapping to be either geometrically irreducible, or the dimension of that fiber to be $< N$. By a shrinking of S, we may furthermore assume:

(c') For every morphism

$$t : \mathrm{Spec}(\mathbb{F}_q) \to S$$

the dimensions of all fibers of the induced morphisms

$$h_t = h \otimes \overline{F}_q : \quad X_t = X \otimes \overline{\mathbb{F}}_q \longrightarrow \mathbb{A}_t = \mathbb{A}_S \otimes \overline{\mathbb{F}}_q ,$$

are $\leq N$. The general fiber is geometrically irreducible or its dimension is $< N$.

Claim. The weight is always $\leq 2N$ for $b \leq 2N,\ a \leq 1$.

By the given assumptions, we only have to show that the weight of $E_2^{1,2N}$ is $\leq 2N$. If the dimension of the general fiber of h_t is less than N then $\mathrm{R}^{2N} h_{t!}(\overline{\mathbb{Q}}_t)$ is concentrated on finitely many points and $E_2^{1,2N}$ is zero. Let the general geometric fiber of h_t be irreducible of dimension N.

We now make use of the trace morphism. It is available for flat finitely generated compactifiable morphisms $f : X \to S$ between noetherian schemes of fiber dimension $\leq d$. It is a canonical morphism of sheaves

$$\mathrm{R}^{2d} f_!(\overline{\mathbb{Q}}_l) \to \overline{\mathbb{Q}}_l(-d)$$

with certain natural properties. In particular, if all fibers of the morphism f are geometrically irreducible of dimension d, this trace map is an isomorphism. See

Chap. II, §8 and III, §7 or [SGA4], expose XVIII, p. 553 for further details. Let us therefore – for the moment – assume, that the morphism $h_t : X_t \to \mathbb{A}_t$ is flat. Under this additional assumption, we have the trace map

$$\mathrm{R}^{2N} h_{t!}(\overline{\mathbb{Q}}_l) \;\to\; \overline{\mathbb{Q}}_l(-N)$$

at our disposition. By our assumptions, this trace map is an isomorphism at the generic point of $\mathbb{A}_t$. Consider the exact sequences of sheaves, which is obtained after tensoring with $\mathscr{L}(\psi)$

$$0 \to \mathscr{K} \to \mathrm{R}^{2N} h_{t!}(\overline{\mathbb{Q}}_l) \otimes \mathscr{L}(\psi) \to \mathscr{B} \to 0$$

$$0 \to \mathscr{B} \to \overline{\mathbb{Q}}_l(-N) \otimes \mathscr{L}(\psi) \to \mathscr{Q} \to 0$$

for some sheaf $\mathscr{B}$. The sheaves $\mathscr{K}$ and $\mathscr{Q}$ are skyscraper sheaves concentrated on finitely many points and the weight of $\mathscr{Q}$ obviously satisfies $w(\mathscr{Q}) \leq 2N$.

Now $H_c^*(\mathbb{A}_t, \mathbb{Q}_l(-N) \otimes \mathscr{L}(\psi)) = 0$ as shown in Chap. I, §5. Thus

$$H_c^1(\mathbb{A}_t, \mathrm{R}^{2N} h_{t!}(\overline{\mathbb{Q}}_l) \otimes \mathscr{L}(\psi)) \;\cong\; H_c^1(\mathbb{A}_t, \mathscr{B}) \;\cong\; H_c^0(\mathbb{A}_t, \mathscr{Q}).$$

Hence the weight of $E_2^{1,2N}$ is $\leq 2N$.

Let us finally remove our temporary assumption, that h_t was a flat morphism. For this we have to reduce the case of a general morphism h_t to the case of a flat morphism $\tilde{h}_t$. Without restriction of generality we can replace X_t by $(X_t)_{red}$. Whence we can assume, that X_t is reduced. Let $\mathbb{A}_t = Spec(\overline{\mathbb{F}}_q[z])$, let $J \subset O_{X_t}$ be the ideal of $\overline{\mathbb{F}}_q[z]$ torsion elements of the structure sheaf O_{X_t} of X_t and let $\tilde{X}_t \to X_t$ be the closed subscheme defined by the ideal sheaf J

$$\begin{array}{ccc} \tilde{X}_t & \hookrightarrow & X_t \\ & \searrow^{\tilde{h}_t} & \downarrow h_t \\ & & \mathbb{A}_t \end{array}$$

Then, of course, $\tilde{h}_t : \tilde{X}_t \to \mathbb{A}_t$ is a flat morphism. The closed embedding $\tilde{X}_t \hookrightarrow X_t$ is an isomorphism over the generic point of $\mathbb{A}_t$. Hence it is still an isomorphism, if we discard a certain finite number of points of $\mathbb{A}_t$. But this implies, that the kernel and the cokernel of the map

$$\mathrm{R}^{2N} \tilde{h}_{t!}(\overline{\mathbb{Q}}_l) \otimes \mathscr{L}(\psi) \;\to\; \mathrm{R}^{2N} h_{t!}(\overline{\mathbb{Q}}_l) \otimes \mathscr{L}(\psi)$$

are concentrated on a finite number of closed points. Hence the morphism

$$H_c^1(\mathbb{A}_t, \mathrm{R}^{2N} \tilde{h}_{t!}(\overline{\mathbb{Q}}_l) \otimes \mathscr{L}(\psi)) \;\to\; H_c^1(\mathbb{A}_t, \mathrm{R}^{2N} h_{t!}(\overline{\mathbb{Q}}_l) \otimes \mathscr{L}(\psi))$$

is surjective! Now we can easily deduce the desired results from the case of the flat morphism $\tilde{h}_t$, that was already considered above.

Now let R be $\mathbb{Z}$. The uniform estimates of weight and dimension derived so far will suffice for the following arguments. In particular, together with Theorem V.3.1 they imply the estimates for the trigonometric sums, that were stated in Theorem V.2.2. □

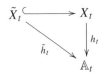

V.10 Uniform Estimates

In this section we give a review of the results of Katz and Laumon [Ka-L] on uniform estimates.

As before, let $R \subseteq \mathbb{C}$ be a finitely generated ring, l a prime that is invertible in R, and $S = \mathrm{Spec}(R)$.

Let $\mathbb{A}^m$ be the m-dimensional affine space over S and let K be a complex in $D_c^b(\mathbb{A}^m, \overline{\mathbb{Q}}_l)$. For a morphism

$$t : \mathrm{Spec}(\mathbb{F}_q) \to S$$

let K_t be the pullback of K to the affine space of dimension m over $\mathbb{F}_q$

$$\mathbb{A}_t^m = \mathbb{A}^m \times_S \mathrm{Spec}(\mathbb{F}_q).$$

Theorem 10.1 *There exists an open nonempty subscheme $S' \subseteq S$ and an open nonempty subscheme $U \subseteq \mathbb{A}^m$ with the following properties:*

Let $t : \mathrm{Spec}(\mathbb{F}_q) \to S'$ be a morphism and $\psi : \mathbb{F}_q \to \overline{\mathbb{Q}}_l^*$ a non-trivial character. We consider the Fourier transform (see III.8.1)

$$T_\psi : D_c^b(\mathbb{A}_t^m, \overline{\mathbb{Q}}_l) \to D_c^b(\mathbb{A}_t^m, \overline{\mathbb{Q}}_l).$$

Then $U_t = U \times \mathrm{Spec}(\mathbb{F}_q) \subseteq \mathbb{A}_t^m$ is nonempty. All cohomology sheaves $\mathscr{H}^\nu(T_\psi(K_t))$ are smooth on U_t. The ranks of $\mathscr{H}^\nu(T_\psi(K_t))|U_t$ can be estimated independently of the choice of t and ψ. The number

$$\sum_\nu (-1)^\nu \mathrm{Rank}\, \mathscr{H}^\nu(T_\psi(K_t))|U_t$$

does not depend on the choice of t and ψ.

Proof. See [Ka-L], theorem 4.1 and corollary 4.2. □

Remark 10.2 One can choose U to be homogeneous. This means, there exists a homogeneous nonvanishing polynomial $F(z_1, \dots, z_m)$ with coefficients in R so that U is the complement in $\mathbb{A}^m$ of the subvariety of codimension one defined by the zero locus of F. This property comes from the fact that the theorem also holds for all characters

$$\psi_\alpha(x) = \psi(\alpha x)\,,\ \forall \alpha \in \mathbb{F}_q^*$$

of $\mathbb{F}_q$.

Remark 10.3 Let K_t be a perverse sheaf. Then the Fourier transform $T_\psi(K_t)$ is perverse again. Therefore the restriction to U_t vanishes

$$\mathscr{H}^\nu(T_\psi(K_t))|\,U_t = 0 \quad \text{for } \nu \neq -m\,,$$

if U respectively F was suitably chosen. Suppose K_t to be pure of weight w. Then $T_\psi(K_t)$ is pure of weight $w + m$.

Let X be a smooth affine scheme over S of relative dimension

$$d = d(X/S)$$

over S and consider an S-morphism

$$\mathbf{h} = (h_1, \dots, h_m) : X \to \mathbb{A}^m ,$$

which we assume to be finite. In addition to these fixed data, we suppose that the following additional data are given: an S-morphism into the multiplicative group G_m

$$g : X \longrightarrow \mathbf{G_{m}}_{,S} ,$$

over S, a specialization morphism

$$t : \mathrm{Spec}(\mathbb{F}_q) \longrightarrow S ,$$

as above, furthermore a non-trivial character $\psi : \mathbb{F}_q \to \overline{\mathbb{Q}}_l^*$, an arbitrary character $\chi : \mathbb{F}_q^* \to \overline{\mathbb{Q}}_l^*$, an m-tuple $\mathbf{a} = (a_1, \dots, a_m) \in \mathbb{F}_q^m$. These extra data will play the role of variables in the following. For each six-tuple $(g, \mathbb{F}_q, t, \mathbf{a}, \psi, \chi)$ of these variables we define a trigonometric sum:

$$f(g, \mathbb{F}_q, t, \mathbf{a}, \psi, \chi) = \sum_{x \in X_t(\mathbb{F}_q)} \psi\left(\sum_{i=1}^{m} a_i \cdot h_i(x)\right) \chi(g(x)).$$

Here $X_t = X \times_S \mathrm{Spec}(\mathbb{F}_q)$.

Theorem 10.4 *([Ka-L], theorem 5.2) There exists a homogeneous nonvanishing polynomial $F(z_1, \dots, z_m) \in R[z_1, \dots, z_m]$, a non-zero open subscheme $S' \subseteq S$, and an integer $A \geq 0$ such that for every six-tuple $(g, \mathbb{F}_q, t, \mathbf{a}, \psi, \chi)$ with $t : \mathrm{Spec}(\mathbb{F}_q) \to S' \subseteq S$ and $0 \neq F(a_1, \dots, a_m) \in \mathbb{F}_q$*

$$|f(g, \mathbb{F}_q, t, \mathbf{a}, \psi, \chi)| \leq A \cdot q^{d/2}$$

holds.

Remark. Katz and Laumon prove that A can be chosen, up to the choice of sign, to be the difference of the **Euler-Poincaré** characteristic of $X \otimes \mathbb{C}$ and the Euler-Poincaré characteristic of the preimage in $X \otimes \mathbb{C}$ of a general hyperplane in $\mathbb{A} \otimes \mathbb{C}$.

Proof. We restrict ourselves to a proof in the special case $\chi = 1$.

We apply Theorem V.10.1 for the complex $K = \mathbf{h}_*(\overline{\mathbb{Q}}_l)[d]$. Let $S' \subseteq S$, $U \subseteq \mathbb{A}^m$, and $F(z_1, \dots, z_m)$ be as in Theorem V.10.1 and Remark V.10.2. Then $K_t = (\mathbf{h} \otimes \mathbb{F}_q)_*(\overline{\mathbb{Q}}_l)[d]$ is a perverse sheaf on $\mathbb{A}_t^m$, which is pure of weight d. Let be $F(a_1, \dots, a_m) \neq 0$. Then

$$\mathbf{a} = (a_1, \dots, a_r) \in U_t(\mathbb{F}_q).$$

Let $\mathscr{L}_{\mathbf{a}}(\psi^{-1})$ be the pullback of the Artin-Schreier sheaf $\mathscr{L}(\psi^{-1})$ under the morphism

$$\mathbb{A}_t^m \to \mathbb{A}_t = \mathbb{A}_t^1$$

defined by

$$(x_1, \ldots , x_m) \mapsto a_1 x_1 + \ldots + a_m x_m \,.$$

The Remarks V.10.2 and V.10.3 made after Theorem V.10.1 now imply

$$H_c^\nu(\mathbb{A}^m \otimes \overline{\mathbb{F}}_q, K_t \otimes \mathscr{L}_{\mathbf{a}}(\psi^{-1})) = 0 \quad \text{for } \nu \neq 0 \,,$$

by the choice of $\mathbf{a}$, $F(\mathbf{a}) \neq 0$. Furthermore $H_c^0(\mathbb{A}^m \otimes \overline{\mathbb{F}}_q, K_t \otimes \mathscr{L}_{\mathbf{a}}(\psi^{-1}))$ is pure of weight d. Finally, the dimension $A = \dim_{\overline{\mathbb{Q}}_l} H_c^0(\mathbb{A}^m \otimes \overline{\mathbb{F}}_q, K_t \otimes \mathscr{L}_{\mathbf{a}}(\psi^{-1}))$ of the cohomology group is independent of the choice of the variables $\mathbb{F}_q,\ t,\ \psi$ and $\mathbf{a}$.

The cohomological interpretation of trigonometric sums then implies the desired estimate using Theorem V.3.1 and the subsequent Remark V.3.2. Apply this for $h : X \to \mathbb{A}$ defined by $h = h_{\mathbf{a}} = \sum_i a_i \cdot h_i$. Note that $h^*(\mathscr{L}(\psi^{-1})) = \mathbf{h}^*(\mathscr{L}_{\mathbf{a}}(\psi^{-1}))$. □

V.11 An Application

Let l be a prime and let $a \neq 0$ be an integer. Let $R = \mathbb{Z}[l^{-1}]$ be localization of $\mathbb{Z}$ with respect to the powers of l and put $S = Spec(R)$. The Kloosterman manifold

$$X = \{(x_1, ..., x_m) \in \mathbb{A}^m \mid x_1 \cdots x_m = 1\}$$

is an affine smooth scheme over S; X is equidimensional of pure relative dimension $d = m - 1$ over S. Let $g = 1 : X \to \mathbf{G}_{\mathbf{m},S}$ be the constant map with image 1. Let $\chi : \mathbb{F}_q^* \to \overline{\mathbb{Q}}_l^*$ be the trivial character. For the six-tuple $(g, \mathbb{F}_q, t, \mathbf{a}, \psi, \chi) = (1, \mathbb{F}_q, t, \mathbf{a}, \psi, 1)$ consider the trigonometric sum

$$f(\mathbb{F}_q, t, \mathbf{a}, \psi) = f(1, \mathbb{F}_q, t, \mathbf{a}, \psi, 1) = \sum_{x \in X(\mathbb{F}_q)} \psi(\sum_{i=1}^m a_i \cdot x_i) \,.$$

We want to apply Theorem V.10.4 of Katz and Laumon. The constant A mentioned in this theorem can be explicitly evaluated by the remark following the theorem. This gives (see [Bry], p. 118)

$$A = m \,.$$

Let $F(z_1, \ldots, z_m) \neq 0$ be the homogeneous polynomial from the theorem. Then the theorem gives a lower bound c such that for every triple $(\mathbb{F}_q, t, \psi)$ and for every point $\mathbf{a} = (a_1, \ldots, a_m) \in (\mathbb{F}_q)^{*m}$ satisfying

$$char(\mathbb{F}_q) \geq c \quad , \quad F(a_1, \ldots, a_m) \neq 0$$

the following estimate $(*)$ holds $|f(\mathbb{F}_q, t, \mathbf{a}, \psi)| \leq m \cdot q^{\frac{m-1}{2}}$.

Counting points, we get for $char(\mathbb{F}_q) \geq c$ and for every $a \in \mathbb{F}_q^*$

$$\#\{\lambda(a_1, \ldots, a_m) \in (\mathbb{F}_q)^{*m} \mid \lambda, a_\nu \in (\mathbb{F}_q)^*, \prod_\nu a_\nu = a\} \geq \frac{(q-1)^m}{m}$$

$$\#\{(z_1, \ldots, z_m) \in (\mathbb{F}_q)^{*m} \mid F(z_1, \ldots, z_m) = 0\} \leq \text{const} \cdot q^{m-1}$$

We choose c big enough. If $char(\mathbb{F}_q) \geq c$, there exists for every $a \in (\mathbb{F}_q)^*$ a point $\mathbf{a} \in (\mathbb{F}_q)^{*m}$, such that

$$F(a_1, \ldots, a_m) \neq 0 \quad , \quad \prod_\nu a_\nu = a$$

The estimate $(*)$ holds for such a point $\mathbf{a}$. Furthermore we have

$$\{(x_1, \ldots, x_m) \in (\mathbb{F}_q)^m \mid \prod_\nu x_\nu = a\}$$

$$= \{(x_1 a_1, \ldots, x_m a_m) \in (\mathbb{F}_q)^m \mid x_\nu \in \mathbb{F}_q, \prod_\nu x_\nu = 1\}$$

and therefore $Kloos_m(q, \psi, a) = f(\mathbb{F}_q, t, \mathbf{a}, \psi)$.

We Obtain the Following Conclusion. Let $char(\mathbb{F}_q) \geq c$ and let

$$\psi : \mathbb{F}_q \to \overline{\mathbb{Q}}_l$$

be a nontrivial character. Then for all a in $\mathbb{F}_q^*$ the following holds

$$|\ Kloos_m(q, \psi, a)\ | \leq m \cdot q^{\frac{m-1}{2}} .$$

This is Deligne's estimate V.4.1 for the generalized Kloosterman sums, and the statement holds for all $q = p^r$ such that $p \geq c$ is large enough.

Remark. On the Kloosterman manifold X there is a natural action of the group

$$G = \{(g_1, .., g_m) \in (\mathbf{G_m})^m \mid \prod_i g_i = 1\} .$$

Therefore the set of points $\mathbf{a} = (a_1, ..., a_m) \in (\mathbb{F}_q)^{*m}$, satisfying the inequality $(*)$, is stable under the action of $G(\mathbb{F}_q)$. We conclude:

The inequality $(*)$ holds for all quadruples $(\mathbb{F}_q, t, \mathbf{a}, \psi)$ satisfying $char(\mathbb{F}_q) \geq c$ and $\mathbf{a} \in (\mathbb{F}_q)^{*m}$.

Bibliography for Chapter V

[Bry] J.L. Brylinski, Transformations Canoniques, Dualite Projective, Transformations de Fourier et Sommes Trigonometriques, in Geometrie et Analyse Microlocales, astérisque, 140–141 (1986), 3–134

[De] P. Deligne, Applications de la formule des traces aux sommes trigonométriques, in [SGA4$\frac{1}{2}$], Springer Lecture Notes 569

[Gr] A. Grothendieck, Formule d'Euler-Poincaré en Cohomologie étale, Expose X in SGA5, Springer Lecture Notes 589

[Ka1] N. Katz, Théorème d'uniformité pour la structure cohomologique des sommes exponentielles. Astérisque 79, 1980 (Sommes exponentielles), 83–146

[Ka2] N. Katz, Gauss Sums, Kloosterman Sums and Monodromy Groups. Annals of Mathematics, Studies, Princeton University Press 1988

[Ka3] N. Katz, Perversity and Exponential Sums, Advanced Studies in Pure Mathematics 17, 1989

[Ka4] N. Katz, Independence of l and weak Lefschetz, Proc. of Symp. pure math, vol. 55, part 1 Motives 1994

[Ka5] N. Katz, Affine cohomological transforms, perversity, and monodromy, J. Amer. Math. Soc. vol. 6, No. 1, 1993, 149–172

[Ka-L] N. Katz and G. Laumon, Transformation de Fourier et Majoration de Sommes Exponentielles, Publ. Math. I.H.E.S. 62, 1986

[Lau1] G. Laumon, Majoration de sommes exponentielles, Exp. 10 in caractéristique d' Euler-Poincaré. Astérisque 82–83

[Lau2] G. Laumon, Semi-continuité du conducteur de Swan, Exp. 9 in Charactéristique d'Euler-Poincaré, Astérisque 82–83

[Se1] J-P. Serre, Corps Locaux, Publications d'Institut de Mathematiques de l'Universite de Nancago VIII, Hermann Paris 1962

[Se2] J-P. Serre, Représentations lineaires des groupes finis. Hermann Paris 1978

[SGA1] Exposé XIII, Propriéte Cohomologique des Faisceaux d'Ensembles et de Groupes non commutatifs, Springer Lecture Notes 224

VI. The Springer Representations

Looking at the myriad approaches to Springer's representations in the literature, one is at first reminded of the proverbial blind men attempting to describe an elephant.

J.E. Humphreys

VI.1 Springer Representations of Weyl Groups of Semisimple Algebraic Groups

This chapter has its own bibliography.

The classification of the finite dimensional irreducible characters of a finite group of Lie type has been established by Lusztig in a series of papers culminating in [Lu1]. The unipotent irreducible characters play an important role in this classification theory. The unipotent characters are strongly related to the unipotent conjugacy classes of the algebraic group on one hand and to the irreducible characters of the Weyl group on the other hand. See also [Ca1], [Ca2] for an overview of these results.

Now T.A. SPRINGER [S6] had previously discovered representations of the Weyl group of a semisimple algebraic group, which are related to the unipotent conjugacy classes of the algebraic group. These Springer representations in fact are important for the understanding of the unipotent characters of a finite simple group of Lie type.

Besides the original method used by Springer there are several constructions for the Springer representations in the literature. One of the basic ideas of the more recent constructions is due to SLODOWY [Sl2], who was the first to view the Springer representations in terms of monodromy representations (in characteristic 0).

In the following our main concern is not the original construction of Springer. We emphasize the elegant methods developed by BORHO, MACPHERSON, LUSZTIG, KAZHDAN, BRYLINSKI [Bry2] and KASHIWARA. These more recent approaches use the theory of perverse sheaves. One of them applies DELIGNE's Fourier transform and is described in the article [Bry2] of Brylinski. Although in loc. cit. Brylinski mentions the influence of Kashiwara, for simplicity we will refer to this approach – which seems to be due to ideas of both Kashiwara and Brylinski – as Brylinski's approach. The representations of the Weyl group obtained in this way will be called Brylinski's representations. There is a closely related approach proposed by Lusztig and used by Borho and MacPherson, which we also present. This second approach leads to different representations of the Weyl groups W. However it turns out, that the difference is just a twist with the sign character of the Coxeter group W.

Let k be an algebraically closed base field of characteristic $p > 0$.

In the following let G be a connected semisimple algebraic group over k. We fix a Borel subgroup B of G. Then every Borel group $B' \subseteq G$ defined over k is conjugate to B, i.e. $B' = gBg^{-1}$ for some $g \in G(k)$. Furthermore $gBg^{-1} = B$ implies $g \in B(k)$. Hence the flag manifold

$$\mathscr{B} = G/B$$

is the manifold of all Borel groups of G. The orbit gB in $\mathscr{B}$ corresponds to the Borel subgroup gBg^{-1}.

The unipotent conjugacy classes in $G(k)$ can be classified in terms of the Bala-Carter theorem, as long as the characteristic p is a good prime for G. See Bala-Carter [B-C] for large primes p and Pommerening [Po] for the more general case.

Let $u \in G(k)$ be a unipotent element of G. Define

$$\mathscr{B}_u \subseteq \mathscr{B}$$

to be the reduced closed subscheme of $\mathscr{B}$, corresponding to all Borel subgroups of G, which contain u. In other words $\mathscr{B}_u(k) = \{g \in G(k)/B(k) : u \in gB(k)g^{-1}\}$. Let G_u be the centralizer of u in G and let $G_u^0 \subseteq G_u$ denote the connected component of the unit element. The group

$$G_u/G_u^0 = A(u)$$

is a finite group. Its structure has been determined by Alekseevski and Mizuno for fields, whose characteristic is large enough compared to the Dynkin graph of the group. See [Hu], 7.17 and 9.4 and [Hu], p. 136–137. One immediately reduces to the case, where G is a simple group. E.g, if G is an adjoint group of the type A_r then $A(u)$ is trivial. For the other types of Dynkin diagrams the following holds: Either $A(u)$ is an elementary two-abelian group of the form $(\mathbb{Z}/2\mathbb{Z})^l$ or it is one of the symmetric groups S_3, S_4, S_5

$$A(u) \in \{(\mathbb{Z}/2\mathbb{Z})^l, S_3, S_4, S_5\} .$$

It seems not to be known or at least it seems not to be explicitly stated in the literature, whether this classification holds for all very good primes[1].

The algebraic group G_u acts on the scheme $\mathscr{B}_u$, hence on its l-adic cohomology groups $H^\nu(\mathscr{B}_u, \overline{\mathbb{Q}}_l)$. Here l will always be an auxiliary chosen prime number different from p. Since G_u^0 is connected, G_u^0 acts trivially on these cohomology groups. This induces an action of the finite group $A(u)$ on the cohomology groups $H^\nu(\mathscr{B}_u, \overline{\mathbb{Q}}_l)$.

Now something surprising happens. Despite the fact that the Weyl group W of G does not act itself on the scheme $\mathscr{B}_u$, Springer was able to prove that there is a "natural" action of W on each of the cohomology groups $H^\nu(\mathscr{B}_u, \overline{\mathbb{Q}}_l)$. This will be shown in Theorem VI.13.4. But let us formulate the consequences

Theorem 1.1 (Springer) *Assume G to be a connected semisimple algebraic group over k. Under certain assumptions on the characteristic p of the base field k, for instance if p is large enough with respect to a bound depending only on the Dynkin diagram of G, the following holds:*

There is a "natural" action of the Weyl group W on the cohomology groups $H^\nu(\mathscr{B}_u, \overline{\mathbb{Q}}_l)$ which commutes with the action of $A(u)$. Put $d = d(u) = \dim(\mathscr{B}_u)$.

[1] As pointed out to us by R. Carter the method of E. Sommers (IMRN 1998, no. 11) should be considered to try to solve this problem.

Let χ be an irreducible character of $A(u)$ with values in $\overline{\mathbb{Q}}_l$, i.e. the character of an irreducible representation of W in a finite-dimensional vector space over $\overline{\mathbb{Q}}_l$. Then the cohomology groups decompose into irreducible representations of $A(u)$. Each χ-isotypic component is a W-module. In particular for the highest cohomology degree, the corresponding isotypic component $V_{u,\chi}$ defined by

$$H^{2d}(\mathscr{B}_u, \overline{\mathbb{Q}}_l) \cong \bigoplus_{\chi \in \hat{A}(u)} \chi \otimes V_{u,\chi}$$

defines an irreducible representation space $V_{u,\chi}$ of the Weyl group, if it is nonzero. This irreducible representation of W on $V_{u,\chi}$ will be called $\phi_{u,\chi}$. The following statements hold:

(1) Every irreducible character of W with values in $\overline{\mathbb{Q}}_l$ is obtained as one of the characters $\phi_{u,\chi}$.
(2) Furthermore the following statements are equivalent:
  (a) $\phi_{u,\chi} = \phi_{u',\chi'}$.
  (b) u and u' are conjugate in $G(k)$ (this allows to identify $A(u)$ with $A(u')$). With the obvious identifications made $\chi = \chi'$ holds.

The representations of W in the highest degree cohomology groups $H^{2d}(\mathscr{B}_u, \overline{\mathbb{Q}}_l)$ resp. in $V_{u,\chi}$ are the **Springer representations** of the Weyl group W, mentioned in the introduction.

An Example. Let $G = \mathrm{SL}_n(k)$ be the special linear group of rank n. Over the algebraically closed field k very unipotent matrix A in G is conjugate to its (essentially unique) Jordan normal form B:

$$B = \begin{pmatrix} B_1 & 0 & & \dots & 0 \\ 0 & B_2 & & \dots & 0 \\ & & \ddots & & \\ 0 & 0 & & \dots & B_r \end{pmatrix} \in G\,.$$

Here, B_ν is the matrix of rank n_ν whose matrix elements are 1 in the diagonal and right to it and 0 otherwise. Obviously

$$n_1 + \ldots + n_r = n, \quad 1 \le n_1 \le \ldots \le n_r\,.$$

Hence the unipotent conjugacy classes of G are in one to one correspondence with the set of all partitions of the number n. The set of these partitions of n corresponds to the set of all conjugacy classes of the permutation group S_n of n elements – or alternatively – corresponds one to one with Young tableaus, i.e. with the isomorphism classes of irreducible representations of S_n on a finite-dimensional vector space over a fixed algebraically closed base field k of characteristic zero. Since S_n is the Weyl group of the group $G = Sl_n$ we find in this special case, rather by chance, a naive description of a one-to-one correspondence between the unipotent conjugacy classes of G and the isomorphism classes of irreducible finite representations over K of the Weyl group of G. This is due to the fact, that for the linear group the groups $A(u)$, defined for the unipotent elements u, are all trivial groups modulo the center of G. See [Hu], 7.17.

Remark 1.2 Let u be a regular unipotent element in $G(k)$, where G is of adjoint type. Then $\mathscr{B}_u = Spec(k)$ and $A(u) = 1$, hence necessarily $\chi = 1$. Then a natural guess is, that the "canonical" action of the Weyl group on $H^0(\mathscr{B}_u, \overline{\mathbb{Q}}_l) = \overline{\mathbb{Q}}_l$ should be the trivial representation $\phi_{u,\chi} = 1_W$ of W. In the formulation of Theorem VI.1.1 we might have postulated this. But this does not give Springer's original construction.

Nevertheless this postulate seems natural. It is satisfied, if one follows Lusztig's approach. Actually Lusztig's and Borho–MacPherson's construction gives the Springer representations only up to a character twist by the sign character of the Coxeter group W. This was shown by Hotta (see [Sl2], 4.6). In fact, we show in the Theorem VI.15.1 below, that Brylinski's construction differs from Lusztig's construction by the same character twist. Therefore Brylinski's representations and the original Springer representations coincide.

Let us consider the other extreme $u = 1$. Then $\mathscr{B}_u = \mathscr{B}$ is the flag variety of G. Its highest cohomology group is one dimensional, and the "canonical" action of W on it should be given by the sign character ϵ_W of the Coxeter group W. In fact we show in the next section, how the Weyl group acts on the other cohomology groups $H^\mu(\mathscr{B}, \overline{\mathbb{Q}}_l)$ of the flag space $\mathscr{B}$ in a natural way. The sheaf theoretic version of Theorem VI.1.1, proved later, indeed provides analogous "natural" actions of the Weyl group W on the cohomology groups $H^\mu(\mathscr{B}_u, \overline{\mathbb{Q}}_l)$ for all $\mu \leq 2d(u)$.

Concerning the Characteristic p. The assumptions of Theorem VI.1.1 on the characteristic p of the base field are the following. The theorem above will be proved under the assumption, that p is a **very good** prime for the group G in the sense of [Sl1], 3.13. See Definition VI.1.6 below.

Remark 1.3 If G is not simply connected, an analogous result holds. The scheme $\mathscr{B}_u$ depends only on the isogeny class of G. For the proof of Theorem 1 it therefore suffices to choose an arbitrary group G for a given Dynkin diagram. An appropriate choice is the adjoint group.

Remark 1.4 In general, although all characters of W occur, not all irreducible characters χ of $A(u)$ need to occur. In other words it can happen, that $V_{u,\chi} = 0$ for certain characters χ of $A(u)$.

Remark 1.5 The Springer representations also exist – now unconditionally – if the construction is done over an algebraically closed base field of characteristic 0 (i.e. without loss of generality over the base field $\mathbb{C}$). In particular the method of BRYLINSKI et al., described in the following, can be carried through over the base field $\mathbb{C}$. Only DELIGNE's Fourier transform, used in the arguments, has to be replaced by another Fourier transform defined by MALGRANGE (see also [Gi]).

Definition 1.6 *Let G be a connected semisimple algebraic group defined over the algebraically closed base field k. A prime p is called* **good** *resp.* **very good** *for G iff it is good resp. very good for all simple components. Let G be a simple group. Then these properties only depend on the Dynkin diagram of G:*

a) If G is Type A_r then every prime is good. If G is Type B_r, C_r or D_r, then p is good if $p \neq 2$. If G is Type E_6, E_7, F_4 or G_2, then p is good if p is $\neq 2, 3$. If G is Type E_8 then p is good if p is $\neq 2, 3, 5$.
b) A prime p is very good for the simple group G if p is good for G and if, in addition, $p \not| r+1$ in case of the type A_r.
c) A prime p is called **bad** *for the semisimple group G, if p is not good for G. (See later Theorem VI.4.1 for another, more instructive definition.)*

Remark 1.7 Let $L(\Sigma^*) \subseteq X(T) \otimes \mathbb{Q}$ be the lattice generated by the weights. Then $L(\Sigma) \subseteq X(T) \subseteq L(\Sigma^*)$. Here $X(T)$ denotes the character group and $L(\Sigma)$ denotes the root lattice. Then the primes p, that are very good for G, do not divide the order of the group $L(\Sigma^*)/L(\Sigma)$. If

$$\tilde{G} \to G$$

is the simply connected cover of G then $\tilde{G} \to G$ is separable if the characteristic p is very good for G. In this case, the Lie algebras of G and $\tilde{G}$ are canonically isomorphic. So the Lie algebra of G depends only on its Dynkin diagram. See also [Sl1], 3.6.

To construct SPRINGER's representations of the Weyl group, we need further ingredients:

(1) Properties of DELIGNE's Fourier transform.
(2) Properties of the Lie algebra of a semisimple algebraic group over a base field with positive characteristic.
(3) A resolution of the singularities of the nilpotent variety in the Lie algebra of a semisimple algebraic group respectively GROTHENDIECK's simultaneous resolution of singularities in this case.

VI.2 The Flag Variety $\mathscr{B}$

In this section, we describe the cohomology ring of the flag variety $\mathscr{B} = G/B$. Moreover, we define an operation of the Weyl group W of G on the cohomology ring $H^\bullet(\mathscr{B}, \overline{\mathbb{Q}}_l)$. This is due to A.Borel and we briefly describe these well known results. For further details see [Sl2], chap. IV and [Bb] and [De1], [De2].

Let $X(T) = Hom(T, \mathbb{G}_m)$ be the character group of a maximal torus T in B. The Weyl group W of the torus T acts on T, hence canonically on $X(T)$. Each character

$\alpha \in X(T)$ gives rise to a line sheaf $\mathscr{L}_\alpha$ of the flag manifold $\mathscr{B}$. Since $T = B/U$, the character α can be viewed as a character of the Borel group B. In other words, there exists a one dimensional representation E_α of B. As a vector space $E_\alpha = k$. The line sheaf $\mathscr{L}_\alpha$ is defined as the sheaf of sections of the associated vector bundle

$$G \overset{B}{\times} E_\alpha = (G \times E_\alpha)/B ,$$

and similarly define $\mathscr{L}'_\alpha$ as the pullback of $\mathscr{L}_\alpha$ to G/T. (For the notation $G \overset{B}{\times} E_\alpha$ see also VI.8.1 in this chapter). Obviously $\mathscr{L}_{\alpha+\beta} = \mathscr{L}_\alpha \otimes \mathscr{L}_\beta$. The first Chern class $c(\mathscr{L}_\alpha) = c(\alpha) \in H^2(\mathscr{B}, \overline{\mathbb{Q}}_l)$ therefore defines a homomorphism

$$c : \ X(T) \ \to \ H^2(\mathscr{B}, \overline{\mathbb{Q}}_l) ,$$

which becomes an isomorphism $X(T) \otimes \overline{\mathbb{Q}}_l \cong H^2(\mathscr{B}, \overline{\mathbb{Q}}_l)$ after a tensoring with $\overline{\mathbb{Q}}_l$. Extended to the symmetric algebra $S^\bullet(V)$ of V, this gives rise to a ring homomorphism

$$S^\bullet \to H^{2\bullet}(\mathscr{B}, \overline{\mathbb{Q}}_l) .$$

Let $J^\bullet$ denote the homogeneous ideal of $S^\bullet(V)$, generated by the W-invariant homogeneous elements of positive degree in $S^\bullet(V)$. Then

Theorem 2.1 (Borel) *The homomorphism above induces a ring isomorphism*

$$S^\bullet(V)/J^\bullet \ \cong \ H^{2\bullet}(\mathscr{B}, \overline{\mathbb{Q}}_l) .$$

By transport of structure this makes W act on the graded cohomology ring $H^\bullet(\mathscr{B}, \overline{\mathbb{Q}}_l)$. *All odd degree cohomology groups* $H^{2\nu+1}(\mathscr{B}, \overline{\mathbb{Q}}_l)$ *of* $\mathscr{B}$ *vanish.*

Proof. The proof of this theorem can be reduced to the characteristic zero case. It is enough to consider a simple Chevalley group, which is connected and simply connected or of adjoint type. See [Bb], V.5.3. See also Demazure [De1] for a direct approach in characteristic $p > 0$. □

The representation of W on $S^\bullet(V)/J^\bullet$ turns out to be the regular representation of W. See proposition 5 of [Bb], V.5.4. Therefore

Theorem 2.2 *The representation of W defined in the last theorem is isomorphic to the regular representation of W. On the highest one dimensional cohomology group* $H^{2dim(\mathscr{B})}(\mathscr{B}, \overline{\mathbb{Q}}_l)$ *the action is given by the sign character* ε_W *of the Coxeter group W.*

By this theorem the regular representation of W on the group algebra $\overline{\mathbb{Q}}_l[W]$ is endowed with a natural grading induced by the cohomology degrees. To emphasize this we also write

$$\overline{\mathbb{Q}}_l[W]^\bullet = \bigoplus_\mu H^{2\mu}(\mathscr{B}, \overline{\mathbb{Q}}_l)[-2\mu]$$

for this graded W-module.

For any irreducible character ϕ of W let $H^{\mu}(\mathscr{B}, \overline{\mathbb{Q}}_l)_\phi$ be the ϕ-isotypic component of the cohomology group in degree μ. We have the corresponding Poincare polynomials $P_\phi(t) = deg(\phi)^{-1} \sum_\nu dim_{\overline{\mathbb{Q}}_l} H^{2\nu}(\mathscr{B}, \overline{\mathbb{Q}}_l)_\phi \cdot t^\nu$.

There is a second approach via the

Lemma 2.3 *Let $f : Z \to S$ be a smooth morphism of schemes. Let the fibers of f be isomorphic to affine spaces of dimension N. Then $f_*(\overline{\mathbb{Q}}_{l\,Z}) = \overline{\mathbb{Q}}_{l\,S}$ and $R^\nu f_*(\overline{\mathbb{Q}}_{l\,Z}) = 0$ for $\nu > 0$. In particular $R\Gamma(Z, \overline{\mathbb{Q}}_{l\,Z}) = R\Gamma(S, \overline{\mathbb{Q}}_{l\,S})$. Similarly $Rf_!\overline{\mathbb{Q}}_{l\,Z} = \overline{\mathbb{Q}}_{l\,S}[-2N](-N)$ by the trace map.*

We apply this lemma in the following two cases:

1. $f : Z \to S$ is a vector bundle.
2. The morphism
$$f : G/T \to G/B = \mathscr{B}\ .$$
The fibers of this morphism are isomorphic to the unipotent radical U of the Borel group B. As a scheme U is isomorphic to affine space of dimension $\#\Sigma^+$ (the number of positive roots).

This second case is of particular interest for us, since G/T and G/B have canonically the same cohomology groups by the last lemma. On the other hand, G/T carries visibly a canonical left action of W defined by

$$w(g \cdot T) = g n_w^{-1} \cdot T \quad , \quad n_w \in N(T)\ ,$$

where n_w is a representative of W in the normalizer $N(T)$ of the torus T. This defines a canonical action of W on the complex level, since

$$\mathrm{R}\Gamma(\mathscr{B}, \overline{\mathbb{Q}}_l) = \mathrm{R}\Gamma(G/T, \overline{\mathbb{Q}}_l)\ .$$

The induced W-action on the cohomology ring $H^\bullet(\mathscr{B}, \overline{\mathbb{Q}}_l)$ of the flag manifold $\mathscr{B}$ was studied by Slodowy. To distinguish it from the previous construction due to Borel we call it the Slodowy action. Both these actions coincide. This is seen as follows: There is a cartesian diagram

$$\begin{array}{ccc} G \times^T E_\alpha & \longrightarrow & G \times^T E_\beta \\ \downarrow & & \downarrow \\ G/T & \xrightarrow{\ w\ } & G/T \end{array} ,$$

where

$$\beta(t) = \alpha(n_w^{-1} t n_w)$$

and where $n_w \in N(T)$ is a representative of $w \in W$. The upper arrow is defined by $(g, \lambda) \mapsto (g n_w^{-1}, \lambda)$. Observe $(gt, \alpha(t^{-1})\lambda) \mapsto (g t n_w^{-1}, \alpha(t^{-1})\lambda) =$

$(gn_w^{-1}t', \alpha(n_w^{-1}(t')^{-1}n_w)\lambda)$, which is $(gn_w^{-1}t', \beta((t')^{-1})\lambda)$. So this is well defined. $\mathscr{L}'_\alpha$ is the pullback of $\mathscr{L}'_\beta$, or equivalently $(w^{-1})^*(\mathscr{L}'_\alpha) = \mathscr{L}'_\beta$ resp. $w^*(\mathscr{L}'_\beta) = \mathscr{L}'_\alpha$. Hence

$$(w^{-1})^*(c(\alpha)) = c(\beta) .$$

This completely determines the induced left action of W on the cohomology ring $H^\bullet(G/T, \overline{\mathbb{Q}}_l)$. Observe, that we converted the induced right action of W on the cohomology of G/T – defined by the pullback w^* – into a left action $w(.) = (w^*)^{-1}(.)$. This left action is uniquely determined by $w(c(\alpha)) = c(\beta)$ or $w(\alpha)(t) = \alpha(n_w^{-1}tn_w)$.

In particular

$$\overline{\mathbb{Q}}_l[W] \longrightarrow End_{D(Spec(k))}(R\Gamma(\mathscr{B}, \overline{\mathbb{Q}}_l))$$

is an injective ring homomorphism.

VI.3 The Nilpotent Variety $\mathscr{N}$

For the following we describe the scheme $\mathscr{B}_u$. This uses the Lie algebra $\mathfrak{g} = Lie(G)$ of G. Let $U \subseteq B$ be the unipotent radical of the fixed Borel subgroup B of G.

Definition 3.1 *Consider algebraic representations $\varrho : G \to GL_n(k)$ of G and the corresponding Lie algebra representations $Lie(\varrho) : \mathfrak{g} = Lie(G) \to Lie(GL_n(k)) = M_n(k)$, where M_n is the matrix algebra. An element $x \in Lie(G)$ is called* **nilpotent** *iff one of the following two equivalent properties holds:*

(1) For every representation ϱ the matrix $Lie(\varrho)(x)$ is nilpotent.
(2) There is a faithful representation such that $\mathrm{Lie}(\varrho)(x)$ *is nilpotent. Then ϱ is an isomorphism onto a closed subscheme of* GL_n.

In [S5], 4.4.20 it is shown, that both definitions are equivalent.

Let $\mathscr{U}$ denote the **unipotent variety** of G, i.e. the Zariski closed subset of all unipotent elements of G. Similar let $\mathscr{N}$ denote the **nilpotent variety** of $Lie(G)$, i.e. the Zariski-closed subset of $Lie(G)$ of all its nilpotent elements. We endow the Zariski closed subsets $\mathscr{U}$ resp. $\mathscr{N}$ with the reduced subscheme structure of G resp. $Lie(G)$. The Lie algebra $\mathfrak{u} = Lie(U)$ of U is contained in $\mathscr{N}$

$$Lie(U) \subseteq \mathscr{N} \quad , \qquad U \subseteq \mathscr{U}.$$

The algebraic group G acts both on $\mathscr{U}$, by inner automorphisms, and on $\mathscr{N}$, by the adjoint representation. We write $g(v) = Ad(g)(v)$ for this adjoint action. Since all Borel groups are conjugate, we get

$$\mathscr{U} = \bigcup_{g \in G(k)} gUg^{-1}.$$

This implies

$$\mathcal{N} = \bigcup_{g\in G(k)} g\big(Lie(U)\big)$$

under the following important

Assumption (∗). Assume, that there exists an G-equivariant isomorphism of schemes

$$\sigma : \mathcal{N} \overset{\cong}{\to} \mathcal{U} ,$$

such that $\sigma(Lie(U)) \subseteq U$. ($\sigma$ should be viewed as a substitute for the exponential map, which is not a priori defined in case of characteristic $p > 0$).

Assumption (∗) implies $\sigma(Lie(U)) = U$, since $Lie(U)$ and U are irreducible, smooth and of the same dimension. Therefore the claim above follows immediately. Furthermore for an arbitrary Borel group B' of G and its unipotent radical U', also $Lie(U') \subseteq \mathcal{N}$ and

$$\sigma(Lie(U')) = U'$$

holds by conjugation.

Now suppose, we have been given an isomorphism σ as in the assumption (∗) above. The unipotent elements u of $G(k)$ then uniquely correspond to the nilpotent elements v of $Lie(G)$ via

$$\sigma(v) = u .$$

This allows to describe the scheme $\mathcal{B}_u$ in another way.

For this we may a priori assume that $u \in U(k)$, hence $v \in Lie(U)(k)$, since we know from assumption (∗) that $\mathcal{N} = \cup_{g\in G(k)} g(Lie(U))$ and $\mathcal{B}_u \cong \mathcal{B}_{gug^{-1}}$. For $u \in U(k)$ define $H = \{g \in G : g^{-1}ug \in U\}$ or $H = \{g \in G : u \in gBg^{-1}\}$. H is a Zariski closed subset of G. We endow H with the reduced subscheme structure. B acts on H from the right. Furthermore H/B exists and, viewed as a subset of $G/B = \mathcal{B}$, can be identified with $\mathcal{B}_u$

$$\begin{array}{ccc} H & \hookrightarrow & G \\ \downarrow & & \downarrow \\ \mathcal{B}_u \cong H/B & \hookrightarrow & G/B = \mathcal{B} \end{array}$$

Now we use assumption (∗). It implies, that $H = \{g \in G : g^{-1}(v) \in Lie(U)\}$ can be described in terms of the Lie algebra. Furthermore the conjugation action $G \times U \to \mathcal{U}$ defined by $(g, u) \mapsto gug^{-1}$ may be replaced by $(g, v) \mapsto g(v)$.

$$\begin{array}{ccc} G \times U & \longrightarrow & \mathcal{U} \\ {\scriptstyle id\times\sigma}\uparrow{\scriptstyle\cong} & & {\scriptstyle\sigma}\uparrow{\scriptstyle\cong} \\ G \times Lie(U) & \longrightarrow & \mathcal{N} \end{array}$$

On $G \times Lie(U)$ the Borel group B acts by $b(g, z) = (gb^{-1}, bz)$, where $b \in B$ and $g \in G, z \in Lie(U)$. The quotient scheme

$$G \overset{B}{\times} Lie(U) = \big(G \times Lie(U)\big)/B$$

exists and defines a vector bundle over $\mathscr{B} = G/B$ with fiber $Lie(U)$. The adjoint action $(g, z) \mapsto g(z)$ induces a morphisms

$$q_{\mathscr{N}} : G \overset{B}{\times} Lie(U) \longrightarrow Lie(G) ,$$

whose image is $\mathscr{N}$, which follows from the assumption $(*)$ and the remarks above. This morphism q is a proper morphism. This is shown after VI.8.1. Let us assume this for the moment.

Consider the fibers $q_{\mathscr{N}}^{-1}(v)$ for $v \in \mathscr{N}(k)$, endowed with the reduced subscheme structure. They can be described in terms of the set $H = \{g \in G : g^{-1}(v) \in Lie(U)\}$ defined above. In fact, the map $\varphi : H \to G \overset{B}{\times} Lie(U)$ induced by $h \mapsto (h, h^{-1}(v))$ defines an isomorphism of the quotient H/B with the fiber $q_{\mathscr{N}}^{-1}(v)$. Set theoretically this is clear, since obviously $q_{\mathscr{N}}^{-1}(v)(k) = H(k)/B(k)$. However this is enough, since the morphism $G \to \mathscr{B}$ locally has enough etale sections. This shows $\mathscr{B}_u \cong H/B \cong q_{\mathscr{N}}^{-1}(v)$ provided $\sigma(v) = u$.

Theorem 3.2 (Springer Fibers) *Assume that there exists a G-equivariant isomorphism $\sigma : \mathscr{N} \overset{\cong}{\to} \mathscr{U}$, such that $\sigma(Lie(U)) \subseteq U$. Consider the morphism*

$$q_{\mathscr{N}} : G \overset{B}{\times} Lie(U) \twoheadrightarrow \mathscr{N} \subseteq Lie(G)$$

induced by the adjoint representation. Let u be an unipotent element of G and $v = \sigma^{-1}(u)$ be the corresponding nilpotent element of $Lie(G)$. Then there is a natural isomorphism

$$\mathscr{B}_u \cong q_{\mathscr{N}}^{-1}(v) \quad , \quad \sigma(v) = u \in \mathscr{U} ,$$

both schemes being endowed with the reduced subscheme structures.

Concerning Assumption $(*)$. The existence of an isomorphism $\sigma : \mathscr{N} \overset{\cong}{\to} \mathscr{U}$ with the desired properties is not a priori clear. However σ – if it exists at all – is usually not uniquely defined. Let us give examples:

The Case $G = \mathrm{SL}_n$. In this case the Lie algebra is the vector space M_n of all $n \times n$-matrices with trace zero: $Lie(\mathrm{SL}_n) = M_n$. Let E be the unity matrix. Then $\sigma : \mathscr{N} \to \mathscr{U}$ defined by $\sigma(Y) = E + Y$ is a well defined G-equivariant map. Its inverse is the map $u \mapsto u - E$. So σ has the desired properties.

To give a different choice of σ in this example assume, that the characteristic p of k be different from 2. Then for every nilpotent matrix A, the Cayley transform

$$C(A) = (E + A)(E - A)^{-1}$$

or $C(A) = -E + 2(E - A)^{-1}$ is defined. Conversely for every unipotent matrix B, the inverse Cayley transform

$$F(B) = (B - E)(B + E)^{-1}$$

or $F(B) = E - \left(\frac{B+E}{2}\right)^{-1}$ is defined. These Cayley transforms provide a SL_n-equivariant isomorphism between nilpotent and unipotent matrices and transform upper triangular matrices into corresponding upper triangular matrices. Hence C has the desired properties

$$C : \mathcal{N} \overset{\cong}{\to} \mathcal{U} .$$

The Classical Groups. Again assume $\mathrm{char}(k) = p \neq 2$. Let $G \subseteq \mathrm{SL}_n$ be either the special orthogonal group

$$G = \{A \in \mathrm{SL}_n : A \cdot {}^t\!A = E\},$$

or in case $2|n$, the symplectic group

$$G = \{A \in \mathrm{SL}_n : A \cdot I \cdot {}^t\!A = I\}.$$

Here, ${}^t\!A$ denotes the transposed matrix to the matrix A, and I is the standard symplectic matrix $I = \left(\begin{smallmatrix} 0 & -E \\ E & 0 \end{smallmatrix}\right)$.

The Lie algebra $\mathfrak{g} = Lie(G)$ is $\{A \in M_n : A + {}^t\!A = 0\}$ for the orthogonal respectively $\{A \in M_n : AI + I\,{}^t\!A = 0\}$ for the symplectic group. In both cases, it is easy to verify that for every nilpotent matrix $A \in Lie(G)(k)$ the Cayley transform defined above has the property

$$C(A) \in G(k) .$$

If B is a Borel group of G with the unipotent radical U, then one can find an element $g \in \mathrm{SL}_n$ such that gBg^{-1} is in the subgroup of upper triangular matrices in Sl_n. Due to the equivariance of C therefore $C(Lie(U)) \subseteq U$. So for $p \neq 2$ the Cayley transform has the desired properties also for the orthogonal and symplectic groups.

A Third Example. Consider subgroups G of Sl_n and assume $\mathrm{char}(k) = p \geq n$.

For a nilpotent matrix $a \in M_n$ respectively a unipotent matrix $b \in \mathrm{SL}_n$ the equation $a^\nu = (b - E)^\nu = 0$ holds for all $\nu \geq n$. Therefore the exponential

$$\exp(a) \;=\; \sum_{\nu=0}^{\infty} \frac{a^\nu}{\nu!} \;=\; \sum_{\nu=0}^{n-1} \frac{a^\nu}{\nu!}$$

is well defined. Similar for its inverse, the logarithm $\log(b) = \log(E + (b - E))$

$$\log(b) \;=\; \sum_{\nu=1}^{\infty} (-1)^{\nu-1} \frac{(b-E)^\nu}{\nu} \;=\; \sum_{\nu=1}^{n-1} (-1)^{\nu-1} \frac{(b-E)^\nu}{\nu} .$$

The function $a \mapsto \exp(a)$ is equivariant and provides an isomorphism between nilpotent and unipotent matrices. It transforms upper triangular matrices into upper triangular matrices.

Now let $G \subseteq \mathrm{SL}_n$ be a semisimple algebraic subgroup of SL_n. Let d be an upper bound of all degrees of polynomials of a defining equation system in the matrix coefficients for G. Recall that there exists an intrinsic definition of the exponential function on the Lie algebra of an algebraic group ([D-G], see the proof of proposition 3.1 from chap. II, §3). We immediately deduce

$$p \geq d(n-1) \Rightarrow \exp(\mathcal{N}) \subseteq G\,.$$

As in the previous case this exponential map has all desired properties

$$\exp : \mathcal{N} \to \mathcal{U}\,.$$

Obviously $X \in \mathcal{N}_G$ iff $X \in Lie(G) \cap \mathcal{N}_{Sl(n)}$ by VI.3.1(2). Since the Borel group $B = B_G$ of G can be assumed to be contained in the group $B_{Sl(n)}$ of upper triangular matrices (use conjugation in $Sl(n)$), we get $B_G = B_{Sl(n)} \cap G$. Recall B_G is a maximal solvable connected subgroup of G. It follows that exp maps $\mathfrak{u}_G = Lie(U_G)$ into the unipotent radical U_G of B_G.

Springer's Result. From the last example it is clear, that for a given Dynkin diagram, the assumption (∗) holds for all primes large enough. However there is a more refined result due to Springer [S1]. The proof of this result is rather involved. For instance it uses the fact that the unipotent variety and also the nilpotent variety $\mathcal{N}$ are normal varieties. This latter fact follows from results of Demazure [De1]. See also [Hu], p. 117 for further information. Depending on the above mentioned results Springer ([S1]) proved the following theorem

Theorem 3.3 *Let the characteristic p of k be a very good prime for the semisimple algebraic group G over k. Then there is an G-equivariant isomorphism*

$$\sigma : \mathcal{N} \overset{\cong}{\to} \mathcal{U}$$

with $\sigma(\mathfrak{u}) \subseteq U$.

Springer reduces the proof to the construction of an B-equivariant isomorphism $\sigma^{-1} : U \cong Lie(U)$. This induces a G-equivariant isomorphism

$$G \overset{B}{\times} U \cong G \overset{B}{\times} Lie(U)$$

and an isomorphism $\mathcal{U} \cong \mathcal{N}$ by proper descent via the Stein factorization.

Remark 3.4 A completely different approach can be found in [B-R] p. 315. These authors start with a G-equivariant map $\varphi : G \to Lie(G)$, such that $\varphi(e) = 0$ and $d\varphi_e = id_{Lie(G)}$. Hence φ is etale in a neighborhood of the unit element $e \in G(k)$. Such maps exist in the case of a very good characteristic. Since e is a closed orbit, it follows from a characteristic p version of Luna's etale slice theorem, that φ induces an isomorphism $\varphi : \mathcal{U} \to \mathcal{N}$ of the closed fiber over e resp 0, by considering

$$
\begin{array}{ccc}
G & \xrightarrow{\varphi} & Lie(G) \\
\downarrow & & \downarrow \\
G/Ad(G) & \longrightarrow & Lie(G)/Ad(G) \\
\downarrow \cong & & \downarrow \cong \\
T/W & \longrightarrow & Lie(T)/W \\
\Cup & & \Cup \\
e & \longmapsto & 0
\end{array}
$$

VI.4 The Lie Algebra in Positive Characteristic

As before let G be a semisimple connected group defined over an algebraically closed field k of positive characteristic p. Let $B \subseteq G$ be a Borel group of G. Let $U \subseteq B$ be its unipotent radical, and let $T \subseteq B$ be a maximal torus of G. The corresponding Lie algebras are

$$\mathfrak{t} = Lie(T), \quad \mathfrak{u} = Lie(U), \quad \mathfrak{b} = Lie(B) = \mathfrak{t} \oplus \mathfrak{u} \subseteq \mathfrak{g} = Lie(G) .$$

Let $X(T)$ be the group of all characters $T \to G_m$, let $\Sigma \subseteq X(T)$ be the set of all roots of G. Our choice of a Borel group defines a subset $\Sigma^+ \subseteq \Sigma$ of positive roots, such that $\Sigma = \Sigma^+ \cup \Sigma^-$, $\Sigma^- = -\Sigma^+$. G acts on its Lie algebra $\mathfrak{g}$ by the adjoint representation, and so does T. This defines the decomposition of $\mathfrak{g}$ into root spaces $\mathfrak{g}^\alpha$

$$\mathfrak{g} = \mathfrak{t} \oplus \bigoplus_{\alpha \in \Sigma} \mathfrak{g}^\alpha = \mathfrak{t} \oplus \mathfrak{u}^+ \oplus \mathfrak{u}^- ,$$

where $\mathfrak{g}^\alpha = \{x \in Lie(G) : t(x) = \alpha(t) \cdot x \ \forall t \in T\}$ and

$$\mathfrak{u} = \mathfrak{u}^+ = \bigoplus_{\alpha \in \Sigma^+} \mathfrak{g}^\alpha, \qquad \mathfrak{u}^- = \bigoplus_{\alpha \in \Sigma^-} \mathfrak{g}^\alpha .$$

We now give a better characterization of the "bad" primes then the one previously given in Definition VI.1.6. Let $S = (\alpha_1, \ldots, \alpha_r) \subseteq \Sigma^+$ be a system of simple roots in Σ^+. The sublattice $L(\Sigma)$ in $X(T)$ generated by all roots has the simple roots as a basis

$$L(\Sigma) = \bigoplus_{\nu=1}^{r} \mathbb{Z}\alpha_\nu .$$

Every root $\alpha \in \Sigma^+$ can be uniquely written in the form $\alpha = \sum_{i=1}^r m_\nu \alpha_\nu$ with nonnegative integers m_ν. The root $\tilde{\alpha} \in \Sigma^+$ of maximal height $ht(\tilde{\alpha}) = \sum_\nu m_\nu$ is called the longest root. See [Bb], chap. VI, §1.8, prop. 25. It is well known, that the coefficients $m_1, \ldots, m_r$ of the longest root in Σ^+ have the following property

Theorem 4.1 ([S-St], chap. 1, 4.3) *The following properties of a prime p are equivalent:*

(a) p is a bad prime in the sense of Definition VI.1.6.
(b) There is an index i such that $p = m_i$.
(c) There is an index i such that $p|m_i$.
(d) There is an index i such that $p \leq m_i$.

Proof. This follows from a case by case inspection of the tables for the irreducible root systems. See also ([Bb, VI §4]). □

A character $\varphi : T \to \mathbb{G}_m$ induces a linear map $\mathrm{Lie}(\varphi) : \mathfrak{t} \to k$. So there is a natural homomorphism

$$X(T) \to \mathfrak{t}^* ,$$

where φ maps to $Lie(\varphi)$. This defines an isomorphism $X(T) \otimes_{\mathbb{Z}} k \overset{\cong}{\to} \mathfrak{t}^*$.

Let the characteristic p of K be a very good prime for G (see Definition VI.1.6). Then Remark VI.1.7 implies $p \not| [X(T) : L(\Sigma)]$. Therefore also

$$L(\Sigma) \otimes_{\mathbb{Z}} k \cong \mathfrak{t}^*.$$

Consider the induced map $\Sigma \longrightarrow \mathfrak{t}^*$, called "reduction" and denoted by $\alpha \mapsto \overline{\alpha}$. Obviously $[t, x] = \overline{\alpha}(t) \cdot x$ for $t \in \mathfrak{t}$ and $x \in \mathfrak{g}^{\alpha}$.

Theorem 4.2 *Let the characteristic p be a very good prime for G. Then the images under the reduction map of the roots* $\alpha \in \Sigma$ *are nonzero in* $\mathfrak{t}^*$

$$\Sigma \to \mathfrak{t}^* \setminus \{0\} .$$

If furthermore p is odd, the reduction map is injective.

Remark. So if p is an odd very good prime we can identify Σ with its image in $\mathfrak{t}^*$. In other words we need not distinguish between $\alpha \in \Sigma$ and its image $\overline{\alpha} \in \mathfrak{t}^*$. Hence in particular, for all $\alpha \in \Sigma$

$$\mathfrak{g}^{\alpha} = \{x \in \mathfrak{g} : [t, x] = \alpha(t) \cdot x \ \forall t \in \mathfrak{t}\}.$$

Proof. Recall $L(\Sigma) = \oplus_{\nu=1}^{r} \mathbb{Z} \cdot \alpha_\nu$. Furthermore $\mathfrak{t}^* = \oplus_{\nu=1}^{r} k \cdot \overline{\alpha}_\nu$. Every root $\alpha \in \Sigma$ is conjugate to a simple root in S under the Weyl group. Therefore we can assume $\alpha \in S$ without restriction of generality, say $\alpha = \alpha_1$. But then $\overline{\alpha} \neq 0$ is obvious. This proves the first claim.

To proof injectivity assume p also odd. Let $\beta = \sum_{\nu=1}^{r} c_\nu \alpha_\nu$ be a second root. Suppose to the contrary

$$\overline{\alpha_1} = \overline{\alpha} = \overline{\beta} = \sum_{\nu} \overline{c}_\nu \overline{\alpha}_\nu .$$

Here, $\bar{c}$ is the image of the integer c in k. Then $p \mid c_\nu$ for $\nu \geq 2$. It is well known that $|c_\nu| \leq m_\nu$ holds for all roots β, if $\sum m_\nu \alpha_\nu \in \Sigma^+$ is the longest root. We thus obtain $p \leq |c_\nu| \leq m_\nu$ for all $\nu \geq 2$ such that $c_\nu \neq 0$. Since p is a very good prime, Theorem VI.4.1 implies $c_\nu = 0$ for all $\nu \geq 2$. Hence $\beta = \pm\alpha$. Therefore $2\overline{\alpha_1} = 0$ or $\overline{\alpha_1} = 0$, since $p \neq 2$. This is a contradiction and completes the proof. □

VI.5 Invariant Bilinear Forms on $\mathfrak{g}$

If p is a very good prime for G, then the Lie algebra of a group, which is isogenious to the semisimple group G, is isomorphic to $\mathfrak{g} = Lie(G)$. See Remark VI.1.7. Hence lemma 5.3 from chap. 1 of [S-St] holds without restrictions on the isogeny type of the semisimple group G. Therefore

Theorem 5.1 *Let the characteristic p of the base field be a very good prime for G. Then on the Lie algebra $\mathfrak{g}$ of G there is a G-invariant nondegenerate symmetric k-bilinear form $b(x, y)$ with values in k.*

Remark. If $G = \mathrm{SL}_n$ is the special linear group, then b can be chosen to be

$$b(x, y) = \mathrm{trace}(x \cdot y), \qquad x, y \in \mathfrak{g} \subseteq M_n.$$

If $p \neq 2$ and $G \subset \mathrm{Sl}_n$ is either the special orthogonal group or the symplectic group with the natural realization defined earlier, then the restriction of $\mathrm{trace}(x \cdot y)$ to $Lie(G)$ remains nondegenerate.

Lemma 5.2 *Assume p is a very good prime for G and $b(x, y)$ the bilinear form as in Theorem VI.5.1. Then*

$$\mathfrak{g} = \mathfrak{t} \oplus \bigoplus_{\alpha \in \Sigma^+} (\mathfrak{g}^\alpha \oplus \mathfrak{g}^{-\alpha})$$

is an orthogonal decomposition of $\mathfrak{g}$. In particular, the restrictions $b|\mathfrak{t}$ and $b|\mathfrak{g}^\alpha \oplus \mathfrak{g}^{-\alpha}$ are nondegenerate. Furthermore $\mathfrak{g}^\alpha \perp \mathfrak{g}^\alpha$. The bilinear form b induces a nondegenerate pairing between $\mathfrak{g}^\alpha$ and $\mathfrak{g}^{-\alpha}$ for all $\alpha \in \Sigma^+$.

Corollary 5.3 *In the situation of the last lemma $\mathfrak{u}^\perp = \mathfrak{t} \oplus \mathfrak{u} = \mathfrak{b}$ and $\mathfrak{u} = \mathfrak{u}^+ = \bigoplus_{\alpha \in \Sigma^+} \mathfrak{g}^\alpha$ holds.*

Proof. Let φ be a character of T and let be $\mathfrak{g}^\varphi = \{g \in \mathfrak{g} : t(g) = \varphi(t)g \ \forall t \in T\}$ be its root space. Suppose given two characters φ, ψ of T such that $\varphi \cdot \psi$ is not the trivial character. Then

$$\mathfrak{g}^\varphi \perp \mathfrak{g}^\psi \quad , \quad \varphi \cdot \psi \neq 1 .$$

This is clear, since otherwise choose an element $t \in T(k)$ with $\varphi(t) \cdot \psi(t) \neq 1$. For $x \in \mathfrak{g}^\varphi$ and $y \in \mathfrak{g}^\psi$ we have $\varphi(t) \cdot b(x, y) = b(t(x), y) = b(x, t^{-1}(y)) = \psi(t)^{-1} \cdot b(x, y)$. Hence $b(x, y) = 0$. □

VI.6 The Normalizer of $Lie(B)$

The scheme $\mathscr{B} = G/B$ is the moduli space of the Borel subgroups of G. Later we need the related fact, that $\mathscr{B}$ may also be considered as the moduli space of all Borel subalgebras of the Lie algebra $\mathfrak{g}$, i.e. of the Lie subalgebras of G which are the Lie algebra of a Borel subgroup of G. This result can be deduced from the next theorem.

Theorem 6.1 *Let p be a very good prime for G. Then the Borel group B is the normalizer of the Borel algebra $\mathfrak{b} = Lie(B)$ in G.*

Proof. Let $N = \{\sigma \in G : \sigma(\mathfrak{b}) = \mathfrak{b}\}$ be the normalizer of $\mathfrak{b} = Lie(B)$ in G. N is an algebraic subgroup scheme of G. It contains the Borel group B of G, but need not be reduced. $P = N_{red} \subseteq N$ contains B, hence is a parabolic subgroup of G. Hence it is smooth and connected.

Suppose $B \neq P$. Choose an element $\sigma \in P(k)$ in the normalizer of the torus T, that represents a reflection $\sigma = s_\alpha$ in the Weyl group of G with respect to a root $\alpha \in \Sigma^+$. Then

$$\sigma(\mathfrak{g}^\alpha) = \mathfrak{g}^{-\alpha} ,$$

hence $\sigma(\mathfrak{b}) \neq \mathfrak{b}$. Thus σ is not in the normalizer of $\mathfrak{b}$, which is a contradiction. Thus $P = N_{red} = B$.

Suppose $B \neq N$, in other words suppose N is not reduced. Let m be the maximal ideal of the local ring of the scheme B at the unit element and let $\tilde{m}$ be the corresponding maximal ideal of N of the local ring at the unit element. The surjection

$$\tilde{m}/\tilde{m}^2 \longrightarrow m/m^2$$

is not an isomorphism by our assumptions. Otherwise $\dim \tilde{m}/\tilde{m}^2 = \dim m/m^2$ is $\dim B = \dim N$. Then N would be smooth at the unit element, hence smooth everywhere. Thus $B = N$, contradicting our assumptions. The dual statement for the Lie algebras now gives a proper inclusion

$$\mathfrak{b} = Lie(B) \hookrightarrow Lie(N) \subseteq \mathfrak{g} .$$

In particular, there exists a non-zero element $x \in Lie(N) \cap \mathfrak{u}^-$. Since p is very good, the first part of Theorem VI.4.2 implies

$$0 \neq [\mathfrak{t}, x] \subseteq \mathfrak{u}^- ,$$

hence

$$[x, \mathfrak{b}] \not\subseteq \mathfrak{b} .$$

In other words $x \notin \mathrm{Lie}(N)$, which is a contradiction. This proves $N = B$. □

VI.7 Regular Elements of the Lie Algebra $\mathfrak{g}$

Again, assume that the characteristic p of k is a very good prime for the group G. Assume that G is a connected, semisimple algebraic group over the algebraically closed field k. Let t be an element in the Lie algebra of the torus $\mathfrak{t} = \mathrm{Lie}(T)$. Then, according to a theorem of Steinberg, the centralizer G_t of t in G is a connected, reductive, algebraic subgroup of G. See [S-St], chap. II, 3.19, if G is simply connected.

A very good prime is not a torsion prime [Sl1], 3.3. Since by assumption p is a very good prime, we do not need to assume, that G is simply connected. See Remark VI.1.7.

Definition 7.1 *The element t in the Lie algebra $\mathfrak{t} = Lie(T)$ of the torus T is called* **regular** *iff its stabilizer G_t under the adjoint action is $G_t = T$.*

In particular, the stabilizer

$$\{w \in W : w(t) = t\}$$

of a regular element $t \in \mathfrak{t}$ in the Weyl group W of T is trivial. Therefore $s_\alpha(t) = t - \alpha(t)\alpha^\vee \neq t$ for all roots $\alpha \in \Sigma$, hence $\alpha(t) \neq 0$. Recall that we simply write α for the linear map $Lie(\alpha) : \mathfrak{t} \to k$. Therefore $t \in \mathscr{O}$ holds for regular elements, where $\mathscr{O}$ is the Zariski open subset

$$\mathscr{O} = \{t \in \mathfrak{t} : \alpha(t) \neq 0 \; \forall \alpha \in \Sigma\}$$

of the Lie algebra $\mathfrak{t}$. Obviously W acts on $\mathscr{O}$. By the first part of Theorem VI.4.2 $\mathscr{O}$ is the complement of finitely many proper linear subspaces of $\mathfrak{t}$, so in particular nonempty.

Theorem 7.2 *Let p be a very good prime. Then the set $\mathfrak{t}_{reg}$ of the regular elements $t \in \mathfrak{t}$ is the non-empty open subset*

$$\mathfrak{t}_{reg} = \{t \in \mathfrak{t} : \alpha(t) \neq 0 \; \forall \alpha \in \Sigma\}$$

of $\mathfrak{t}$.

Proof. Suppose $t \in \mathscr{O} \setminus \mathfrak{t}_{reg}$. Then G_t is a connected reductive group and T is also a maximal torus of G_t. Since $G_t \neq T$ there exists a root α of G_t in $X(T)$. This root α is also a root of G. In other words $\alpha \in \Sigma$. The root space $\mathfrak{g}^\alpha$ is contained in the Lie algebra $Lie(G_t) \subseteq \mathfrak{g}$. By definition $[t, x] = 0$ holds for all elements $x \in \mathrm{Lie}(G_t)$. Applied to any nonzero element $x \in \mathfrak{g}^\alpha$ this gives

$$0 = [t, x] = \alpha(t) \cdot x \neq 0 \, .$$

A contradiction! □

Remark 7.3 Due to a lack of reference, let us observe that Theorem VI.7.2 also holds in the schematic sense: G_t is a reduced scheme for any element $t \in \mathfrak{t}_{reg}$. This follows from

$$\mathfrak{t} \subseteq Lie(G_t) \subseteq \{x \in \mathfrak{g} : [t, x] = 0\} = \mathfrak{t} \, .$$

See also [SGA3], III, exp. XXII 5.

VI.8 Grothendieck's Simultaneous Resolution of Singularities

The general reference for this section is [Sl1], chap. II.

Suppose, that a representation of the Borel group B of G on a finite-dimensional vector space V is given. We can interpret V as an affine scheme. The group B acts on the scheme $G \times V$ via

$$b(g, v) = (gb^{-1}, bv) .$$

On $G \times V$ there is a second action, the action of the affine line by multiplication on the second factor V. This operation commutes with the action on B. $G \times V$ is a scheme over $\mathscr{B} = G/B$

$$G \times V \longrightarrow \mathscr{B} .$$

We want to construct the quotient scheme $(G \times V)/B$ with respect to the action on B as a scheme over $\mathscr{B}$. To do this, we view a scheme over $\mathscr{B}$ as a contravariant functor on the category of the schemes over $\mathscr{B}$. The functor corresponding to such a scheme is then a sheaf for the faithfully flat, quasi-compact topology.

Now we construct the quotient sheaf of the sheaf $G \times V$ with respect to the action of the sheaf B. If this quotient sheaf is representable by a scheme over $\mathscr{B}$, we call this scheme the quotient scheme of the scheme $G \times V$ to the operation of B. Notation:

$$(G \times V)/B = G \overset{B}{\times} V .$$

For our purposes, we may use the etale topology rather than the faithfully flat quasicompact topology. The quotient space $\mathscr{B} = G/B$ admits a Zariski open covering $C \to \mathscr{B}$, such that there exists a section σ

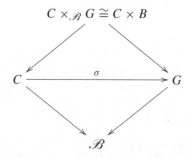

To show this use translates of the large Bruhat cell. The principal homogeneous space G for B over $\mathscr{B} = G/B$ becomes trivial on the Zariski open subsets, which are left translates gV_0 of the dense open Bruhat cell $V_0 = Uw_0B$, where w_0 is the longest element in the Weyl group. Therefore the same holds for the associated vector bundle $G \overset{B}{\times} V$. Hence

Theorem 8.1 *The quotient scheme*

$$(G \times V)/B = G \overset{B}{\times} V \to \mathscr{B}$$

exists in the sense above. It is in a natural way a locally trivial vector bundle over $\mathscr{B}$ (with respect to the Zariski topology).

The Adjoint Representation. In the following, we apply this to the adjoint representation of G on the Lie algebra $\mathfrak{g}$ of G, respectively its restriction to the Borel group. This action of B on $\mathfrak{g}$ contains the subrepresentation, where B acts on $\mathfrak{b} = Lie(B) = \mathfrak{t} \oplus \mathfrak{u}$. The factor Lie algebra $\mathfrak{t} = \mathfrak{b}/\mathfrak{u}$ is a trivial B-module. As before, $\mathfrak{t} = \mathrm{Lie}(T)$ denotes the Lie algebra of a maximal torus $T \subseteq B \subseteq G$, and $\mathfrak{u} = Lie(U)$ is the Lie algebra of the unipotent radical U of B.

The vector bundle $X = G \overset{B}{\times} \mathfrak{b}$ over $\mathscr{B}$ can be embedded into a trivial vector bundle. The map defined by $g \overset{B}{\times} x \mapsto (gB, g(x))$ defines a homomorphism of vector bundles

$$G \overset{B}{\times} \mathfrak{b} \to \mathscr{B} \times \mathfrak{g}\,.$$

Here $g \overset{B}{\times} x$ denotes a representative of an element in $G \overset{B}{\times} \mathfrak{b}$. Furthermore gB for $g \in G$ denotes an element of $\mathscr{B}$. The vector bundle homomorphism considered is injective. It identifies $G \overset{B}{\times} \mathscr{B}$ with a locally split subbundle of the trivial vector space bundle $\mathscr{B} \times \mathfrak{g}$ over $\mathscr{B}$

$$\begin{array}{ccc} G \overset{B}{\times} \mathfrak{b} & \hookrightarrow & \mathscr{B} \times \mathfrak{g} \\ \downarrow & & \downarrow \\ \mathscr{B} & = & \mathscr{B}\,. \end{array}$$

Since $\mathscr{B}$ is complete, the projection $\mathscr{B} \times \mathfrak{g} \to \mathfrak{g}$ is proper. Hence also the composition map

$$q : G \overset{B}{\times} \mathfrak{b} \to \mathfrak{g}\,.$$

Since $\mathfrak{b} = \mathfrak{t} \oplus \mathfrak{u}$, and since B acts trivially on $\mathfrak{t} = \mathfrak{b}/\mathfrak{u}$, there is also a morphism

$$p : G \overset{B}{\times} \mathfrak{b} \to \mathfrak{t}\,,$$

defined by $g \overset{B}{\times} (t + x) \mapsto t$ for $t \in \mathfrak{t}$ and $x \in \mathfrak{u}$. This morphism is a smooth morphism.

The Weyl Group. For the following let $p = \mathrm{char}(k)$ be a very good prime for G. The affine algebra of the affine vector scheme $\mathfrak{g}$ is the symmetric algebra $S(\mathfrak{g}^*)$ of the dual space $\mathfrak{g}^*$ of the Lie algebra $\mathfrak{g}$; the affine algebra of $\mathfrak{t}$ is the symmetric algebra

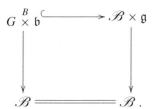

$S(\mathfrak{t}^*)$ of the dual $\mathfrak{t}^*$ of $\mathfrak{t}$. Let $W = \mathrm{Norm}_G(T)/T$ be the Weyl group $W = W(k)$ of G. Then G acts on $S(\mathfrak{g}^*)$ and the Weyl group W acts on $S(\mathfrak{t}^*)$.

Theorem 8.2 *Assume that the characteristic p of k is a very good prime for G. Then the natural restriction map $S(\mathfrak{g}^*) \to S(\mathfrak{t}^*)$ induces a homomorphism between the rings of invariants $S(\mathfrak{g}^*)^G \to S(\mathfrak{t}^*)^W$. This map is an isomorphism*

$$S(\mathfrak{g}^*)^G \overset{\cong}{\to} S(\mathfrak{t}^*)^W .$$

Let $r = \dim T = \dim \mathfrak{t}$ be the semisimple rank of G. In $R = S(\mathfrak{g}^*)^G = S(\mathfrak{t}^*)^W$ there are r over k algebraic independent homogeneous elements, which generate the ring R over k. Hence R is a polynomial ring over k.

Proof. See [Sl1], chap. II, 3.12 and [De1]. □

Adjoint Quotients. Consider the affine quotient scheme $\theta : \mathfrak{t} \to \mathfrak{t}/W$, which is defined by the ring inclusion $S(\mathfrak{t}^*)^W \hookrightarrow S(\mathfrak{t}^*)$

$$\mathfrak{t}/W = \mathrm{Spec}(S(\mathfrak{t}^*)^W).$$

The ring homomorphisms

$$S(\mathfrak{g}^*) \hookleftarrow S(\mathfrak{g}^*)^G \overset{\cong}{\to} S(\mathfrak{t}^*)^W$$

induce a morphism of schemes, the so called *adjoint quotient*

$$\gamma : \mathfrak{g} \to \mathfrak{t}/W .$$

Theorem 8.3 (See [Sl1], chap. II, 4.7) *Let the characteristic p be a very good prime for G. Then the diagram*

$$\begin{array}{ccc} X = G \overset{B}{\times} \mathfrak{b} & \overset{q}{\longrightarrow} & \mathfrak{g} \\ \Big\downarrow p & & \Big\downarrow \gamma \\ \mathfrak{t} & \overset{\theta}{\longrightarrow} & \mathfrak{t}/W \end{array}$$

is commutative. The morphisms of this diagram define, what is called the simultaneous resolution of singularities of the morphism

$$\mathfrak{g} \to \mathfrak{t}/W$$

in the sense of Grothendieck. This means, that the following four properties hold:

(1) p is a smooth morphism.
(2) θ is a finite and surjective morphism.

(3) *q is proper and surjective.*
(4) *For $t \in \mathfrak{t}(k)$ the morphism between the fibers*

$$p^{-1}(t) \longrightarrow \gamma^{-1}(\theta(t))_{red}$$

is a resolution of singularities.

The properties (1)–(3) have been discussed already. The proof of property (4) can be reduced to the case where $t = 0$ by considering the centralizer of t. See loc. cit. In the case $t = 0$ one has to find an open dense subset, where the morphism is separable, such that the fiber has cardinality one. Elements x in $\mathfrak{u}$ with fibers of cardinality one exist. Choose $x \in \mathfrak{u}$ such that the components x_α with respect to the decomposition into root spaces $x = \sum_{\Sigma^+} x_\alpha$ are nonzero for all simple roots: $x_\alpha \neq 0$ for all simple roots α in Σ^+. For separability see [St], section 6.

VI.9 The Galois Group W

Assume that we are in the situation of the last section. In particular assume that the characteristic is a very good prime for G. Consider the simultaneous resolution of singularities as in Theorem VI.8.3. We show that the upper horizontal map q in the simultaneous resolution diagram defines an etale Galois covering on some Zariski open dense subset.

Consider the open subscheme $\mathfrak{t}_{reg}$ of $\mathfrak{t}$ of the regular elements. $\mathfrak{t}_{reg}$ is non-empty, hence Zariski dense in $\mathfrak{t}$. See Theorem VI.7.2. On $\mathfrak{t}_{reg}$ the group W acts freely. Hence $\mathfrak{t}_{reg}$ is the inverse image of an open subset $(\mathfrak{t}/W)_{reg}$ in $\mathfrak{t}/W$. Consider the inverse images of $(\mathfrak{t}/W)_{reg}$ in the other schemes of the simultaneous resolution diagram of Theorem VI.8.3. This gives a corresponding diagram of these open subschemes:

$$\begin{array}{ccc}
X_{reg} = (G \overset{B}{\times} \mathfrak{b})_{reg} & \overset{q}{\longrightarrow} & \mathfrak{g}_{reg} \\
\Big\downarrow p & & \Big\downarrow \gamma \\
\mathfrak{t}_{reg} & \overset{\theta}{\longrightarrow} & \mathfrak{t}_{reg}/W
\end{array}$$

The elements of $\mathfrak{g}_{reg}$ will be called the **regular** elements of the Lie algebra $\mathfrak{g}$. (This differs from the usual terminology).

Theorem 9.1 *Let the characteristic p be a good prime for G. Then*

$$q : X_{reg} = (G \overset{B}{\times} \mathfrak{b})_{reg} \longrightarrow \mathfrak{g}_{reg}$$

is an unramified Galois covering of $\mathfrak{g}_{reg}$, whose Galois group is W. The Weyl group W acts on X_{reg} consistent with the morphism $p : X_{reg} \to \mathfrak{t}_{reg}$ and the action of W on $\mathfrak{t}_{reg}$, such that

$$X_{reg}/W \cong \mathfrak{g}_{reg}.$$

Proof. We make some definitions first. Recall, that T acts trivially on $\mathfrak{t} \subseteq \mathfrak{b}$. Put

$$\Upsilon = G/T \times \mathfrak{t}_{reg} .$$

Consider the commutative diagram, with the right arrow $\tilde{q}$ defined by $\Upsilon \ni (gT, t) \mapsto g(t)$ and the left arrow f defined by $f(gT, t) = g \overset{B}{\times} t$:

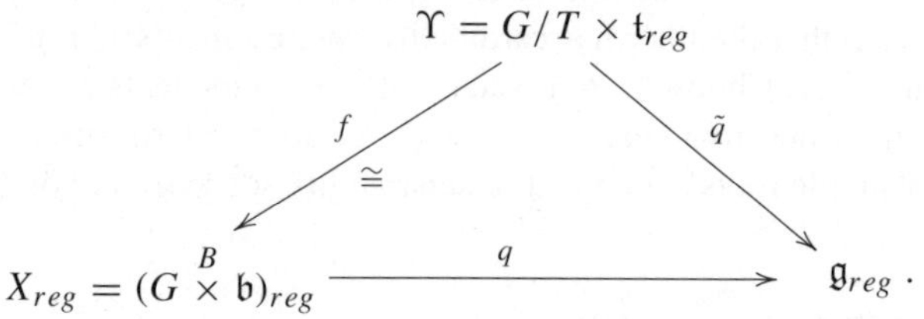

Obviously $p \circ f$ is the projection $\Upsilon \to \mathfrak{t}_{reg}$ onto the second factor.

In the first two steps of the proof we will show, that the morphism f defines an isomorphism. This allows to replace q by the upper right arrow $\tilde{q}$. We first prove

(1) $f : \Upsilon(k) \longrightarrow X_{reg}(k)$ *is Surjective.* Jordan decomposition: In Theorem resp. Definition VI.3.1, nilpotent elements of the Lie algebra of an affine algebraic group were defined. It is easy to define in a similar way semisimple elements, such that every element x of the Lie algebra has a unique Jordan decomposition – i.e. x is a sum of a semisimple element x_s and a nilpotent element x_n, such that x_s and x_n commute (see [S5], 4.4.20):

$$x = x_s + x_n, \qquad [x_s, x_n] = 0 .$$

Let $g \overset{B}{\times} x$ be given in $(G \overset{B}{\times} \mathfrak{b})_{reg}$. Then $x = t + u \in \mathfrak{b}(k)$ is in the Lie algebra of the Borel group, with the regular component $t \in \mathfrak{t}_{reg}(k)$ and $u \in \mathfrak{u}(k)$. We want show

$$x \in B(k)(\mathfrak{t}_{reg}(k)) .$$

For this consider the Jordan decomposition $x = x_s + x_n$ of x in $Lie(B)$. According to [Bo] (chap. IV, prop. 11.8) there exists an element $\sigma \in B(k)$, which moves x_s into $\mathfrak{t}$

$$\sigma(x_s) \in \mathfrak{t}(k) \quad , \quad \sigma \in B(k)$$

(this uses, that all maximal tori in B are conjugate). The corresponding decomposition of $\sigma(x)$ in $\mathfrak{b}(k)$ is

$$\sigma(t) + \sigma(u) = \sigma(x) = \sigma(x_s) + \sigma(x_n) .$$

Note that $\sigma(x_n), \sigma(u)$ are both in $\mathfrak{u}(k)$, and $B(k)$ acts trivially on $\mathfrak{t}(k) = \mathfrak{b}(k)/\mathfrak{u}(k)$. Hence

$$t \equiv \sigma(t) \equiv \sigma(x_s)$$

modulo $\mathfrak{u}$ and therefore $t = \sigma(x_s)$. In particular, t commutes with $\sigma(x_n)$, i.e. $[t, \sigma(x_n)] = 0$. Since $t \in \mathfrak{t}_{reg}(k)$ is regular, we have $\alpha(t) \neq 0$ in k for all $\alpha \in \Sigma$ by the characterization of $\mathfrak{t}_{reg}$ given in Theorem VI.7.2. This implies, that $\sigma(x_n)$ is zero and $\sigma(x) = t$. Hence $x = \sigma^{-1}(t) \in B(k)(\mathfrak{t}_{reg}(k))$ as desired. Therefore it is shown, that every element of $X_{reg}(k)$ is in the image of $\Upsilon(k)$.

(2) *f is an Isomorphism.* Let A be a (finite) commutative k-algebra with unit element. Suppose given $t \in \mathfrak{t}_{reg}(k)$. Let $\sigma \in B(A)$ be an A-valued point of the Borel group B with the particular property, that

$$\sigma(t) \in \mathfrak{t}_{reg}(A) \subseteq \mathfrak{t}(A) .$$

Here t also denotes the image of t under the map $\mathfrak{t}_{reg}(k) \to \mathfrak{t}_{reg}(A)$.

We will show below, that these assumptions imply $\sigma \in T(A)$. In other words, f is (infinitesimally) injective. Therefore the morphism

$$f : \Upsilon \to X_{reg}$$

is unramified. Hence f is etale, since Υ and X_{reg} are both smooth and have the same dimension. Furthermore $f : \Upsilon(k) \to X_{reg}(k)$ is injective ($A = k$) and therefore, due to step (1), bijective. Hence $f : \Upsilon \to X_{reg}$ is an isomorphism

$$\Upsilon \cong X_{reg}.$$

Now the postponed proof of the fact, that $\sigma \in T(A)$ holds under the assumptions above. The centralizer of $\sigma(t)$ in $G/A = G \times_{Spec(k)} Spec(A)$ is

$$(G/A)_{\sigma(t)} = \sigma \cdot (T/A) \cdot \sigma^{-1} ,$$

where $T/A = T \times_{Spec(k)} Spec(A)$. Since $\sigma(t) \in \mathfrak{t}(A)$

$$T/A \subseteq (G/A)_{\sigma(t)} = \sigma \cdot (T/A) \cdot \sigma^{-1} .$$

Both these tori are smooth over $Spec(A)$, hence

$$T/A = \sigma \cdot (T/A) \cdot \sigma^{-1}$$

and thus $\sigma \in N(T)(A)$ is an A-valued point of the normalizer $N(T)$ of T in G. On the other hand $\sigma \in B(A)$, therefore $\sigma \in T(A)$ using $N(T) \cap B = T$.

(3) *Action of W.* On $(G/T) \times \mathfrak{t}_{reg}$, the Weyl group W acts in a natural way. Let $n \in N(T)$ represent an element $w \in W$. Then

$$(gT, t) \cdot w = (gTn, n^{-1}(t)) = (gnT, n^{-1}(t))$$

defines a right action of W on $G/T \times \mathfrak{t}_{reg}$. Since W acts freely on $\mathfrak{t}_{reg}$, W acts freely on $\Upsilon = (G/T) \times \mathfrak{t}_{reg}$. With a similar argument as in the previous steps of the proof, we will now show

$$\Upsilon/W \cong \mathfrak{g}_{reg}.$$

(4) *Step Four.* Let (gT,t) and (g_1T,t_1) be two elements in $\Upsilon(k) = (G/T)(k) \times \mathfrak{t}_{reg}(k)$ such that $g(t) = g_1(t_1)$. Then we show $(gT,t)\cdot w = (g_1T,t_1)$ for some $w \in W$. The argument is as follows:

$$g_1^{-1}g(t) = t_1, \quad n = g_1^{-1}g \in G(k),$$

hence

$$G_{n(t)} = n \cdot G_t \cdot n^{-1}\,.$$

Both t and t_1 are regular, hence their centralizers are both T. This implies

$$T = n \cdot T \cdot n^{-1}\,.$$

Hence n is in the normalizer $N(T)$ of T. Let $w \in W$ be the image of n^{-1} in the Weyl group. Then $(g,t)\cdot w = (g_1,t_1)$, which proves the claim.

(5) *Step Five.* In the same way we can conclude infinitesimally: Let A be a k-algebra as in step (2). Assume $(gT,t) \in ((G/T)\times\mathfrak{t}_{reg})(k)$ as in step (4), but now let (g_1T,t_1) be an arbitrary A-valued point in $((G/T)\times\mathfrak{t}_{reg})(A)$. Assume $g(t) = g_1(t_1)$. Put $n = g_1^{-1}g \in G(A)$. Then $T/A \subseteq n(T/A)n^{-1}$, hence $T/A = n(T/A)n^{-1}$ and $n \in N(T)(A)$. Thus $(\Upsilon(k))/W = \mathfrak{g}_{reg}(k)$ and, due to the infinitesimal statement at hand,

$$\Upsilon/W = \mathfrak{g}_{reg}\,.$$

□

VI.10 The Monodromy Complexes Φ and Φ'

Assume, that the characteristic is a very good prime for G.

The Regular Representation. Since $q : X_{reg} \to \mathfrak{g}_{reg}$ is an etale Galois covering with deck-transformation group W, the direct image of the constant sheaf defines a smooth sheaf on $\mathfrak{g}_{reg}$. It corresponds to a representation of the fundamental group $\pi_1(\mathfrak{g}_{reg})$, that factorizes over the quotient group W. In fact, the underlying representation is the regular representation of W. We get from Remark III.15.3 part d) the next

Corollary 10.1 *The direct image of the constant sheaf $\overline{\mathbb{Q}}_l$ on X_{reg} is a smooth $\overline{\mathbb{Q}}_l$-sheaf $\mathcal{G} = q_!\overline{\mathbb{Q}}_{l\,X_{reg}}$ on $\mathfrak{g}_{reg}$ of rank #W. Furthermore*

$$\mathcal{G} \;=\; \bigoplus_{\phi\in\hat{W}} \mathcal{G}_\phi \otimes_{\overline{\mathbb{Q}}_l} V_\phi \quad , \quad \mathcal{G} = q_!\overline{\mathbb{Q}}_{l\,X_{reg}}\,.$$

The sum is over all irreducible characters ϕ of W. The sheaves $\mathcal{G}_\phi$ are smooth irreducible $\overline{\mathbb{Q}}_l$-sheaves on $\mathfrak{g}_{reg}$ of rank equal to the degree $deg(\phi)$, and V_ϕ are

$\overline{\mathbb{Q}}_l$-vectorspaces of dimension $deg(\phi)$. Two sheaves $\mathcal{G}_\phi$, $\mathcal{G}_{\phi'}$ are isomorphic iff the characters ϕ and ϕ' are equal. Furthermore the endomorphism ring of the perverse sheaf $\Phi_{reg} := q_!\overline{\mathbb{Q}}_{l\,X_{reg}}[dim(X)]$

$$\Phi_{reg} = \mathcal{G}[dim(X)]$$

is isomorphic to the group ring $\overline{\mathbb{Q}}_l[W]$ of the Weyl group

$$End_{Perv(\mathfrak{g}_{reg})}(\Phi_{reg}) = \bigoplus_\phi End(V_\phi) \cong \overline{\mathbb{Q}}_l[W]\,.$$

In fact, the perverse sheaf Φ_{reg} on $\mathfrak{g}_{reg}$ is a W-equivariant complex. This action of the Weyl group W preserves the isotypic components. As a special case of the Lemma III.15.4, this defines homomorphisms

$$\phi_L : W \longrightarrow Gl(V_\phi) = End_{\overline{\mathbb{Q}}_l}(V_\phi)^*\,.$$

From the structure of $End_{Perv(\mathfrak{g}_{reg})}(q_!\overline{\mathbb{Q}}_{l\,X_{reg}}[dim(X)]) \cong \overline{\mathbb{Q}}_l[W]$ it is clear, that among the endomorphisms the centralizer of W has dimension equal to $\#\hat{W}$. Hence the representations ϕ_L are irreducible and each irreducible character of W occurs exactly once. This explains the previous notation $\mathcal{G}_\phi$, once we assume that $\phi = \phi_L$. So assume this. In particular we now often drop the index L.

Now we will make use of the following remarkable property of the map

$$q : X = G \overset{B}{\times} \mathfrak{b} \to \mathfrak{g}\,,$$

which we simply quote from **Springer**'s Bourbaki article ([S4], prop. 4.3). For details see [St], theorem 4.6, p. 217 or the discussion in [Hu], p. 100ff and p. 112ff.

Theorem 10.2 (Dimension Formula) *irreducible components of the Springer fibers $q^{-1}(\xi)$ have the dimension*

$$d(\xi) = \frac{1}{2}(\dim(G_\xi) - r)\,.$$

Here $r = \dim(T) = \dim(\mathfrak{t})$ denotes the semisimple rank of G and G_ξ the stabilizer of ξ in G.

Remark. If $\xi \in \mathfrak{g}_{reg}(k)$ then $d(\xi) = 0$.

Lemma 10.3 *The Lie algebra $\mathfrak{g}$ can be stratified $\mathfrak{g} = \coprod_{i\geq 0} X_i$ into locally closed subsets $X_i \subseteq \mathfrak{g}$, such that*

(a) $X_0 = \mathfrak{g}_{reg}$.
(b) For $i > 0$ and $\xi \in X_i$ we have $\dim(G_\xi) - r < \operatorname{codim}(X_i)$.

Proof. See [S4], lemma 4.5 and [B-K]. □

We give a description of the stratification $\mathfrak{g} = \coprod X_i$, following **Springer** [S4], lemma 4.5: Let $P \subsetneq G$ be a proper parabolic subgroup, M a Levi group of P and S the centre of M. Furthermore

$$\mathfrak{m} = \mathrm{Lie}(M), \qquad \mathfrak{s} = \mathrm{Lie}(S),$$

$$\mathfrak{s}_{reg} = \{\xi \in \mathfrak{s} : G_\xi = M\}.$$

We choose a nilpotent element $\eta \in \mathfrak{m}$ and set $Z = \cup_{g \in G}\, g(\mathfrak{s}_{reg} + \eta)$. Now vary P and η. This gives a family $\{Z\}$, which defines the desired partition.

Due to the last two facts Theorem VI.10.2 and Lemma VI.10.3 we derive

Corollary 10.4 *The fiber dimensions of the morphism* $q : X = G \overset{B}{\times} \mathfrak{b} \to \mathfrak{g}$ *satisfy*

$$\dim q^{-1}(\xi) \leq \frac{1}{2}(\mathrm{codim}(X_i) - 1) \quad , \quad i > 0, \quad \xi \in X_i \, .$$

In particular, q is a **small** *morphism.*

Since X is smooth of dimension $n = dim(G)$, the sheaf $\overline{\mathbb{Q}}_{l\,X}[n]$ is a perverse on X. Since q is a small map, also the derived direct image complex

$$\Phi := \mathrm{R}q_!\overline{\mathbb{Q}}_l[n]_X$$

is a G-equivariant perverse sheaf. Smallness also implies that the perverse sheaf Φ can be recovered from its restriction Φ_{reg} to the top stratum by intermediate extension. With this information at hand, one can endow the perverse sheaf Φ with additional structure, namely a W-action. Consider

$$\begin{array}{ccc} X_{reg} & \xrightarrow{\ q\ } & \mathfrak{g}_{reg} \\ \big\downarrow & & \big\downarrow{\scriptstyle j} \\ X & \xrightarrow{\ q\ } & \mathfrak{g} \end{array}$$

The restriction of this perverse sheaf Φ on $\mathfrak{g}$ to the regular locus $\mathfrak{g}_{reg}$ is the perverse sheaf Φ_{reg} studied in Corollary VI.10.1.

This sheaf was shown to be a W-equivariant sheaf. Therefore, by the characterization of $j_{!*}$ from Chap. III, §6, we get from Remark III.15.3 and Exercise III.15.6

Corollary 10.5 *For $n = dim(X)$ the direct image complex*

$$\Phi := Rq_!\overline{\mathbb{Q}}_l[n]_X$$

is a perverse sheaf on the Lie algebra $\mathfrak{g}$. Let $j : \mathfrak{g}_{reg} \hookrightarrow \mathfrak{g}$ be the open embedding defined by the inclusion of the regular elements. Then

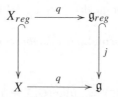

$$\boxed{\Phi \;=\; j_{!*}\Phi_{reg} \;\in\; Perv(\mathfrak{g})}$$

is a G-left invariant and W-right invariant perverse sheaf on $\mathfrak{g}$ *with respect to the trivial action of W on* $\mathfrak{g}$ *and with respect to the adjoint action of G on* $\mathfrak{g}$*. Furthermore, the perverse sheaves* $\Phi_\phi = j_{!*}(\mathscr{G}_\phi[n])$ *are the irreducible constituents of* Φ

$$\boxed{\Phi = \bigoplus_{\phi\in\hat{W}} \Phi_\phi \otimes_{\overline{\mathbb{Q}}_l} V_\phi \quad , \qquad \Phi_\phi \;=\; j_{!*}\mathscr{G}_\phi[dim(X)]\,.}$$

Each vectorspace V_ϕ *counts the multiplicity of* Φ_ϕ*. the W-action on* Φ *is via representations* $\phi : W \to Gl(V_\phi)$*, such that*

$$End_{Perv(\mathfrak{g})}(\Phi) \;\cong\; \overline{\mathbb{Q}}_l[W]\,.$$

Remark 10.6 Since intermediate extensions commute with Verdier duality, it is immediately clear, that there exists an isomorphism of W-equivariant complexes $D(\Phi) \cong \Phi(n)$. Since $D(\Phi_\phi \otimes_{\overline{\mathbb{Q}}_l} V_\phi) \cong D(\Phi_\phi) \otimes V_{\phi^*}$ for the dual representation ϕ^* of ϕ, one obtains

$$D(\Phi_\phi) \cong \Phi_{\phi^*}(n)\,.$$

In fact, we will show later $\phi^* \cong \phi$, hence $D(\Phi_\phi) = \Phi_\phi(n)$.

Remark 10.7 (The Coadjoint Version) Theorem VI.8.3 admits a dual version. Since we assumed the characteristic p of k to be very good for G, the Lie algebra $\mathfrak{g}$ has a nondegenerate, symmetric, G-invariant bilinear form (Theorem VI.5.1 and Lemma VI.5.2). Let $\mathfrak{g}^*,\ \mathfrak{t}^*$ etc. denote the dual spaces of $\mathfrak{g},\ \mathfrak{t}$. Using this invariant bilinear form, we get a G-equivariant identification

$$\mathfrak{g} \cong \mathfrak{g}^*$$

between the Lie algebra and its dual. This induces identifications

$$\begin{array}{ccc} \mathfrak{b} & \hookrightarrow & \mathfrak{g} \\ \downarrow\cong & & \downarrow\cong \\ \mathfrak{u}^\perp & \hookrightarrow & \mathfrak{g}^* \end{array} \qquad\qquad \begin{array}{ccc} \mathfrak{t} & \hookrightarrow & \mathfrak{g} \\ \downarrow\cong & & \downarrow\cong \\ \mathfrak{t}^* & \twoheadleftarrow & \mathfrak{g}^* \end{array}$$

(Lemma VI.5.2). The left diagram is B-equivariant, the right diagram is W-equivariant. Using these isomorphisms, we get from the simultaneous resolution of singularities – the diagram of Theorem VI.8.3) – a dual, but isomorphic diagram

$$\begin{array}{ccc} G \times^B \mathfrak{u}^{\perp} & \xrightarrow{\quad q' \quad} & \mathfrak{g}^* \\ \Big\downarrow{\scriptstyle p'} & & \Big\downarrow{\scriptstyle \gamma'} \\ \mathfrak{t}^* & \xrightarrow{\quad \theta' \quad} & \mathfrak{t}^*/W \end{array}$$

Since we can also identify

$$\begin{aligned} \mathfrak{t}^*_{reg} &\cong \mathfrak{t}_{reg}, \\ (\mathfrak{t}^*/W)_{reg} &\cong (\mathfrak{t}/W)_{reg}, \end{aligned}$$

we obtain in the obvious way a corresponding diagram for the regular points. For these dual diagrams, the statements of Theorem VI.8.3 and Theorem VI.9.1 and also the statements of Lemma VI.10.2 and VI.10.3 obviously carry over. In particular

Corollary 10.8

$$\Phi' = Rq'_!\overline{\mathbb{Q}}_l[n] = Rj'_{!*}\Phi'_{reg} \in Perv(\mathfrak{g}^*)$$

is a G-(left)equivariant and W-(right)equivariant perverse sheaf on $\mathfrak{g}'$. *Its decomposition in irreducible perverse constituents is*

$$\boxed{\Phi' = \bigoplus_{\phi \in \hat{W}} \Phi'_\phi \otimes_{\overline{\mathbb{Q}}_l} V_\phi \quad , \quad \Phi'_\phi = \mathscr{G}'_\phi[dim(X)]\,.}$$

W acts on Φ' *via irreducible representations*

$$\phi : W \to Gl(V_\phi)\,,$$

such that $End_{Perv(\mathfrak{g}^*)}(\Phi') \cong \overline{\mathbb{Q}}_l[W]$.

VI.11 The Perverse Sheaf Ψ

Consider the nilpotent variety

$$\mathscr{N} \subseteq \mathfrak{g}$$

contained in $\mathfrak{g}$ and its resolution of singularities obtained in Theorem VI.8.3. Using the notations of this theorem it is immediately clear, that $p^{-1}(0) = G \times^B \mathfrak{u} = q_{\mathscr{N}}^{-1}(\mathscr{N})$. Furthermore $\theta^{-1}(0) = 0$ and $\gamma^{-1}(0) = \mathscr{N}$ hold. This gives the picture

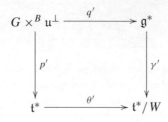

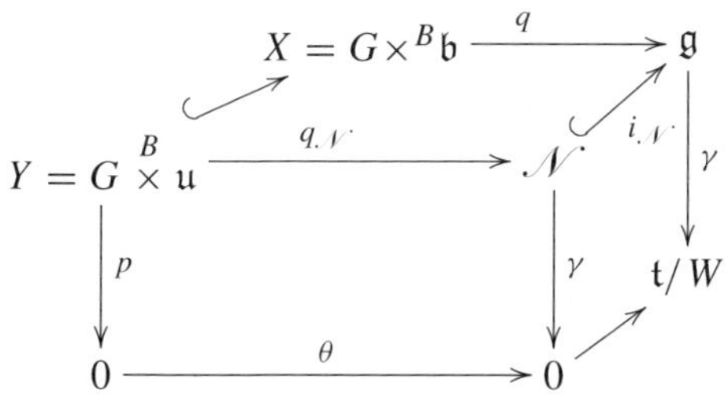

Therefore by part (4) of Theorem VI.8.3 the restriction $q_{\mathcal{N}^\cdot}$ of q to the inverse image of $\mathcal{N}^\cdot$ defines a resolution of the singularities of the nilpotent variety $\mathcal{N}^\cdot$

$$q_{\mathcal{N}^\cdot} : \; Y = G \overset{B}{\times} \mathfrak{u} \longrightarrow \mathcal{N}^\cdot .$$

The restriction of a small map, in our case q, need not be small. But in the present case the restriction $q_{\mathcal{N}^\cdot}$ of the small map q still is semi-small. This is shown as follows: $q_{\mathcal{N}^\cdot}$ is a proper birational map, which is G-equivariant. In fact, there are only finitely many G-orbits $\mathcal{N}_\xi^\cdot$ in $\mathcal{N}^\cdot$. This follows from the existence of a G-invariant isomorphism $\sigma : \mathcal{N}^\cdot \tilde{\to} \mathcal{U}$ (Theorem VI.3.3), since there are only finitely many unipotent conjugacy classes in G (see [Hu], 3.9 for further references). Each orbit is of the form $\mathcal{N}_\xi^\cdot = G(\xi')$ for an arbitrary nilpotent representative $\xi' \in \mathcal{N}_\xi^\cdot$ and defines an irreducible, smooth and locally closed submanifold of $\mathcal{N}^\cdot$. Note $\dim(\mathcal{N}_\xi) = \dim(G) - \dim(G_\xi)$. Due to the dimension formula (Theorem VI.10.2) we have $d(\xi) = \dim(q_{\mathcal{N}^\cdot}^{-1}(\xi)) = \frac{1}{2}(\dim(G_\xi) - r)$, hence $d(\xi') = \dim(q_{\mathcal{N}^\cdot}^{-1}(\xi')) = \frac{1}{2}\big(\dim(G) - \dim(\mathcal{N}_\xi) - r\big)$. But this is the same as $\frac{1}{2}\big(\dim(G\times^B\mathfrak{u}) - \dim(\mathcal{N}_\xi)\big) = \frac{1}{2}\big(\dim(\mathcal{N}^\cdot) - \dim(N_\xi)\big)$. This proves

Lemma 11.1 *Let* $\mathrm{char}(k) = p$ *be a very good prime for the group* G*. Consider the resolution of singularities of the nilpotent variety*

$$q_{\mathcal{N}^\cdot} : \; Y = G \overset{B}{\times} \mathfrak{u} \to \mathcal{N}^\cdot \quad , \quad \mathfrak{u} = Lie(U) .$$

The stratification of the nilpotent variety $\mathcal{N}^\cdot$ *into* G*-orbits* $\mathcal{N}_\xi^\cdot = G(\xi)$ *under the adjoint action of* G

$$\mathcal{N}^\cdot = \coprod_{\mathcal{N}^\cdot \bmod G} \mathcal{N}_\xi^\cdot$$

is a finite stratification and it has the property

$$\dim(q_{\mathcal{N}^\cdot}^{-1}(\xi')) = \frac{1}{2} \cdot codim(\mathcal{N}_\xi^\cdot, \mathcal{N}^\cdot) \quad , \quad \xi' \in N_\xi .$$

In particular $q_{\mathcal{N}^\cdot}$ *is* **semi-small**.

Note that $m = \dim(G \overset{B}{\times} \mathfrak{u}) = \dim(\mathcal{N}^\cdot)$ is an even number. Note $m = 2N$, where N is the number of positive roots. We abbreviate $\langle m\rangle = [m](\frac{m}{2})$. This combined shift and twist does not change weights.

Corollary 11.2 *The complex $\overline{\mathbb{Q}}_l\langle m\rangle$ is a G-equivariant perverse sheaf on the smooth manifold $G \overset{B}{\times} \mathfrak{u}$. Since $q_{\mathcal{N}} : G \overset{B}{\times} \mathfrak{u} \to \mathcal{N}$ is semi-small and G-equivariant, the direct image complex*

$$\boxed{\Psi \;=\; R(q_{\mathcal{N}})_*\overline{\mathbb{Q}}_l\langle m\rangle \;\in\; Perv(\mathcal{N})}$$

is a G-equivariant perverse sheaf on the nilpotent variety $\mathcal{N}$. Since the complex $\overline{\mathbb{Q}}_l\langle m\rangle$ is pure of weight zero, and because $q_{\mathcal{N}}$ is a proper morphism, the perverse sheaf Ψ is a again pure of weight zero. Obviously $D\Psi \cong \Psi$.

Remark: To speak of weights makes sense, since all terms are defined over a finite base field.

VI.12 The Orbit Decomposition of Ψ

Let the assumptions be as in the last section. Recall, that there are only finitely many G-orbits in $\mathcal{N}$. This is not true on $\mathfrak{g}$ of course. Therefore, contrary to the situation for the G-equivariant perverse sheaf Φ on $\mathfrak{g}$, the G-equivariant perverse sheaf Ψ on $\mathcal{N}$ admits a decomposition with respect to these nilpotent orbits.

The G-equivariant perverse sheaf Ψ was defined by

$$\Psi = \mathrm{R}(q_{\mathcal{N}})_!(\overline{\mathbb{Q}}_{lY}\langle m\rangle) \in Perv(\mathcal{N}) \quad , \quad m = dim(\mathcal{N})\ .$$

The pure perverse sheaf Ψ has a direct sum decomposition into irreducible perverse sheaves according to Gabber's decomposition theorem. We will see later, that the use of Gabber's comparatively deep theorem can be avoided. The decomposition for the perverse sheaf Ψ could be directly obtained also from VI.13.6 and the trivial decomposition of Φ' (Corollary VI.10.8).

Since Ψ is a G-equivariant perverse sheaf on $\mathcal{N}$ and G is connected, each of its irreducible constituents is G-equivariant. See Exercise III.15.6. The irreducible constituents are of the form $i_*(g_{!*}\mathcal{F}[dim(U)])$ for smooth $\overline{\mathbb{Q}}_l$-sheaves $\mathcal{F}$ on open dense smooth subschemes $g : U \hookrightarrow Z$, where $i : Z \hookrightarrow \mathcal{N}$ are irreducible closed subschemes of $\mathcal{N}$. Since each of the irreducible perverse constituent sheaves is G-equivariant, the schemes U can be chosen to be among the finitely many adjoint orbits $U = \mathcal{N}_\xi$ of $\mathcal{N}$, and Z can be chosen to be the Zariski closure of that orbit. Note that all orbits U have even dimension. The sheaves $\mathcal{F} = \mathcal{F}_\chi(\frac{dim(U)}{2})$ – for χ in a certain index set depending on the orbit $U = \mathcal{N}_\xi$ – are G-equivariant smooth irreducible $\overline{\mathbb{Q}}_l$-sheaves on the orbit $\mathcal{N}_\xi$. We write

$$\Psi_\xi(\mathcal{F}_\chi) = i_*g_{!*}\mathcal{F}_\chi\langle dim(\mathcal{N}_\xi)\rangle$$

for the constituents. We thus obtain $\Psi = \bigoplus_{\xi\in\mathcal{N}/G}\bigoplus_\chi \Psi_\xi(\mathcal{F}_\chi)\otimes_{\overline{\mathbb{Q}}_l} V(\xi,\chi)$. In this formula $V(\xi,\chi)$ denotes certain finite dimensional $\overline{\mathbb{Q}}_l$-vectorspaces, which count the multiplicities of constituents. Since $End_{Perv(\mathcal{N})}(\Psi_\xi(\chi)) = \overline{\mathbb{Q}}_l$, we obtain

Lemma 12.1 (Orbit Decomposition) *The G-W-equivariant perverse sheaf* $\Psi = R(q_{\mathcal{N}^\cdot})_!(\overline{\mathbb{Q}}_{lY})[m](\frac{m}{2})$ *on* $\mathcal{N}^\cdot$ *admits a direct sum decomposition*

$$\boxed{\Psi = \bigoplus_{\xi}\bigoplus_{\chi} \Psi_\xi(\mathcal{F}_\chi) \otimes_{\overline{\mathbb{Q}}_l} V(\xi,\chi)\,,}$$

where the $\Psi_\xi(\mathcal{F}_\chi)$ *are the irreducible perverse sheaves from above, such that*

$$End_{Perv(\mathcal{N}^\cdot)}(\Psi) = \bigoplus_{\xi\in\mathcal{N}^\cdot/G}\bigoplus_{\mathcal{F}_\xi} End_{\overline{\mathbb{Q}}_l}\big(V(\xi,\mathcal{F}_\xi)\big)\,.$$

The sum is over all G-orbits ξ of the nilpotent variety $\mathcal{N}^\cdot$. For each ξ the summation over χ is with respect to finitely many G-equivariant smooth sheaves $\mathcal{F}_\chi$, for each orbit respectively $\mathcal{N}_\xi^\cdot$.

Fibers. Next we are interested in the stalks Ψ_v of the perverse sheaf Ψ. Let $v \in \mathcal{N}^\cdot$ be nilpotent. The stalk Ψ_v will be computed from the cohomology of the fiber of the morphism $q_{\mathcal{N}^\cdot}$. The fiber $q_{\mathcal{N}^\cdot}^{-1}(v)$ can be identified with $\mathcal{B}_u$, where

$$\sigma(u) = v$$

using a fixed chosen exponential σ, provided by Theorem VI.3.2. Thus the stalk Ψ_v of Ψ at v is $\Psi_v^\bullet = R\Gamma^\bullet(\mathcal{B}_u, \overline{\mathbb{Q}}_{lY}\langle m\rangle) = R\Gamma^{\bullet+dim(\mathcal{N}^\cdot)}(\mathcal{B}_u, \overline{\mathbb{Q}}_l(\frac{m}{2}))$. Let $IC(\mathcal{F}_\chi)$ for the smooth $\overline{\mathbb{Q}}_l$-sheaves $\mathcal{F}_\chi$ be the intermediate extension up to twists and shifts

$$IC(\mathcal{F}_\chi) := \Psi_\xi(\mathcal{F}_\chi)[-dim(\mathcal{N}_\xi^\cdot)]\,.$$

Then by a comparison with the orbit decomposition we get

$$R\Gamma^{\bullet+dim(\mathcal{N}^\cdot)}(\mathcal{B}_u, \overline{\mathbb{Q}}_l(\frac{m}{2})) = \bigoplus_{\overline{\mathcal{N}^\cdot_\xi}\ni v}\bigoplus_{\chi} \mathcal{H}^{\bullet+dim(\mathcal{N}_\xi^\cdot)} IC(\mathcal{F}_\chi)_v \otimes_{\overline{\mathbb{Q}}_l} V(\xi,\chi)\,.$$

The first sum is over all the orbits $\mathcal{N}_\xi^\cdot$, whose closure contains the nilpotent element v. The inner sums are over all sheaves $\mathcal{F}_\chi$ belonging to the fixed orbit $\mathcal{N}_\xi^\cdot$.

The dimension formula of Lemma VI.11.1 states $dim(\mathcal{N}^\cdot) - dim(\mathcal{N}_\xi^\cdot) = 2\cdot d(\xi) = 2\cdot dim(q_{\mathcal{N}^\cdot}^{-1}(\xi))$. Therefore

$$H^\mu(\mathcal{B}_u, \overline{\mathbb{Q}}_l(\frac{m}{2})) = \bigoplus_{\overline{\mathcal{N}^\cdot_\xi}\ni v}\bigoplus_{\chi} \mathcal{H}^{\mu-2d(\xi)} IC(\mathcal{F}_\chi)_v \otimes_{\overline{\mathbb{Q}}_l} V(\xi,\chi)\,.$$

The left side vanishes unless $\mu \leq 2\cdot d(u)$, where $d(u) = dim(\mathcal{B}_u)$. Since $codim(\mathcal{N}_v^\cdot, \overline{\mathcal{N}^\cdot_\xi}) = 2d(u) - 2d(\xi)$, we obtain for the highest possible degree

$$H^{2d(u)}(\mathcal{B}_u, \overline{\mathbb{Q}}_l(\frac{m}{2})) = \bigoplus_{\overline{\mathcal{N}^\cdot_\xi}\ni v}\bigoplus_{\chi} \mathcal{H}^{codim(\mathcal{N}_v^\cdot,\overline{\mathcal{N}^\cdot_\xi})} IC(\mathcal{F}_\chi)_v \otimes_{\overline{\mathbb{Q}}_l} V(\xi,\chi)\,.$$

The left side of this formula does not depend on the choice of the particular point v in the G-orbit $\mathcal{N}_v$, which contains v. So we can choose v to be a generic point of this stratum $\mathcal{N}_v$.

Vanishing Conditions. If $\mathcal{N}_v \neq \mathcal{N}_\xi$ is not the top stratum, then by definition of the intermediate extension all stalks $\mathcal{H}^\mu IC(\mathcal{F}_\chi)_v = 0$ vanish for the cohomology degrees $\mu \geq codim(\mathcal{N}_v, \overline{\mathcal{N}}_\xi)$. This implies, that ξ with $\mathcal{N}_\xi = \mathcal{N}_v$ contributes to the formula above, only if $\mu = 2d(u)$. Hence $H^{2d(u)}(\mathcal{B}_u, \overline{\mathbb{Q}}_l(\frac{m}{2}))$ is

$$\bigoplus_\chi \mathcal{H}^0 IC(\mathcal{F}_\chi)_v \otimes_{\overline{\mathbb{Q}}_l} V(\xi,\chi) = \bigoplus_\chi (\mathcal{F}_\chi)_v(\frac{dim(\mathcal{N}_v)}{2}) \otimes_{\overline{\mathbb{Q}}_l} V(\xi,\chi) .$$

The sheaves $\mathcal{F}_\chi$ are those attached to the G-orbit of v.

Consider the G-orbit $\mathcal{N}_\xi$ of v. Varying v in the orbit the formula above implies, that the smooth G-equivariant sheaves $\mathcal{F}_\chi$ with support $\mathcal{N}_\xi$, that appear in the orbit decomposition of Ψ (Lemma VI.12.1), arise as follows: The sheaves $\mathcal{F}_\chi$ on the corresponding orbit $\mathcal{N}_\xi$ of v are obtained by decomposing the higher direct image sheaf

$$\mathrm{R}^{2d(u)} q_!(\overline{\mathbb{Q}}_{l\,Y_\xi})(d(u)) = \bigoplus_\chi \mathcal{F}_\chi \otimes_{\overline{\mathbb{Q}}_l} V(\xi,\chi)$$

into irreducible smooth sheaves over $\mathcal{N}_\xi$. Here Y_ξ denotes the inverse image of the orbit $\mathcal{N}_\xi$ of v under the proper morphism

$$Y \supset Y_\xi \xrightarrow{q_{\mathcal{N}}} \mathcal{N}_\xi \subset \mathcal{N} ,$$

obtained by the restriction of $q_{\mathcal{N}}$.

The Sheaves $\mathcal{F}_\chi$. Recall $\mathcal{N}_\xi \cong G/G_\xi$. Let G_ξ^0 be the connected component (in the schematic sense) of the stabilizer G_ξ of v. Then $G/G_\xi^0 \to G/G_\xi$ is an etale covering with Galois group $A(u) = G_u/G_u^0 \cong G_\xi/G_\xi^0$. Consider the pullback

$$\tilde{q} : \tilde{Y}_\xi = Y_\xi \times_{\mathcal{N}_\xi} G/G_\xi^0 \to G/G_\xi^0 .$$

Choose $\tilde{\xi}$ to be the coset of the unit element in G/G_ξ^0. Then $\tilde{\xi}$ maps to the unit coset $e \in G/G_\xi$, which corresponds to $\xi \in \mathcal{N}_\xi$. Let $\tilde{Y}(1), \tilde{Y}(2), ...$ be the irreducible components of $\tilde{Y} = \tilde{Y}_\xi$ of highest dimension $d(\xi) = d(u)$. We claim

$$\mathrm{R}^{2d(u)} \tilde{q}_!(\overline{\mathbb{Q}}_{l\,\tilde{Y}_\xi})(d(u)) \cong \bigoplus_j \mathrm{R}^{2d(u)} \tilde{q}_!(\overline{\mathbb{Q}}_{l\,\tilde{Y}(j)}(d(u)) \cong \bigoplus_j \overline{\mathbb{Q}}_{l\,G/G_\xi^0} .$$

The inverse of the first isomorphism is induced by the obvious maps

$$\bigsqcup_j \tilde{Y}(j) \to \bigcup_j \tilde{Y}(j) \to \tilde{Y} .$$

The second isomorphism is induced by the trace map ([SGA4], expose XVIII, p. 553). See also the discussion following Theorem III.6.2. For this we need, that the fibers of $\tilde{q} : \tilde{Y}(j) \to G/G_\xi^0$ are irreducible and nonempty. This is clear, since the components $\tilde{Y}(j)$ are obtained from the irreducible components of the fiber $\tilde{q}^{-1}(\tilde{\xi})$ by translation with G. Note that G_ξ^0 acts on the fiber, but stabilizes each component being a connected group. Thus, using base change, the pullback of each sheaf $\mathscr{F}_\chi$ from $\mathscr{N}_\xi$ to the etale cover G/G_ξ^0 becomes trivial. This implies that $\mathscr{F}_\chi$ is isomorphic to one of the smooth $\overline{\mathbb{Q}}_l$-sheaf $\mathscr{F}_\chi$ attached to an irreducible representation χ of the covering group $A(u)$ on a finite dimensional vectorspace over $\overline{\mathbb{Q}}_l$. So we have proved, that the smooth $\overline{\mathbb{Q}}_l$-sheaves $\mathscr{F}_\chi$ arising in orbit decomposition of Ψ for the orbit $\mathscr{N}_\xi$ are of the form $\mathscr{F}_\chi$ for some $\chi \in \hat{A}(u)$. From the description given, it is clear that the sheaves $\mathscr{F}_\chi$ for the trivial character always occur. This implies

Lemma 12.2 *Suppose $\sigma(u) = v$ for v in some nilpotent G-orbit $\mathscr{N}_\xi$. Then the smooth $\overline{\mathbb{Q}}_l$-sheaves $\mathscr{F}_\xi$, which arise in the orbit decomposition of Ψ (see Lemma VI.12.1), are obtained from the "highest" direct image sheaf*

$$\boxed{R^{2d(u)}q_!(\overline{\mathbb{Q}}_{l\,Y_\xi})(d(u)) = \bigoplus_{\chi \in \hat{A}(u)} \mathscr{F}_\chi \otimes_{\overline{\mathbb{Q}}_l} V(\xi, \chi) .}$$

For the trivial character $\chi = 1$ of $A(u)$ the vectorspace $V(\xi, \chi) \neq 0$ is always nontrivial. The $\overline{\mathbb{Q}}_l$-sheaves $\mathscr{F}_\chi$ are smooth sheaves on $\mathscr{N}_\xi$. They are attached to the irreducible representations χ of the Galois group $A(u)$ of the etale Galois covering

$$\pi : G/G_\xi^0 \to G/G_\xi = \mathscr{N}_\xi$$

(as in Remark III.15.3), such that

$$\pi_*(\overline{\mathbb{Q}}_l) = \bigoplus_{\chi \in \hat{A}(u)} \mathscr{F}_\chi \otimes_{\overline{\mathbb{Q}}_l} \overline{\mathbb{Q}}_l^{\,deg(\chi)} .$$

VI.13 Proof of Springer's Theorem

We want to prove Springer's Theorem VI.1.1 in this section using the Deligne-Fourier transform. Doing this we actually prove more, namely we even construct a representation of the Weyl group W on a perverse sheaf. So in fact we construct a perverse Springer representation. Decomposing the underlying perverse sheaf into irreducible constituents via Gabber's decomposition theorem, one obtains from this perverse Springer representation the Springer representations without difficulty. The perverse construction may be outlined as follows: The morphisms $q' : X' \to \mathfrak{g}^*$ and the similar morphism $q_{\mathscr{N}} : Y \to \mathfrak{g}$ have been considered already. Recall, that $\mathscr{N}$ is the nilpotent variety in $\mathfrak{g}$. By the semi-smallness of the morphism $q_{\mathscr{N}}$, the complex $\mathrm{R}(q_{\mathscr{N}})_*\overline{\mathbb{Q}}_{l\,Y}$ is a perverse sheaf, up to a shift. This complex is the candidate for the perverse Springer representation to be constructed. We have to find a homomorphism $W \to End_{D(\mathcal{N})}(\mathrm{R}(q_{\mathcal{N}})_*\overline{\mathbb{Q}}_{l\,Y})$. For this one observes, that the complex $\mathrm{R}(q_{\mathcal{N}})_*\overline{\mathbb{Q}}_{l\,Y}$ turns out to be the Deligne-Fourier transform – computed on $\mathfrak{g} \times \mathcal{B}$ – of the complex $\mathrm{R}q'_*\overline{\mathbb{Q}}_{l\,X'}$; this is true at least up to shifts and twists. Using this fact, it is therefore enough to find the representation of W on the complex $\mathrm{R}q'_*\overline{\mathbb{Q}}_{l\,X'}$. Now by the smallness of the morphism q' this complex is not only perverse, but also the intermediate extension of its restriction to the regular locus $\mathfrak{g}^*_{reg}$ – again up to shifts and twists. Since over the regular locus the morphism q' is an unramified Galois extension with covering group W (see Theorem VI.9.1 resp. its dual version), the restriction of $\mathrm{R}q'_*\overline{\mathbb{Q}}_{l\,X'}$ to the regular locus is a smooth sheaf on $\mathcal{F}$ on $\mathfrak{g}^*_{reg}$. The monodromy group W of the covering $X'_{reg} \to \mathfrak{g}^*_{reg}$ acts on $\mathcal{F}$ in a natural way. Furthermore $End_{D(\mathfrak{g}^*_{reg})}(\mathcal{F}) \cong \overline{\mathbb{Q}}_l[W]$. By the perverse continuation principle (see Corollary III.5.11) this geometrically "visible" natural action of W on $\mathcal{F}$ extends to an action on the complex $\mathrm{R}q'_*\overline{\mathbb{Q}}_{l\,X'}$ as desired. By a Fourier transform it gives the perverse Springer representation. This approach for the construction of the Springer representations can be found in [Bry2], §11. We also consider other constructions later.

Let the situation be as in the previous section. In particular assume, that the characteristic p of the base field is a very good prime for the semisimple connected group G. Fix some unipotent element u of G. There exists a nilpotent element $v \in \mathfrak{g}$ such that $\sigma(v) = u$, where $\sigma : \mathcal{N} \cong \mathcal{U}$ is a fixed chosen isomorphism between the nilpotent and the unipotent variety. Let $\mathcal{N}_\xi$ be the nilpotent G-orbit of v.

In order to prove Springer's theorem, we have to improve upon the Lemma VI.12.2 of the last section in the following way: In Springer's theorem one has to decompose $H^{2d(u)}(\mathcal{B}_u, \overline{\mathbb{Q}}_l)$ with respect to the action of the characters χ of the group G/G^0_u. By the proper base change theorem we may therefore start from the decomposition

$$\mathrm{R}^{2d(u)}q_!(\overline{\mathbb{Q}}_{l\,Y_\xi})(d(u)) = \bigoplus_{\chi \in \hat{A}(u)} \mathcal{F}_\chi \otimes_{\overline{\mathbb{Q}}_l} V(\xi, \chi)$$

stated in this lemma. It is therefore necessary to relate the vector spaces $V(\xi, \chi)$ to irreducible representation spaces $\phi_{u,\chi}$ of the Weyl group W, as stated in Springer's Theorem VI.1.1.

How did the vector spaces $V(\xi, \chi)$ arise? They were defined as multiplicity spaces in the orbit decomposition VI.12.1 of the perverse sheaf Ψ on the nilpotent variety $\mathcal{N}$, attached to the orbit ξ such that $v \in \mathcal{N}_\xi$

$$\Psi = \bigoplus_{\xi} \bigoplus_{\chi \in \hat{A}(\xi)} \Psi_\xi(\mathcal{F}_\chi) \otimes_{\overline{\mathbb{Q}}_l} V(\xi, \chi) .$$

Here we wrote

$$A(\xi) = A(\sigma(\xi))$$

with respect to the fixed isomorphism $\sigma : \mathcal{N} \to \mathcal{U}$.

By the Proposition VI.13.6 at the end of this section we obtain an unexpected W-action on the perverse sheaf Ψ. It is an action compatible with the trivial action of W on the underlying space $\mathcal{N}$. This surprising W-action is induced from the natural W-action on the monodromy sheaf Φ' – on the dual of the Lie algebra of G – and the surprising Fourier transform identity $T_\psi(i_{\mathcal{N}*}\Psi) = \Phi'$. So Proposition VI.13.6 below allows to compare the two decompositions of Φ' and Ψ. One obtains

$$\begin{aligned} T_\psi\big(i_{\mathcal{N}*}\Psi\big) &= \bigoplus_\xi \bigoplus_\chi T_\psi\big(i_{\mathcal{N}*}\Psi_\xi(\mathcal{F}_\chi)\big) \otimes_{\overline{\mathbb{Q}}_l} V(\xi,\chi) \\ &= \Phi' = \bigoplus_{\phi\in\hat{W}} \Phi'_\phi \otimes_{\overline{\mathbb{Q}}_l} V'_\phi \end{aligned}$$

from Corollary VI.10.8. The monodromy perverse sheaf Φ' is equipped with a natural W-right action, which makes the vectorspaces V'_ϕ into irreducible representation spaces of W with respect to the irreducible representation $\phi \in \hat{W}$

$$\phi : W \to Gl(V'_\phi) .$$

Since the perverse sheaves Φ'_ϕ and the perverse sheaves $T_\psi(i_{\mathcal{N}*}\Psi_\xi(\mathcal{F}_\chi))$ are pairwise nonisomorphic by definition, and since $V'_\phi \neq 0$ for all $\phi \in \hat{W}$ by monodromy reasons, the comparison of the two decompositions defines an injective map

$$B : \hat{W} \hookrightarrow \{\xi \in \mathcal{N}/G, \chi \in \hat{A}(\xi)\} .$$

This map $B(\phi) = (\xi, \chi)$, for irreducible characters ϕ of W, is defined by requiring

Definition 13.1 *Define the Brylinski-Springer correspondence $B(\phi) = (\xi, \chi)$ by requiring*

$$\Phi'_\phi \cong T_\psi\big(i_{\mathcal{N}*}\Psi_\xi(\mathcal{F}_\chi)\big) \quad , \quad V'_\phi \cong V(\xi,\chi) .$$

Conversely, if (ξ, χ) is in the image of this map B put

$$\phi_{\xi,\chi} = \phi \in \hat{W} ,$$

provided $B(\phi) = (\xi, \chi)$. We then also write $\phi_{u,\chi}$ instead of $\phi_{\xi,\chi}$, if $v \in \mathcal{N}_\xi$ and v corresponds to u.

Remark 13.2 To consider a Fourier transform T_ψ as above we have to fix an auxiliary finite subfield $\kappa = \mathbb{F}_q$ of k and a nontrivial character ψ of the additive group of κ. We choose κ such that G, B and T are defined over κ, unless stated otherwise.

Remark 13.3 The Brylinski-Springer correspondence defined above does not depend on the choice of the nontrivial additive character ψ, used to define the Fourier transform. Since any two Artin-Schreier sheaves $\mathcal{L}_\psi$, $\mathcal{L}_{\psi'}$ are related by a rescaling of the underlying affine line, this follows from the following fact: Let $h : \mathbb{G}_m \times \mathfrak{g}^* \to \mathfrak{g}^*$ denote the rescaling map $h(t, v) = t \cdot v$ of the vectorspace $\mathfrak{g}^*$. Then we claim $h^*(\Phi') \cong \Phi'$ and also for its constituents $h^*(\Phi'_\phi) \cong \Phi'_\phi$ for $\phi \in \hat{W}$. This is an easy consequence of the definition of the monodromy sheaf Φ'. Use, that rescaling on the second factor $\mathfrak{t}_{reg}$ defines a $\mathbb{G}_m$ action on Υ (see VI.9.1). It commutes with the action of W and makes the morphism $\tilde{q}$ of Theorem VI.9.1 into a $\mathbb{G}_m$-equivariant map. Here we use, that the arguments of VI.9.1 can be copied in the coadjoint situation.

If we put things together, we have obtained a sheaf theoretic version of Theorem VI.1.1. This sheaf theoretic version implies Theorem VI.1.1 by specialization.

Theorem 13.4 *Let $u = \sigma(v) \in \mathscr{U}$ be a unipotent element of G. Let v be contained in the G-orbit $\mathscr{N}_\xi$ of $\mathscr{N}$. Let $q : Y_\xi \to \mathscr{N}_\xi$ be the restriction of the simultaneous resolution to the inverse image of $\mathscr{N}_\xi$, a morphism with fiber dimension $d(u)$. Then*

$$R^{2d(u)}q_!(\overline{\mathbb{Q}}_{l\,Y_\xi})(d(u)) \cong \bigoplus_{\chi \in \hat{A}(u)} \mathscr{F}_\chi \otimes_{\overline{\mathbb{Q}}_l} V'_\phi .$$

The vector space $V'_{\phi_{u,\chi}}$ has a natural structure as a representation space of W. It defines an irreducible representation $\phi_{u,\chi}$ of W. If $\phi_{u,\chi} \cong \phi_{u',\chi'}$, then u, u' are in the same G-orbit and $\chi = \chi'$, if $A(u)$ is identified with $A(u')$ in the obvious way. The representation $\phi_{u,\chi}$ is completely determined by $T_\psi\left(\Psi_\xi(\mathscr{F}_\chi)\right) \cong \Phi_{\phi_{u,\chi}}$.

We have seen, that the isomorphism $T_\psi(i_{\mathscr{N}*}\Psi) = \Phi'$ gives Ψ the structure of a W-equivariant sheaf. This structure has defined the Springer correspondence B, as explained above. However it turns out, that there is more then one "natural" choice for such a W-action on Ψ. In fact, these different possible W-actions give rise to different representations $\phi_{u,\chi}$ of the Weyl group W.

Remark 13.5 The relation between the nilpotent element v and the unipotent element u comes from a choice of an isomorphism $\sigma : \mathscr{N} \to \mathscr{U}$. For the consideration made above this isomorphism was fixed. It is reasonable to ask, how the characters $\phi_{u,\chi}$ of the Weyl group depend on this choice?

Now let us prove the Proposition VI.13.6, which is stated below: Consider the simultaneous Grothendieck resolution of singularities in the coadjoint setting. Let $\mathfrak{b}' = \mathfrak{u}^\perp$. Then there is the following commutative diagram

$$\begin{array}{ccccc} X' = G \times^B \mathfrak{b}' & \xhookrightarrow{i'} & \mathfrak{g}^* \times \mathscr{B} & \xrightarrow{pr'} & \mathfrak{g}^* \\ \downarrow & & \downarrow & & \downarrow \\ \mathscr{B} & \overset{id}{=\!=\!=} & \mathscr{B} & \longrightarrow & Spec(k) \end{array} \quad (q' = pr' \circ i' : X' \to \mathfrak{g}^*).$$

Put $\delta_{X'}[dim(X')] = i'_*\overline{\mathbb{Q}}_{l\,X'}[dim(X')]$. This is a perverse sheaf on the vector bundle $\mathfrak{g}^* \times \mathscr{B}$. The morphism i' on the upper left is an inclusion of vector bundles over

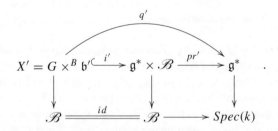

$.\mathscr{B}$. Hence this perverse sheaf has a linear support in each fiber. The morphism pr' has the property, that the trivial vector bundle $\mathfrak{g}^* \times .\mathscr{B}$ is the pullback of the k-vector space $\mathfrak{g}^*$. Therefore, using Theorem III.13.3, the relative Fourier transform of the perverse sheaf $\delta_{X'}[dim(X')]$ on $\mathfrak{g}^* \times .\mathscr{B}$ satisfies

$$Rpr_! \, T_\psi^{\mathfrak{g}^* \times .\mathscr{B}/.\mathscr{B}}(\delta_{X'}[n]) = T_\psi^{\mathfrak{g}^*/Spec(k)}(\, Rq'_! \overline{\mathbb{Q}}_{l\,X'}[n]\,)$$

$$\Phi' \;=\; Rq'_! \overline{\mathbb{Q}}_{l\,X'}[n] \in Perv(\mathfrak{g}') \,.$$

On the other hand the relative Fourier transform of "Dirac" sheaves of subvector bundles can be easily computed (Corollary III.13.4). This computes the perverse sheaf

$$T_\psi^{\mathfrak{g}^* \times .\mathscr{B}/.\mathscr{B}}\big(i'_* \overline{\mathbb{Q}}_{l\;G \overset{B}{\times} \mathfrak{b}'}[n]\big)\big(dim(B)\big) \;=\; i_* \overline{\mathbb{Q}}_{l\;G \overset{B}{\times} \mathfrak{u}}[m] \,,$$

on $\mathfrak{g} \times .\mathscr{B}$. With our previous notations $Y = G \overset{B}{\times} \mathfrak{u}$ and $m = dim(Y), n = dim(G)$ this can be written as

$$T_\psi^{\mathfrak{g}^* \times .\mathscr{B}/.\mathscr{B}}\big(\delta_{X'}[n]\big) \;= i_* \overline{\mathbb{Q}}_{l\,Y}[dim(Y)](-dim(B)) \,.$$

Since furthermore $dim(G) - dim(B) = dim(U) = \frac{1}{2} dim(Y)$, we obtain from the definition $\Psi = Rpr_! i_* \overline{\mathbb{Q}}_{l\,Y}\langle dim(Y)\rangle$ the next

Proposition 13.6 *Let $i_{\mathscr{N}^\cdot}$ denote the inclusion $\mathscr{N}^\cdot \hookrightarrow \mathfrak{g}$. Then the Fourier transform $T_\psi(\Phi')$ of the perverse sheaf Φ' on the vector space $\mathfrak{g}^*$ is the perverse sheaf $i_{\mathscr{N}^\cdot *}\Psi$ on the dual vectorspace $\mathfrak{g}$ up to a twist*

$$\boxed{T_\psi(\Phi') = i_{\mathscr{N}^\cdot *}\Psi(-n) \quad , \quad T_\psi\big(i_{\mathscr{N}^\cdot *}\Psi\big) = \Phi' \,.}$$

In particular $End_{Perv(\mathscr{N}^\cdot)}(\Psi) \;\cong\; End_{Perv(\mathfrak{g}^*)}(\Phi') \;\cong\; \overline{\mathbb{Q}}_l[W]$.

Duality. For the moment suppose, that the characteristic p of the base field k is large enough. Then $A(u)$ is one of the groups $(\mathbb{Z}/2\mathbb{Z})^e$, S_3, S_4, S_5. The irreducible characters χ of these groups are self dual $\chi^* \cong \chi$. Furthermore, by the definition of the Brylinski-Springer correspondence $T_\psi(i_{\mathscr{N}^\cdot *}\Psi_\xi(\mathscr{F}_\chi)) \cong \Phi'_\phi$ for $\phi = \phi_{\xi,\chi}$. Furthermore $D(\Phi'_\phi) \cong \Phi'_{\phi^*}(n)$ by VI.10.6 and $DT_\psi(K) \cong T_{\psi^{-1}} D(K)(n)$ by III.12.2. This implies $\Phi'_{\phi^*}(n) \;\cong\; D\Phi'_\phi \;\cong\; DT_\psi(i_{\mathscr{N}^\cdot *}\Psi_\xi(\mathscr{F}_\chi)) \;\cong\; T_{\psi^{-1}} D(i_{\mathscr{N}^\cdot *}\Psi_\xi(\mathscr{F}_\chi))(n) \;\cong\; T_{\psi^{-1}} i_{\mathscr{N}^\cdot *}\Psi_\xi(\mathscr{F}_{\chi^*}))(n) \;=\; T_{\psi^{-1}} i_{\mathscr{N}^\cdot *}\Psi_\xi(\mathscr{F}_\chi))(n) \;=\; \Phi'_\phi(n)$ using VI.13.3. Hence $\phi^* \cong \phi$, since $\Phi'_{\phi_1} \cong \Phi'_{\phi_2}$ implies $\phi_1 \cong \phi_2$ (and conversely).

Corollary 13.7 *The irreducible characters ϕ of the Weyl group W are self dual $\phi^* = \phi$. Any irreducible representation of W is isomorphic to its contragredient.*

In particular, this implies that an arbitrary W-right action of W on Ψ has the property, that the Verdier dual $D(\Psi)$ of the W-equivariant complex Ψ is isomorphic to Ψ in the equivariant sense.

VI.14 A Second Approach

In this section we will describe a second construction, which defines another W-action on the perverse sheaf Ψ. p is assumed to be a very good prime. This W-action is different from the one obtained in the last section. It is obtained by restriction from the action of W on Φ, by means of a natural restriction map $\Phi \to i_{\mathcal{N}*}\Psi[r](N)$. This new W-action on Ψ is in some sense dual to the one obtained in the last section. In this section we will only consider this new action, the Lusztig action.

We remind the reader, that we use the following abbreviations, where N is the number of positive roots and $r = dim(T)$, $n = dim(G)$, $m = n - r = 2N$. Further recall $K\langle 2s\rangle = K[2s](s)$ and see VI §10,11 for the definition of the complexes Φ and Ψ. Consider the commutative diagram

$$\begin{array}{ccccccccc} & & & & & & \overset{q}{\frown\!\!\!\longrightarrow} & & \\ G\times^B\{0\} & \overset{i_0}{\hookrightarrow} & Y = G\times^B\mathfrak{u} & \overset{i_Y}{\hookrightarrow} & X = G\times^B\mathfrak{b} & \overset{i}{\hookrightarrow} & \mathfrak{g}\times\mathcal{B} & \overset{pr}{\longrightarrow} & \mathfrak{g} \\ \downarrow & & \downarrow & & \downarrow & & \downarrow & & \downarrow \\ \mathcal{B} & \overset{id}{=\!=\!=} & \mathcal{B} & \overset{id}{=\!=\!=} & \mathcal{B} & \overset{id}{=\!=\!=} & \mathcal{B} & \longrightarrow & Spec(k) \end{array}.$$

Also recall the restriction $q_{\mathcal{N}} : Y \to \mathcal{N}$ of the map $q : X \to \mathfrak{g}$ to the inverse image Y of the nilpotent variety $i_{\mathcal{N}} : \mathcal{N} \to \mathfrak{g}$. With these notations we obtain the following complexes on $\mathfrak{g}$:

$$\Phi = Rq_!\overline{\mathbb{Q}}_{lX}[dim(X)] = Rpr_!i_*\overline{\mathbb{Q}}_{lX}[n]$$

$$\Psi(-N) = Rq_{\mathcal{N}!}\overline{\mathbb{Q}}_{lY}[dim(Y)] = Rpr_!(i\circ i_Y)_*\overline{\mathbb{Q}}_{lY}[m]$$

$$K^\bullet := Rpr_!\ \overline{\mathbb{Q}}_l[n] \cong \overline{\mathbb{Q}}_{l\mathfrak{g}}[n]\otimes_{\overline{\mathbb{Q}}_l} R\Gamma(\mathcal{B},\overline{\mathbb{Q}}_l)\ .$$

Φ and Ψ are perverse sheaves. The complex $K^\bullet$ is not a perverse sheaves; it is a direct sum of its perverse cohomology sheaves

$$K^\bullet \cong \bigoplus_{\mu=0}^{dim(\mathcal{B})} {}^pH^{2\mu}K[-2\mu] = \bigoplus_{\mu=0}^{dim(\mathcal{B})} H^{2\mu}(\mathcal{B},\overline{\mathbb{Q}}_l(\mu))\otimes_{\overline{\mathbb{Q}}_l}\overline{\mathbb{Q}}_{l\mathfrak{g}}[n]\langle -2\mu\rangle\ .$$

This is a trivial case of Gabber's decomposition theorem. Note that the "bullet" does not refer to K being a complex! It refers to this direct sum decomposition into translates of perverse sheaves. $K^\bullet$ is a sum of twisted right translates of the constant perverse sheaf $\overline{\mathbb{Q}}_{l\mathfrak{g}}[n]$; recall $n = dim(\mathfrak{g})$. Furthermore $K^\bullet$ canonically carries a graded right W-action, i.e each of the translated perverse sheaves has a canonical structure as W-right equivariant perverse sheaf, compatible with the trivial right action of W on the underlying affine space $\mathfrak{g}$. This action is called the Slodowy action and is defined via Theorem VI.2.2. See also the discussion following Lemma VI.2.3.

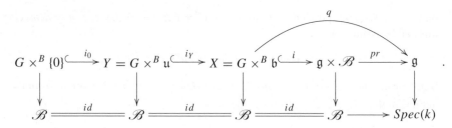

The adjunction maps $id \to i_* i^*$, $id \to (i_Y)_*(i_Y)^*$ and $id \to (i_0)_*(i_0)^*$ induce natural restriction maps between the complexes defined above, by applying $Rq_!$. We abbreviate δ_0 to be the constant skyscraper sheaf on $\mathfrak{g}$ with support in the origin and stalk $\overline{\mathbb{Q}}_l$. Put $\delta_{\mathfrak{g}}[n] = \overline{\mathbb{Q}}_{l\,\mathfrak{g}}[n]$. Then we obtain a sequence of restriction maps, which we call the

Control Sequence

$$K^\bullet = \delta_{\mathfrak{g}}[n] \otimes_{\overline{\mathbb{Q}}_l} \overline{\mathbb{Q}}_l[W]^\bullet \longrightarrow \Phi \longrightarrow i_{\mathcal{N}^\cdot,*}\Psi[r](-N) \longrightarrow \delta_0[n] \otimes_{\overline{\mathbb{Q}}_l} \overline{\mathbb{Q}}_l[W]^\bullet$$

This is not an exact sequence! It is a sequence of maps in the derived category between sheaf complexes. These sheaf complexes are supported on $\mathfrak{g}$ respectively $\mathfrak{g}$ respectively $\mathcal{N}^\cdot$ respectively $\{0\}$. The maps are induced from the inclusions $\{0\} \subset \mathfrak{u} \subset \mathfrak{b} \subset \mathfrak{g}$. The monodromy action of W on Φ induces a W action on Ψ. This follows from the usual permanence properties of W-equivariant sheaves. See Remark III.15.3. We also view the other two complexes in the control sequence as W-equivariant sheaves on $\mathfrak{g}$, with respect to the trivial action of W on $\mathfrak{g}$. For this we use the Slodowy action on $H^{2\bullet}(\mathcal{B}, \overline{\mathbb{Q}}_l) = \overline{\mathbb{Q}}_l[W]^\bullet$ discussed in VI.2.2.

With these four W-action specified, we claim

Proposition 14.1 *The maps of the control sequence are W-equivariant homomorphisms of sheaf complexes.*

Recall. In this section the action of W on $\Psi = \Phi[-r](N)|_{\mathcal{N}^\cdot}$ is the action induced from the action of W on Φ. This is not the action of W defined in VI §13 by Fourier transform.

Proof. $\Phi \to i_{\mathcal{N}^\cdot *}\Psi[r](-N)$ is W-equivariant by definition. Hence it is enough to prove, that the first restriction map $res : K^\bullet \to \Phi$ is W-equivariant and that the induced action on the stalk $\Phi_0 = \overline{\mathbb{Q}}_l[W]^\bullet[n]$ is the Slodowy action. Assume the W-equivariance of the first map. To see that the induced action on the stalk at the origin $0 \in \mathfrak{g}$ is the same as the Slodowy action, consider the stalks of the control sequence

$$(K^\bullet)_0 = \overline{\mathbb{Q}}_l[W]^\bullet[n] \xrightarrow{\cong} \Phi_0 \xrightarrow{\cong} \Psi[r](-N)_0 \xrightarrow{\cong} \overline{\mathbb{Q}}_l[W]^\bullet[n] \quad (\text{composite} = id).$$

Since the composed map $(K^\bullet)_0 \to \overline{\mathbb{Q}}_l[W]^\bullet[n]$ is the identity and since all restriction maps are W-equivariant, the induced W-module structure of the stalk Ψ_0 is given by the Slodowy action (up to a shift) and is isomorphic to the regular representation.

So it remains to verify the W-equivariance of the first map of the control sequence.

Step 1. We show equivariance on the regular locus $\mathfrak{g}_{reg}$.

For this consider the diagram

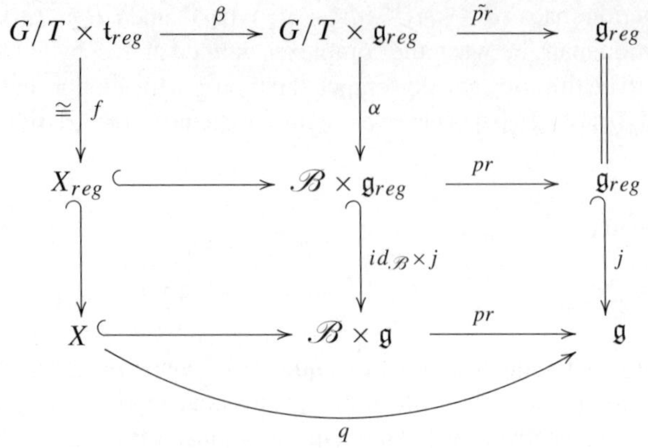

The upper left arrow β is defined by $(gT, t) \mapsto (gT, g(t))$. This is obviously a W-equivariant morphism, if W acts on G/T from the right by $gT \mapsto gnT$, on $\mathfrak{t}_{reg}$ by the natural right action $t \mapsto w^{-1}(t)$ and on $\mathfrak{g}$ by the trivial right action. Here $n \in N(T)$ denotes a representative of $w \in W$.

The map pr is proper. Hence $\mathrm{R}pr_! = \mathrm{R}pr_*$. Therefore the restriction $K^\bullet_{reg}$ of the complex $K^\bullet = \mathrm{R}pr_! \overline{\mathbb{Q}}_{l,\mathscr{B} \times \mathfrak{g}}$ to the regular locus is

$$K^\bullet_{\mathfrak{g}_{reg}} \cong \mathrm{R}pr_* \overline{\mathbb{Q}}_{l,\mathscr{B} \times \mathfrak{g}_{reg}}[n] \cong \mathrm{R}\tilde{pr}_* \overline{\mathbb{Q}}_{l\,G/T \times \mathfrak{g}_{reg}}[n] ,$$

since $\mathrm{R}\alpha_* \overline{\mathbb{Q}}_{l\,G/T \times \mathfrak{g}_{reg}} = \overline{\mathbb{Q}}_{l,\mathscr{B} \times \mathfrak{g}_{reg}}$ by Lemma VI.2.3. Thus the restriction map $K^\bullet \to \Phi$ – when considered over the open subset $\mathfrak{g}_{reg}$ – can be canonically identified with the complex map

$$\mathrm{R}pr_* \left(\overline{\mathbb{Q}}_{l\,G/T \times \mathfrak{g}_{reg}}[n] \to \beta_* \beta^* \overline{\mathbb{Q}}_{l\,G/T \times \mathfrak{g}_{reg}}[n] \right)$$

via the isomorphism f. This morphism is W-equivariant.

Step 2. We now show, that the W-equivariance proved over $\mathfrak{g}_{reg}$ in step 1 already implies W-equivariance over the whole Lie algebra $\mathfrak{g}$. We have to show the commutativity of the next diagram

$$
\begin{array}{ccc}
K^\bullet & \xrightarrow{\ res\ } & \Phi \\
{\scriptstyle \varphi_w^{K^\bullet}} \big\uparrow & & \big\uparrow {\scriptstyle \varphi_w^\Phi} \\
w^*(K^\bullet) & \xrightarrow{w^*(res)} & w^*(\Phi) .
\end{array}
$$

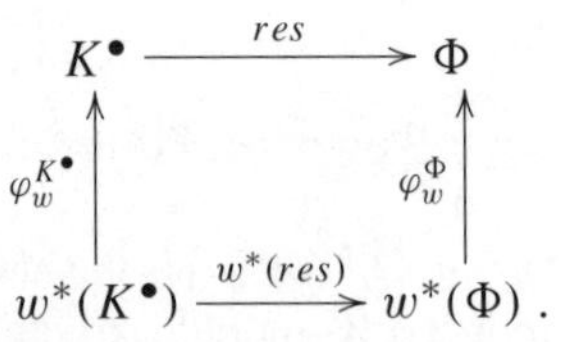

Note $w^*(K^\bullet) = K^\bullet$ etc., since W acts trivially on $\mathfrak{g}$. To show the commutativity of the diagram $res \circ \varphi_w^{K^\bullet} - \varphi_w^\Phi \circ res = 0$, it is enough to show injectivity of the natural map

$$Hom_{D^b_c(\mathfrak{g}, \overline{\mathbb{Q}}_l)}(K^\bullet, \Phi) \hookrightarrow Hom_{D^b_c(\mathfrak{g}_{reg}, \overline{\mathbb{Q}}_l)}(K^\bullet|_{\mathfrak{g}_{reg}}, \Phi|_{\mathfrak{g}_{reg}}) .$$

Here homomorphisms are understood to be homomorphisms in the derived category. To proof this injectivity, one may replace the complex $K^\bullet$ – ignoring W-equivariance – by a direct sum of translates of the constant sheaf. Because $Hom_{D^b_c(Z,\overline{\mathbb{Q}}_l)}(\overline{\mathbb{Q}}_{l\,Z}[-\mu], -) = H^\mu(Z, -)$ the injectivity statement would follow from the statements

$$res: \ H^\mu(\mathfrak{g}, \Phi) \hookrightarrow H^\mu(\mathfrak{g}_{reg}, j^*(\Phi)) \quad , \quad \mu \in \mathbb{Z}\,.$$

Now $H^\mu(\mathfrak{g}, \Phi) = H^{\mu+dim(X)}(X, \overline{\mathbb{Q}}_l)$ and $H^\mu(\mathfrak{g}_{reg}, j^*(\Phi)) = H^{\mu+dim(X)}(X_{reg}, \overline{\mathbb{Q}}_l)$. But X is a vector bundle over $\mathscr{B}$ with fibers isomorphic to $\mathfrak{b}$. It admits the trivial vector bundle $\mathscr{B} \times \mathfrak{t}$ as a quotient bundle. Thus X can be viewed as a vector bundle over $\mathscr{B} \times \mathfrak{t}$. This defines a smooth morphisms with affine fibers isomorphic to $\mathfrak{u}$. Now, to proof injectivity of the restriction map in cohomology, we may restrict to the cohomology of the basis (Lemma VI.2.3). So it is enough to prove, that the restriction maps

$$H^\nu(\mathscr{B} \times \mathfrak{t}, \overline{\mathbb{Q}}_l) \ \to \ H^\nu(\mathscr{B} \times \mathfrak{t}_{reg}, \overline{\mathbb{Q}}_l)$$

are injective. But this is an immediate consequence of (a trivial case of) the Künneth formulas. □

Isotypic Components. Let $i_{\mathscr{N}^\cdot} : \mathscr{N}^\cdot \to \mathfrak{g}$ be the inclusion map. Recall, that by definition of the new W-action obtained on Ψ, the restriction map

$$\begin{array}{ccc} \Phi & \xrightarrow{\ res\ } & i_{\mathscr{N}^\cdot,*}\Psi[r](-N) \\ \Big\downarrow{\scriptstyle id} & & \Big\downarrow{\scriptstyle \cong} \\ \Phi & \xrightarrow{\ adj\ } & i_{\mathscr{N}^\cdot,*}(i^*_{\mathscr{N}^\cdot}\Phi) \end{array}$$

is W-equivariant. By III.15.4, since W acts trivially on $\mathfrak{g}$, the W-right action on Ψ gives rise to a decomposition $\Psi = \bigoplus_{\phi\in\hat{W}} \Psi_\phi \otimes_{\overline{\mathbb{Q}}_l} V_\phi$ into isotypic ϕ-components analogous to the decomposition $\Phi = \bigoplus_{\phi\in\hat{W}} \Phi_\phi \otimes_{\overline{\mathbb{Q}}_l} V_\phi$. The restriction map

$$i^*_{\mathscr{N}^\cdot}\Phi[-r] \to \Psi(-N)$$

is equivariant, hence decomposes into maps

$$i^*_{\mathscr{N}^\cdot}\Phi_\phi[-r] \to \ \Psi_\phi(-N)\,.$$

These maps are isomorphisms. This follows from the proper base change theorem applied to the proper map $q : X \to \mathfrak{g}$ with respect to the base restriction map $i_{\mathscr{N}^\cdot} : \mathscr{N}^\cdot \to \mathfrak{g}$.

Theorem 14.2

(a) For each irreducible character ϕ of the Weyl group, the restriction maps

$$i^*_{\mathscr{N}^\cdot}\Phi_\phi[-r] \ \cong \ \Psi_\phi(-N)$$

are isomorphisms between irreducible perverse sheaves.

(b) *The stalks at the origin of* Φ_ϕ *and* Ψ_ϕ *are nonzero*

$$(\Phi_\phi)_0 \cong (\Psi_\phi[r](-N))_0 \cong \overline{\mathbb{Q}}_l[W]^\bullet_\phi[n] .$$

These are W*-equivariant isomorphisms with respect to the Lusztig-action.*

(c) *For all* $\phi_1, \phi_2 \in \hat{W}$ *the* $\overline{\mathbb{Q}}_l$*-vectorspace* $Hom_{D^b_c(\mathfrak{g})}(\Phi_{\phi_1}, i_{\mathcal{N}*}\Psi_{\phi_2}[r])$ *has dimension one, if* $\phi_1 \cong \phi_2$*, and is zero else.*

Proof. For the proof of (b), it is enough to consider the perverse sheaves Ψ_ϕ. The stalk of each Ψ_ϕ at the origin is the ϕ-isotypic component with respect to the W-action of the stalk Ψ_0 of Ψ. By Proposition VI.14.1 the direct sum of the cohomology groups of the stalk Ψ_0 define a representation, which is isomorphic to the regular representation of W. More precisely $\Psi_0 \cong \overline{\mathbb{Q}}_l[W]^\bullet[2N](N)$ (proper base change theorem applied for q). This proves (b).

By (b) the perverse sheaves Ψ_ϕ are nonzero. Recall $End_{Perv(\mathcal{N})}(\Psi) \cong \overline{\mathbb{Q}}_l[W]$ from VI.13.6. Hence $\sum_\phi dim_{\overline{\mathbb{Q}}_l}\left(End_{Perv(\mathcal{N})}(\Psi_\phi)\right) \cdot deg(\phi)^2 = \#W$. Therefore $End_{Perv(\mathcal{N})}(\Psi_\phi) \cong \overline{\mathbb{Q}}_l$. Thus all perverse sheaves Ψ_ϕ are irreducible perverse sheaves on $\mathcal{N}$ by the decomposition theorem. This proves (a). (c) is an immediate consequence of (a) and

$$dim_{\overline{\mathbb{Q}}_l} Hom_{D^b_c(\mathfrak{g})}\left(\Phi_{\phi_1}, i_{\mathcal{N}*}\Psi_{\phi_2}[r]\right) = dim_{\overline{\mathbb{Q}}_l} Hom_{D^b_c(\mathcal{N})}\left(i^*_{\mathcal{N}}\Phi_{\phi_1}[-r], \Psi_{\phi_2}\right) .$$

□

Corollary 14.3 *The natural restriction map*

$$End_{Perv(\mathfrak{g})}(\Psi) \to End_{\overline{\mathbb{Q}}_l}(\Psi_0)$$

induced by the stalk functor $res_0 : \Psi \to \Psi_0$ *is injective. Similar for* Φ.

Example 14.4 By definition $\Phi_1 \cong \overline{\mathbb{Q}}_{l\,\mathfrak{g}}[n]$ holds for the trivial representation $\phi = 1$ of W. Therefore we get $\Psi_1(-N) \cong i^*_{\mathcal{N}}\Phi_1[-r] \cong \overline{\mathbb{Q}}_{l,\mathcal{N}}[dim(\mathcal{N})]$ or $\Psi_1 = \overline{\mathbb{Q}}_{l,\mathcal{N}}\langle dim(\mathcal{N})\rangle$. In particular

$$\overline{\mathbb{Q}}_{l,\mathcal{N}}[dim(\mathcal{N})] \in Perv(\mathcal{N})$$

is a perverse sheaf on $\mathcal{N}$.

VI.15 The Comparison Theorem

Consider the "control sequence" as defined in the last section. All its maps are W-equivariant, where Ψ was endowed with the Lusztig-action (as defined by the second approach). We write $\Psi = \Psi_L$ to indicate this. Similarly we write Ψ_B, Φ_B

for the W-equivariant complexes Ψ, Φ defined by the Brylinski action of W (see IV §10,11,12).

The Verdier dual of this control sequence defines again a sequence of W-equivariant maps, since we have W-equivariant isomorphisms $D(\Phi) \cong \Phi(n)$ (Remark VI.10.6) and $D(\Psi_L) \cong \Psi_L$ (Corollary VI.13.7). If one furthermore uses $D(\delta_{\mathfrak{g}}[n]) \cong \delta_{\mathfrak{g}}[n](n)$ and $D(\delta_0) \cong \delta_0$, the dual of the control sequence yields the following

Twisted Dual of the Control Sequence. The dual of the control sequence twisted by $(-n)$ gives the following W**-equivariant** maps

$$\delta_{\mathfrak{g}}[n] \otimes_{\overline{\mathbb{Q}}_l} H^{-2\bullet}(\mathscr{B}, \overline{\mathbb{Q}}_l)^* \longleftarrow \Phi_L \longleftarrow i_{\mathscr{N}*}\Psi_L[-r](-n+N) \longleftarrow \cdots .$$

Fourier Transform of the Control Sequence. There is another way to invert arrows. Start in the coadjoint situation and consider the perverse sheaves Φ' and Ψ' on $\mathfrak{g}^*$. Here Φ' is endowed with the monodromy W-action and Ψ' inherits this W-action by restriction. There is an analogous control sequence for these sheaves on $\mathfrak{g}'$. The Deligne-Fourier transform applied to it gives – after a twist and a shift by $[-r](N)$ – new W**-equivariant** maps

$$\delta_{\mathfrak{g}}[n] \otimes_{\overline{\mathbb{Q}}_l} H^{2\bullet}(\mathscr{B}, \overline{\mathbb{Q}}_l)[2N](N) \longleftarrow \Phi_B \longleftarrow i_{\mathscr{N}*}\Psi_B[-r](-n+N) \longleftarrow .$$

Here we used $T_\psi(\Phi') = i_{\mathscr{N}*}\Psi_B(-n)$ and $T_\psi(i_{\mathscr{N}*}\Psi') = \Phi_B$ as well as $T_\psi(\delta'_0[n]) = \delta_{\mathfrak{g}}[2n]$ and $2n - r = n + 2N$. The last map on the left is $cores = T_\psi(res : i_{\mathscr{N}'*}\Psi' \to \delta'_0[n-r](N)))$ and is called corestriction for simplicity.

The index $_B$ indicates, that the W-action on Φ_B resp. Ψ_B may be different from the one on Φ_L resp. Ψ_L considered above, although the underlying perverse sheaves are the same. We refer to the action on Φ_B as the "Brylinski"-action, in contrast to the "Lusztig"-action on Φ_L used above.

Poincare Duality. The group W acts on the highest cohomology group $H^{2N}(\mathscr{B}, \overline{\mathbb{Q}}_l)(N) \cong \overline{\mathbb{Q}}_l$ by the sign character ε_W of the Coxeter group W. Poincare duality for the flag variety $\mathscr{B}$ gives the following W**-equivariant** isomorphism

$$PD: \ H^{-2\bullet}(\mathscr{B}, \overline{\mathbb{Q}}_l[2N](N)) = \varepsilon_W \otimes_{\overline{\mathbb{Q}}_l} H^{2\bullet}(\mathscr{B}, \overline{\mathbb{Q}}_l)^* .$$

We claim, that the twisted dual of the control sequence and the Fourier transform of the control sequence essentially yield the same maps by identifying the targets via the isomorphism PD. In particular, we claim that there exists a unique isomorphism ρ of perverse sheaves (ignoring W-actions !), which makes the diagram

$$\begin{array}{ccc} \delta_{\mathfrak{g}}[n] \otimes_{\overline{\mathbb{Q}}_l} H^{-2\bullet}(\mathscr{B}, \overline{\mathbb{Q}}_l)^* & \longleftarrow & \Phi_L \\ {\scriptstyle id\otimes_{\overline{\mathbb{Q}}_l} PD}\uparrow\cong & & \uparrow\rho \\ \delta_{\mathfrak{g}}[n] \otimes_{\overline{\mathbb{Q}}_l} H^{2\bullet}(\mathscr{B}, \overline{\mathbb{Q}}_l)[2N](N) & \xleftarrow{cores} & \Phi_B \end{array}$$

commutative.

Uniqueness of ρ. Apply Fourier inversion and use, that $T_{\psi^{-1}}(cores)$ essentially is the restriction map res_0. So uniqueness follows from Corollary VI.14.3.

Theorem 15.1 $\rho : \Phi_B \cong \varepsilon_W \otimes_{\overline{\mathbb{Q}}_l} \Phi_L$ *is an isomorphism of* W*-equivariant perverse sheaves.*

Corollary 15.2 $\Psi_B \cong \varepsilon_W \otimes_{\overline{\mathbb{Q}}_l} \Psi_L$ *as* W*-equivariant perverse sheaves.*

Proof. Assume for the moment that ρ exists, making the last diagram commutative (but without assuming any kind of W-equivariance). The two horizontal maps in the last diagram are W-equivariant. The map $id \otimes_{\overline{\mathbb{Q}}_l} PD$ is W-equivariant up to a twist by the character ε_W. Then uniqueness of ρ implies the W**-equivariance** of

$$\rho : \Phi_B \longrightarrow \varepsilon_W \otimes_{\overline{\mathbb{Q}}_l} \Phi_L \,.$$

Recall, that each $w \in W$ acts on Φ_B by some isomorphism $\varphi_w^B : \Phi_B \to \Phi_B$ and similar $\varphi_w^L : \Phi_L \to \Phi_L$. The statement follows, since both $\varepsilon_W(w) \cdot \varphi_w^L \circ \rho \circ (\varphi_w^B)^{-1}$ and ρ make the diagram commute. This proves the statement modulo the existence of ρ.

Existence of ρ. We have to find a vertical isomorphism ρ (not necessarily W-equivariant), which makes the following diagram commutative

$$\begin{array}{ccc}
DRpr_!(\delta_{\mathfrak{g}\times\mathscr{B}}[n]) = D(\delta_{\mathfrak{g}}[n]) \otimes_{\overline{\mathbb{Q}}_l} DR\Gamma_c(\mathscr{B}, \overline{\mathbb{Q}}_l) & \xleftarrow{\;D(res_X)\;} & D(\Phi_L) \\
\uparrow \cong & & \uparrow \cong \\
\delta_{\mathfrak{g}}[n](n) \otimes_{\overline{\mathbb{Q}}_l} H^{-2\bullet}(\mathscr{B}, \overline{\mathbb{Q}}_l)^* & \longleftarrow & \Phi_L(n) \\
id\otimes PD \uparrow \cong & & \vdots\, \rho \\
Rpr_! D(\delta_{\mathfrak{g}\times\mathscr{B}}[n]) = \delta_{\mathfrak{g}}[n](n) \otimes_{\overline{\mathbb{Q}}_l} H^{2\bullet}(\mathscr{B}, \overline{\mathbb{Q}}_l\langle 2N\rangle) & \xleftarrow{\;cores\;} & \Phi_B(n) \\
\uparrow \cong & & \| \\
T_\psi(\delta_0\langle 2N\rangle(n)) \otimes_{\overline{\mathbb{Q}}_l} R\Gamma_c(\mathscr{B}, \overline{\mathbb{Q}}_l) & \xleftarrow{\;Rpr_! T_\psi^{E'/S}(res_0)[2N]\;} & T_\psi(i_{\mathscr{N}'*}\Psi')(n)
\end{array}$$

Recall that $T_\psi^{E/S}(\delta_0) = \delta_{\mathfrak{g}}[n]$ and $R\Gamma_c(\mathscr{B}, \overline{\mathbb{Q}}_l) \cong H^{2\bullet}(\mathscr{B}, \overline{\mathbb{Q}}_l)$ gives the lower left isomorphism.

Recall that the diagram at the beginning of section VI §14 defines subbundles X and Y of the vector bundle $E = \mathfrak{g} \times \mathscr{B}$ over $\mathscr{B}$. Similar subbundles X' and Y' were defined for the dual bundle E' over $\mathscr{B}$; see the diagram before Proposition VI.13.6.

Now use the functoriality $\mathrm{R}pr_! T_\psi^{E'/S} = T_\psi \mathrm{R}pr_!$ of the Fourier transform, where $E' = \mathfrak{g}^* \times \mathscr{B}$ is the trivial vector bundle over the base $S = \mathscr{B}$. To find ρ, it is enough to complete the next diagram to a commutative diagram with the dotted isomorphisms

$$\begin{array}{ccccc}
D\mathrm{R}pr_!\big(\delta_{\mathfrak{g}\times\mathscr{B}}[n]\big) & \xleftarrow{D(res_X)} & D\mathrm{R}pr_!\big(\delta_X[n]\big) & = \!= & D(\Phi_L) \\
{\scriptstyle rel.PD}\ \uparrow \cong & & \cong \uparrow\ {\scriptstyle rel.PD} & & \vdots \cong \\
\mathrm{R}pr_! D\big(\delta_{\mathfrak{g}\times\mathscr{B}}[n]\big) & \xleftarrow{D(res_X)} & \mathrm{R}pr_! D\big(\delta_X[n]\big) & & \vdots \\
\uparrow \cong & & \vdots & & \vdots \\
\mathrm{R}pr_! T_\psi^{E'/S}\big(\delta'_{0\times\mathscr{B}}\langle 2N\rangle(n)\big) & \longleftarrow & \mathrm{R}pr_! T_\psi^{E'/S}\big(\delta_{Y'}\langle 2N\rangle(n)\big) & = \!= & T_\psi\big(i_{\mathcal{N}'*}\Psi'\big)(n)
\end{array}$$

The lower left arrow is given by the map $\mathrm{R}pr_! T_\psi^{E'/S}\big(res_0\big)[2N]$. The left upper square commutes, since the relative Poincare duality isomorphism $rel.PD : D\mathrm{R}pr_! \cong \mathrm{R}pr_! D$ defines a natural transformation.

Of course, one can now drop $\mathrm{R}pr_!$ and restrict oneself to the lower left square of the last diagram. So – after a shift by $[n]$ – one has to complete the diagram

$$\begin{array}{ccc}
D\big(\delta_{\mathfrak{g}\times\mathscr{B}}\big) & \xleftarrow{\quad D(res_X)\quad} & D\big(\delta_X\big) \\
\uparrow \cong & & \vdots \\
T_\psi^{E'/S}\big(\delta'_{0\times\mathscr{B}}\big)\langle 2N\rangle[n](n) & \xleftarrow{T_\psi^{E'/S}\big(res_0\big)[2N+n]} & T_\psi^{E'/S}\big(\delta_{Y'}\big)\langle 2N\rangle[n](n)\,.
\end{array}$$

But $D(\delta_X) = \delta_X\langle 2n\rangle$, $D(\delta_{\mathfrak{g}\times\mathscr{B}}) = D(\delta_E) = \delta_E\langle 2n + 2N\rangle$. Furthermore by Corollary III.13.4

$$T_\psi^{E'/S}(\delta_{Y'})\langle 2N\rangle[n](n) = \delta_X\langle 2n\rangle$$

$$T_\psi^{E'/S}(\delta_{0\times\mathscr{B}})\langle 2N\rangle[n](n) = \delta_E\langle 2n + 2N\rangle\,.$$

Note that Y' is a subbundle W of E'/S of rank N, whose annihilator V in E/S is the subbundle X of E. So we end up in the more abstract situation, where $W \hookrightarrow E'$ is a locally split subvectorbundle of rank N of the vector bundle E'/S of rank n with the corresponding subvectorbundle $V = W^\perp \hookrightarrow E$ over the smooth base $S = \mathscr{B}$. So – up to a shift and twist by $\langle 2dim(S)\rangle = \langle 2N\rangle$ – we have to complete the diagram

$$\begin{array}{ccc} \delta_E\langle 2n\rangle & \xleftarrow{D_{E/S}(res_{E,V})} & \delta_V\langle 2n-2N\rangle \\ \uparrow \cong & & \uparrow \exists\cong \\ \delta_E\langle 2n\rangle & \xleftarrow{cores_{V,E}} & \delta_V\langle 2n-2N\rangle \ . \end{array}$$

This is possible by III.13.6. □

Corollary 15.3 $\sum_\nu dim_{\overline{\mathbb{Q}}_l} H_c^\nu\big(\mathcal{N}, (\Psi_L)_\phi\big)\cdot t^{\nu/2} = P_{\phi\otimes\varepsilon_W}(t)$.

Proof. This follows, since $\bigoplus_\phi H_c^*(\mathcal{N}, (\Psi_L)_\phi)\otimes\phi$ is equal to $H_c^*(\mathfrak{g}, i_{\mathcal{N}*}\Psi_L) = \mathcal{H}^{*-n}T_\psi(i_{\mathcal{N}*}\Psi_L)_0$, which equals $\mathcal{H}^{*-n}(\Phi'_B)_0$ by VI.13.6. By VI.15.1 and VI.14.2 this is isomorphic as W-module to $\varepsilon_W\otimes\overline{\mathbb{Q}}_l[W]^\bullet$. □

The polynomials $P_\phi(t)$, defined after VI.2.2, satisfy $P_\phi(1) = deg(\phi)$ and $P_{\phi\otimes\varepsilon_W}(t) = t^N P_\phi(t^{-1})$, a consequence of Poincare duality and $\phi^*\cong\phi$. Therefore the degree $b_\phi = deg_t(P_\phi)$ and the leading order a_ϕ of the polynomials are related by $a_\phi + b_{\phi\otimes\varepsilon} = N$. The dimensional bounds of the cohomological dimension imply $H_c^\nu(\mathcal{N}, \Psi_\xi(\mathcal{F}_\chi)) = 0$ for $\nu > dim(\mathcal{N}_\xi) = m - 2\cdot d(\xi)$. Therefore $b_{\phi\otimes\varepsilon}\leq 1/2\cdot dim(\mathcal{N}_\xi) = N - d(u)$ for $\phi = \varepsilon_W\otimes\phi_{u,\chi}$ or

$$d(u)\ \leq\ a_\phi \quad , \quad \phi = \varepsilon_W\otimes\phi_{u,\chi}\ .$$

Equality holds iff $\mathcal{F}_\chi$ is constant, i.e. iff $\chi = 1$ is the trivial character of $A(u)$.

As a special case one obtains

Corollary 15.4 *Let p be a very good prime. Then*

$$H_c^\nu(\mathcal{N}, \overline{\mathbb{Q}}_l) = 0 \quad , \quad \nu\neq 2m$$

and $H_c^{2m}(\mathcal{N}, \overline{\mathbb{Q}}_l)\cong\overline{\mathbb{Q}}_l(-m)$.

Recall, that $\mathcal{N}$ is affine and of dimension m. Thus $\mathcal{N}$ behaves like affine space of dimension m. On the other hand $\mathcal{N}$ has singularities. To illustrate this consider the simplest case $G = Sl(2)$. Then $\mathcal{N} = Spec\ k[a,b,c]/(a^2+bc)$, which has a singularity at the origin. Note $\mathcal{U}\cong\mathcal{N}$ via $g\mapsto g-1$. A matrix with entries a, b, c, d and trace $a+d=0$ is nilpotent iff $a^2+bc=0$.

For very good primes p the computation of the cohomology of $\mathcal{N}$ strengthens Steinberg's theorem

Theorem 15.5 (Steinberg) *The number of unipotent elements in $G(\mathbb{F}_q)$ equals q^m, the square of the number of elements in a p-Sylow group (if G is defined over the finite field $\mathbb{F}_q$).*

VI.16 Regular Orbits

The method used in the previous section can be carried over to study more general situations as well. In particular it is of interest to have information on the Fourier transform of (co)adjoint orbits. This is the starting point of Springer's fundamental papers [S6] and [S2], and it appears in [S6] in disguised form as a problem on trigonometric sums. The case already studied in the last sections was the case of nilpotent orbits. The case of regular orbits, studied in Springer's papers, is the other extreme.

General Reference. [Bry2, §11]

Let p be a very good prime for G. Let T be a maximal torus of G and B a Borel subgroup of G containing T. We fix a nondegenerate G-equivariant symmetric bilinear form K on $\mathfrak{g}$. From our assumptions on p the existence of K is guaranteed by VI.5.1.

For any $y \in \mathfrak{t}^*$ define a morphism $\nu_y : X \to \mathbb{A}$ on $X = G \times^B \mathfrak{b}$ by

$$\nu_y(g \times^B b) = y(\overline{b}) \quad , \quad (y \in \mathfrak{t}^*, \overline{b} \in \mathfrak{t})$$

where $\overline{b}$ is the image of b in $\mathfrak{t} = \mathfrak{b}/\mathfrak{u}$. The map ν_y is well defined. We get the diagram

$$\begin{array}{ccc} X & \xrightarrow{\quad q \quad} & \mathfrak{g} \\ {\scriptstyle \nu_y}\downarrow & & \\ \mathbb{A} & , \ \mathscr{L}(\psi) & \end{array}$$

where $q : X \to \mathfrak{g}$ is defined by the Grothendieck simultaneous resolution of singularities, and $\nu_y : X \to \mathbb{A}$ is defined as above. Let $\kappa = \mathbb{F}_q$ be a finite field contained in k. Let ψ be a nontrivial character of the additive group of κ. Let $\mathscr{L}(\psi)$ be the corresponding Artin-Schreier sheaf on $\mathbb{A}$.

Definition 16.1 *For* $y \in \mathfrak{t}^*$ *define the* **Springer sheaf** $\mathscr{S}(\psi, y)$

$$\mathscr{S}(\psi, y) \;=\; Rq_!\Big(\nu_y^*(\mathscr{L}(\psi))[n]\Big) .$$

Obviously $D\mathscr{S}(\psi, y) \cong \mathscr{S}(\psi^{-1}, y)(n)$.

The Springer sheaves are perverse sheaves on $\mathfrak{g}$. Namely q is a small proper morphism, X is smooth and $\nu_y^*(\mathscr{L}(\psi))$ is a smooth sheaf on X. If G, B and T and y are defined over κ, then we obtain a canonical Weil sheaf structure on the Springer sheaf.

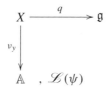

Remark 16.2 The morphism ν_y is trivial on the closed subset $Y = G \times^B \mathfrak{u}$ of $X = G \times^B \mathfrak{b}$. Therefore, since $Y = q^{-1}(\mathscr{N})_{red}$, we have the following consequence of the proper base change theorem: Let $y \in \mathfrak{t}^*$ and let $\mathscr{S}(\psi, y) \in Perv(\mathfrak{g})$ be the

corresponding Springer sheaf. Then the restriction of $\mathscr{S}(\psi, y)$ to the subvariety $i_{\mathscr{N}^\cdot} : \mathscr{N}^\cdot \to \mathfrak{g}$ is isomorphic to the perverse sheaf $\Psi[r](-N)$ on $\mathscr{N}^\cdot$. For the definition of Ψ see VI.11.2. In particular, the restriction of $\mathscr{S}(\psi, y)$ to $\mathscr{N}^\cdot$ does not depend on ψ and y. There are natural maps

$$res : \mathscr{S}(\psi, y) \longrightarrow i_{\mathscr{N}^\cdot *} i^*_{\mathscr{N}^\cdot}(\mathscr{S}(\psi, y)) \cong i_{\mathscr{N}^\cdot *}\Psi[r](-N) .$$

Since $\Psi \cong D(\Psi) = D\big(i^*_{\mathscr{N}^\cdot}[-r](N)\mathscr{S}(\psi, y)\big) = i^!_{\mathscr{N}^\cdot}[r](-N)\mathscr{S}(\psi^{-1}, y)(n)$ this implies (using the ψ-independence)

$$i^!_{\mathscr{N}^\cdot}\mathscr{S}(\psi, y)[2r](r) \cong i^*_{\mathscr{N}^\cdot}\mathscr{S}(\psi, y) .$$

Now we consider regular (co)adjoint orbits $\mathscr{O}$ of G. For this we use the fixed G-equivariant bilinear form K as in Remark VI.10.7 to identify

$$\mathfrak{t}^* \cong \mathfrak{t} \hookrightarrow \mathfrak{g} \cong \mathfrak{g}^* .$$

Then by definition a **regular orbit** $\mathscr{O}$ of G in $\mathfrak{g}^*$ contains an element y in $\mathfrak{t}^*_{reg}$. Since in fact $\mathfrak{t}^*$ is not contained in $\mathfrak{g}^*$, we have to proceed as follows to give sense to this statement: $y \in \mathfrak{t}^*$ is called regular, if y corresponds to a regular element $\tilde{y} \in \mathfrak{t}_{reg}$ under the isomorphism $\mathfrak{t}^* \cong \mathfrak{t}$. Then $\mathfrak{t}^*_{reg} \cong \mathfrak{t}_{reg}$ and $\mathscr{O}_y$ is the image of the orbit $G(\tilde{y}) \subset \mathfrak{g}$ under the isomorphism $\mathfrak{g} \cong \mathfrak{g}^*$. The orbit $\mathscr{O}_y$ is closed in $\mathfrak{g}^*$ and has dimension m. This is a well known consequence of the assumption, that y is regular. See Lemma VI.16.5 below and its proof. Let $i_{\mathscr{O}} : \mathscr{O} \hookrightarrow \mathfrak{g}^*$ be the inclusion map and let

$$\delta_{\mathscr{O}} = i_{\mathscr{O}*}\big(\overline{\mathbb{Q}}_{l\mathscr{O}}\big)$$

be the constant sheaf on this orbit viewed as a sheaf on $\mathfrak{g}^*$, where $\langle m\rangle = [2N](N)$. Since $\mathscr{O}_y$ with the reduced subscheme structure is smooth

$$\delta_{\mathscr{O}}\langle m\rangle \in Perv(\mathfrak{g}^*) .$$

The next proposition compares the Springer sheaves with the Deligne-Fourier transform of regular orbits. Formulas of that type first arose in the work of HARISH-CHANDRA on the representations of real reductive Lie groups.

Proposition 16.3 *For regular elements $y \in \mathfrak{t}^*_{reg}$ the Fourier transform $T_\psi\big(\delta_{\mathscr{O}_y}\langle m\rangle\big)$ of the perverse sheaf $\delta_{\mathscr{O}_y}\langle m\rangle$ on $\mathfrak{g}^*$, defined by the m-dimensional smooth regular (co)adjoint orbit $\mathscr{O}_y$, and the* **Springer sheaf** *$\mathscr{S}(\psi, y)$ are isomorphic perverse sheaves on $\mathfrak{g}$*

$$\boxed{h_y : \mathscr{S}(\psi, y) \cong T_\psi\big(\delta_{\mathscr{O}_y}\langle m\rangle\big) \quad , \quad y \in \mathfrak{t}^*_{reg} .}$$

Remark 16.4 Even for nonregular $y \in \mathfrak{t}^*$ both sides remain defined. This is obvious for the left side. The right side can be defined by some extension of the constant sheaf to the closure of an orbit. However one can not expect to obtain an isomorphism any longer. E.g. for $y = 0$ we see that $T_\psi(\delta_{\mathcal{O}_0}\langle m\rangle) = \delta_{\mathfrak{g}}[n]\langle m\rangle$ is not even a perverse sheaf, whereas the left side specializes to the perverse sheaf

$$\mathcal{S}(\psi, 0) = \Phi \, .$$

Now we come to the proof of the proposition

Proof. Step 1. $y \in \mathfrak{t}^*_{reg}$ has stabilizer $G_y = T$, the orbit $\mathcal{O}_y$ is a closed orbit in $\mathfrak{g}^*$. With respect to the diagram of the Grothendieck simultaneous resolution of singularities

$$\begin{array}{ccccccc}
(p')^{-1}(y) & \hookrightarrow & X'_{reg} & \xrightarrow{q'} & \mathfrak{g}^*_{reg} & \hookleftarrow & \mathcal{O}_y \\
\downarrow & & \downarrow{\scriptstyle p'} & & \downarrow{\scriptstyle \gamma'} & & \\
\{y\} & \hookrightarrow & \mathfrak{t}^*_{reg} & \xrightarrow{\theta'} & \mathfrak{t}^*_{reg}/W & &
\end{array}$$

the following holds

Lemma 16.5 *For regular* $y \in \mathfrak{t}^*_{reg}$ *the morphisms* $q' : (p')^{-1}(y) \to \mathcal{O}_y$ *is an isomorphism, hence*

$$\delta_{\mathcal{O}_y}\langle m\rangle = Rq'_!\Big(\delta_{(p')^{-1}(y)}\Big)\langle m\rangle \, .$$

Proof. The morphism $q' : X'_{reg} \to \mathfrak{g}^*_{reg}$ is a finite etale Galois covering map (Theorem VI.9.1 respectively its coadjoint version). Since p' is smooth, its fiber $F = (p')^{-1}(y)$ is a smooth closed subscheme of X' (of "constant" dimension). Indeed $F \subset X'_{reg}$. The orbit $\mathcal{O}_y = G(y)$ is smooth and

$$(q')^{-1}(\mathcal{O}_y) = X'_{reg} \times_{\mathfrak{g}^*_{reg}} \mathcal{O}_y \to \mathcal{O}_y$$

is a finite etale covering map of smooth schemes. The smooth scheme F is contained in the smooth scheme $(q')^{-1}(\mathcal{O}_y)$ and has the same "constant" dimension. Therefore F is open and closed, as a subscheme of $(q')^{-1}(\mathcal{O}_y)$. Therefore

$$q' : F \to \mathcal{O}_y$$

is a finite etale covering map onto the irreducible scheme $\mathcal{O}_y$, so it is enough to show injectivity of this morphism in the set theoretic sense: Use the isomorphism $G/T \times \mathfrak{t}^*_{reg} \to X'_{reg}$ (as in the proof of Theorem VI.9.1). If $g_1T \times y = g_2T \times y$, then $g_1 = g_2t$ with $t \in T(k)$ by the regularity of y. Hence q' is injective on $F = (p')^{-1}(y)$. This completes the proof of the lemma.

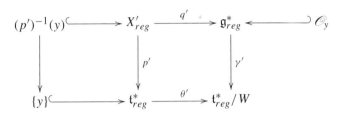

Noteworthy the closed subset $F = (p')^{-1}(y)$ of X' maps to a closed subset of $\mathfrak{g}^*$, since $q' : X' \to \mathfrak{g}^*$ is a proper morphism. Hence by the proof above the regular orbits $\mathscr{O}_y = q'(F)$ are closed in $\mathfrak{g}^*$ – a fact already mentioned and used earlier. □

Step 2. The vector bundle $X' = G \times^B \mathfrak{b}'$ over $\mathscr{B}$ is the middle term of an exact sequence of vector bundles over $\mathscr{B}$.

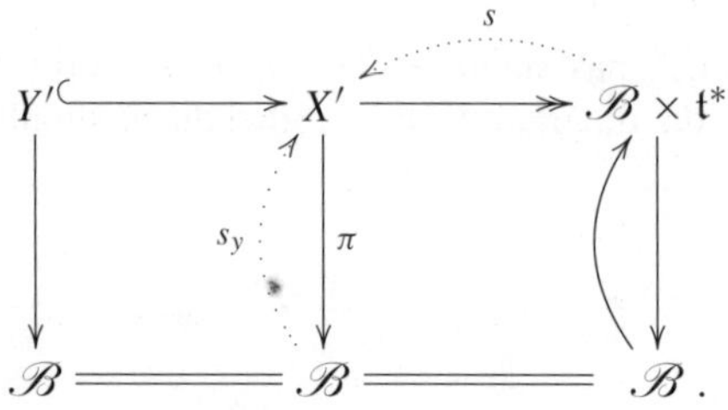

The trivial vector bundle $\mathscr{B} \times \mathfrak{t}^*$ is the quotient bundle, and the kernel is the vector bundle $Y' = G \times^B \mathfrak{u}'$. Note $\mathfrak{b}' = \mathfrak{u}^\perp$, $\mathfrak{u}' = \mathfrak{b}^\perp$ and $\mathfrak{b}' = \mathfrak{t}^* \oplus \mathfrak{u}'$. This exact sequence of vector bundles over $\mathscr{B}$ does not split. However sections s, as indicated in the diagram, exist locally over the base $\mathscr{B}$. Any local section s of the exact sequence induces a local section s_y of π in the middle, by composing s with the global section on the right defined by y via $gB \mapsto (gB, y)$. For a local section s_y over a suitable small open subset $U \subset \mathscr{B}$ one obtains

Corollary 16.6 *Suppose* $y \in \mathfrak{t}^*_{reg}$. *Then*

$$(p')^{-1}(y) \cap \pi^{-1}(U) \xleftarrow{\sim} s_y(U) + (Y'|U) \subset X'_{reg} \; .$$

Step 3. To obtain a global section, we make a base scheme extension with respect to the quotient map

$$\tilde{pr} : \tilde{\mathscr{B}} = G/T \longrightarrow \mathscr{B} = G/B \; .$$

Working over the new base $\tilde{\mathscr{B}}$ is advantageous since

a) Over this new base $\tilde{\mathscr{B}}$ the pull back of the exact sequence of vector bundles, considered in step 2, splits.
b) The map $\tilde{pr}$ is acyclic in the sense, that

$$R\tilde{pr}_!(\tilde{pr}^* K)\langle m\rangle \;=\; K \quad , \quad K \in D^b_c(\mathscr{B}) \; .$$

See also VI.2.3.

To verify property a) it is enough to remark, that the pullback $\tilde{X}' = \tilde{\mathscr{B}} \times_{\mathscr{B}} X'$ of $X' = G \times^B \mathfrak{b}'$ can be canonically identified with $G \times^T \mathfrak{b}'$ by an isomorphism λ of vector bundles over $\tilde{\mathscr{B}} = G/T$

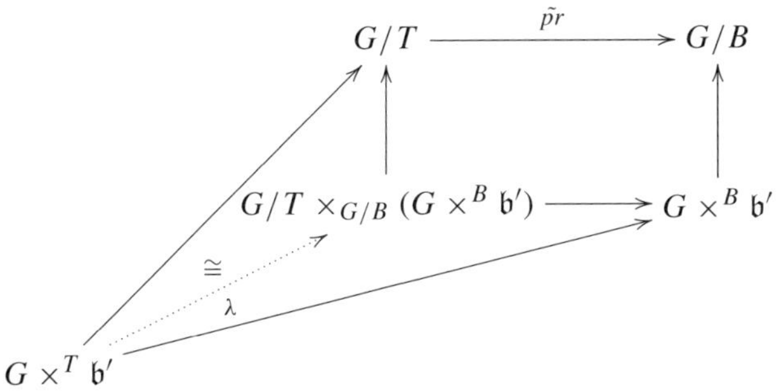

induced from the maps $g \times^T x \mapsto gT$ and $g \times^T x \mapsto g \times^B x$ from $G \times^T \mathfrak{b}'$ to G/T respectively $G \times^B \mathfrak{b}'$. Similarly $\tilde{Y}' = Y' \times_{\mathscr{B}} \tilde{\mathscr{B}} \cong G \times^T \mathfrak{u}'$. Therefore the sequence of vector bundles, considered in step 2), splits over the base $\tilde{\mathscr{B}}$ with the desired global sections $s(gT, y) = g \times^T (y, 0)$ and $s_y(gT) = g \times^T (y, 0) \in G \times^T (\mathfrak{t} \oplus \mathfrak{u}')$

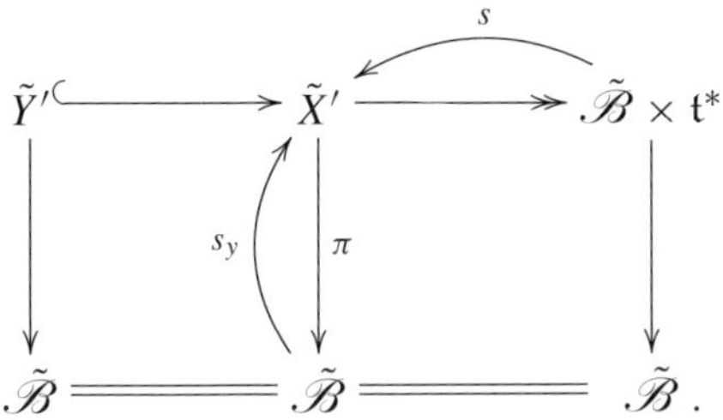

Step 4. Consider the trivial vector bundle

$$\tilde{\mathscr{B}} \times \mathfrak{g}^* \to \tilde{\mathscr{B}}$$

over $\tilde{\mathscr{B}}$ and its subbundle $\tilde{X}'$. The global section s_y of the subvectorbundle $\tilde{X}'$ in particular defines a global section of the morphism $\tilde{\mathscr{B}} \times \mathfrak{g}^* \to \tilde{\mathscr{B}}$. This section will also be denoted s_y. Then let T_{-s_y} denote the translation with the negative of this section in the fibers of the vector bundle $\tilde{\mathscr{B}} \times \mathfrak{g}^* \to \tilde{\mathscr{B}}$.

Now consider the diagram

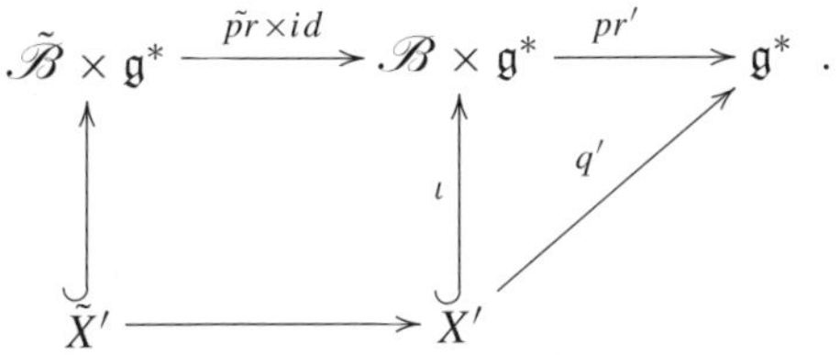

By Lemma III.13.3 and by Lemma VI.16.5 we have

$$T_\psi(\delta_{\mathscr{O}_y}\langle m\rangle) \;=\; T_\psi(Rpr'_!(K)) \;=\; Rpr_!\big(T_\psi^{\tilde{\mathscr{B}}\times\mathfrak{g}^*/\tilde{\mathscr{B}}}(K)\big)$$

for $K = \iota_*\delta_{(p')^{-1}(y)}\langle m\rangle$. Put

$$\tilde{K} = (\tilde{pr} \times id)^*(K)\langle m\rangle = T^*_{-s_y}(\delta_{\tilde{Y}'})\langle 2m\rangle .$$

The second equality holds, since the inverse image of $(p')^{-1}(y)$ in $\tilde{\mathscr{B}} \times \mathfrak{g}^*$ is equal to $s_y + \tilde{Y}' = T_{s_y}(\tilde{Y}')$. Once more by Lemma III.13.3 we have for any complex $\tilde{K} \in D_c^b(\tilde{\mathscr{B}} \times \mathfrak{g}^*, \overline{\mathbb{Q}}_l)$

$$T_\psi^{\mathscr{B}\times\mathfrak{g}^*/\mathscr{B}}\big(\mathrm{R}(\tilde{pr} \times id)_!\tilde{K}\big) = \mathrm{R}(\tilde{pr} \times id)_!\big(T_\psi^{\tilde{\mathscr{B}}\times\mathfrak{g}^*/\tilde{\mathscr{B}}}(\tilde{K})\big) .$$

Since

$$\mathrm{R}(\tilde{pr} \times id)_!(\tilde{K}) = K$$

by step 3 b), this gives $T_\psi^{\mathscr{B}\times\mathfrak{g}^*/\mathscr{B}}(K) = \mathrm{R}(\tilde{pr} \times id)_! T_\psi^{\tilde{\mathscr{B}}\times\mathfrak{g}^*/\tilde{\mathscr{B}}}(\tilde{K})$, hence

$$T_\psi(\delta_{\mathscr{O}_y}\langle m\rangle) = \mathrm{R}pr_!\big(\mathrm{R}(\tilde{pr} \times id)_! T_\psi^{\tilde{\mathscr{B}}\times\mathfrak{g}^*/\tilde{\mathscr{B}}}(\tilde{K})\big) .$$

Under Fourier transform translates become twists with the Artin-Schreier character sheaf. In other words, the relative Fourier transform of the complex $\tilde{K}$ can be computed by Lemma III.14.1

$$T_\psi^{\tilde{\mathscr{B}}\times\mathfrak{g}^*/\tilde{\mathscr{B}}}(\tilde{K}) = \tilde{\nu}_y^*(\mathscr{L}_\psi) \otimes_{\overline{\mathbb{Q}}_l} T_\psi^{\tilde{\mathscr{B}}\times\mathfrak{g}^*/\tilde{\mathscr{B}}}(\delta_{\tilde{Y}'}\langle 2m\rangle) .$$

The relative Fourier transform of the sheaf complex $\delta_{\tilde{Y}'}\langle m\rangle$ is $\delta_{\tilde{X}}[n]$. Here $\delta_{\tilde{X}} = \overline{\mathbb{Q}}_{l\,\tilde{X}}$ and similar for $\tilde{X}'$, $\tilde{Y}'$. Hence we get

$$T_\psi^{\tilde{\mathscr{B}}\times\mathfrak{g}^*/\tilde{\mathscr{B}}}\big(\tilde{K}\big) = i_*(L)\langle m\rangle \quad , \quad L = \tilde{\nu}_y^*(\mathscr{L}_\psi) \otimes_{\overline{\mathbb{Q}}_l} \delta_{\tilde{X}}[n] \in D_c^b(\tilde{X}, \overline{\mathbb{Q}}_l) ,$$

where $i : \tilde{X} \hookrightarrow \mathfrak{g} \times \tilde{\mathscr{B}}$ denotes the subbundle obtained from pulling back the subbundle $X \hookrightarrow \mathfrak{g} \times \mathscr{B}$. Furthermore $\tilde{\nu}_y : \tilde{X} \to \mathbb{A}$ is the characteristic function on $\tilde{X}$ associated to the section s_y as in III.14.1. By definition $\tilde{\nu}_y$ is the pullback under $pr : \tilde{X} \to X$ of the function ν_y defined on X by $\nu_y(g \times^B b) = K(b, y)$. So

$$T_\psi(\delta_{\mathscr{O}_y}\langle m\rangle) = \mathrm{R}pr_!\big(\mathrm{R}(\tilde{pr} \times id)_! i_*(L)\langle m\rangle\big) .$$

Since $L = (\tilde{pr} \times id)^*(M)$ holds for $M = \nu_y^*(\mathscr{L}_\psi) \otimes_{\overline{\mathbb{Q}}_l} \delta_X[n]$, the projection formula $\mathrm{R}(\tilde{pr} \times id)_!(\tilde{pr} \times id)^*\langle m\rangle = id$ – which is analogous to the one in step 3, assertion b) – implies

$$T_\psi(\delta_{\mathscr{O}_y}\langle m\rangle) = \mathrm{R}\tilde{pr}_!(M) = \mathrm{R}q_!(M) = \mathscr{S}(\psi, y) .$$

Proposition VI.16.3 is proved. □

Remark 16.7 We mention without proof a result, which immediately follows from III.14.1. Let $\iota : Y' \to \mathfrak{g}^* \times \mathscr{B}$ be the inclusion map. Then the Fourier transform

$$T_\psi(res) : T_\psi(\delta_{\tilde{X}'}) \to T_\psi(\delta_{\tilde{Y}'_y})$$

of the restriction map $res : \delta_{\tilde{X}'} \to \delta_{\tilde{Y}'_y}$ uniquely factorizes over the adjunction map

$$\iota_! \iota^! T_\psi(\delta_{\tilde{Y}'_y}) \to T_\psi(\delta_{\tilde{Y}'_y})$$

and induces an isomorphism

$$T_\psi(\delta_{X'}) \cong \iota_! \iota^! T_\psi(\delta_{Y'_y}) \, .$$

VI.17 W-actions on the Universal Springer Sheaf

The Springer sheaf $\mathscr{S}(\psi, y)$ only depends on the conjugacy class of the element $y \in \mathfrak{t}^*_{reg}$. Thus y could be replaced by a conjugate under the Weyl group W. Furthermore it is very natural to consider the parameter $y \in \mathfrak{t}^*$ as a variable. To do so, enlarge the base by a base scheme extension with $\mathfrak{t}^*_{reg}$. Then the statements made in the previous section carry over to this generic situation. Since W acts on $\mathfrak{t}^*_{reg}$, this allows to define W-actions on the universal Springer sheaf and also on Ψ and Φ by restrictions. As before these actions are "visible" on some dense open subsets, and are then extended to actions over the total space by perverse analytic continuation. We discuss this and show, that in this way one can reproduce the W-actions already considered before.

Our intention is to equip the Springer sheaves with a natural right action of the Weyl group W. For this it will be necessary to enlarge the base scheme. So we consider $\mathfrak{g} \times \mathfrak{t}^*$ instead of $\mathfrak{g}$. Recall that Grothendieck's simultaneous resolution of singularities provided maps $q : X \to \mathfrak{g}$ and $p : X \to \mathfrak{t}$ from $X = G \times^B \mathfrak{b}$ to the Lie algebras $\mathfrak{g}$ and $\mathfrak{t}$. So in analogy with the situation of the last section we consider the morphisms

$$\begin{array}{ccc} X \times \mathfrak{t}^* & \xrightarrow{q \times id} & \mathfrak{g} \times \mathfrak{t}^* \, . \\ \downarrow{\scriptstyle p \times id} & & \\ \mathfrak{t} \times \mathfrak{t}^* & & \\ \downarrow{\scriptstyle \langle\, ,\, \rangle} & & \\ \mathbb{A} & & \end{array}$$

(with the composite $\nu : X \times \mathfrak{t}^* \to \mathbb{A}$)

We define the **universal Springer sheaf**

$$\boxed{\mathscr{S}(\psi) \;=\; \mathrm{R}(q \times id)_! \Big(\nu^*\big(\mathscr{L}(\psi)\big) \Big)[n + r] \, .}$$

Here $n = dim(\mathfrak{g})$ and $r = dim(\mathfrak{t})$. Smallness of q implies smallness of $q \times id$. By the smallness of the map $q \times id$ the universal Springer sheaf is a perverse sheaf

$$\mathscr{S}(\psi) \in Perv(\mathfrak{g} \times \mathfrak{t}^*) \, .$$

Lemma 17.1 *Let* $j : \mathfrak{g}_{reg} \times \mathfrak{t}^*_{reg} \hookrightarrow \mathfrak{g} \times \mathfrak{t}^*$ *be the inclusion map of the regular locus. Then*

$$\mathscr{S}(\psi) \;=\; j_{!*}\big(\mathscr{S}(\psi)|\mathfrak{g}_{reg} \times \mathfrak{t}^*_{reg}\big)\,.$$

In particular for $j' : \mathfrak{g} \times \mathfrak{t}^*_{reg} \hookrightarrow \mathfrak{g} \times \mathfrak{t}^*$ *we get* $\mathscr{S}(\psi) = j'_{!*}\big(\mathscr{S}(\psi)|\mathfrak{g} \times \mathfrak{t}^*_{reg}\big)$.

Proof. Consider

$$\begin{array}{ccc}
\mathfrak{g}_{reg} \times \mathfrak{t}^*_{reg} & \xrightarrow{\;j_1\;} & \mathfrak{g}_{reg} \times \mathfrak{t}^* \\
\big\downarrow & \searrow^{j} & \big\downarrow{\scriptstyle j_2} \\
\mathfrak{g} \times \mathfrak{t}^*_{reg} & \xrightarrow{\;j'\;} & \mathfrak{g} \times \mathfrak{t}^*\,.
\end{array}$$

The $(j_1)_{!*}$-extension of $\mathscr{S}(\psi)|\mathfrak{g}_{reg} \times \mathfrak{t}^*_{reg}$ is $\mathscr{S}(\psi)|\mathfrak{g}_{reg} \times \mathfrak{t}^*$, since the latter is a smooth sheaf by Theorem VI.9.1.

The $(j_2)_{!*}$-extension of $\mathscr{S}(\psi)|\mathfrak{g}_{reg} \times \mathfrak{t}^*$ is $\mathscr{S}(\psi)$, since $q \times id$ is a small map with top stratum $\mathfrak{g}_{reg}$ (Lemma VI.10.3(a)) and since $X \times \mathfrak{t}^*$ and $\nu^*(\mathscr{L}(\psi))$ are smooth. The claim now follows, since intermediate extensions are functorial: $j_{!*} = j_{2!*} \circ j_{1!*}$. □

Notation. For $K \in Perv(\mathfrak{g})$ we write $K/\mathfrak{t}^* \in Perv(\mathfrak{g} \times \mathfrak{t}^*)$ for the pullback of $K[r]$ under the projection $\mathfrak{g} \times \mathfrak{t}^* \to \mathfrak{g}$.

With this notation we get similar as in VI.16.2

$$i^*_{\mathscr{N} \times \mathfrak{t}^*}\big(\mathscr{S}(\psi)\big)[-r](N) \;=\; \Psi/\mathfrak{t}^*\,,$$

where $r = dim(\mathfrak{t})$ and $2N + r = n = dim(\mathfrak{g})$, and where

$$i_{\mathscr{N} \times \mathfrak{t}^*} : \mathscr{N} \times \mathfrak{t}^* \hookrightarrow \mathfrak{g} \times \mathfrak{t}^*$$

denotes the inclusion map of the universal nilpotent locus into $\mathfrak{g} \times \mathfrak{t}^*$.

Let y be in $\mathfrak{t}^*$. For the inclusion

$$i_y : \mathfrak{g} \hookrightarrow \mathfrak{g} \times \mathfrak{t}^*$$

defined by $i_y(x) = (x, y)$ the pullback $i_y[-r]^*(\mathscr{S}(\psi))$ of the universal Springer sheaf is the Springer sheaf $\mathscr{S}(\psi, y)$ considered in the last section. It is interesting to note, that the most singular specialization gives the perverse sheaf Φ introduced earlier

$$i_0[-r]^*(\mathscr{S}(\psi)) = \Phi\,.$$

Right W-actions. We now fix the following W-right actions:

1. the trivial action on $\mathfrak{g}$, $\mathfrak{g}^*$ and $\mathbb{A}$
2. the canonical right actions on $\mathfrak{t}$ and $\mathfrak{t}^* \cong \mathfrak{t}$

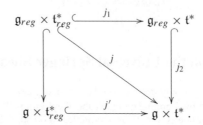

3. the action on $X_{reg} \cong G/T \times \mathfrak{t}_{reg}$ defined by $w(gT, x) = (gTn_w, n_w^{-1}(x))$, where n_w is a representative in $N(T)$ of $w \in W$
4. the similar action on X'_{reg}.

We always use the induced right W-actions on cartesian products. The schemes $\mathfrak{g}_{reg} \times \mathfrak{t}^*$ and $\mathfrak{g}_{reg} \times \mathfrak{t}^*_{reg}$ are W-stable. The morphism

$$q \times id | X_{reg} \times \mathfrak{t}^* : X_{reg} \times \mathfrak{t}^* \to \mathfrak{g}_{reg} \times \mathfrak{t}^*$$

and the morphism $\nu : X_{reg} \times \mathfrak{t}^* \to \mathbb{A}$ are W-(right)-equivariant. Therefore $\nu^*(\mathscr{L}(\psi))$ is a W-right equivariant sheaf on $X_{reg} \times \mathfrak{t}^*$. Its direct image sheaf $\mathscr{S}(\psi)|\mathfrak{g}_{reg} \times \mathfrak{t}^*$ under $R(q \times id)_!$ therefore carries a natural W-right action. Using the last Lemma VI.17.1 and perverse analytic continuation, this action uniquely extends to a W-right-action on the whole universal Springer sheaf $\mathscr{S}(\psi)$. We call this action the **Lusztig action** on $\mathscr{S}(\psi)$ and write

$$\mathscr{S}(\psi)_L$$

to emphasize the specified W-action on $\mathscr{S}(\psi)$. Note, that there exists an W-equivariant isomorphism $D\big(\mathscr{S}(\psi)_L\big) \cong \mathscr{S}(\psi^{-1})_L(n+r)$.

Universal Regular Orbit. Consider the universal regular orbit

$$\mathscr{O} \subset \mathfrak{g}^*_{reg} \times \mathfrak{t}^*_{reg}$$

defined by $\mathscr{O} = \{(x, y) \in \mathfrak{g}^*_{reg} \times \mathfrak{t}^*_{reg} \mid x \in G(y)\}$. Define

$$\Gamma \subset X' \times \mathfrak{t}^*_{reg} = G \times^B \mathfrak{b}' \times \mathfrak{t}^*_{reg}$$

by $\Gamma = \{(x, y) \in X' \times \mathfrak{t}^*_{reg} \mid p'(x) = y\}$. Here X' is the Grothendieck simultaneous resolution of singularities with the maps $q' : X' \to \mathfrak{g}^*$ and $p' : X' \to \mathfrak{t}^*$. The universal regular orbit $\mathscr{O}$ is obtained as the image of Γ under

$$q' \times id : X' \times \mathfrak{t}^*_{reg} \to \mathfrak{g}^* \times \mathfrak{t}^*_{reg}$$

$$(q' \times id)(\Gamma) = \mathscr{O} .$$

Clearly Γ is the restriction of $\overline{\Gamma} = \{(x, y) \in X' \times \mathfrak{t}^* \mid p'(x) = y\} = X' \times_{\mathfrak{t}^*} \mathfrak{t}^*$ (viewed as a subset of $X' \times \mathfrak{t}^*$) to the open subset $X' \times \mathfrak{t}^*_{reg}$ (the cartesian product $\overline{\Gamma}$ is defined via the maps $p' : X' \to \mathfrak{t}^*$ and $id : \mathfrak{t}^* \to \mathfrak{t}^*$). In particular $\overline{\Gamma} \cong X'$.

In fact $\mathscr{O} \subset \mathfrak{g}^*_{reg} \times \mathfrak{t}^*_{reg}$, hence $\Gamma \subset X'_{reg} \times \mathfrak{t}^*_{reg}$. Using Theorem VI.9.1

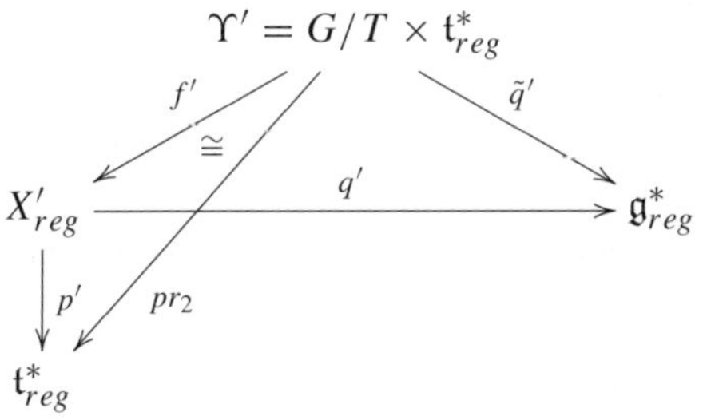

we see that $(f')^{-1}(\Gamma) = \{gT \times y \times y \mid g \in G, y \in \mathfrak{t}^*_{reg}\}$. Hence the morphism

$$q' \times id_{\mathfrak{t}^*} : \Gamma \xrightarrow{\sim} \mathscr{O} \ ,$$

or equivalently the morphism

$$\tilde{q}' \times id_{\mathfrak{t}^*} : (f' \times id_{\mathfrak{t}^*})^{-1}(\Gamma) \xrightarrow{\sim} \mathscr{O} \ ,$$

is an isomorphism. In fact $\tilde{q}' \times id(g_1T, y_1, y_1) = \tilde{q}' \times id(g_2T, y_2, y_2)$ implies $y_1 = y_2$ and $g_1(y_1) = g_2(y_2)$, hence $g_1T = g_2T$ as $y = y_1 = y_2$ is regular.

Put $\delta_{\mathscr{O}} = \overline{\mathbb{Q}}_{l\mathscr{O}}$. Then suitably normalized it defines an irreducible perverse sheaf $K = \delta_{\mathscr{O}}\langle m\rangle[r]$ viewed as a sheaf on $\mathfrak{g}^* \times \mathfrak{t}^*_{reg}$ by zero extension. As before we abbreviated $\langle m\rangle = [m](m/2)$. Similar as in the proof of proposition VI.16.3 we now prove the next Lemma VI.17.3. We skip the details and only give a short summary of the necessary arguments in the present setting over the base $\mathfrak{t}^*$.

17.2 A Summary. Recall that $\tilde{\mathscr{B}} = G/T$ and $\mathscr{B} = G/B$. Put $S = \mathscr{B} \times \mathfrak{t}^*$ and $V = \mathfrak{g} \times S$. Furthermore put $\tilde{S} = \tilde{\mathscr{B}} \times \mathfrak{t}^*$ and $\tilde{V} = \mathfrak{g} \times \tilde{S}$. We view V as a vector bundle V/S over S. Its dual bundle is V'/S, where $V' = \mathfrak{g}^* \times S$. Similarly for $\tilde{V}$. Base change with respect to the morphism $\tilde{pr} : \tilde{S} \to S$ – induced by the natural map $G/T \to G/B$ – gives the vector bundles $\tilde{V}/\tilde{S}$ as pullback of V/S. We define subvectorbundles $\tilde{Y}' \hookrightarrow \tilde{X}' \hookrightarrow \tilde{V}'$ over $\tilde{S}$ as pullback of Y' and X' and similar for $\tilde{V}/\tilde{S}$. Then there is an isomorphism

$$\tilde{X}' \cong G \times^T \mathfrak{b}' \times \mathfrak{t}^* \ .$$

Using this we define a section $s : \tilde{S} \to \tilde{X}'$ via $s(gT, y) = (g \times^T y, y)$.

Consider the complex

$$\tilde{K} = \delta_{T_s(\tilde{Y}')}\langle 2m\rangle[r] \ ,$$

with support on the translate $T_s(\tilde{Y}')$ of $\tilde{Y}'$ defined by the section s. According to Lemma III.14.1 the Fourier transform of $\tilde{K}$ on the vectorbundle $\tilde{V}'$ is

$$T_\psi^{\tilde{V}'/\tilde{S}}(\tilde{K}) = \tilde{\mathfrak{v}}^*(\mathscr{L}_\psi) \otimes_{\overline{\mathbb{Q}}_l} \delta_{\tilde{X}}[2n](N)$$

viewed as a complex on $\tilde{V}$ with support in $\tilde{X}$. The proof of this identity is essentially the same as the proof of proposition VI.16.3.

Under the natural projection $pr : \tilde{V} \to V$ we obtain from this Fourier transform the universal Springer sheaf

$$R\mu_! Rpr_!\big(T_\psi^{\tilde{V}'/\tilde{S}}(\tilde{K})\big) = \mathscr{S}(\psi) \ .$$

For the dual vector bundles we have the similar projection $pr' : \tilde{V}' \to V'$. Here we get the following canonical isomorphism

$$R\mu'_! Rpr'_!(\tilde{K}) = R(q' \times p')_! \overline{\mathbb{Q}}_{l X'}\langle m\rangle[r]$$

for the direct image. This follows from considering the following commutative diagram

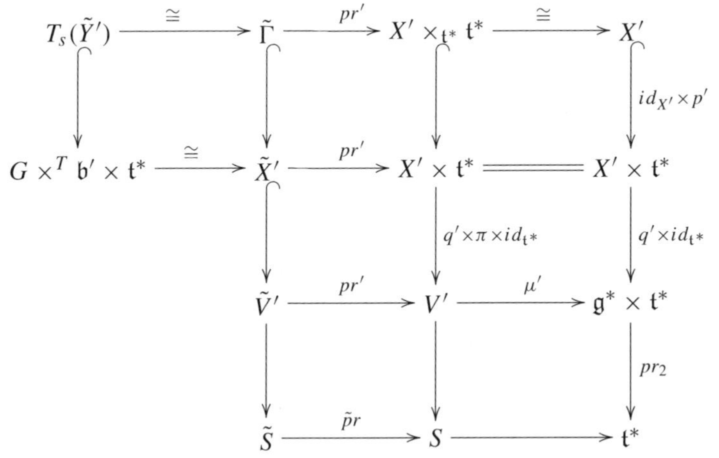

where $\overline{\Gamma} = X' \times_{\mathfrak{t}^*} \mathfrak{t}^* \hookrightarrow X' \times \mathfrak{t}^*$ and $\tilde{\Gamma} = (pr')^{-1}(\overline{\Gamma})$. The cartesian product is with respect to the maps $p' : X' \to \mathfrak{t}^*$ and $id : \mathfrak{t}^* \to \mathfrak{t}^*$ and $\overline{\Gamma} \cong X'$. Furthermore $\tilde{\Gamma} = (pr')^{-1}(\overline{\Gamma})$ is isomorphic to $T_s(\tilde{Y}')$. One therefore concludes that

$$\mathrm{R}pr'_!\big(\overline{\mathbb{Q}}_{l\,\tilde{\Gamma}}\big)\langle 2m\rangle[r]$$

$$= \mathrm{R}pr'_!(pr')^*\big(\overline{\mathbb{Q}}_{l\,X'\times_{\mathfrak{t}^*}\mathfrak{t}^*}\big)\langle 2m\rangle[r]$$

$$\cong \overline{\mathbb{Q}}_{l\,X'\times_{\mathfrak{t}^*}\mathfrak{t}^*}\langle m\rangle[r] \cong \overline{\mathbb{Q}}_{l\,X'}\langle m\rangle[r]\,,$$

using Lemma VI.2.3(2) with $m = 2N$ and $\langle m\rangle = [2N](N)$.

Furthermore, since $\mathscr{S}(\psi)$ is a perverse sheaf and since Fourier transform preserves perversity, also $\mathrm{R}pr'_!(\tilde{K})$ is a perverse sheaf. Over the regular locus $\mathfrak{t}^*_{reg} \subset \mathfrak{t}^*$ the restriction Γ of $\overline{\Gamma}$ maps isomorphically to $\mathscr{O}$. If we restrict the base to its regular open subset $\mathfrak{t}^*_{reg} \subset \mathfrak{t}^*$, therefore the following complexes can be canonically identified

$$\left(\mathrm{R}(q' \times p')_!\overline{\mathbb{Q}}_{l\,X'}\langle m\rangle[r]\right)\Big|\mathfrak{g}^* \times \mathfrak{t}^*_{reg} = \delta_{\mathscr{O}}\langle m\rangle[r]\,.$$

This immediately implies

Lemma 17.3 *For the relative Fourier transform on $\mathfrak{g}^* \times \mathfrak{t}^*_{reg}/\mathfrak{t}^*_{reg}$ we have*

$$T_\psi^{\mathfrak{g}^*\times\mathfrak{t}^*_{reg}/\mathfrak{t}^*_{reg}}\left(\delta_{\mathscr{O}}\langle m\rangle[r]\right) \cong \mathscr{S}(\psi)\Big|\mathfrak{g} \times \mathfrak{t}^*_{reg}\,.$$

The last two lemmas together immediately imply

Proposition 17.4 *For the inclusion* $j' : \mathfrak{g} \times \mathfrak{t}^*_{reg} \hookrightarrow \mathfrak{g} \times \mathfrak{t}^*$ *we have*

$$\mathcal{S}(\psi) \cong j'_{!*}\left(T_\psi^{\mathfrak{g}^*\times\mathfrak{t}^*_{reg}/\mathfrak{t}^*_{reg}}(\delta_{\mathcal{O}}\langle m\rangle[r])\right).$$

From this new description of the universal Springer sheaf it is clear, that we do obtain a new W-right action on $\mathcal{S}(\psi) \in Perv(\mathfrak{g} \times \mathfrak{t}^*)$, which is compatible with the W-right action on $\mathfrak{g} \times \mathfrak{t}^*$. It is inherited from the canonical right action of W on the universal regular orbit $\mathcal{O} \subset \mathfrak{g}^* \times \mathfrak{t}^*_{reg}$, respectively the corresponding action on $\delta_{\mathcal{O}}$ via Fourier transform, via perverse analytic continuation with respect to the open embedding j'. This transport of structure defines a W-(right)-action on the universal Springer sheaf $\mathcal{S}(\psi)$, which is compatible with the natural right action of W on $\mathfrak{g} \times \mathfrak{t}^*$. This action will be called the **Brylinski action**. We write

$$\mathcal{S}(\psi)_B$$

for the corresponding W-equivariant perverse sheaf. Obviously there exists a W-equivariant isomorphism $D(\mathcal{S}(\psi)_B) \cong \mathcal{S}(\psi^{-1})_B(n+r)$.

Corollary 17.5 *The universal Springer sheaf* $\mathcal{S}(\psi)$ *is an irreducible perverse sheaf on* $\mathfrak{g} \times \mathfrak{t}^*$. *It is a* W*-right equivariant perverse sheaf. Any two* W*-right actions on* $\mathcal{S}(\psi)$, *compatible with the fixed* W*-right action on the underlying space* $\mathfrak{g} \times \mathfrak{t}^*$, *differ by a twist with a character* $\varepsilon : W \to \overline{\mathbb{Q}}_l^{\,*}$.

We also have the following

Corollary 17.6 *Let* $j' : \mathfrak{g}^* \times \mathfrak{t}^*_{reg} \hookrightarrow \mathfrak{g}^* \times \mathfrak{t}^*$ *denote the base change map to the regular open subset. Then the relative Fourier transform* T_ψ *on* $\mathfrak{g}^* \times \mathfrak{t}^*_{reg}/\mathfrak{t}^*_{reg}$ *of the irreducible perverse sheaf* $j'_{!*}(\delta_{\mathcal{O}}\langle m\rangle[r])$ *is isomorphic to the universal Springer sheaf* $\mathcal{S}(\psi)$.

$$\boxed{h : \mathcal{S}(\psi) \cong T_\psi^{\mathfrak{g}^*\times\mathfrak{t}^*/\mathfrak{t}^*}\left(j'_{!*}\delta_{\mathcal{O}}\langle m\rangle[r]\right).}$$

Proof. The relative Fourier transform commutes with base change defined by j'. Hence the assertion follows from the last proposition VI.17.4, since both perverse sheaves are irreducible and generically supported on $\mathfrak{g} \times \mathfrak{t}^*_{reg}$. □

Since the universal Springer sheaf $\mathcal{S}(\psi)$ was shown to be an irreducible perverse sheaf on $\mathfrak{g} \times \mathfrak{t}^*$, the inverse Fourier transform of it is an irreducible perverse sheaf on $\mathfrak{g}^* \times \mathfrak{t}^*$. The Fourier transform is functorial with respect to base change (Theorem III.13.3). By the summary VI.17.2 above this implies, that the inverse Fourier transform of $\mathcal{S}(\psi)$ can be canonically identified with the direct image complex

$$\mathrm{R}(q' \times p')_! \overline{\mathbb{Q}}_{l\,X'} \langle m \rangle [r] \ ,$$

where $q' \times p' : X' \to \mathfrak{g}^* \times \mathfrak{t}^*$ is defined by completing the commutative diagram obtained from the coadjoint version of the Grothendieck simultaneous resolution of singularities

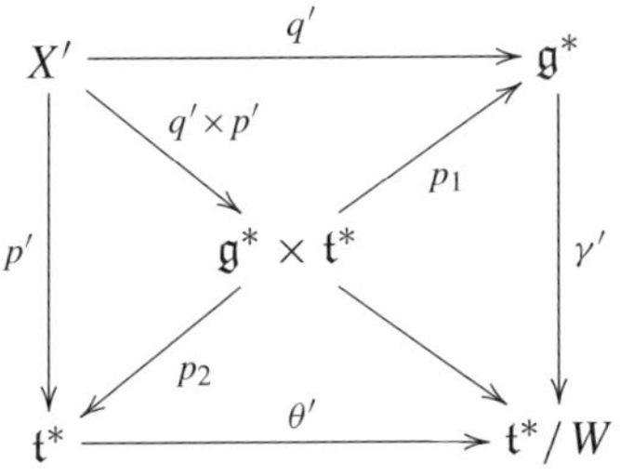

Restricted from $\mathfrak{g}^* \times \mathfrak{t}^*$ to $\mathfrak{g}^* \times \mathfrak{t}^*_{reg}$, this direct image complex coincides with $\delta_{\mathscr{O}} \langle m \rangle [r]$. We therefore conclude

Corollary 17.7 *There exists a canonical isomorphism of perverse sheaves*

$$\boxed{\; j'_{!*}(\delta_{\mathscr{O}}) \; = \; R\big(q' \times p'\big)_! \overline{\mathbb{Q}}_{l\,X'} \;} \ .$$

By specialization we obtain in particular

$$j'_{!*}\big(\delta_{\mathscr{O}} \langle m \rangle [r]\big) \, \Big| \mathfrak{g}^* \times \{0\} \; = \; i_{\mathscr{N}^* *}(\Psi')[r] \ .$$

(Tacitly again the G-equivariant identification of $\mathfrak{g}^*$ and $\mathfrak{g}$ was used in the definition of the inclusion $i_{\mathscr{N}^* *}$!)

Remark 17.8 We emphasize, that the identifications made in the last corollary are induced by a canonical map. Therefore we preferred to write equality signs. We call the second isomorphism of the last corollary the **restriction isomorphism**. By construction it is clear, that this restriction isomorphism is W-equivariant, provided Ψ' is endowed with the **Lusztig** action.

17.9 Corestriction map. There is a natural W-equivariant map defined by "restriction to the orbit $\mathscr{O}$"

$$res_{\mathscr{O}} : \quad \Phi'/\mathfrak{t}^*_{reg} \to \delta_{\mathscr{O}}[n+r] \quad , \qquad \text{in } Perv(\mathfrak{g}^* \times \mathfrak{t}^*_{reg}) \ .$$

Up to a shift it is deduced from the adjunction morphism adj between the constant sheaf $(\overline{\mathbb{Q}}_l)_{X' \times \mathfrak{t}^*_{reg}}$ to the constant sheaf $(\overline{\mathbb{Q}}_l)_\Gamma$

$$\begin{array}{ccccc}
\overline{\mathbb{Q}}_l[n+r] & X'_{reg}\times \mathfrak{t}^*_{reg} & \xrightarrow{q'\times id} & \mathfrak{g}^*_{reg}\times \mathfrak{t}^*_{reg} & \Phi'/\mathfrak{t}^*_{reg} \\
\downarrow{\scriptstyle adj} & \cup & & \cup & \downarrow{\scriptstyle res_{\mathscr{O}}} \\
\delta_\Gamma[n+r] & \Gamma & \xrightarrow{\sim} & \mathscr{O} & \delta_{\mathscr{O}}[n+r]
\end{array}$$

such that $res_{\mathscr{O}} = \mathrm{R}(q'\times id)_!(adj)$. Its Fourier transform will be called the **corestriction**

$$cores:\ (i_{\mathscr{N}*}\Psi(-n))/\mathfrak{t}^*_{reg} \longrightarrow \mathscr{S}(\psi)(-N)[r]\Big|\mathfrak{g}\times\mathfrak{t}^*_{reg}$$

$$cores = T_\psi^{\mathfrak{g}^*\times \mathfrak{t}^*_{reg}/\mathfrak{t}^*_{reg}}(res_{\mathscr{O}})\ .$$

Since this Brylinski-action on $\mathscr{S}(\psi)_B$ has been described by means of a Fourier transform, the corestriction map is W-equivariant by definition, provided Ψ and $\mathscr{S}(\psi)$ are both equipped with the Brylinski action. Similar to VI.16.7 one obtains the next

Lemma 17.10 *If Ψ and $\mathscr{S}(\psi)$ are equipped with the Brylinski action, the corestriction is a W-equivariant map. It canonically factorizes into an isomorphism and the adjunction morphism*

$$\begin{array}{ccc}
(i_{\mathscr{N}*}\Psi_B)/\mathfrak{t}^*_{reg} & \xrightarrow{cores} & \mathscr{S}(\psi)_B[r](-N+n)\big|\mathfrak{g}\times\mathfrak{t}^*_{reg} \\
{\scriptstyle\cong}\searrow & & \nearrow{\scriptstyle adj} \\
& (i_{\mathscr{N}\times\mathfrak{t}^*_{reg}})_!(i_{\mathscr{N}\times\mathfrak{t}^*_{reg}})^!\,\mathscr{S}(\psi)_B[r](-N+n)\big|\mathfrak{g}\times\mathfrak{t}^*_{reg}\ . &
\end{array}$$

The corestriction and obviously the adjunction morphism, hence also the left isomorphism, are W-equivariant.

Induced W-actions on Ψ. There are now several ways to induce a W-right action, say, on Ψ from the W-right actions on the universal Springer sheaf respectively the universal regular orbit described above. There are various possibilities to produce such W-actions. However we will see that all these induced actions on Ψ coincide either with the Lusztig or the Brylinski action, that where previously defined. Nevertheless this gives a new approach to define these actions. In the following we describe four different methods, called a)-d) to obtain a W right action on Ψ. These are

a) (Corestriction to $\mathscr{N}$) Use the lower left isomorphism of the last diagram giving the isomorphism

$$i^!_{\mathscr{N}\times\mathfrak{t}^*_{reg}}\mathscr{S}(\psi)[r](-N+n)\ \cong\ \Psi/\mathfrak{t}^*_{reg}\ .$$

b) (Restriction to $\mathcal{N}$) Use $i^*_{\mathcal{N}\times\mathfrak{t}^*}\mathcal{S}(\psi)[-r](N) \cong \Psi/\mathfrak{t}^*$.
c) (Corestriction to $\mathfrak{g} \times \{0\} \subset \mathfrak{g} \times \mathfrak{t}^*$) Use $\mathcal{S}(\psi)|\mathfrak{g} \times \{0\} \cong \Phi[r]$ and then apply Fourier transform (up to twists and shifts).
d) (Restriction to $\mathfrak{g}^* \times \{0\} \subset \mathfrak{g}^* \times \mathfrak{t}^*$) Use $j'_{!*}(\delta_{\mathcal{O}}\langle m\rangle[r])\Big|\mathfrak{g}^* \times \{0\} = i_{\mathcal{N}*}(\Psi')$.

Concerning a) and b) one has to take into consideration, that any W-action on the perverse sheaf $\Psi/\mathfrak{t}^*_{reg}$ – or $\Psi/\mathfrak{t}^*$ – compatible with the underlying W-(right)-action on $\mathfrak{g} \times \mathfrak{t}^*_{reg}$ – resp. $\mathfrak{g} \times \mathfrak{t}^*$ – descends to a unique W-action on Ψ. This follows from Lemma III.15.4. So each of the four methods defines a W-action on Ψ compatible with the underlying W action on $\mathfrak{g}$.

This being said, we discuss the four cases a),b),c) and d) and compare them with the W-actions defined on Ψ in Chap. VI §13 and §14.

In case a) the isomorphism

$$i^!_{\mathcal{N}\times\mathfrak{t}^*_{reg}}[r](-N+n)\mathcal{S}(\psi)_B \cong \Psi_B/\mathfrak{t}^*_{reg}$$

is W-equivariant with respect to the **Brylinski** actions by the last Lemma VI.17.10.

In case b) and c) it is clear from the definitions, that $i^*_{\mathcal{N}\times\mathfrak{t}}[-r](N)\mathcal{S}(\psi)_L \cong \Psi_L/\mathfrak{t}^*$ and $\mathcal{S}(\psi)_L|\mathfrak{g} \times \{0\} \cong \Phi_L[r]$ are W-equivariant isomorphisms with respect to the **Lusztig** actions.

In case d) the natural W-action on $\delta_{\mathcal{O}}$ induces the **Lusztig** action on Ψ'. This was already observed in VI.17.8.

To clarify the picture we finally show

Proposition 17.11 *The identity map induces a W-equivariant isomorphism comparing the Brylinski action and the Lusztig action on the universal Springer sheaf*

$$\mathcal{S}(\psi)_B = \varepsilon_W \otimes_{\overline{\mathbb{Q}}_l} \mathcal{S}(\psi)_L$$

on $\mathfrak{g} \times \mathfrak{t}^*$. Here ε_W is the sign character of the Coxeter group W.

Proof. We already know from Corollary VI.17.5, that this assertion holds for some character

$$\varepsilon : W \to \overline{\mathbb{Q}}_l^*$$

of the Weyl group W. It remains to determine ε. It can be determined by restriction to a suitable subscheme of $\mathfrak{g}\times\mathfrak{t}^*$. So restrict to $\mathfrak{g}^*\times\{0\}$. Using $\mathcal{S}(\psi)_L|\mathfrak{g}\times\{0\} \cong \Phi_L[r]$ as in method c) above, we see the the right side specializes to $\varepsilon\otimes_{\overline{\mathbb{Q}}_l}\Phi_L$. Since Fourier transform commutes with base change, the restriction of the left side $S(\psi)_B$ becomes the Fourier transform of the corestriction of $j'_{!*}(\delta_{\mathcal{O}}\langle m\rangle[r])$. By method d) this is the Fourier transform of Ψ'_L, hence Φ_L. We therefore get a W-(right)-equivariant isomorphism

$$\Phi_B \cong \varepsilon \otimes_{\overline{\mathbb{Q}}_l} \Phi_L \ .$$

Note, that there exists at most one character ε of W with this property (restrict to $\mathfrak{g}_{reg}$ and use that in the regular representation there is contained at least one irreducible representation ϕ such that $\phi \cong \varepsilon' \otimes \phi$ implies $\varepsilon' = 1$; e.g. the trivial representation ϕ. Further note that $\Phi_L = \bigoplus_{\phi \in \hat{W}} \Phi_\phi \otimes_{\overline{\mathbb{Q}}_l} V_\phi$ such that all Φ_ϕ are pairwise nonisomorphic). This remark now implies

$$\varepsilon = \varepsilon_W$$

using the existence of the W-equivariant isomorphism shown in VI.15.1. □

Corollary 17.12 *The identity map induces W-equivariant isomorphisms between Φ_B and $\varepsilon_W \otimes_{\overline{\mathbb{Q}}_l} \Phi_L$ resp. between Ψ_B and $\varepsilon_W \otimes_{\overline{\mathbb{Q}}_l} \Psi_L$.*

Proof. It is enough to prove the first statement. Since the Fourier transform commutes with base change, it follows from VI.17.11 and the method c) and d) above via the base change

$$\{0\} \hookrightarrow \mathfrak{t}^* .$$

□

We think, that it is reasonable to apply the method of §15 in the generality of working with the universal Springer sheaves. Combined with arguments as above this probably allows to improve upon VI.15.1, to obtain a canonical isomorphism between the twisted dual of the entire control sequence and its Fourier transform. However the authors have not checked this.

VI.18 Finite Fields

Let κ be a finite field with algebraic closure k, let G_0 be a connected semisimple group over κ. Let $\mathbf{B}_0$ be a Borel subgroup over κ and let $\mathbf{T}_0 \subset \mathbf{B}_0 = \mathbf{T}_0 \cdot \mathbf{U}_0$ be a maximal torus defined over κ. Suppose p is a very good prime for G, and suppose that the fixed G-equivariant nondegenerate bilinear form $\langle .,. \rangle$ on $\mathfrak{g}$ is defined over κ. We also fix a nondegenerate additive character ψ of κ.

Weil Complexes. In this section it is quite crucial to consider etale $\overline{\mathbb{Q}}_l$-sheaves on schemes X_0 over $Spec(\kappa)$ and the corresponding F_X-equivariant $\overline{\mathbb{Q}}_l$-sheaves on X. Recall that by our conventions $X = X_0 \times_{Spec(\kappa)} Spec(k)$ is the scheme obtained from X_0 by extending the base to k. Furthermore F_X denotes the Frobenius automorphism of X. This Frobenius automorphism was introduced in Chap. I, §1. It is not a morphism over $Spec(k)$, but only a morphism over $Spec(\kappa)$. In Chap. I, §1 we also introduced the concept of Weil sheaves. By definition, a Weil sheaf on X_0 is a F_X-equivariant etale $\overline{\mathbb{Q}}_l$-sheaf on X. Similarly we can define a Weil complex on X_0 to be a F_X-equivariant complex $K \in D_c^b(X, \overline{\mathbb{Q}}_l)$ equipped with an F_X-action in the sense of Chap. III, §15 (although F_X is not of finite order, the situation of III §15 obviously carries over). Of course any complex K_0 in $D_c^b(X_0, \overline{\mathbb{Q}}_l)$ gives by pullback a Weil complex K, i.e. a F_X-equivariant complex on the scheme X, and most Weil complexes arise in this way. To simplify notation, it is sometimes useful to speak of Weil sheaves K (or complexes K on X) suppressing indices $_0$, if the underlying arithmetic structure of X – defined by X_0 – is understood to be fixed.

Let $\mathcal{S}(\mathbf{T}, \psi)$ denote the universal Springer sheaf on $\mathfrak{g} \times \mathfrak{t}^*$, where $\mathfrak{t}$ denotes the Lie algebra of $\mathbf{T}$. Note that we now keep track of the underlying torus. Let $\mathbf{F} = \mathbf{F}_{\mathfrak{g}\times\mathfrak{t}^*}$ be the Frobenius automorphism of $\mathfrak{g} \times \mathfrak{t}^*$. Then $\mathbf{X}_0 = G_0 \times^{\mathbf{B}_0} \mathfrak{b}_0$ and the maps $\nu_0 : \mathbf{X}_0 \to \mathbb{A}_0$ and $q_0 : \mathbf{X}_0 \to \mathfrak{g}_0$ are defined over κ. So this defines a **"standard" Weil sheaf structure**

$$F^* : \mathbf{F}^*(\mathcal{S}(\mathbf{T}, \psi)) \xrightarrow{\sim} \mathcal{S}(\mathbf{T}, \psi)$$

on $\mathcal{S}(\mathbf{T}, \psi)$. In fact it is obtained from the arithmetically defined complex $\mathcal{S}(\mathbf{T}_0, \psi) = R(q_0 \times id_0)_! \nu_0^*\big(\mathcal{L}(\psi)_0[n + r]\big)$ on $\mathfrak{g}_0 \times \mathfrak{t}_0^*$ by base change; see Definition I.1.2.

In the case of arbitrary maximal κ-tori T_0 of G_0 we can not directly apply this procedure to define an arithmetic structure on the corresponding Springer sheaf. Indeed, we have to modify the approach, since T_0 is not necessarily contained in a κ-rational Borel subgroup of G_0. Therefore the corresponding X and the corresponding morphism q is only defined over $Spec(k)$.

Suppose T_0 is a maximal torus in G_0 defined over κ, but not necessarily maximal split over κ. The fixed bilinear form identifies the Lie algebra $\mathfrak{g}_0$ and its dual $\mathfrak{g}_0^*$ and also identifies $\mathfrak{t}_0$ and $\mathfrak{t}_0^*$. The group

$$\Omega(T_0) = \langle W(T), F\rangle ,$$

generated by the Weyl group $W(T)$ over k and the Frobenius automorphism $F = F_{\mathfrak{g}\times\mathfrak{t}^*}$ acts on $\mathfrak{g} \times \mathfrak{t}^*$ from the right. These actions are actions over $Spec(\kappa)$

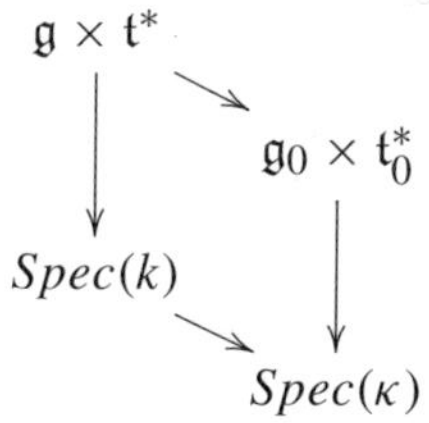

In fact, $W(T)$ is a normal subgroup of $\Omega(T_0)$, hence $\Omega(T_0)$ is a semidirect product of the Weyl group $W(T)$ and the Weil group $W(k/\kappa)$ generated by F.

Adjoint Construction. The isomorphism $f : \Upsilon \to X_{reg}$ of Theorem VI.9.1 cannot be realized over the base field κ. However, if we restrict to the regular locus $\mathfrak{g}_{reg} \times \mathfrak{t}^*$

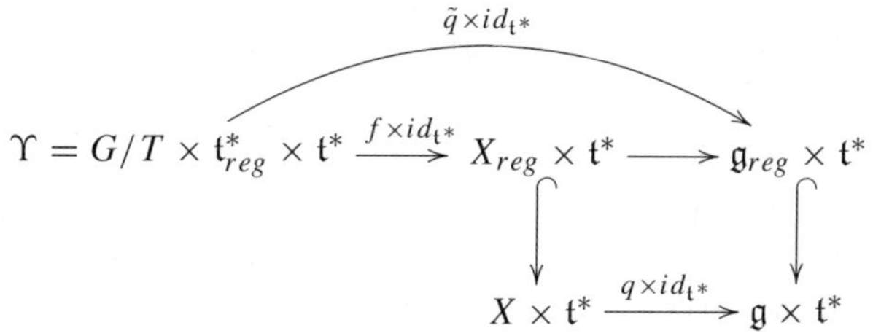
q̃×id_t*
ϒ = G/T × t*_reg × t*
f×id_t*
X_reg × t*
g_reg × t*
X × t*
q×id_t*
g × t*

we can identify the restriction $\mathscr{S}(T,\psi)\,|\mathfrak{g}_{reg}\times\mathfrak{t}^*$ of the universal Springer sheaf with the complex

$$\mathrm{R}(\tilde{q}\times id_{\mathfrak{t}^*})_!(f\times id_{\mathfrak{t}^*})^*\nu^*(\mathscr{L}(\psi)[n+r])\,.$$

All ingredients of this complex are defined over the base field κ! There is $\tilde{q}_0:\Upsilon_0=G_0/T_0\times(\mathfrak{t}_{reg})_0\to\mathfrak{g}_{reg0}$ as well as $\mathscr{L}(\psi)_0$. Also the morphism $\nu\circ(f\times id_{\mathfrak{t}^*})$ is defined over κ. Hence in this way we can define a Weil sheaf on $\mathfrak{g}_{reg0}\times(\mathfrak{t}_0)^*$. By perverse analytic continuation (see III.5.11) this puts a Weil sheaf structure on the universal Springer sheaf $\mathscr{S}(T,\psi)$, from which we started. This is done by extending the cocycle defining the action to the whole space $\mathfrak{g}\times\mathfrak{t}^*$. (The same argument more generally defines an action of the group $\Omega(T_0)$ on $\mathscr{S}(T,\psi)$ compatible with the action of $\Omega(T_0)$ on the underlying space, where the Weyl group $W(T)$ acts via the Lusztig action).

Coadjoint Construction. Since T_0 is defined over κ, also its universal regular orbit $\mathscr{O}_0=\mathscr{O}(T_0)$

$$\mathscr{O}(T_0)\subset\mathfrak{g}_0^*\times(\mathfrak{t}_0)^*_{reg}$$

is defined over κ. The group $\Omega(T_0)$ acts on $\mathscr{O}(T)$ from the right, and also acts canonically on the trivial sheaf $\delta_{\mathscr{O}(T)}=\overline{\mathbb{Q}}_{l\,\mathscr{O}(T)}$; this sheaf is a pull back from $Spec(\kappa)$, so we can apply III.15.3a). Hence $\Omega(T_0)$ acts on the perverse sheaf $\delta_{\mathscr{O}(T)}\langle m\rangle[r]$ and its intermediate extension K, which is the base extension of the perverse sheaf

$$K_0\;=\;j'_{0!*}\big(\delta_{\mathscr{O}(T_0)}\langle m\rangle[r]\big)\;\in\;Perv(\mathfrak{g}_0^*\times\mathfrak{t}_0^*)\,.$$

In other words K is a $\Omega(T_0)$-equivariant perverse sheaf on $\mathfrak{g}^*\times\mathfrak{t}^*$.

Consider the relative Deligne-Fourier transform on $\mathfrak{g}^*\times\mathfrak{t}^*/\mathfrak{t}^*$. The relative Fourier transform of K carries an induced $\Omega(T_0)$-structure, say with cocycles ψ_σ. Now we use the isomorphism h of Corollary VI.17.6. h is a priori only defined over k. Nevertheless we can use h as in the diagram

$$\begin{array}{ccc}
\mathscr{S}(T_0,\psi) & \xrightarrow[\sim]{\;h\;} & T_\psi(K)\\
\uparrow{\scriptstyle \varphi_\sigma\,\cong} & & \uparrow{\scriptstyle \cong\,\psi_\sigma}\\
\sigma^*(\mathscr{S}(T_0,\psi)) & \xrightarrow[\sim]{\;\sigma^*(h)\;} & \sigma^*(T_\psi(K))
\end{array}$$

to define the cocycles φ_σ for $\sigma\in\Omega(T_0)$. These cocycles φ_σ put a $\Omega(T_0)$-structure on the perverse sheaf

$$\mathscr{S}(T,\psi)\in Perv(\mathfrak{g}\times\mathfrak{t}^*)\,.$$

Now the action of $W(T)$ is the *Brylinski action.*

Since $\mathscr{S}(T,\psi)$ is an irreducible perverse sheaf and since for $T_0=\mathbf{T}_0$ all constructions are defined over κ, we obtain

Lemma 18.1 *Let T_0 be a maximal torus in G_0. Then the universal Springer sheaf $\mathscr{S}(T,\psi)$ carries two different structures as $\Omega(T_0)$-equivariant sheaf, which are defined by the adjoint respectively coadjoint construction such that*

(i) *Both structure are compatible with the canonical right action of $\Omega(T_0)$ on the underlying space $\mathfrak{g}\times\mathfrak{t}^*$.*
(ii) *They coincide up to a twist with a character*

$$\varepsilon:\Omega(T_0)\to\overline{\mathbb{Q}}_l^{\,*}\ .$$

Define: $\varepsilon_T=\varepsilon\big(F_{\mathfrak{g}\times\mathfrak{t}^*}\big)\ \in\overline{\mathbb{Q}}_l^{\,*}$.
(iii) *On the Coxeter group $W(T)$ the character ϵ is the sign character $\varepsilon_{W(T)}$.*
(iv) *For the maximal split torus $\mathbf{T}_0$, the Weil sheaf structures on $\mathscr{S}(\mathbf{T},\psi)$ induced by both structures coincide with the standard Weil sheaf structure of $\mathscr{S}(\mathbf{T},\psi)$.*

Transport of Structure. For a scheme X_0 over $Spec(\kappa)$ let F_X denote the Frobenius automorphism of X obtained by scalar extension $F_X=id_{X_0}\times F$ from the geometric Frobenius automorphism F of k/κ. For X_0,Y_0 over $Spec(\kappa)$ and a morphism $f:Y\to X$ over $Spec(k)$ put $f^F=F_X\circ f\circ F_Y^{-1}:Y\to X$, which is a morphism over $Spec(k)$. Suppose $f:X\cong X'$ is an isomorphism over $Spec(k)$. Then pullback defines the following automorphism of X

$$F':=f^*(F_{X'})=f^{-1}\circ F_{X'}\circ f=F_X\circ w=wF_X\ .$$

Here w is a uniquely determined k-automorphism of X. For a Weil complex K' on X' with structure map $\phi_{F_{X'}}:F_{X'}^*(K')\cong K'$, the pullback $(K,(F')^*)=(f^*(K'),f^*(\phi_{F_{X'}}))$ defines a F'-equivariant sheaf complex K on X. For a closed fixed point x, $F'(x)=x$ of X with image $x'=f(x)$ the stalk endomorphisms $((F')^*,K_{\overline{x}})$ and $(F_{X'}^*,K'_{\overline{x}'})$ can be identified.

Suppose T_0',T_0 are two maximal κ-tori in G_0. Then there exist elements $g\in G(k)$ such that $T'=g^{-1}Tg$. If we identify T with T' over k in this way, this induces an isomorphism

$$\omega_g:\Omega(T_0')\xrightarrow{\ \sim\ }\Omega(T_0)\ .$$

By this isomorphism the Weyl group $W(T')$ is identified with the Weyl group $W(T)$. The Frobenius automorphism $F_{T'}$ of T' maps to an automorphism

$$F':=wF_T\ \in\Omega(T_0)$$

for some $w=w(T,T',g)\in W(T)$. Note $F'(x)=F_T(w(x))$ due to the right action.

Notations. Fix an isomorphism $\tau:\overline{\mathbb{Q}}_l\cong\mathbb{C}$. This allows to view ψ as a $\mathbb{C}$-valued character. For maximal tori T,T' in G put $N(T,T')=\{g\in G\mid g(T)=T'\}$. Recall that the Grothendieck simultaneous resolution $\mathbf{X}_0\to\mathfrak{g}_0$ defines the Springer fibers $\mathscr{B}_x$. For a maximal torus T_0 define $w=w(T,g)\in W(\mathbf{T})$ as above, such that $T=g^{-1}\mathbf{T}g$. Let $\mathfrak{t}$ denote the Lie algebra of T. For $y\in(\mathfrak{t}_{reg}^*)^F$ put $\tilde{y}=Ad_g(y)\in\mathfrak{g}$.

Recall $y \in (\mathscr{O}(T)_y)^F \subset (\mathfrak{g}^*)^F \cong (\mathfrak{g})^F$. Recall $n = dim(\mathfrak{g}) = m + r = 2N + r$, where $r = dim(\mathfrak{t})$.

Then – as an application of the previous results – we obtain

Theorem 18.2 (Springer) *([S6], theorem 3.15 and 4.4)*
For $y \in (\mathfrak{t}_{reg})^{*F}$ and for $x \in \mathfrak{g}^F$ the following values (a) and (b) are equal

(a) *The function $f^{L_0}(x, y)$ defined for the Weil complex $L_0 = \mathscr{S}(T_0, \psi)$ (coadjoint construction) or by transport of structure*

$$\tau(trace(w\mathbf{F}, S(\mathbf{T}, \psi)_{(x,\tilde{y})}))\,,$$

where w acts by the Brylinski action.

(b) *The function $f^{K_0}(x, y)$ defined for the Weil complex $K_0 = T_\psi(M(T_0))$, where $M(T_0) = (j_0')_{!*}(\delta_{\mathscr{O}(T_0)}\langle m\rangle[r])$. Explicitly*

$$q^{-N} \cdot \sum_{z \in (\mathscr{O}_y)^F} \psi^{-1}\big(\langle x, z\rangle\big)\,.$$

Furthermore:

(c) *If $x \in \mathfrak{g}^F$ is* **regular***, then (a) is also equal to*

$$\varepsilon_T \cdot \sum_{w \in N(T',T)^F/T^F} \psi^{-1}\big(\langle x, y^w\rangle\big)\,.$$

This value is zero unless the centralizer T_0' of x in G_0 is conjugate to T_0 under G^F.

(c′) *If $x \in \mathscr{N}^F \subset \mathfrak{g}^F$ is* **nilpotent***, then (a) is also equal to*

$$\sum_{\nu \geq 0} (-1)^\nu \cdot \tau\; trace\big(w\mathbf{F}, H_c^\nu(\mathscr{B}_x, \overline{\mathbb{Q}}_l)\big)\,,$$

where $\mathscr{B}_x = q_{\mathscr{N}}^{-1}(x)$ is the Springer fiber over x and w acts by the Brylinski action. This value depends only on w but not on the particular choice of y.

Proof. (a)$\Longleftrightarrow$(b) holds by the definition of the coadjoint construction. Since $n + m + r = 2n$ is even, the explicit formula in (b) follows from the dictionary in Chap. III §12. It gives

$$f^{T_\psi(j_{!*}'(\delta_{\mathscr{O}})\langle m\rangle[r]}(x, y) = (-1)^n \sum_{z \in \mathfrak{g}^F} (-1)^{m+r} q^{-m/2} f^{j_{!*}'(\delta_{\mathscr{O}_y})}(z) \cdot \psi^{-1}(\langle x, z\rangle)\,,$$

or equivalently $q^{-N} \cdot \sum_{z \in (\mathscr{O}_y)^F} \psi^{-1}\big(\langle x, z\rangle\big)$.

Claim (c) follows from comparison with the adjoint construction using Lemma VI.18.1. This gives the alternative expression

$$\varepsilon_T \cdot f^{\mathrm{R}(\tilde{q}\times id)_!(\nu^*(\mathscr{L}(\psi))[n+r]}(x,y) = \varepsilon_T \cdot \sum_{(gT,\tilde{x})} (-1)^{n+r}\psi^{-1}(\langle \tilde{x}, y\rangle) .$$

The sum is over all $(gT,\tilde{x}) \in (G/T \times \mathfrak{t}_{reg})^F$ such that $g(\tilde{x}) = x$. Since $(-1)^{n+r} = (-1)^{2(N+r)} = 1$, the sum is equal to $\sum_{w\in N(T',T)^F/T^F} \psi^{-1}(\langle x, y^w\rangle)$.

Claim (c′) follows from $\mathscr{S}(T,\psi)|(\mathscr{N}^{\cdot}\times\overline{y}) \cong \mathrm{R}q_{\mathscr{N}^{\cdot}!}(\overline{\mathbb{Q}}_l[n+r])$. Again since $(-1)^{n+r} = 1$ we get

$$\tau(trace(w\mathbf{F}, S(\mathbf{T},\psi)_{(x,\tilde{y})})) = \sum_{\nu\geq 0}(-1)^{\nu}\cdot\tau\ trace(w\mathbf{F}, H_c^{\nu}(\mathscr{B}_x, \overline{\mathbb{Q}}_l)) .$$

□

From VI.19.2 will give $\varepsilon_T = \varepsilon_W(w)$. Multiplying the expression in VI.18.2(c′) by ε_T has therefore the same result as a change from the Brylinski to the Lusztig action. With this modification (w is supposed to act now via the Lusztig action)

$$Q_w(x) = \sum_{\nu\geq 0}(-1)^{\nu}\cdot\tau\ trace\big(w\mathbf{F}, H_c^{\nu}(\mathscr{B}_x,\overline{\mathbb{Q}}_l)\big) \quad , \quad x\in\mathscr{N}^{\cdot F}$$

defines for each $w \in W(T)$ a function

$$Q_w : \mathscr{N}^{\cdot F} \to \mathbb{C}$$

on the nilpotent κ-rational elements. It is called the **Green function**. We may also write Q_T instead of Q_w. The Green functions satisfy the following remarkable **orthogonality relations** ([S6], theorem 5.6).

Theorem 18.3 (Springer) *Suppose p and q are sufficiently large. Let Q_T and $Q_{T'}$ be Green functions attached to κ-rational maximal tori T_0 and T_0' of G_0. Then*

$$\sum_{x\in\mathscr{N}^{\cdot F}} Q_T(x)Q_{T'}(x) = |G^F||N(T,T')^F||T^F|^{-1}|(T')^F|^{-1} .$$

We briefly outline, how these orthogonality relations can be deduced from the Plancherel formula for the finite abelian group $\mathfrak{g}^F$. It states for Weil complexes K_0, L_0 on $\mathfrak{g}$, that

$$\sum_{x\in\mathfrak{g}^F} f^{T_\psi(K_0)}(x)\cdot f^{L_0}(x) = \sum_{x\in\mathfrak{g}^F} f^{K_0}(x)\cdot f^{T_\psi(L_0)}(x) .$$

Here $\mathfrak{g}$ and $\mathfrak{g}^*$ were tacitly identified via the fixed bilinear form on $\mathfrak{g}$, so all complexes are viewed as complexes on $\mathfrak{g}\times\mathfrak{t}^*$.

Proof. Assume p and q are large enough. Then $\mathfrak{t}_{reg}^F$ and $(\mathfrak{t}')_{reg}^F$ are nonempty. Choose $y \in \mathfrak{t}_{reg}^{*F}$. Consider the Weil complex $M(T_0) = (j_0')_{!*}(\delta_{\mathscr{O}(T_0)}\langle m\rangle[r])$ attached to the universal regular orbit $\mathscr{O}(T_0)$ for the torus T_0. Base change by the closed embedding $\{y\} \hookrightarrow \mathfrak{t}^*$ defines

$$K_0 = i_y^*\big(M(T_0)\big)\ .$$

Similarly the Weil complex $M(T_0') = (j_0')_{!*}(\delta_{\mathscr{O}(T_0')})\langle m\rangle[r]$ on $\mathfrak{g} \times \mathfrak{t}'^*$ defines

$$L_0 = i_0^*\big(M(T_0')\big)$$

by base change for the closed embedding $\{0\} \hookrightarrow \mathfrak{t}'^*$. Then $T_\psi(K_0)$ and $\mathscr{S}(T,\psi)|\mathfrak{g}\times y$ are isomorphic Weil complexes by the definition of the coadjoint construction. Furthermore $L_0 \cong \Psi_L[r] = (i_{\mathscr{N}_0^\cdot})_*\mathrm{R}(q_{\mathscr{N}_0^\cdot})_*(\overline{\mathbb{Q}}_l\langle m\rangle[r])$. This uses the restriction isomorphism VI.17.7. Finally $T_\psi(L_0) \cong \mathscr{S}(T_0',\psi)|\mathfrak{g}\times 0$. This follows from the coadjoint construction, functoriality of T_ψ (Theorem III.13.2.1) and perverse analytic continuation (Corollary VI.17.6). Using these facts we obtain from the "dictionary"

- $f^{T_\psi(K_0)}|\mathscr{N}^{\cdot F} = \varepsilon_T \cdot Q_T$ (Theorem VI.18.2c′)
- $f^{L_0} = (-1)^r q^{-N} \cdot Q_{T'}$ (with support in $\mathscr{N}^{\cdot F}$)
- f^{K_0} (up to a constant $(-1)^r q^{-N}$ the characteristic function of the orbit $\mathscr{O}_y^F$; in particular it is a function with *regular* support in $\mathfrak{g}_{reg}^F$)
- $f^{T_\psi(L_0)}(x) = \varepsilon_{T'} \cdot \sum_{w\in N(T',T'')^F/(T')^F} 1$ for all regular $x \in \mathfrak{g}^F$, where T_0'' denotes the centralizer of x in G_0 (a variant of Theorem VI.18.2c).

The theorem follows immediately from this, since $\varepsilon_T = \varepsilon_{T'}$ for $N(T,T')^F \neq \emptyset$ (Lemma VI.19.2).

Concluding Remarks

Both $\mathbf{T} \subset \mathbf{B}$ are stable under the Frobenius automorphism F_G of G. Suppose that the isomorphism $\sigma : \mathscr{N}^\cdot \to \mathscr{U}$ is defined over κ, i.e. commutes with F_G. Suppose $\sigma(v) = u \in \mathbf{U}^F$. Then – more or less by the definition of an induced representation – the trace

$$R(u) = trace\Big(u, Ind_{\mathbf{B}^F}^{G^F}(1_{\mathbf{B}^F})\Big)$$

is equal to the number of the cosets $g\mathbf{B}^F \in G^F/\mathbf{B}^F$, for which $ug\mathbf{B}^F = g\mathbf{B}^F$ or equivalently $u \in g\mathbf{B}^F g^{-1}$ holds. Since $\mathbf{B}$ is connected we have $(G/\mathbf{B})^F = G^F/\mathbf{B}^F$. Thus $\mathscr{B}_u^F$ is the number of cosets $g \in G^F/\mathbf{B}^F$, for which $u \in g\mathbf{B}^F g^{-1}$. Hence the Grothendieck-Lefschetz fixed point formula gives

$$\sum_\nu (-1)^\nu trace\Big(\mathbf{F}, H_c^\nu(\mathscr{B}_u, \overline{\mathbb{Q}}_l)\Big) = trace\Big(u, Ind_{\mathbf{B}^F}^{G^F}(1_{\mathbf{B}^F})\Big)\ .$$

In terms of the Green function $Q_w(u)$

$$Q_1(u) = R(u) \quad , \quad u \in \mathscr{U}^F\ .$$

Using the orthogonality relations of Green functions it was shown by Kazhdan [K] for sufficiently large characteristic p and sufficiently large q (and later by Lusztig for all very good primes p), that the following much deeper result holds: The last formula remains valid for arbitrary $w \in W$, if $R(u)$ is replaced by a trace $R_w(u)$ on the right

$$Q_w(u) = R_w(u) \quad , \quad u \in \mathscr{U}^F .$$

Here $R_w = R^{\theta}_{T_w,G}$ is one of the cohomologically induced representations of G^F defined by Deligne and Lusztig [81]. It is instructive, that even for regular unipotent elements u this formula is nontrivial; in this case it is equivalent to $R_w(u) = 1$, because $Q_w(u) = 1$ holds as mentioned in Remark VI.1.2.

VI.19 Determination of ε_T

Let the situation be as in the previous section. In this section we determine the constant ε_T defined in Lemma VI.18.1. For this we discuss the behaviour of the adjoint and the coadjoint construction of Weil sheaf structures – given in the last section – under transport of structure.

Adjoint Situation. We have a commutative diagram

$$\begin{array}{ccc}
X(B') \times (\mathfrak{t}')^* & \xrightarrow{q(B')\times id} & \mathfrak{g} \times (\mathfrak{t}')^* \\
{\scriptstyle R_g\times Ad_g}\downarrow\cong & & \cong\downarrow{\scriptstyle id\times Ad_g} \\
X(B) \times \mathfrak{t}^* & \xrightarrow{q(B)\times id} & \mathfrak{g} \times \mathfrak{t}^*
\end{array}$$

where $gB'g^{-1} = B$. Here $R_g : X(B') = G \times^{B'} \mathfrak{b}' \to G \times^B \mathfrak{b}$ is induced by right translation $R_g(h) = hg^{-1}$ on G and by $Ad_g : \mathfrak{b}' \to \mathfrak{b}$ on the second factor. We have a similar diagram

$$\begin{array}{ccc}
G/T' \times \mathfrak{t}'_{reg} \times (\mathfrak{t}')^* & \xrightarrow{\tilde{q}(T')\times id} & \mathfrak{g}_{reg} \times (\mathfrak{t}')^* \\
{\scriptstyle R_g\times Ad_g}\downarrow\cong & & \cong\downarrow{\scriptstyle id\times Ad_g} \\
G/T \times \mathfrak{t}_{reg} \times \mathfrak{t}^* & \xrightarrow{\tilde{q}(T)\times id} & \mathfrak{g}_{reg} \times \mathfrak{t}^*
\end{array}$$

Since transport of structure by pullback with respect to the vertical arrows commutes with direct images of compact support, the pullback of the $\Omega(T'_0)$-equivariant structure on $\mathscr{S}(T', \psi)$ gives the corresponding $\Omega(T_0)$-equivariant structure on $\mathscr{S}(T, \psi)$.

Coadjoint Situation. The two universal regular orbits $\mathscr{O}(T)$ and $\mathscr{O}(T')$ attached to the tori T and T' can be identified over k via the automorphism $id_{\mathfrak{g}^*} \times Ad_g$.

$$\begin{array}{ccccc}
\mathcal{O}(T') & \hookrightarrow & \mathfrak{g}^* \times (\mathfrak{t}')^*_{reg} & \xleftarrow{F} & \mathfrak{g}^* \times (\mathfrak{t}')^*_{reg} \\
\downarrow \cong & & \cong \downarrow id \times Ad_g & & \cong \downarrow id \times Ad_g \\
\mathcal{O}(T) & \hookrightarrow & \mathfrak{g}^* \times \mathfrak{t}^*_{reg} & \xleftarrow{F'} & \mathfrak{g}^* \times \mathfrak{t}^*_{reg}
\end{array}$$

Via this isomorphism the Frobenius automorphism of $\mathfrak{g}^* \times (\mathfrak{t}')^*_{reg}$ becomes the automorphism

$$F' = (id_{\mathfrak{g}^*} \times w)F_{\mathfrak{g}^* \times \mathfrak{t}^*}$$

on $\mathfrak{g}^* \times (\mathfrak{t})^*_{reg}$. By transport of structure as in III.15.3a (although $\Omega(T'_0)$ is not a finite group) the "standard" $\Omega(T'_0)$-action on the constant $\overline{\mathbb{Q}}_l$-sheaf on $\mathcal{O}(T')$ pulls back to an $\Omega(T'_0)$-action on the "standard" constant $\overline{\mathbb{Q}}_l$-sheaf on $\mathcal{O}(T)$, which lies over the right action of $\Omega(T'_0)$ on $\mathcal{O}(T)$ defined by the isomorphism ω_g. Similar one has an action on the perverse sheaves K' and by transport of structure an induced action on K. (This follows from III.15.3a, where we take $S = Spec(\kappa)$. All morphisms like Ad_g, F_X, w etc. are morphisms over $Spec(\kappa)$). Fourier transform is functorial with respect to base change, and also the isomorphism h of Corollary VI.17.6 is compatible with base change maps like Ad_g. Therefore the pullback of the $\Omega(T'_0)$-action on the universal Springer sheaf $\mathcal{S}(T', \psi) \in Perv(\mathfrak{g} \times \mathfrak{t}'^*)$ to an action on the universal Springer sheaf $\mathcal{S}(T, \psi) \in Perv(\mathfrak{g} \times \mathfrak{t}^*)$ gives the canonical coadjoint $\Omega(T_0)$-action defined via the cocycles ϕ_σ, once $\Omega(T'_0)$ and $\Omega(T_0)$ are identified via the isomorphism ω_g.

This implies

Lemma 19.1 *Both the adjoint and the coadjoint construction are compatible with respect to transport of structure.*

Therefore the character ε defined in the last lemma can be evaluated by transport of structure. Specialize to the case $T_0 = \mathbf{T}_0$ and rename T'_0 to be T_0. Define $w \in W(\mathbf{T})$ with $w = w(T_0, g)$ as above via an isomorphism $Ad_g(T) = \mathbf{T}$. Then the characters ε^T for T and $\varepsilon^{\mathbf{T}}$ for $\mathbf{T}$ are related by $\varepsilon^T(\sigma) = \varepsilon^{\mathbf{T}}(\omega_g(\sigma))$ for all $\sigma \in \Omega(T_0)$. Since $\omega_g(F_{\mathfrak{g}\times\mathfrak{t}^*}) = w\mathbf{F}_{\mathfrak{g}\times\mathfrak{t}^*}$, this gives

Lemma 19.2 *Let T_0 be a maximal torus of G_0. Define $w = w(T, g)$ as above. Then the two Weil sheaf structures on the universal Springer sheaf $\mathcal{S}(T, \psi)$ on $\mathfrak{g} \times \mathfrak{t}^*$, defined by the adjoint respectively the coadjoint construction, have structure maps differing by the sign*

$$\varepsilon_T = \varepsilon^{\mathbf{T}}(w) = (-1)^{l(w)} .$$

The pullback $(F')^*$ of the cocycle map corresponding to the Frobenius automorphism $F_{\mathfrak{g}\times\mathfrak{t}'^*}$ gives the cocycle

$$\varphi_{F'} = \varphi_{(id\times w)\mathbf{F}} .$$

The cocycle relation for a right action therefore give the following commutative diagram

$$(*) \qquad \begin{array}{ccc} (id \times w)^*(\mathbf{F}^*(\mathscr{S}(\psi)) & \xrightarrow[\sim]{(F')^*} & \mathscr{S}(\psi)\,. \\ {\scriptstyle (id\times w)^*(F^*)} \downarrow \cong & \nearrow {\scriptstyle \varphi_w} & \\ (id \times w)^*(\mathscr{S}(\psi)) & & \end{array}$$

which expresses the cocycle map φ_w in terms of the Frobenius cocycle map $F^* = \varphi_{\mathbf{F}}$ for $\mathbf{F} = \mathbf{F}_{\mathfrak{g}\times\mathfrak{t}^*}$ and the pullback $(F')^* = \varphi_{F'}$ of the corresponding Frobenius cocycle map for T_0. Note, both cocycle maps F^* and $(F')^*$ were essentially obtained from the canonical Weyl sheaf structures on the constant sheaves on $\mathscr{O}$ respectively $\mathscr{O}'$ by transport of structure.

For every element w of the Weyl group $W = W(\mathbf{T})$ of the torus $\mathbf{T}_0$ there exists a torus T_0 defined over κ and a suitable choice of Ad_g, such that $F' = w\mathbf{F}$. The Galois structure on the κ-rational tori therefore uniquely determines the Brylinski action φ_w via the diagrams $(*)$. It is the underlying idea – more or less – of Springer's original approach in [S6] to define φ_w by the diagrams $(*)$. So in this sense the *Brylinski* action coincides with the **Springer** action, which is defined through the various Frobenius actions and their pullbacks.

Looked at it the other way round, for any κ-torus T_0 in G_0 this determines the canonical Weil sheaf structure on

$$h : \mathscr{S}(T, \psi) \cong T_\psi(j'_{!*}\delta_{\mathscr{O}}\langle m\rangle[r])\ ,$$

which was induced from the natural Weil sheaf structure on $\overline{\mathbb{Q}}_{l\,\mathscr{O}}$:

After transport of structure it can be obtained from the corresponding Weil sheaf structure on $\mathscr{S}(\mathbf{T}, \psi)$ for a maximal split torus $\mathbf{T}_0$, by "twisting" with the Brylinski action of a suitable element w in the Weyl group.

Bibliography for Chapter VI

[A] Alekseevski A.V.: Component groups of centralizers of unipotent elements in semisimple algebraic groups, Trudy Tbiliss. Math. Inst. Razmadfze Akad. Gruzin SSR, 62 (1979), 5–27

[BBD] Analyse et Topologie sur les espaces singuliers I, Astérisque 100, 1982

[Bb] Bourbaki N.: Groupes et Algèbres de Lie, Chap. 4, 5 et 6, Éléments de Mathematique 1968 Hermann, Paris; 1982 Masson, Paris

[B-C] Bala P., Carter R.W.: Classes of unipotent elements in simple algebraic groups, Math. Proc. Cambridge Philos. Soc. I, 79 (1976), 401–425; II, 80 (1976), 1–18

[B-K] Borho W., Kraft H.: Über Bahnen und deren Deformationen bei linearen Aktionen reduktiver Gruppen. Comm. Meth. Helv. 54 (1976), 61–104

[B-M1] Borho W., MacPherson R.: Representation de groupes de Weyl et homology d'intersection pour la varietes nilpotent, C.R.A.S. t. 292 (27-4-1981), 707–710

[B-M2] Borho W., MacPherson R.: Partial resolutions of nilpotent varieties, Analyse et Topologie sur les espaces singuliers II–III, Astérisque 101–102, 1983

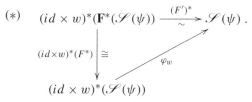

[Bo] Borel A.: Linear Algebraic Groups, Benjamin 1969

[B-R] Bardsley P., Richardson R.W.: Etale slices for algebraic transformation groups in characteristic p, Proc. London Math. Soc. 51 (1985), 295–317

[Bry1] Brylinski J.L.: (Co-) homology d'intersection et faisceaux pervers, Sem. Bourbaki Exp. 585, Vol. 1981/82

[Bry2] Brylinski J.L.: Transformations canoniques, Duality Projective Theorie de Lefschetz, Transformations de Fourier et Sommes Trigonometriques in Geometrie et Analyse Microlocales, Astérisque 140–141, 1986

[Ca1] Carter : On the Representation Theory of finite groups of Lie Type, Encyclopedia of Math. Sciences Vol. 77, Algebra IX, Springer-Verlag

[Ca2] Carter R.W.: Finite Groups of Lie Type: Conjugacy Classes and Complex Characters, John Wiley, London 1985

[CG] Chriss N., Ginzburg V.: Representation theory and complex geometry, Birkhäuser 1977

[Cu] Curtis C.W.: Representations of Hecke algebras, asterisque 173 (1989)

[De1] Demazure M.: Invariants symmetriques entiers de groupes de Weyl et torsion, Invent. Math. 21 (1973), 103–161

[De2] Demazure M.: Desingularisation des Varietes de Schubert generalisées, Ann. scient. Ec. Norm. Sup. 4e serie t7 (1974)

[DG] Demazure M., Gabriel P.: Groupes Algébriques, Tome I, Masson Paris, North Holland Amsterdam

[Gi] Ginzburg V.: Geometric methods in the representation theory of Hecke algebras and quantum groups. (Notes by Vladimir Baranovsky). In: Broer A. (ed.) et al., Representation theories and algebraic geometry. Proceedings of the NATO Advanced Study Institute, Montreal, Canada, July 28–August 8, 1997. Dordrecht: Kluwer Academic Publishers. NATO ASI Ser., Ser. C, Math. Phys. Sci. 514, 127–183 (1998)

[Hd-Str] Séminaire Heidelberg-Strasbourg, 1965/66; Groupes Algébriques Linéaires, Publ. I.R.M.A. Strasbourg

[Hu] Humphreys J.P.: Conjugacy Classes in semisimple algebraic groups, A.M.S. Math. Surveys and Monographs, Vol. 43

[IL] Illusie L.: Deligne's l-adic Fourier Transform, Proc. of Symposia in pure Math. Vol. 46, Part 2 (1987)

[K] Kazhdan D., Proof of Springer's hypothesis, Israel J. Math. 28 (1977), 272–286

[Ka-L] Katz M., Laumon G.: Transformation de Fourier et Majoration de Sommes Exponentielles, Publ. Math. I.H.E.S. 62, 1986

[Ka-Lu] Kazhdan, Lusztig G.: A topological approach to Springer representations, Adv. in Math. 38 (1980)

[Lu1] Lusztig G.: Characters of Reductive Groups over a finite field, Ann. of Math. stud. (1984), Princeton University Press

[Lu2] Lusztig G.: Green polynomials and singularities of unipotent classes, Adv. in Math. 42 (1982), 169–178

[MV] Mirkovic I., Vilonen K.: Characteristic varietes of character sheaves, Invent. Math. 93, No. 2, (1988), 405–418

[Po] Pommerening K.: Über die unipotenten Klassen reduktiver Gruppen. J. Algebra 49 (1977), 525–536, Teil II, J. Algebra 65 (1980), 373–398

[S1] Springer T.A.: The unipotent variety of semisimple groups, Proc. Colloq. Algebr. Geometry Bombay 1968 (Tata Institute) 1969

[S2] Springer T.A.. Construction of representations of Weyl groups, Invent. Math. 44, 1978

[S3] Springer T.A.: On representations of Weyl groups, Proc. Hyderabad Conf. Algebraic Groups (1991), 517–536

[S4] Springer T.A.: Quelques applications de la Cohomologie d'Intersection, Sem. Bourbaki Exp. 589, 1981/82

[S5] Springer T.A.: Linear Algebraic Groups, Sec. Ed. Birkhäuser

[S6] Springer T.A.: Trigonometrical sums, Green functions of finite groups and representations of Weyl groups, Inv. Math. 36, (1976), 173–203

[SGA3] SGA 3, Schémas en Groupes I–III, Springer lecture Notes volume 151–153

[Sl1] Slodowy P.: Simple singularities and simple algebraic groups, Lect. Notes in Math. Vol. 815, Springer 1980

[Sl2] Slodowy P.: Four lectures on simple groups and singularities, Comm. Math. Inst. Utrecht 1980

[S-St] Springer T.A., Steinberg R.: Conjugacy classes. In Borel et alii: Seminar on algebraic groups and related finite groups, Lecture Notes in Math. 1311 (1970), Springer Verlag

[St] Steinberg R.: On the Desingularization of the unipotent variety, Inventiones math. 36, 209–224 (1976)

Appendix

A. $\overline{\mathbb{Q}}_l$-Sheaves

All references in this section refer to the first two chapters of the book [FK] "Etale Cohomology and the Weil Conjecture" by E. Freitag and R. Kiehl, if not stated otherwise. In this book the theory of etale sheaves of $\mathbb{Z}/n\mathbb{Z}$-modules, of l-adic sheaves and $\mathbb{Q}_l$-sheaves on noetherian schemes was developed. Without difficulty it can be shown, that the essential results on $\mathbb{Q}_l$-sheaves carry over to the case of sheaves over finite extension fields of the field $\mathbb{Q}_l$. We want to indicate the small changes, that are necessary for the proof of Poincaré duality and Grothendieck's trace formula in this setting.

Let l be a prime and let $\Lambda = \mathbb{Z}/l^n\mathbb{Z}, n \geq 1$. We will consider noetherian schemes, on which the prime l is invertible. Let R be a finite Λ-algebra. In this part of the appendix $M(X, \Lambda)$ will denote the category of etale sheaves of R-modules on the noetherian scheme X. Without difficulty all definitions and results of [FK], chap. I, §1–§11 carry over from the category $M(X, \Lambda)$ to the category $M(X, R)$. Some remarks on that:

The category $M(X, R)$ is an abelian category and contains enough injectives (see [FK], chap. I, §2, p. 20). For sheaves $\mathscr{F}$ in $M(X, R)$ the abelian group of sections $\Gamma(X, \mathscr{F})$ and the derived groups $H^\nu(X, \mathscr{F})$ constructed in the category $M(X, R)$ carry a natural structure of R-modules. The derived functors $H^\nu(X, \mathscr{F})$ defined in the category $M(X, R)$, or defined in the category $M(X, \Lambda)$, or defined in the category of all abelian sheaves on X, all coincide. Similar statements are valid for the direct image functor and its higher derivatives

$$f_*, \mathrm{R}^i f_* : M(X, R) \to M(Y, R)$$

with respect to a morphism $f : X \to Y$, and also for the derived functors $H_A^\nu(X, \mathscr{F})$ of the functor of sections $\Gamma_A(X, \mathscr{F})$ with supports in the closed subset A of X ([FK], chap. I, 4.13 and 4.14). Of essential importance are finiteness properties of sheaves ([FK], chap. I, §4). A sheaf $\mathscr{F}$ in $M(X, R)$ is said to be constructible, if it is constructible as a sheaf of sets ([FK], chap. I, 4.3). It follows, that $\mathscr{F}$ is constructible as a sheaf of R-modules iff it is constructible as a sheaf of Λ-modules. A sheaf $\mathscr{F}$ from $M(X, R)$ is constructible iff there exists an etale morphism $j : U \to X$ and a surjective homomorphism of sheaves of R-modules

$$j_!(R_U) = j_!(\Lambda_U) \otimes_\Lambda R \to \mathscr{F}$$

([FK], chap. I, proposition 4.10, remark). Furthermore, the category $M(X, R)$ is locally noetherian. The noetherian objects are the constructible sheaves of R-modules

(Chap. I, Proposition 4.8, Corollary). Using these remarks essentially all results from Chap. I, §1–§11 carry over to the category $M(X, R)$ of R-modules. Especially one should mention the two base change theorems (Chap. I, 6.1 resp. 8.7) and the finiteness theorems (Chap. I, 8.9 and 8.10); furthermore the theorems of Künneth type (Chap. I, 8.14 and 8.17). The important results proved in the second chapter are the Poincare duality theorem and the Grothendieck trace formula.

We consider smooth schemes X of pure dimension $d(X)$ over an algebraically closed base field. We remind the reader of the definition of the sheaves $\mu_{l^n,X}$ of l^n roots of unity on X and the positive and negative tensor powers $\mu_{l^n,X}^{\otimes r}$. Put $R_X(r) = R \otimes_{\Lambda_X} \mu_{l^n,X}^{\otimes r}$. For a sheaf $\mathscr{F}$ of R-modules we have by definition

$$\mathscr{F}(r) = \mathscr{F} \otimes_{R_X} R_X(r) = \mathscr{F} \otimes_{\Lambda_X} \Lambda_X(r) .$$

In [FK], chap. II (theorem 1.6) a Λ-homomorphism

$$H_c^{2d(X)}(X, \Lambda_X(d(X))) \to \Lambda$$

was constructed for a smooth scheme X of pure dimension $d(X)$ over k, the so called trace map. By tensoring one obtains the trace homomorphism

$$H_c^{2d(X)}(X, R_X(d(X))) = H_c^{2d(X)}(X, \Lambda_X(d(X))) \otimes_\Lambda R \to R .$$

This new trace map over R inherits its properties from the original trace map. Let us state two such properties. Let $j : Y \to X$ be an etale map. Then the natural diagram

$$\begin{array}{ccccc} H_c^{2d(X)}(X, j_! R_Y(d(Y))) & = & H_c^{2d(Y)}(Y, R_Y(d(Y))) & \longrightarrow & R \\ \downarrow & & & & \| \\ H_c^{2d(X)}(X, R_X(d(X))) & & \longrightarrow & & R \end{array}$$

is commutative. Furthermore for irreducible X, the trace map

$$H_c^{2d(X)}(X, R_X(d(X))) \to R$$

is an isomorphism.

Next let us assume, that R is an injective R-module. For example, let E be a finite extension field of the field $\mathbb{Q}_l$ of l-adic integers with valuation ring $\mathfrak{o}$ and generator π of the maximal ideal. Then for all natural numbers $j \geq 1$ the ring $\mathfrak{o}/\pi^j\mathfrak{o}$ is an injective module over itself. Using the trace map ([FK], chap. II, §1, p. 143) available for sheaves from the category $M(X, R)$, one constructs a natural transformation

$$\mathrm{R}Hom(\mathscr{F}, R_X(d(X)))[2d(X)] \to Hom_R(\mathrm{R}\Gamma_c(X, \mathscr{F}), R) ,$$

the duality map. This duality map induces natural transformations of δ-functors

$$Ext_R^\nu(\mathscr{F}, R_X(d(X))) \to Hom_R(H_c^{2d(X)-\nu}(X, \mathscr{F}), R) \qquad \nu = 0, .., 2d(X) .$$

Theorem A.1 (Poincare Duality in the Category $M(X, R)$). *Under the assumptions made above the duality map is an isomorphism.*

Proof. The proof is based on lemma 1.15 from [FK], chap. II. Two things need to be verified in order to carry over the argument of this lemma 1.15:

(1) Let $j : Y \to X$ be an etale map, then the composition

$$H^0(Y, R_Y(d(Y))) = \!=\!= Hom_R(j_! R_Y, R_X(d(X)))$$
$$\swarrow$$
$$Hom_R(H_c^{2d(X)}(X, j_! R_Y), R) = \!= Hom_R(H_c^{2d(Y)}(Y, R_Y), R)$$

is an isomorphism.
It is enough to verify this in case of an irreducible Y. But then (1) is a consequence of the fact, that under this hypothesis the trace is an isomorphism.

(2) Let U be a smooth scheme of pure dimension d and let a be a geometric point of U. Then there exists an etale neighborhood

$$j : V \to U$$

of a, such that the natural map

$$H^\nu(U, j_! R_V) = H_c^\nu(V, R_V) \longrightarrow H_c^\nu(U, R_U)$$

is the zero map for $\nu \neq 2d$. (Chap. II, 1.14)

The verification of (2) is easily reduced to the corollary of lemma 1.14 ([FK], chap. I). Use that the ring R is, as Λ-module, a direct sum of copies of modules of the form $\mathbb{Z}/l^m\mathbb{Z}$. So one simply applies Lemma 1.14 to the various rings $\mathbb{Z}/l^m\mathbb{Z}$. □

This being said, let us discuss the Grothendieck trace formula ([FK], chap. II, 3.14 and 3.15). Let k be the algebraic closure of the finite field κ, X_0 be a finitely generated scheme over κ and $\mathcal{G}_0$ be an etale sheaf on X_0. Consider the pullback $\mathcal{G}$ of $\mathcal{G}_0$ to $X = X_0 \otimes k$. Let $|X_0|$ be the set of closed points of X_0. For $x \in |X_0|$ let $\overline{x}$ be a geometric point of X_0 over x with values in the field k; $\overline{x}$ defines a geometric point of X with values in k, also denoted $\overline{x}$. Obviously $\mathcal{G}_{0\overline{x}} = \mathcal{G}_{\overline{x}}$. We then have the Frobenius morphism

$$Fr_{X_0} : X_0 \to X_0$$

for X_0, and also for the sheaf $\mathcal{G}_0$ the (geometric) Frobenius homomorphism

$$Fr_{\mathcal{G}_0} : Fr_{X_0}^*(\mathcal{G}_0) \to \mathcal{G}_0 .$$

See [FK], chap. I, 3.8. Base field extension yields the Frobenius morphisms

$$Fr_X : X \to X$$

$$Fr_{\mathcal{G}} : Fr_X^*(\mathcal{G}) \to \mathcal{G} .$$

For every fixed point $\overline{x}$ in the fixed point set $Fix(Fr_X) = X_0(\kappa) \subset X_0(k)$ of Fr_X, one obtains a homomorphism of the stalks $\mathcal{G}_{\overline{x}}$

$$F_x : \mathcal{G}_{\overline{x}} \to \mathcal{G}_{\overline{x}} .$$

If $\mathcal{G}_0$ is a sheaf of R-modules, then $Fr_{\mathcal{G}}$ is a homomorphism in the category $M(X, R)$ and F_x is a homomorphism of R-modules. For a sheaf $\mathcal{G}_0$ of R-modules the pair $(Fr_X, Fr_{\mathcal{G}})$ induces a morphism in the derived category of $M(X, R)$

$$F : \mathrm{R}\Gamma_c(X, \mathcal{G}) \to \mathrm{R}\Gamma_c(X, \mathcal{G}) .$$

Next let R be a free Λ-module. Then projective R-module are again projective Λ-modules; a perfect complex of R-modules is therefore also a perfect complex of Λ-modules. For a definition see [FK], chap. II, definition 3.1. For any endomorphism h of a perfect complex in the derived category of the category of R-modules the trace

$$Tr_R(h) \in R$$

is well defined ([FK], chap. II, 3.4). Let furthermore λ be an element of the ring R. Then the pair $(Fr_{X_0}, \lambda Fr_{\mathcal{G}_0})$ is universal, i.e. defined for all finitely generated schemes X_0 over κ and all sheaves $\mathcal{G}_0 \in M(X_0, R)$, and satisfies the obvious compatibility relations. All fixed points of the Frobenius morphism Fr_X have multiplicity 1. These two properties of the pair $(Fr_{X_0}, Fr_{\mathcal{G}_0})$ were used in the proof of [FK], theorem 3.10 resp. 3.14 following Grothendieck and Nielsen-Wecken. So the proof carries over for the new ring Λ, of course using now the trace Tr_Λ with values in Λ.

Theorem A.2 *Let R be a free Λ-module and λ be an element of R. Let X_0 be a finitely generated scheme over κ and let $\mathcal{G}_0$ be a constructible sheaf of R-modules, all whose stalks are projective R-modules. Then $R\Gamma_c(X, \mathcal{G})$ is represented by a perfect complex of R-modules. For the homomorphism*

$$\lambda F : R\Gamma_c(X, \mathcal{G}) \to R\Gamma_c(X, \mathcal{G})$$

induced by the pair $(Fr_{X_0}, \lambda Fr_{\mathcal{G}_0})$ we have

$$Tr_\Lambda(\lambda F) = \sum_{\overline{x} \in Fix(Fr_X)} Tr_\Lambda(\lambda F_x) .$$

A corresponding formula holds for powers of Fr_{X_0} resp. $Fr_{\mathcal{G}_0}$ ([FK], chap. II, 3.15).

Now consider a finite extension field E of the field of l-adic numbers $\mathbb{Q}_l$. Let $\mathfrak{o}$ be the valuation ring of E and let π be a generator of the maximal ideal of $\mathfrak{o}$.

We want to define the category of π-adic sheaves resp. the category of E-sheaves (see [FK], chap. I, §12). For all $j \geq 1$ the ring $\mathfrak{o}/\pi^j\mathfrak{o}$ is an injective module over itself. The ring $\mathfrak{o}/l^n\mathfrak{o}$ furthermore is a free $\mathbb{Z}/l^n\mathbb{Z}$-module. A constructible π-adic sheaf $\mathscr{G}$ on a scheme X, or in other terminology a constructible $\mathfrak{o}$-sheaf on X, is a projective system

$$\mathscr{G} = (\mathscr{G}_i)_{i\geq 1} .$$

There all $\mathscr{G}_i$ are assumed to be constructible sheaves of $R_i = \mathfrak{o}/\pi^i\mathfrak{o}$-modules such that

$$\mathscr{G}_i = \mathscr{G}_{i+1} \otimes_{R_{i+1}} R_i .$$

The π-adic constructible sheaves define an abelian category in the obvious way. We write

$$\mathscr{G}_i = \mathscr{G} \otimes_{\mathfrak{o}} R_i \qquad \mathscr{G} = \lim_{\substack{\leftarrow \\ i}} \mathscr{G}_i .$$

Let $l\mathfrak{o} = \pi^e\mathfrak{o}$. Without loss of information we can consider the subsystem

$$\left(\mathscr{G}_{e\cdot i}\right)_{i\geq 1} .$$

Hence the category of constructible π-adic sheaves is equivalent to the category of projective systems

$$\mathscr{G} = \left(\mathscr{G}_i\right)_{i\geq 1} ,$$

where the $\mathscr{G}_i$ are constructible sheaves of $\mathfrak{o}/l^i\mathfrak{o}$ modules, such that

$$\mathscr{G}_i = \mathscr{G}_{i+1}/l^i\mathscr{G}_{i+1} .$$

We again write: $\mathscr{G}_i = \mathscr{G}/l^i\mathscr{G}$ and $\mathscr{G} = lim \mathscr{G}_i$. Hence, in that sense a π-adic sheaf is a special case of an l-adic sheaf.

Remark. The category of π-adic sheaves is a full exact subcategory of a larger abelian category, the Artin-Rees category. This Artin-Rees category is the quotient category of the abelian category of all projective systems $(\mathscr{G}_n)_{n\geq 1}$ of constructible π-torsion sheaves $\mathscr{G}_n$ with respect to the full exact subcategory of null systems $(\mathscr{N}_n)$. A system $\mathscr{N}_n$ is called a null system, if there exists a natural number $t \geq 1$ such that the maps

$$\mathscr{N}_{n+t} \to \mathscr{N}_n$$

are zero for all n. Many constructions do not directly yield π-adic sheaves, but only projective systems which become isomorphic to π-adic sheaves in the Artin-Rees quotient category. These are called A-R π-adic sheaves. It is therefore often convenient to replace the category of π-adic sheaves by the larger category of A-R π-adic sheaves.

See [FK], chap. I, §12 for the analogous l-adic (instead of π-adic) situation.

Without difficulties, all basic notions and properties of the category of l-adic sheaves carry over to the category of π-adic sheaves ([FK], chap. I, §12). A π-adic sheaf is said to be smooth, if all the sheaves $\mathscr{G} \otimes \Lambda_i$ are locally constant sheaves of Λ_i-modules. See [FK], chap. I, §12. Let $\mathscr{F}$ be a sheaf of constructible $\mathfrak{o}/\pi^m\mathfrak{o}$ modules. Then $\mathscr{F}$ defines a π-adic sheaf $\mathscr{G} = (\mathscr{G}_i)_{i\geq 1}$, namely a torsion sheaf given by

$$\mathscr{G}_i = \begin{cases} \mathscr{F}/\pi^i\mathscr{F} & i \le m \\ \mathscr{F} & i \ge m \end{cases}$$

The category $M(X, E)$ of constructible E-sheaves on X, or simply the category of E-sheaves on X, is defined to be the quotient category of the category of π-adic sheaves divided by the category of torsion sheaves. If $\mathscr{F}$ is a π-adic sheaf, then $\mathscr{F} \otimes_{\mathfrak{o}} E$ denotes the sheaf $\mathscr{F}$ viewed as an E-sheaf. The morphisms of this category are defined by

$$Hom(\mathscr{F} \otimes_{\mathfrak{o}} E, \mathscr{G} \otimes_{\mathfrak{o}} E) = Hom(\mathscr{F}, \mathscr{G}) \otimes_{\mathfrak{o}} E .$$

The composition of such morphisms is defined in the obvious way. A sheaf belonging to $M(X, E)$ can always be represented in the form $\mathscr{F} \otimes_{\mathfrak{o}} E$ with a torsionfree sheaf $\mathscr{F}$, i.e. with a π-adic sheaf $\mathscr{F}$ for which the stalks of the sheaves $\mathscr{F} \otimes_{\mathfrak{o}} \mathfrak{o}/\pi^i\mathfrak{o}$ are projective $\mathfrak{o}/\pi^i\mathfrak{o}$-modules.

One can show the following: The category of E-sheaves is equivalent to the category of $\mathbb{Q}_l$-sheaves endowed with an action of E on $\mathscr{F}$. This means, that there exists an embedding

$$E \hookrightarrow Hom(\mathscr{F}, \mathscr{F}) .$$

Either by the limiting procedure and then by passing to the quotient, or by this second point of view, one carries over all important notions to the category $M(X, E)$ and proves their essential properties. Especially one finds, that $M(X, E)$ is a noetherian abelian category. The inverse image functor and the direct image functor with compact support are defined and they coincide with the corresponding notions for $\mathbb{Q}_l$-sheaves. A sheaf in $M(X, E)$ is said to be smooth, if it can be represented in the form

$$\mathscr{F} \otimes_{\mathfrak{o}} E$$

for some smooth π-adic sheaf $\mathscr{F}$ (locally constant in the terminology of [FK], chap. I, §12). The sheaf of homomorphisms between two sheaves will be defined only, if the first sheaf is smooth: In that case

$$\mathscr{H}om(\mathscr{F} \otimes_{\mathfrak{o}} E, \mathscr{G} \otimes_{\mathfrak{o}} E) = \varprojlim \mathscr{H}om_{\mathfrak{o}/\pi^i\mathfrak{o}}(\mathscr{F}/\pi^i\mathscr{F}, \mathscr{G}/\pi^i\mathscr{G}) ,$$

for smooth torsionfree π-adic sheaves $\mathscr{F}$.

Theorem A.1 implies

Theorem A.3 (Poincare Duality) *Let X be a smooth scheme over some algebraically closed field k of pure dimension d and let $\mathscr{F}$ be a smooth E-sheaf on X. Then there are natural isomorphisms*

$$H^{\nu}(X, \mathscr{H}om(\mathscr{F}, E_X(d)) \xrightarrow{\cong} Hom_E(H_c^{2d-\nu}(X, \mathscr{F}), E) .$$

Let us fix notation. As usual let κ denote a finite field with q elements and let k denote its algebraic closure. Let $\mathscr{F}_0$ be an E-sheaf on X_0. Let $\mathscr{F}$ denote the inverse image of $\mathscr{F}_0$ on $X = X_0 \otimes_{\kappa} k$. Then again the Frobenius homomorphisms

$$Fr_{\mathscr{G}_0} : Fr^*_{X_0}(\mathscr{F}_0) \to \mathscr{F}_0$$

and

$$Fr_{\mathscr{G}} : Fr^*_X(\mathscr{F}) \to \mathscr{F}$$

in the category $M(X, E)$ are defined. From Theorem A.2 one deduces a trace formula for the Frobenius multiplied with a constant $\lambda \in E$.

Remark. Let V be a finite dimensional vector space over E and

$$h : V \to V$$

an E-linear map. Then V is also a finite dimensional $\mathbb{Q}_l$-vectorspace and h is also $\mathbb{Q}_l$-linear. This defines the traces $Tr_E(h) \in E$ and $Tr_{\mathbb{Q}_l}(h) \in \mathbb{Q}_l$ of h. They are related by

$$Tr_{\mathbb{Q}_l}(h) = Tr_{E/\mathbb{Q}_l}\big(Tr_E(h)\big) ,$$

where $Tr_{E/\mathbb{Q}_l}$ is the trace map of the field extension $E/\mathbb{Q}_l$. $Tr_{E/\mathbb{Q}_l}(\lambda a) = 0$ for all $\lambda \in E$ implies $a = 0$ for elements $a \in E$. This and Theorem A.2 implies

Theorem A.4 (Lefschetz Fixed Point Formula for Frobenius) *Let $\mathscr{F}_0$ be an E-sheaf on X_0. The homomorphism*

$$F_\nu : H^\nu_c(X, \mathscr{F}) \to H^\nu_c(X, \mathscr{F})$$

induced by the pair $(Fr_X, Fr_{\mathscr{G}})$ satisfies

$$\sum_{\nu=0}^{2dim(X)} (-1)^\nu Tr(F_\nu) = \sum_{x \in Fix(Fr_X)} Tr_E(F_x) .$$

Corresponding formulas also hold for the powers of Fr_X and $Fr_{\mathscr{G}}$.

Let now F be a finite extension field of E. Then there is the obvious functor, extension of coefficients

$$\begin{aligned} M(X, E) &\to M(X, F) \\ \mathscr{F} &\mapsto \mathscr{F} \otimes_E F . \end{aligned}$$

Let $\overline{\mathbb{Q}}_l$ be the algebraic closure of $\mathbb{Q}_l$. We then define the category $M(X, \overline{\mathbb{Q}}_l)$ of $\overline{\mathbb{Q}}_l$-sheaves on the scheme X as the direct limit

$$M(X, \overline{\mathbb{Q}}_l) = \varinjlim_E M(X, E) ,$$

where E ranges over all finite extension fields of $\mathbb{Q}_l$ in $\overline{\mathbb{Q}}_l$.

More concretely: An object in $M(X, \overline{\mathbb{Q}}_l)$ is an E-sheaf $\mathscr{F}$ over a sufficiently large subfield $E \subset \overline{\mathbb{Q}}_l$ of finite degree over $\mathbb{Q}_l$. We write $\mathscr{F} \otimes_E \overline{\mathbb{Q}}_l$ for $\mathscr{F}$ viewed as a $\overline{\mathbb{Q}}_l$-sheaf.

Let $\mathscr{G}$ be a second E'-sheaf on X for another finite extension field E' of $\mathbb{Q}_l$ in $\overline{\mathbb{Q}}_l$. Choose a finite extension field F of $\mathbb{Q}_l$ in $\overline{\mathbb{Q}}_l$ containing both E and E'. Then

$$Hom(\mathscr{F} \otimes_E \overline{\mathbb{Q}}_l, \mathscr{G} \otimes_{E'} \overline{\mathbb{Q}}_l) = Hom(\mathscr{F} \otimes_E F, \mathscr{G} \otimes_{E'} F) \otimes_F \overline{\mathbb{Q}}_l .$$

Without difficulties one can show, that all important notions and properties of the sheaf categories $M(X, E)$ for finite extension fields E of $\mathbb{Q}_l$ are inherited by the category $M(X, \overline{\mathbb{Q}}_l)$. The Poincare duality theorem (Theorem A.3) and the Lefschetz fixed point formula (Theorem A.4) for $\overline{\mathbb{Q}}_l$-sheaves should be especially mentioned.

A $\overline{\mathbb{Q}}_l$-sheaf is said to be smooth, if it is represented by a smooth E-sheaf with respect to some finite extension field E of $\mathbb{Q}_l$ in $\overline{\mathbb{Q}}_l$. Let X be a connected scheme, $\overline{x}$ some geometric point of X with values in some algebraically closed field. Let $\mathscr{F}$ be a smooth $\overline{\mathbb{Q}}_l$-sheaf. Then one can show (see [FK], appendix I.8) that $\pi_1(X, \overline{x})$ operates continuously on the finite dimensional $\overline{\mathbb{Q}}_l$-vectorspace $V = \mathscr{F}_{\overline{x}}$. This means:

There is a finite extension field $E \subset \overline{\mathbb{Q}}_l$ of $\mathbb{Q}_l$ and an E-subspace $W \subset V$, on which $\pi_1(X, \overline{x})$ operates continuously, such that

$$V = W \otimes_E \overline{\mathbb{Q}}_l .$$

The fiber functor $\mathscr{F} \to \mathscr{F}_{\overline{x}}$ defines an equivalence of categories between the category of smooth $\overline{\mathbb{Q}}_l$-sheaves on X and the category of continuous representations of the fundamental group $\pi_1(X, \overline{x})$ on finitely dimensional $\overline{\mathbb{Q}}_l$-vectorspaces.

The Categories $D_c^b(X, E)$ and $D_c^b(X, \overline{\mathbb{Q}}_l)$

For the finite extension field $E \subset \overline{\mathbb{Q}}_l$ of $\mathbb{Q}_l$, let $\mathfrak{o}$ be the valuation ring of E and π be a generating element of the maximal ideal of $\mathfrak{o}$.

In Chap. II §5 and §6 the triangulated category $D_c^b(X, \mathfrak{o})$ was defined together with its standard t-structure. In the following we explain the "localized" categories $D_c^b(X, E)$ and $D_c^b(X, \overline{\mathbb{Q}}_l)$. Also on these categories we have standard t-structures induced from the t-structures on $D_c^b(X, \mathfrak{o})$.

The objects are defined to be the same as for the category $D_c^b(X, \mathfrak{o})$. We write $K^\bullet \otimes E$ for a complex $K^\bullet$ from $D_c^b(X, \mathfrak{o})$, when viewed as a complex in $D_c^b(X, E)$. Furthermore

$$Hom(F^\bullet \otimes E, K^\bullet \otimes E) = Hom(F^\bullet, K^\bullet) \otimes_{\mathfrak{o}} E .$$

Admissible triangles in $D_c^b(X, E)$ are triangles, which are isomorphic in $D_c^b(X, E)$ to admissible triangles in $D_c^b(X, \mathfrak{o})$.

Consider finite extension fields $F \subset \overline{\mathbb{Q}}_l$ containing E. Let $\tilde{\mathfrak{o}}$ denote the valuation ring of F and let $\tilde{\pi}$ be a generator of the maximal ideal. In case of ramification

$$\pi \tilde{\mathfrak{o}} = \tilde{\pi}^e \tilde{\mathfrak{o}}$$

let e be the ramification number. We construct natural functors

$$D_c^b(X, E) \to D_c^b(X, F)$$

in the following way: Since $\tilde{\mathfrak{o}}$ is a free $\mathfrak{o}$-module of rank $[F:E]$,

$$\tilde{\mathfrak{o}}_{re} = \tilde{\mathfrak{o}}/\tilde{\pi}^{re}\mathfrak{o} = \tilde{\mathfrak{o}}/\pi^r\tilde{\mathfrak{o}}$$

is free over $\mathfrak{o}_r = \mathfrak{o}/\pi^r\mathfrak{o}$ for all $r \geq 1$. Consider first the functors

$$D^b_{ctf}(X, \mathfrak{o}_r) \to D^b_{ctf}(X, \tilde{\mathfrak{o}}_{re})$$

$$K^\bullet \mapsto K^\bullet \otimes_{\mathfrak{o}_r} \tilde{\mathfrak{o}}_{re} = K^\bullet \otimes^L_{\mathfrak{o}_r} \tilde{\mathfrak{o}}_{re} .$$

The family of these functors for $r = 1, 2, \ldots$ naturally defines a functor

$$\text{“}\varprojlim_r\text{”} D^b_{ctf}(X, \mathfrak{o}_r) \to \text{“}\varprojlim_r\text{”} D^b_{ctf}(X, \tilde{\mathfrak{o}}_{re}) = \text{“}\varprojlim_{r'}\text{”} D^b_{ctf}(X, \tilde{\mathfrak{o}}_{r'}) ,$$

hence by definition a functor

$$D^b_c(X, \mathfrak{o}) \to D^b_c(X, \tilde{\mathfrak{o}}) .$$

By localization, as above, we get from this the desired functor

$$D^b_c(X, E) \to D^b_c(X, F) .$$

Finally the category $D^b_c(X, \overline{\mathbb{Q}}_l)$ is defined as the direct limit

$$D^b_c(X, \overline{\mathbb{Q}}_l) = \text{“}\varinjlim_r\text{”} D^b_{ctf}(X, E)$$

(in the obvious way) of the categories $D^b_c(X, E)$, where $E \subset \overline{\mathbb{Q}}_l$ ranges over all finite extension fields of $\mathbb{Q}_l$. For all such fields E one has natural functors

$$D^b_c(X, E) \to D^b_c(X, \overline{\mathbb{Q}}_l)$$

$$K^\bullet \mapsto K^\bullet \otimes_E \overline{\mathbb{Q}}_l$$

and

$$Hom(F^\bullet \otimes_E \overline{\mathbb{Q}}_l, K^\bullet \otimes_E \overline{\mathbb{Q}}_l) = Hom(F^\bullet, K^\bullet) \otimes_E \overline{\mathbb{Q}}_l .$$

We skip the obvious definitions for the usual derived functors related to the derived category $D^b_c(X, \overline{\mathbb{Q}}_l)$. The results for $D^b_c(X, \mathfrak{o})$ immediately carry over to the categories $D^b_c(X, E)$ and $D^b_c(X, \overline{\mathbb{Q}}_l)$. From the standard t-structure on $D^b_c(X, \mathfrak{o})$, defined in Chap. II §6, we immediately get t-structures on the categories $D^b_c(X, E)$ and $D^b_c(X, \overline{\mathbb{Q}}_l)$.

The Derived Functors $\mathbf{R}f_!$, $\mathbf{R}f_*$, $\mathbf{R}\mathcal{H}om$

In this last section we finally give a short review on some consequences of the finiteness properties of the functor $\mathrm{R}f_!$ respectively of the functor $\mathrm{R}f_*$ (see [SGA4$\frac{1}{2}$], Th. finitude and also Appendix D of this book):

Let $f : X \to S$ be a finitely generated morphism between finitely generated schemes over the base field, the latter being fixed satisfying our basic assumptions. Consider complexes

$$K^\bullet = (K_r)_{r \geq 1} \in D_c^b(X, \mathfrak{o}) .$$

The finiteness theorem for proper direct images implies for the corresponding projective system of perfect complexes

$$K_r^\bullet \in D_{ctf}^b(X, \mathfrak{o}_r) \qquad r = 1, 2, \ldots ,$$

that also the complexes satisfy

$$Rf_! K_r^\bullet \in D_{ctf}(Y, \mathfrak{o}_r) \quad , \quad Rf_! K_{r+1}^\bullet \otimes_{\mathfrak{o}_{r+1}}^L \mathfrak{o}_r = Rf_! K_r .$$

Hence the projective system $(Rf_! K_r^\bullet)_{r \geq 1}$ is in $D_c^b(Y, \mathfrak{o})$.

Definition. $Rf_! K^\bullet = (Rf_! K_r^\bullet)_{r \geq 1}$

Similarly one can immediately define the pullback $f^* L^\bullet$ of an object $L^\bullet \in D_c^b(Y, \mathfrak{o})$ and the derived tensor product

$$K^\bullet \otimes_{\mathfrak{o}}^L L^\bullet$$

for any two objects $K^\bullet, L^\bullet \in D_c^b(X, \mathfrak{o})$. This is very easy, and only uses the corresponding functors on the "finite level".

However, in order to make the corresponding definitions for the functors

$$Rf_*, f^!, R.\mathcal{H}om ,$$

one requires substantial finiteness theorems. The necessary finiteness theorems for these functors were proven by Deligne (the reference [finitude] is [SGA4$\frac{1}{2}$], p. 233f). For our purposes Deligne's theorem suffices in the case of schemes over a field.

Let $f : X \to Y$ be a morphism between finitely generated schemes over a base field k.

Theorem A.5 ([finitude], theorem 1.1) *Let $\mathcal{G}$ be a constructible sheaf of $\mathfrak{o}_r$-modules on X. Then the derived direct images $R^\nu f_* \mathcal{G}$ are constructible sheaves of $\mathfrak{o}_r$-modules on Y.*

Remark Concerning the Proof. A stronger version of this theorem is formulated and proven in Appendix D of this book. Deligne has used the method of proof for the finiteness theorem later again in the proof of the mixedness theorem for the higher direct image functors (see [Del] and I.9.4 and II.12.2 of this book). The reader will find the proof of Deligne's mixedness theorem in the last section of Chap. I of this book (I.9.4), and is advised to give the proof of the finiteness theorem along these lines by himself.

Corollary A.6 ([finitude], corollaire 1.6ff) *Suppose given* $K^\bullet, L^\bullet \in D^b_{ctf}(X, \mathfrak{o}_r)$ *and* $M^\bullet \in D^b_{ctf}(Y, \mathfrak{o}_r)$. *Then*

$$f^!(M^\bullet)\ ,\ R.\mathscr{H}om(K^\bullet, L^\bullet)\ ,\ Rf_*(M^\bullet)\ \in\ D^b_{ctf}(X, \mathfrak{o}_r)\ .$$

Furthermore

$$f^!(M^\bullet \otimes^L_{\mathfrak{o}_r} \mathfrak{o}_{r-1}) = f^!(M^\bullet) \otimes^L_{\mathfrak{o}_r} \mathfrak{o}_{r-1}$$

$$\mathrm{R.}\mathscr{H}om(K^\bullet \otimes^L_{\mathfrak{o}_r} \mathfrak{o}_{r-1}, L^\bullet \otimes^L_{\mathfrak{o}_r} \mathfrak{o}_{r-1})\ =\ \mathrm{R.}\mathscr{H}om(K^\bullet, L^\bullet) \otimes^L_{\mathfrak{o}_r} \mathfrak{o}_{r-1}$$

$$\mathrm{R}f_*(K^\bullet \otimes^L_{\mathfrak{o}_r} \mathfrak{o}_{r-1})\ =\ \mathrm{R}f_*(K^\bullet) \otimes^L_{\mathfrak{o}_r} \mathfrak{o}_{r-1}\ .$$

We can therefore define, as required

Definition. For $K^\bullet = (K_r), L^\bullet = (L_r) \in D^b_c(X, \mathfrak{o}), M^\bullet = (M_r) \in D^b_c(Y, \mathfrak{o})$ define

$$\mathrm{R}f_*(K^\bullet) = (\mathrm{R}f_* K_r)_{r\geq 1}$$

$$f^!(M^\bullet) = (f^!(M_r))_{r\geq 1}$$

$$R.\mathscr{H}om(K^\bullet, L^\bullet) = (\mathrm{R.}\mathscr{H}om(K^\bullet_r, L^\bullet_r))_{r\geq 1}$$

Properties of these new "derived" functors, resp. identities between them, follow immediately from the corresponding properties and identities on "finite level".

B. Bertini Theorem for Etale Sheaves

Let X be a normal irreducible, finitely generated quasiprojective scheme over an algebraically closed field k. X is a locally closed subscheme of the projective space $\mathbb{P}^N$

$$X \subset \mathbb{P}^N\ .$$

We assume: $dim(X) > 1$.

Consider the Grassmannian G of all linear subspaces in $\mathbb{P}^N$ of codimension $dim(X) - 1$. Then $M = \{(x, L) \in X \times G \mid x \in L\}$ is irreducible. Consider the projections

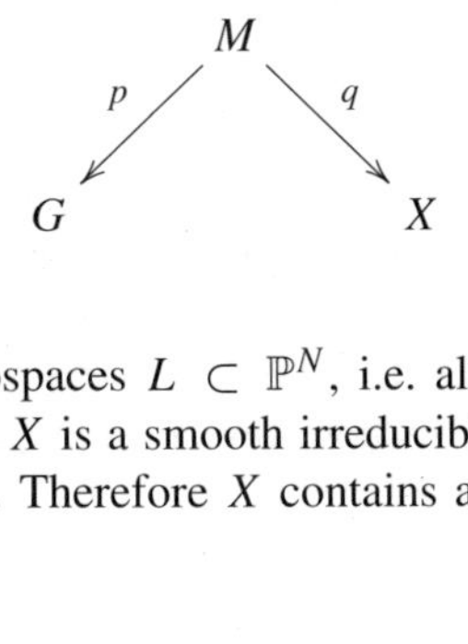

Bertini's theorem says:

For almost all linear subspaces $L \subset \mathbb{P}^N$, i.e. all L in an open dense subset $U \subset G$ the intersection $L \cap X$ is a smooth irreducible curve, which is nonempty. M dominates the scheme X. Therefore X contains an open dense subset V such that $V \subset q(p^{-1}(U))$. For every point $v \in V$ there is a linear subspace L which, considered as a point of the Grassmannian, belongs to U such that $v \in L$.

There is a corresponding statement for irreducible etale coverings of X.

Theorem B.1 *With the preceding assumptions and notations let $\mathscr{F}$ be a smooth irreducible $\overline{\mathbb{Q}}_l$-sheaf on X. Then for almost all linear subspaces $L \subset \mathbb{P}^N$ of codimension $dim(X) - 1$ the intersection $L \cap X$ is a smooth irreducible nonempty curve and the restriction $\mathscr{F}|L \cap X$ of $\mathscr{F}$ to this intersection is an irreducible smooth $\overline{\mathbb{Q}}_l$-sheaf. There is an open nonempty subset V of X, such that every point of V is contained in at least one linear space with the properties mentioned.*

Proof. Only the statement concerning irreducibility needs a proof. There is an extension field $E \subset \overline{\mathbb{Q}}_l$ of finite degree over $\mathbb{Q}_l$ and a smooth E-sheaf $\mathscr{G}$ such that

$$\mathscr{F} = \mathscr{G} \otimes_E \overline{\mathbb{Q}}_l \,.$$

Irreducibility for the E-sheaf $\mathscr{G}$ is weaker then irreducibility of the $\overline{\mathbb{Q}}_l$-sheaf $\mathscr{G} \otimes_E \overline{\mathbb{Q}}_l$.

In the language of representations:
If ρ, belonging to $\mathscr{G}$, is a representation of the fundamental group of X on some finite E-vectorspace, then irreducibility of $\mathscr{G} \otimes_E \overline{\mathbb{Q}}_l = \mathscr{F}$ means **absolute** irreducibility of the corresponding representation ρ.

We first proof a weakened version of Theorem B.1

Lemma B.2 *Let notations and assumptions be as in Theorem B.1. Suppose given a finite field extension $E \subset \overline{\mathbb{Q}}_l$ of $\mathbb{Q}_l$ and a smooth irreducible E-sheaf $\mathscr{G}$ on X. For almost all linear subspaces L of $\mathbb{P}^N$ of codimension $dim(X) - 1$ the sheaf $\mathscr{G}|L \cap X$ is irreducible.*

Proof of Lemma B.2. Let R be the valuation ring of the field E and π be a generating element of the maximal ideal of R. Put $\Lambda_i = R/\pi^{i+1}R$. Let n be the rank of $\mathscr{G}$, $n \neq 0$. The sheaf $\mathscr{G}$ is represented by a smooth π-adic sheaf $\mathscr{H} = (\mathscr{H}_i)_{i \geq 0}$:

$$\mathscr{G} = \mathscr{H} \otimes_R E \,.$$

The sheaves $\mathscr{H}_i = \mathscr{H} \otimes_R \Lambda_i$ are locally constant, locally free etale sheaves of Λ_i-modules of rank n.

The sheaf $\mathscr{H}$ corresponds to a continuous representation of the fundamental group $\pi_1(X, a)$ for a base point a of X in a finite free R-module V of rank n. The representations of $\pi_1(X, a)$ on the quotients $V/\pi^{i+1}V$ describe the sheaves $\mathscr{H}_i$.

The E-sheaf $\mathscr{G}$ is irreducible iff V contains no $\pi_1(X, a)$ module different from 0 and V, which is a direct summand in the sense of R modules, i.e. if $\mathscr{H}$ is irreducible. We then say, that V is an irreducible $\pi_1(X, a)$-module. Similarly $V/\pi^{i+1}V$ is called an irreducible $\pi_1(X, a)$-module, if $V/\pi^{i+1}V$ contains no $\pi_1(X, a)$-submodule different from 0 and $V/\pi^{i+1}V$, which is a direct summand in the sense of Λ_i-modules. We then say $\mathscr{H}_i$ is irreducible.

Lemma B.3 *V is an irreducible $\pi_1(X,a)$ iff there exists an index $i \geq 0$, such that the $\pi_1(X,a)$-module $V/\pi^{i+1}V$ is irreducible.*

Proof of Lemma B.3. Only one direction needs a proof. Let V be an irreducible $\pi_1(X,a)$-module and let M_i be the set of $\pi_1(X,a)$-submodules different from 0 and $V/\pi^{i+1}V$, which are direct summands as Λ_i modules. The sets M_i define a projective system $(M_i)_{i\geq 0}$ in a natural way. Let $M_i \neq \emptyset$ be nonempty for all $i = 0, 1, \ldots$ The sets M_i are finite and nonempty, therefore

$$\lim_{\overleftarrow{i}} M_i \neq \emptyset \,.$$

But this says, that there is a coherent system of $\pi_1(X,a)$-submodules $W_i \subset V/\pi^{i+1}V$ such that $W_i \in M_i$ and $W_j \otimes_{\Lambda_j} \Lambda_i = W_i$ for all $j \geq i \geq 0$. The limit $\lim W_j$ is a $\pi_1(X,a)$-submodule of V and a direct summand as R-module. Furthermore $W \neq 0, V$. This contradicts our assumption on the irreducibility of V and thereby irreducibility of $\mathscr{H}$.

We continue the proof of Lemma B.2. According to Lemma B.3 there is an index $i \geq 0$ such that $V/\pi^{i+1}V$ is an irreducible $\pi_1(X,a)$-module, i.e. such that the sheaf $\mathscr{H}_i$ is irreducible. This is equivalent to the following: There is no locally constant, locally free subsheaf $\mathscr{H}_i'$ of $\mathscr{H}_i$ different from 0 and $\mathscr{H}_i$. Λ_i being injective, all stalks of $\mathscr{H}_i'$ are direct summands of the corresponding stalks of $\mathscr{H}_i$. The sheaf $\mathscr{H}_i$ is represented by an etale covering

$$Y \longrightarrow X$$

of X. Let $Y_1, ..Y_r$ be the irreducible components of Y. Then any locally constant subsheaf of $\mathscr{H}_i$ is represented by a suitable union of components Y_i.

The following is easy to see:
Let $L \subset \mathbb{P}^N$ be a linear subspace of codimension $dim(X) - 1$ such that

1) $X \cap L$ is a nonempty smooth irreducible curve.
2) $(X \cap L) \times_X Y_i$ is irreducible for i=1,..r.

Under these assumptions also the restrictions $\mathscr{H}_i|X \cap L$ are irreducible E-sheaves. The ordinary Bertini theorem together with Lemma B.3 therefore implies Lemma B.2.

Let us return to the proof of Theorem B.1: We had

$$\mathscr{F} = \mathscr{G} \otimes_E \overline{\mathbb{Q}}_l \,.$$

To $\mathscr{G}$ there is attached a representation ρ of the fundamental group $\pi_1(X,a)$ on a vector space of dimension n over E. Let A denote the E-subalgebra of the endomorphism ring $End_E(W)$ generated by $\rho(\pi_1(X,a))$. The E-sheaf $\mathscr{G}$ is semisimple if and only if A is a semisimple algebra.

Lemma B.4 *There is a finite field extension $F \subset \overline{\mathbb{Q}}_l$ of E depending only on E and $n = dimV$, such that the following holds: Let $B \subset End_E(W)$ be any semisimple* *subalgebra of $End_E(W)$. Then the extension algebra $B \otimes_E F \subset End_F(W \otimes_E F)$ is a direct product of matrix rings over F. Especially: Any semisimple representation ρ' of a fundamental group of an algebraic scheme on W decomposes after tensoring with F into a direct sum of absolutely irreducible representations.*

Proof of Lemma B.4. In any case there is a natural number m depending only on E and n, such that every semisimple subalgebra B of $End_E(W)$ has a splitting field $\Omega_B \subset \overline{\mathbb{Q}}_l$ such that $[\Omega_B : E] \leq m$, i.e. $B \otimes_E \Omega_B$ decomposes into a product of matrix rings over Ω_B. But E has only finitely many isomorphism classes of field extensions of bounded degree (Krasner's lemma). This implies Lemma B.4.

Now we apply Lemma B.2 to the E-sheaf $\mathscr{G}$ and to the F-sheaf $\mathscr{G} \otimes_E F$. Here F is chosen to be a finite extension field of E with the properties stated in Lemma B.4. Lemma B.4 then implies Theorem B.1. □

C. Kummer Extensions

Let X_0 be a smooth geometrically irreducible projective curve over κ and let $F(X)$ be the field of meromorphic functions on $X = X_0 \otimes_\kappa k$. Let l be a prime different from the characteristic of κ. Let $\mathscr{D}(X)$ be the group of divisors of degree 0 on X, let $\mathscr{D}(X_0)$ be the subgroup of divisors of degree 0 on X_0 and $\mathscr{H}(X) \subseteq \mathscr{D}(X)$ respectively $\mathscr{H}(X_0) \subseteq \mathscr{D}(X_0)$ be the subgroups of principal divisors on X respectively X_0. Let $A \subseteq \mathscr{D}(X)/\mathscr{H}(X)$ be the l-power torsion subgroup of the group of divisor classes. Let K be the maximal abelian unramified pro-l extension of $F(X)$ and let $G = Gal(K/F(X))$ be the Galois group of K over $F(X)$. The Galois group $Gal(k/\kappa)$ and especially the arithmetic Frobenius automorphism $f \in Gal(k/\kappa)$ operate on A and G. Let us consider the continuous Kummer pairing

$$\begin{array}{ccc} A \times G & \longrightarrow & k^* \\ (x\ ,\ \delta) & \longrightarrow & < x, \delta > \end{array}$$

which is defined as follows: Suppose the divisor D represents the element $x \in A$. Let $n = l^r$ be the order of x. Then $nD = (g)$ is the divisor of a meromorphic function $g \in F(X)^*$. $\sqrt[n]{g}$ is contained in K. For $\delta \in G$

$$\delta(\sqrt[n]{g})/(\sqrt[n]{g}) = \rho \quad ; \quad \rho \in k^* .$$

Then the value $< x, \delta >$ is defined to be ρ.

From Galois theory we draw the following statements: The assignments

$$\begin{array}{ccc} A \supseteq B & \longmapsto & B^\perp = \{\delta \in G \mid < B, \delta >= 1\} \subseteq G \\ G \supseteq H & \longmapsto & H^\perp = \{x \in A \mid < x, H >= 1\} \subseteq A \end{array}$$

define a 1-1 correspondence between the subgroups of A and the closed subgroups of G. For finite subgroups B of A the group $G/B^\perp$ is finite with character group isomorphic to B.

Furthermore we have

$$< f(x), f(\delta) > = f(< x, \delta >) = < x, \delta >^q .$$

Therefore $f(x) = q \cdot x$ iff $< f(x), f(g)/g >= 1$ holds for all $g \in G$. Now put

$$H = \{\delta \in G | \delta = f(g)/g \text{ for some element } g \in G\} .$$

The Frobenius automorphism f acts on H and $H^\perp$. Fix some base point a. Then the group G/H is the l-primary part of the image of $\pi_1(X, a)$ in the factor commutator group of the Weil group $W(X_0, a)$, understood in the sense of locally compact groups. H is a closed subgroup and $x \in H^\perp$ iff $f(x) \in H^\perp$. Therefore $H^\perp$ is the group of all eigenclasses $x \in A$ with $f(x) = q \cdot x$.

Theorem C.1 *The factor group G/H is finite.*

Proof. We use that the Kummer pairing, defined above, is nondegenerate. Hence it is enough to show, that the group $H^\perp$ is finite.

Consider

$$A_n = Ker(A \xrightarrow{l^n} A) .$$

Then we have a canonical isomorphism

$$A_n = H^1(X, \mu_{l^n X}) ,$$

such that again

$$f : H^1(X, \mu_{l^n X}) \to H^1(X, \mu_{l^n X})$$

corresponds to the arithmetic Frobenius homomorphism. Now the orthogonal complement of $A_n \cap H^\perp \subset H^1(X, \mu_{l^n X})$ with respect to Poincare duality pairing

$$H^1(X, \mu_{l^n X}) \otimes H^1(X, \mu_{l^n X}) \to \mu_{l^n X}$$

is given by the group

$$\tilde{A}_n = \{x \in A_n \mid x = y - f(y) \text{ for elements } y \in A_n\} .$$

Counting elements of $\#(A_n \cap H^\perp) = \#(A_n/\tilde{A}_n)$ gives the estimate

$$\#(A_n \cap H^\perp) = \#coker(1 - f : A_n \to A_n) = \#ker(1 - f : A_n \to A_n)$$
$$\leq \#(\mathscr{D}(X_0)/\mathscr{H}(X_0)) ,$$

where we used Hilbert's Theorem 90 in the form

$$\{x \in \mathscr{D}(X)/\mathscr{H}(X) \mid f(x) = x\} = \mathscr{D}(X_0)/\mathscr{H}(X_0) .$$

So we get the estimate

$$\#(H^\perp) \leq \#(\mathscr{D}(X_0)/\mathscr{H}(X_0)) .$$

The class number $\#(\mathscr{D}(X_0)/\mathscr{H}(X_0))$ of divisor classes of degree 0 on X_0 is finite and bounded independently from n. Therefore the claim of the theorem follows. □

D. Finiteness Theorems

The finiteness theorems for the direct image functor $f_!$ with proper support [FK] can be generalized to obtain finiteness theorems for the direct image functor f_* in the case of non-proper maps f. In this section we give a brief report of these results. Our intention is to outline the ideas of the proofs. The original exposition of these theorems can be found in [SGA4$\frac{1}{2}$], expose: Théorems de finitude en cohomology l-adique, briefly [finitude].

Let S be a noetherian regular base scheme with dimension ≤ 1. In the following we consider without further mentioning only schemes and morphisms of the category of the finitely generated schemes over such a base scheme S. For the purposes of this book, it would in fact suffice to assume that S has dimension zero, or even more simple to assume that $S = Spec(k)$ is the spectrum of a field k.

Let Λ be a commutative finite ring, annihilated by an integer invertible on S. In the following it will be enough to assume, that Λ is a factor ring of a discrete valuation ring with a finite residue class field, which is self-injective. Then sheaves are considered to be etale sheaves of Λ-modules. $D(X,\Lambda)$, $D^+(X,\Lambda)$, $D^-(X,\Lambda)$, $D^b(X,\Lambda)$ denote the derived categories of the sheaves on a scheme X resp. the triangulated subcategories of complexes, which are bounded from below, resp. bounded from above, resp. bounded. The index c indicates, that we consider the corresponding triangulated subcategory of the complexes, whose cohomology sheaves are constructible sheaves. Such complexes will be briefly called constructible complexes. The category $D_{tf}(X,\Lambda) \subseteq D^b(X,\Lambda)$ is defined to be the triangulated subcategory of complexes, whose components have finite Λ-Tor dimension, and finally $D_{ctf}(X,\Lambda) = D_{tf}(X,\Lambda) \cap D_c(X,\Lambda)$ is the triangulated category of the constructible complexes with finite Tor dimension. In fact, a complex belongs to $D_{ctf}(X,\Lambda)$ iff it is isomorphic in $D(X,\Lambda)$ to a bounded complex with Λ-flat constructible component sheaves (see II.5.1).

In [SGA4], expose X the following theorem is proved

Theorem D.1 *Let $f : X \to Y$ be a morphism of finitely generated schemes over S. Then the functor Rf_* has finite cohomological dimension, i.e. we have $R^\nu f_* = 0$ for $\nu \geq \nu_0$.*

Proof. (in the case $\dim(S) = 0$): In this case the Theorem D.1 follows from the well known bounds of the cohomological dimension of a finitely generated affine scheme X over a separably closed base field k, e.g. from

$$H^\nu(X,\mathcal{G}) = 0 \qquad \nu > \dim X,\ \mathcal{G} \text{ a sheaf on } X$$

(see [FK], chap. I, theorem 9.1) or from the more elementary statement

$$H^\nu(X,\mathcal{G}) = 0 \qquad \nu > 2\dim X,\ \mathcal{G} \text{ a sheaf on } X\,.$$

(See also [SGA4], expose X, corollaire 4.2 together with theorem 2.1 and proposition 3.2 of loc. cit.) □

Remark D.2 From the finiteness of the cohomological dimension it follows for $f : X \to Y$ and a complex $F^\bullet \in D_c^-(Y, \Lambda)$ of locally constant constructible sheaves and a complex $G^\bullet \in D(X, \Lambda)$, that

$$F^\bullet \otimes^L_\Lambda \mathrm{R}f_*(G^\bullet) = \mathrm{R}f_*(f^*(F^\bullet) \otimes^L_\Lambda G^\bullet) .$$

See [FK], chap. I, proposition 8.14, remark 2. From this one concludes, that $\mathrm{R}f_*$ preserves complexes of finite Tor dimension

$$\mathrm{R}f_* : D_{tf}(X, \Lambda) \to D_{tf}(X, \Lambda) .$$

Theorem D.3 (see [finitude], theorem 1.1) *Let $f : X \to Y$ be a morphism between finitely generated schemes over the base scheme S and $\mathscr{F}$ a constructible sheaf on X. Then the derived direct images*

$$R^\nu f_* \mathscr{F} \qquad \nu = 0, 1, 2, \ldots$$

are constructible sheaves on Y.

Proof. Without loss of generality we may assume that the coefficient ring Λ is a field, e.g. $\Lambda = \mathbb{A}/l\mathbb{Z}$. Then a constructible sheaf is locally free if and only if it is locally constant. In case $\dim(S) = 0$ the proof follows the same pattern as the proof of the mixedness theorem from Chap. I, Theorem I.1.1. Here it can be assumed without restriction of generality, that S is the spectrum of a field. The reader should be aware, that this argument reduces the statement of the theorem to the corresponding finiteness statement for the functors $\mathrm{R}^\nu f_!$ (see [FK], theorem I.8.10) and that one uses a special case of the relative duality theorem for smooth quasi-projective morphisms (see chapter II of this book, in particular II.7.7 and II.7.1 in the section on duality for smooth morphisms.)

If $\dim(S) > 0$ this argument only shows, that the direct image sheaves are constructible over the pullback in Y of an affine dense subscheme of S. So it remains to show the finiteness theorem under the assumption that

$$S = Spec(\mathfrak{o})$$

is the spectrum of a discrete valuation ring $\mathfrak{o}$ in the special fiber. We may for this assume, that the finiteness theorem has been shown, if the base scheme is the spectrum of a field. To complete the proof, we need two additional facts:

A. Let s be the closed point of $S = Spec(\mathfrak{o})$, $\mathscr{F}$ a constructible sheaf on $U = S \setminus \{s\}$ and $j : U \hookrightarrow S$ the open complement. Using Galois cohomology one proves, that the sheaves $\mathrm{R}^\nu j_* \mathscr{F}$ are constructible. The generic point $U = S \setminus \{s\} = \{\eta\}$ is the spectrum of a field. We now assume that the finiteness theorem holds in the case of a base field instead of the base ring $\mathfrak{o}$. Therefore the following holds: Suppose $f : X \to S$ factorizes over η

$$f : X \xrightarrow{g} \eta \overset{j}{\hookrightarrow} S .$$

Let $\mathcal{G}$ be a constructible sheaf or more generally a constructible bounded complex on X. Then the complex $\mathrm{R}f_*(\mathcal{G}) = \mathrm{R}j_*\mathrm{R}g_*(\mathcal{G})$ is constructible.

B. We need a variant of Lemma I.9.7, which is formulated as the next lemma below. We first make some definition. Let $f : Y \to S$ be a finitely generated morphism. A sheaf $\mathcal{F}$ on X, which need not be constructible, is called a **special skyscraper sheaf** (with respect to f), if it is concentrated on the set of closed points of the special fiber $X_s = f^{-1}(s)$. This means: Every constructible subsheaf of $\mathcal{F}$ is concentrated on finitely many points of the special fiber. Such a sheaf $\mathcal{F}$ is acyclic with respect to the functor f_*. Furthermore the following holds: $\mathcal{F}$ is constructible iff $f_*\mathcal{F}$ is constructible.

From this we obtain: Let $K^\bullet$ be a bounded complex on X. Suppose that every cohomology sheaf $\mathcal{H}^\nu(K^\bullet)$ contains a constructible subsheaf, such that the factor sheaf of $\mathcal{H}^\nu(K^\bullet)$ with respect to that subsheaf is a special skyscraper sheaf. Let f be projective and let $\mathrm{R}f_*K^\bullet$ be a complex with constructible cohomology sheaves. Then the sheaves $\mathcal{H}^\nu(K^\bullet)$ are constructible.

Consider commutative diagrams (*) of quasiprojective schemes over S:

$$\begin{array}{ccc} X & \stackrel{j}{\hookrightarrow} & Y \\ & \searrow & \downarrow \\ & & S\,. \end{array}$$

Here j denotes an open embedding.

Let η be the general and s the special point of S.

Lemma D.4 (see also I.9.7) *Fix a diagram of type (*) and a constructible sheaf $\mathcal{G}$ on X. For every diagram of type (*)*

$$\begin{array}{ccc} X' & \stackrel{j'}{\hookrightarrow} & Y' \\ & \searrow & \downarrow \\ & & S' \end{array}$$

with S' the spectrum of an arbitrary discrete valuation ring $\mathfrak{o}'$ with general point η' and with

$$\dim X'_{\eta'} < \dim X_\eta \quad \textit{(generic fibers)}$$

we suppose, that the finiteness theorem for the open embedding j' is already proven. Then each of the higher direct images

$$\mathcal{G}_\nu = R^\nu j_*\mathcal{G} \qquad (\nu = 0, 1, \dots)$$

contains a constructible subsheaf

$$\mathcal{F}_\nu \subseteq \mathcal{G}_\nu\,,$$

such that $\mathcal{G}_\nu/\mathcal{F}_\nu$ is a special skyscraper sheaf.

Proof. The assertion is local with respect to Y. We may suppose, that Y is affine and embedable as a closed subscheme into some affine space $\mathbb{A}^m$ over S.

$$\begin{array}{ccccc} X & \stackrel{j}{\hookrightarrow} & Y & \stackrel{j}{\hookrightarrow} & \mathbb{A}^m \\ & & & \searrow & \downarrow \\ & & & & S\,. \end{array}$$

Consider one of the m projections

$$q : \mathbb{A}^m \to \mathbb{A}$$

onto the affine line $\mathbb{A}$ over S. Let $\mathfrak{o}'$ be the local ring of $\mathbb{A}$ in the generic point of the special fiber of $\mathbb{A}$ over S. $\mathfrak{o}'$ is a discrete valuation ring. Put $S' = Spec(\mathfrak{o}')$ and consider

$$\begin{array}{ccc} X \times_{\mathbb{A}} S' & \stackrel{j'}{\hookrightarrow} & Y \times_{\mathbb{A}} S' \\ & \searrow & \downarrow \\ & & S'\,. \end{array}$$

By assumption the finiteness theorem is true for the open embedding j', since the dimension condition holds. Similar for the open embedding of the generic fibers

$$\begin{array}{ccc} X_\eta & \stackrel{j_\eta}{\hookrightarrow} & Y_\eta \\ & \searrow & \downarrow \\ & & \eta\,, \end{array}$$

η being the spectrum of a field. Since each $\mathscr{G}_\nu = \mathrm{R}^\nu j_* \mathscr{G}$ is a direct limes of its constructible subsheaves, one can find for fixed ν a constructible subsheaf $\mathscr{F} \subseteq \mathscr{G}_\nu$ such that

$$(\mathscr{G}_\nu/\mathscr{F}) \times_S \eta = 0$$

$$(\mathscr{G}_\nu/\mathscr{F}) \times_{\mathbb{A}} S' = 0$$

holds for all projections $\mathbb{A}^m \to \mathbb{A}$. But then the sheaf $\mathscr{G}_\nu/\mathscr{F}$ is a special skyscraper sheaf. □

Having stated the facts A) and B) we suppose, that a diagram

$$\begin{array}{ccc} X & \stackrel{j}{\hookrightarrow} & Y \\ & \searrow & \downarrow f \\ & & S \end{array}$$

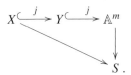

is given, where f is a projective morphism. Let $\mathscr{F}$ be a constructible sheaf on X. Using the facts A) and B) and induction with respect to the dimension of the generic fiber X_η of X, one now proves that the higher direct images $\mathrm{R}^\nu j_*\mathscr{F}$ are constructible sheaves. In the case where the base is a field the proof of the mixedness theorem I.9.4 carries over "word by word" to give the finiteness theorem. In the case where the base is $S = Spec(\mathfrak{o})$ one reduces to the field case. But the argument is now only similar to the argument of the mixedness theorem. We therefore sketch the argument:

We can assume that the finiteness theorem holds in the case of a base field. Therefore the assertion holds, if $\mathscr{G}$ is concentrated on the special fiber X_s or – more generally – the corresponding assertion holds for a complex $N \in D_c^b(X, \Lambda)$, which is concentrated on X_s; by this we mean that all its cohomology sheaves $\mathscr{H}^\nu(N)$ are concentrated on X_s.

Case 1. Suppose that $f \circ j : X \to S$ factorizes over the generic point η of S:

$$f \circ j : X \to \eta \hookrightarrow S .$$

Put $L = \mathrm{R}j_*(\mathscr{F}) \in D^b(X, \Lambda)$. According to A) the direct image complex $\mathrm{R}(f \circ j)_*(\mathscr{F})$ of $\mathscr{F}$ on S is constructible. Hence $\mathrm{R}f_*(L) = \mathrm{R}(f \circ j)_*(\mathscr{F}) \in D_c^b(S, \Lambda)$. By the induction assumption the conditions of Lemma D.4 hold for j. Therefore each cohomology sheaf $\mathscr{H}^\nu(L) = \mathrm{R}^\nu j_*\mathscr{F}$ of the complex L contains a constructible subsheaf $\mathscr{G}_\nu$, such that the factor sheaf $\mathscr{H}^\nu(L)/\mathscr{G}_\nu$ is a special skyscraper sheaf. Moreover the complex $\mathrm{R}f_*(L)$ is constructible and the morphism f is projective. According to B) we conclude that L is constructible.

Case 2. (The general case). The generic fiber X_η is an open subscheme of X. Its complement $X \setminus X_\eta = X_s$ is the special fiber

$$X_\eta \xhookrightarrow{r} X \hookleftarrow X_s .$$

One has a distinguished triangle

$$\mathscr{F} \to \mathrm{R}r_*r^*\mathscr{F} \to N \to \mathscr{F}[1] ,$$

whose third complex N is concentrated on the special fiber X_s. The morphism $X_\eta \to S$ factorizes over the generic point η of S. According to case 1 the complex $\mathrm{R}j_*(\mathrm{R}r_*r^*\mathscr{F}) = \mathrm{R}(j \circ r)_*r^*\mathscr{F}$ and therefore also the complex $\mathrm{R}r_*r^*\mathscr{F} = \mathrm{R}j_*(\mathrm{R}r_*r^*\mathscr{F})|X$ is constructible. Then also the third complex N of the triangle is constructible. Hence the direct image complexes $\mathrm{R}j_*(\mathrm{R}r_*r^*\mathscr{F})$ and $\mathrm{R}j_*N$ are constructible. We conclude, that the direct image $\mathrm{R}j_*\mathscr{F}$ of the first complex of the distinguished triangle is constructible.

This special case of the finiteness theorem for open embeddings $j : X \to Y$, for which $Y \to S$ is proper, and the fundamental Finiteness Theorem for projective morphisms $X \to Y$, (see [FK]), now easily imply the more general finiteness Theorem D.3 over the base $S = Spec(\mathfrak{o})$. For this factorize a given morphism $X \to Y$ into an open embedding, a closed embedding, and a projective map over S. □

From Theorem D.3 respectively Theorem D.1 similar finiteness results can be deduced for the other relevant functors.

Let $i : Y \hookrightarrow X$ be a closed embedding and let $j : U = X/Y \hookrightarrow X$ its open complement. For a sheaf $\mathscr{G}$ on X consider the cohomology sequence of cohomology sheaves with supports in Y

$$0 \to \mathscr{H}_Y^0(X, \mathscr{G}) \to \mathscr{G} \to R^0 j_* j^*(\mathscr{G}) \to \mathscr{H}_Y^1(X, \mathscr{G}) \to 0 \to R^1 j_* j^* \mathscr{G} \to \dots$$

Our results on $\mathrm{R}j_*$ imply, that the functor $\mathscr{H}_Y^0$ has finite cohomological dimension. Furthermore for constructible sheaves $\mathscr{G}$ on X the cohomology sheaves with supports in Y

$$\mathscr{H}_Y^\nu(X, \mathscr{G}) \qquad \nu = 0, 1, 2, \dots$$

are constructible as well. In the section of Chap. II on relative duality we construct, using the functor $\mathrm{R}.\mathscr{H}_Y$, a functor $i^!$ which is right adjoint to the direct image functor

$$\mathrm{R}i_* = i_* = i_! = \mathrm{R}i_! \ ,$$

such that $i^! = i^*\big(\mathrm{R}.\mathscr{H}_Y\big)$. Therefore $i^!$ preserves the categories D^b and D_c^b

$$i^! = i^*\mathrm{R}.\mathscr{H}_Y : \ D^b(X, \Lambda) \to D^b(X, \Lambda)$$

$$i^! = i^*\mathrm{R}.\mathscr{H}_Y : \ D_c^b(X, \Lambda) \to D_c^b(X, \Lambda) \ .$$

The corresponding statements for the right adjoint functor j^* of $j_! = \mathrm{R}j_!$ trivially hold for an open embedding j and, more generally, for the right adjoint functor

$$f^! = f^*[2d](d)$$

of the derived functor $\mathrm{R}f_!$ of a smooth compactifiable morphism $f : X \to Y$ of constant fiber dimension d. See the sections on relative duality theory.

Corollary D.5 (see [finitude], corollary 1.6) *(i) The functor $f^!$, which is the right adjoint of the functor $Rf_!$ for a compactifiable morphism $f : X \to Y$, preserves the categories D_c^b and D_c^+*

$$f^! : \ D_c^+(Y, \Lambda) \to D_c^+(X, \Lambda),$$

$$f^! : \ D_c^b(Y, \Lambda) \to D_c^b(X, \Lambda) \ .$$

Note $D_c^b(Y, \Lambda) \subseteq D_c^+(Y, \Lambda)$ and $D_c^b(X, \Lambda) \subseteq D_c^+(X, \Lambda)$.

(ii) Let $\mathscr{F}, \mathscr{G}$ be two constructible sheaves on X. Then the sheaves $\mathscr{E}xt_\Lambda^\nu(X, \mathscr{F}, \mathscr{G})$ are constructible as well. We obtain the functor

$$\mathrm{R}.\mathscr{H}om(X, -, -) : \ D_c^-(X, \Lambda) \times D_c^+(X, \Lambda) \to D_c^+(X, \Lambda).$$

Proof. (of the second assertion) Let $j : Y \hookrightarrow X$ a locally closed embedding. Then for a constructible sheaf $\mathscr{F}_1$ on Y we have

$$\mathrm{R}.\mathscr{H}om_\Lambda\big(j_!\mathscr{F}_1, \mathscr{G}\big) = \mathrm{R}j_*\mathrm{R}.\mathscr{H}om_\Lambda\big(\mathscr{F}_1, j^!\mathscr{G}\big) \ .$$

With this identity one reduces the proof by noetherian induction to the case of a locally constant sheaf and, since the statement is local with respect to the etale topology, then to the case of a constant sheaf

$$\mathscr{F} = \mathscr{M}_X .$$

Here $\mathscr{M}$ is a finite Λ-module. By a resolution of $\mathscr{M}$ with finite free Λ-modules we can further reduce to the case, where $\mathscr{M}$ is a finite free module $\mathscr{M}$. Hence we can assume $M = \Lambda$. So assume $\mathscr{F} = \Lambda_X$. Then for an arbitrary complex $\mathscr{K}$

$$\mathrm{R}\mathscr{H}om_\Lambda(\mathscr{F}, \mathscr{K}^\bullet) = \mathscr{H}om_\Lambda(\mathscr{F}, \mathscr{K}^\bullet) = \mathscr{K}^\bullet$$

holds. This implies the corollary. □

By Remark D.2 and Theorem D.3 we conclude for $f : X \to Y$ the statements

$$\mathrm{R}f_* : D_{ctf}(X, \Lambda) \to D_{ctf}(Y, \Lambda)$$

$$\mathrm{R}f_*\left(f^*(\mathscr{F}^\bullet) \otimes^L_\Lambda \mathscr{G}^\bullet \right) = \mathscr{F}^\bullet \otimes^L_\Lambda \mathrm{R}f_*(\mathscr{G}^\bullet) .$$

Here $\mathscr{F}^\bullet$ is a complex of locally constant sheaves on Y, which is bounded from above. With the conclusions of the proof of Corollary D.5 the corresponding statements for the functors

$$f^! , \ \mathrm{R}\mathscr{H}om_\Lambda(X, -, -) , \ \mathrm{R}\mathscr{H}_Z ,$$

where $Z \hookrightarrow X$ is a closed embedding, are obtained finally.

Corollary D.6 *The following permanence properties hold: Suppose a morphism $f : X \to Y$ is given. Then $f^!$, $\mathrm{R}f_*$ and $\mathrm{R}\mathscr{H}om$ define functors*

$$f^! : D_{ctf}(Y, \Lambda) \to D_{ctf}(X, \Lambda),$$

$$\mathrm{R}f_* : D_{ctf}(X, \Lambda) \to D_{ctf}(Y, \Lambda) ,$$

$$\mathrm{R}\mathscr{H}om_\Lambda(X, -, -) : D_{ctf}(X, \Lambda) \times D_{ctf}(X, \Lambda) \to D_{ctf}(X, \Lambda) .$$

Let $\Lambda \to \Lambda'$ be a homomorphism into another finite ring Λ' satisfying the same assumptions as Λ. Then the change of ring functor,

$$- \otimes^L_\Lambda \Lambda' : \ D_{ctf}(X, \Lambda) \to D_{ctf}(X, \Lambda)$$

satisfies

$$\mathrm{R}f_*(K^\bullet) \otimes^L_\Lambda \Lambda' \ = \ \mathrm{R}f_*(K^\bullet \otimes^L_\Lambda \Lambda')$$

$$f^!(\mathscr{L}^\bullet) \otimes^L_\Lambda \Lambda' \ = \ f^!(\mathscr{L}^\bullet \otimes^L_\Lambda \Lambda')$$

$$\mathrm{R}\mathscr{H}om_\Lambda(X, K^\bullet, M^\bullet) \otimes^L_\Lambda \Lambda' = \mathrm{R}\mathscr{H}om_{\Lambda'}(X, K^\bullet \otimes^L_\Lambda \Lambda', M^\bullet \otimes^L_\Lambda \Lambda') .$$

Here $K^\bullet, M^\bullet$ are assumed to be in $D_{ctf}(X, \Lambda)$ and $\mathscr{L}^\bullet$ is assumed to be in $D_{ctf}(Y, \Lambda)$. Corresponding statements hold for $\mathrm{R}f_!$.

Vanishing Cycles

Let $S = Spec(R)$ be the spectrum of a strictly henselian discrete valuation ring R, let $\eta = Spec(K) \hookrightarrow S$ be its generic point, let $s = Spec(k) \hookrightarrow S$ be the special point of S, let $\overline{\eta} = Spec(\overline{K})$ be the spectrum of the separable closure $\overline{K}$ of K and let denote $\Gamma = Gal(\overline{K}/K)$ the Galois group of $\overline{K}$ over K. Let π be a generator of the maximal ideal of R. Then

$$\begin{array}{ccccc} \overline{\eta} & & & & \\ \downarrow & \searrow & & & \\ \eta & \hookrightarrow & S & \hookleftarrow & s\,. \end{array}$$

Let X be a scheme over S. Then taking cartesian products over S with the last diagram we obtain a corresponding new diagram

$$\begin{array}{ccccc} X_{\overline{\eta}} & & & & \\ {\scriptstyle f}\downarrow & \overset{\overline{j}}{\searrow} & & & \\ X_\eta & \overset{j}{\hookrightarrow} & X & \overset{i}{\hookleftarrow} & X_s\,. \end{array}$$

The group Γ acts on $X_{\overline{\eta}}$, f is Γ-equivariant with respect to the trivial action of Γ on X_η. We now describe the construction of the complex of vanishing cycles following [SGA7II] expose XIII.

As usual $\mathfrak{o} = \mathfrak{o}_E$ will denote the valuation ring of a finite extension field E of the field $\mathbb{Q}_l$ of l-adic numbers. Let π_E be a generator of its maximal ideal. We assume, that the prime l is invertible in R. We then consider etale sheaves of Λ-modules for the ring $\Lambda = \mathfrak{o}_E/\pi_E^n\mathfrak{o}_E$.

For simplicity we will assume, that the schemes X and $X_{\overline{\eta}}$, hence all other auxiliary schemes in the following are noetherian schemes. For instance this is true, if X was finitely generated over S.

The field $\overline{K}$ is the direct limit (union) of all the finite dimensional extension fields L of K contained in $\overline{K}$. For each such field L the morphism

$$f_L : X_L = X \times_S Spec(L) \longrightarrow X_\eta$$

is a finite etale morphism. Compatibility of etale cohomology with projective limits of schemes, with respect to affine morphisms over a fixed base scheme ([FK], I.4.18), implies for a sheaf $\mathscr{G}$ on X_η

$$\mathrm{R}f_*f^*(\mathscr{G}) = f_*f^*(\mathscr{G}) = \varinjlim_L (f_L)_*(f_L)^*(\mathscr{G})\,.$$

The limit on the right is a direct limit. On $f_*f^*(\mathscr{G})$ the group Γ acts in a natural way. This action is continuous in the following sense:

For each constructible subsheaf $\mathscr{F} \subset f_*f^*(\mathscr{G})$ there exists an open subgroup $H \subset \Gamma$ (i.e. of finite index in Γ), such that

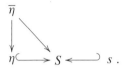

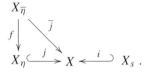

$$\sigma|\mathcal{F} = id_{\mathcal{F}} \quad , \quad \forall \sigma \in H \; .$$

If $\mathcal{G} = \mathcal{J}$ is an injective sheaf, then the sheaves $f^*(\mathcal{J})$ and $f_*f^*(\mathcal{J})$ are again injective. Note, that f_* preserves injectivity, since f^* is an exact functor and we have the (f^*, f_*)-adjunction formula. Similarly, since the etale maps f_L define exact functors $f_L^! = f_L^*$ and $\mathrm{R}(f_L)_! = (f_L)_! = (f_L)_*$ are exact functors, the adjunction formula of II.7.1 shows that f_L^* preserves injectivity. This can be generalized to hold for f^* using a limit theorem. See [FK], III, lemma 3.1 and [FK], I, §4 for further details. Let $K^\bullet$ be a complex of sheaves on X_η, which is bounded from below. Let $K^\bullet \to \mathcal{J}^\bullet$ be an injective resolution of $K^\bullet$, also bounded from below. Then essentially by definition

$$\mathrm{R}\overline{j}_* f^* K^\bullet = \overline{j}_* f^* \mathcal{J}^\bullet = j_* f_* f^* \mathcal{J}^\bullet$$

and the restricted complex $i^*(j_*f_*f^*\mathcal{J}^\bullet)$ is a complex in the category of sheaves on X_s with continuous Γ-action. Let $D(X_s, \Lambda, \Gamma)$ be the corresponding derived category of the category of etale sheaves of Λ-modules on X_s with continuous Γ-action. There is an obvious forget functor

$$forget : D(X_s, \Lambda, \Gamma) \to D(X_s, \Lambda) \; .$$

Now via the construction above, the functor $i^*\mathrm{R}\overline{j}_* f^*$ can be factorized in a natural way over a functor $\mathrm{R}\Psi_\eta$

$$i^*\mathrm{R}\overline{j}_* f^* : D^+(X_\eta, \Lambda) \xrightarrow{\mathrm{R}\Psi_\eta} D^+(X_s, \Lambda, \Gamma) \xrightarrow{forget} D^+(X_s, \Lambda)$$

$$i^*\mathrm{R}j_*f_*f^* = i^*\mathrm{R}\overline{j}_* f^* = forget \circ \mathrm{R}\Psi_\eta \; .$$

For a complex $K^\bullet$ with an injective resolution $\mathcal{J}^\bullet$ as above, the complex $\mathrm{R}\Psi_\eta K^\bullet$ is represented by $i^*j_*f_*f^*\mathcal{J}^\bullet$.

Let $K^\bullet$ be a complex of etale sheaves on X, which is bounded from below. Then we define

$$\mathrm{R}\Psi(K^\bullet) = \mathrm{R}\Psi_\eta j^*(K^\bullet) \; .$$

If $K^\bullet \to \mathcal{J}^\bullet$ is an injective resolution on X, then $j^*(K^\bullet) \to j^*(\mathcal{J}^\bullet)$ is an injective resolution of $j^*(K^\bullet)$ on X_η. Thus

$$\mathrm{R}\Psi(K^\bullet) = i^*j_*f_*f^*j^*\mathcal{J}^\bullet = i^*\overline{j}_*\overline{j}^*\mathcal{J}^\bullet \; .$$

There is a natural Γ-equivariant homomorphism of sheaf complexes with continuous Γ-action, which is defined as the composite map

$$i^*K^\bullet \longrightarrow i^*\mathcal{J}^\bullet \longrightarrow i^*\overline{j}_*\overline{j}^*\mathcal{J}^\bullet = \mathrm{R}\Psi(K^\bullet) \; .$$

Here we assumed Γ to act trivially on the complexes $K^\bullet$ and $\mathcal{J}^\bullet$. We can complete this morphism to the mapping cone, and we obtain a distinguished triangle in the triangulated category $D(X_s, \Lambda, \Gamma)$

$$i^*K^\bullet \to \mathrm{R}\Psi(K^\bullet) \to \mathrm{R}\Phi(K^\bullet) \to i^*K^\bullet[1] .$$

We have now defined the complexes

1. $\mathrm{R}\Psi(K^\bullet)$, the complex of near-by cycles (cycle proche)
2. $\mathrm{R}\Phi(K^\bullet)$, the complex of vanishing cycles (cycle vanescents)

Construction of the Variation map

By construction we have a canonical map

$$can : \mathrm{R}\Psi(K^\bullet) \to \mathrm{R}\Phi(K^\bullet) .$$

The variation maps $var(\sigma)$ for $\sigma \in \Gamma$ are maps in the inverse direction

$$var(\sigma) : \mathrm{R}\Phi(K^\bullet) \to \mathrm{R}\Psi(K^\bullet) \quad , \quad \sigma \in \Gamma .$$

They are morphisms of the category $D(X_s, \Lambda)$. In other words, we forget the structure of $D(X_s, \Lambda, \Gamma)$ and view $\mathrm{R}\Phi(K^\bullet)$ and $\mathrm{R}\Psi(K^\bullet)$ as complexes in $D(X_s, \Lambda)$, on which the group Γ acts, formulated in terms of the category $D(X_s, \Lambda)$. In this sense we have diagrams

$$\begin{array}{ccc} \mathrm{R}\Psi(K^\bullet) & \xrightarrow{can} & \mathrm{R}\Phi(K^\bullet) \\ & \nwarrow^{var(\sigma)} & \uparrow{\scriptstyle \sigma - id} \\ & & \mathrm{R}\Phi(K^\bullet) \end{array}$$

To define $var(\sigma)$ canonically, the considerations made so far have to be refined:

Let $\mathbf{A}$ be the abelian category, whose objects $(\mathscr{F} \to \mathscr{G})$ are Γ-equivariant homomorphisms between sheaves on X_s with continuous Γ-action

$$\mathscr{F} \to \mathscr{G} ,$$

such that Γ acts trivially on the first sheaf $\mathscr{F}$.

For an injective resolution $K^\bullet \to \mathscr{J}^\bullet$ the assignment

$$K^\bullet \mapsto \left(i^*\mathscr{J}^\bullet \to i^*\overline{j}_*\overline{j}^*\mathscr{J}^\bullet \right)$$

defines a functor from the category of complexes $K^\bullet \in D^+(X, \Lambda)$ into the homotopy category $K(\mathbf{A})$ of the complexes of the abelian category $\mathbf{A}$. In the homotopy category $K(\mathbf{A})$ the "complex"

$$\left(i^*\mathscr{J}^\bullet \to i^*\overline{j}_*\overline{j}^*\mathscr{J}^\bullet \right)$$

is naturally isomorphic to a "complex"

$$\left(\varphi : M^\bullet \to N^\bullet \right) ,$$

such that all homomorphisms $\varphi^\nu : M^\nu \to N^\nu$ are monomorphisms (injective) and such that furthermore the short exact sequences

$$0 \to M^\nu \to N^\nu \to C^\nu = Koker(\phi^\nu) \to 0$$

splits in every degree ν. For the construction of $\phi : M^\bullet \to N^\bullet$ we observe, that the mapping cylinder $Cyl(f) = Cone(Cone(f)[-1] \to \tilde{M}^\bullet)$ of a complex map $f : \tilde{M}^\bullet \to \tilde{N}^\bullet$ admits natural maps

$$\begin{array}{ccc} \tilde{M}^\bullet & \xrightarrow{\varphi(f)} & Cyl(f) \\ \| {\scriptstyle id} & & \downarrow {\scriptstyle g} \\ \tilde{M}^\bullet & \xrightarrow{f} & \tilde{N}^\bullet \end{array}$$

such that g is an isomorphism in the homotopy category of complexes of sheaves. Furthermore in each degree $\varphi(f)^\nu$ is an injection onto a direct summand. Namely $\varphi(f)$ and g are defined by $\varphi(f)^\nu(m) = (m, 0, 0) \in \tilde{M}^\nu \otimes \tilde{N}^\nu \otimes \tilde{M}^{\nu+1} \cong Cyl(f)^\nu$ and $g^\nu(m, n, m') = f^\nu(m) + n$. Therefore we put $M^\bullet = \tilde{M}^\bullet$, $N^\bullet = Cyl(f)$ and $\varphi = \varphi(f)$.

The complex $C^\bullet$ represents $\mathrm{R}\Phi(K^\bullet)$ and $N^\bullet$ represents $\mathrm{R}\Psi(K^\bullet)$. For $\sigma \in \Gamma$ the complex map

$$\sigma - id : N^\bullet \to N^\bullet$$

satisfies $(\sigma - id)|M^\bullet = 0$, since Γ acts trivially on $M^\bullet$. Hence for $\sigma \in \Gamma$ there is a uniquely determined complex map

$$var(\sigma) : C^\bullet \to M^\bullet ,$$

which makes the diagram

$$\begin{array}{ccc} M^\bullet & & \\ \uparrow {\scriptstyle \sigma - id} & \nwarrow {\scriptstyle var(\sigma)} & \\ M^\bullet & \longrightarrow & C^\bullet \end{array}$$

commute. This map

$$var(\sigma) : C^\bullet = \mathrm{R}\Phi(K^\bullet) \to M^\bullet = \mathrm{R}\Psi(K^\bullet)$$

represents the variation map. For more details on this construction and further properties see [SGA7II], exp. XIII 1.3.

Constructibility of $\mathbf{R}\Psi(K^\bullet)$ and $\mathbf{R}\Phi(K^\bullet)$

If X is finitely generated over S, then the direct image functor j_* has finite cohomological dimension. Hence we get functors

$$R\Psi\ ,\ R\Phi\ :\ D^b(X,\Lambda)\to D^b(X_s,\Lambda,\Gamma)\ .$$

A complex $K^\bullet\in D(X_s,\Lambda,\Gamma)$ will be called constructible, if its image in $D(X_s,\Lambda)$ under the forget functor is constructible, i.e. has constructible cohomology sheaves. This notion allows to define the category $D_c(X_s,\Lambda,\Gamma)$. A complex $K^\bullet$ has finite Λ-Tor-dimension, if its image in $D(X_s,\Lambda,\Gamma)$ has finite Λ-Tor-dimension. Let

$$D^b_{ctf}(X,\Lambda,\Gamma)\ \subset\ D(X,\Lambda,\Gamma)$$

be the full subcategory of bounded constructible complexes of finite Λ-Tor-dimension.

Theorem D.7 *Let $f:X\to S$ be a finitely generated morphism of schemes. Let $K^\bullet$ be in $D^b_c(X,\Lambda)$. Then also the complexes $R\Phi(K^\bullet)$ and $R\Psi(K^\bullet)$ are constructible.*

Since in the situation of the last theorem the cohomological dimension of j_* is finite, this implies that the cohomological dimension of the functors $R\Phi$ and $R\Psi$ is also finite. By standard arguments therefore the last theorem implies the following

Corollary D.8 *The functors*

$$R\Psi,R\Phi:\ D^+(X,\Lambda)\to D^+(X_s,\Lambda,\Gamma)$$

induce functors between the categories

$$R\Psi,R\Phi:\ D^b_{ctf}(X,\Lambda)\to D^b_{ctf}(X_s,\Lambda,\Gamma)\ .$$

These induced functors are compatible with change of rings

$$\Lambda=o_E/\pi_E^n o_E\ \to\ \Lambda'=o_E/\pi_E^m o_E\quad,\quad n\geq m$$

in the sense that for complexes $K^\bullet\in D^b_{ctf}(X,\Lambda)$ we have

$$R\Psi(K^\bullet\otimes^L_\Lambda\Lambda')\ =\ R\Psi(K^\bullet)\otimes^L_\Lambda\Lambda'$$

$$R\Phi(K^\bullet\otimes^L_\Lambda\Lambda')\ =\ R\Phi(K^\bullet)\otimes^L_\Lambda\Lambda'\ .$$

To prove the last Theorem D.7, it is enough to prove the constructibility of the complex $R\Psi(K^\bullet)$. For this we do not need the Γ-action on the complex $R\Psi(K^\bullet)$. For the proof of the constructibility theorem we therefore consider $R\Psi(K^\bullet)=i^*R\overline{j}_*\overline{j}^*K^\bullet$ as a complex in $D(X_s,\Lambda)$.

Proof. Let $\mathbb{A}=Spec(R[t])$ be the affine line over S, let $S'=Spec(R')$ be the strict Henselization of $\mathbb{A}$ at the generic point of the special fiber of $\mathbb{A}$ over S. Let s' be the special point of S', $\eta'=Spec(K')\subset S'$ the generic point of S'. Here K' denotes the quotient field of R'. Then π also generates the maximal ideal of the discrete valuation ring R'. Let $\overline{K}'$ be the separable closure of K', $\overline{\eta}'=Spec(\overline{K}')$

$$\begin{array}{ccccc} \overline{\eta}' & & & & \\ \downarrow & \searrow & & & \\ \eta' & \hookrightarrow & S' & \hookleftarrow & s' . \end{array}$$

The generic point of the special fiber of $\mathbb{A}$ over S corresponds to the prime ideal $\mathfrak{p} = \pi \cdot R[t]$ in $Spec(R[t]) = \mathbb{A}$. Let $R[t] \hookrightarrow B$ be a integral etale ring extension, let $\mathfrak{P}$ be a prime ideal of B over $\mathfrak{p}$ and let $Q(B)$ be the quotient field of B. Then $Q(B)$ is a separably generated field extension of K. We want to show, that the quotient field K of R is algebraically closed in $Q(B)$. Let L be a finite extension field of K contained in $Q(B)$. Let C be the integral closure of R in L. Then C is a discrete valuation ring, which is a free R-module. Let π' be a generator of the maximal ideal of C. Then

$$R[t] \subset C[t] \subset B$$

$$\mathfrak{P} \cap C[t] = \pi' C[t] = \mathfrak{Q} .$$

Then $B_{\mathfrak{P}}$ is etale over $R[t]_{\mathfrak{p}}$ and also over $C[t]_{\mathfrak{Q}}$, hence $C[t]_{\mathfrak{Q}}$ is etale over $R[t]_{\mathfrak{p}}$. Therefore C is etale over R. Since R is strictly Henselian, we get $R = C$ and $L = K$. Hence

K is algebraically closed in $Q(B)$.

Now vary B. Since obviously R' is the direct limit of the rings $B_{\mathfrak{p}}$ and K' is the direct limit of the fields $Q(B)$, we now get that

$$\tilde{K} = K' \otimes_K \overline{K}$$

is a field. Let p be the characteristic of the residue field of R and let L/K' be a field extension of finite degree say $e = [L : K']$. If p is not zero, let us assume that p is not a divisor of e. Then R' is tamely ramified in L and $L = K'[\pi^{1/e}] = K' \otimes_K K[\pi^{1/e}]$ is therefore contained in $K' \otimes_K \overline{K} = \tilde{K}$. Hence

$L \subset \tilde{K}$ if p does not divide $[L : K']$.

From these two observations we obtain the next

Lemma D.9 *$\tilde{K} = K' \otimes_K \overline{K}$ is a field. In particular the separable closure $\overline{K}'$ of K' is then also the separable closure of $\tilde{K}$. Let $\tilde{P} = Gal(\overline{K}'/\tilde{K})$ be the Galois group of $\overline{K}'$ over $\tilde{K}$ and let p be the characteristic of the residue field $R/\pi R$ of R. If $p = 0$, then $\overline{K}' = \tilde{K}$. If $p > 0$, then $\tilde{P}$ is a pro-p-group.*

Now let

$$Y \to S'$$

be a finitely generated scheme over S'. Then Y is a – not finitely generated – scheme over S via

$$Y \to S' \to S .$$

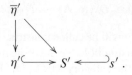

We then also have $Y_\eta = Y_{\eta'}$. Now consider the commutative diagram

$$\begin{array}{ccccccc}
Y_{\overline{\eta}'} & & & & & & \\
\downarrow h & & & & & & \\
Y_{\eta'} \times_{\eta'} Spec(\tilde{K}) & \longrightarrow & Y_{\eta'} & \hookrightarrow & Y & \hookleftarrow & Y_s \\
\| & & \| & & \| & & \| \\
Y_{\overline{\eta}} & \longrightarrow & Y_\eta & \hookrightarrow & Y & \hookleftarrow & Y_s
\end{array}$$

The group $\tilde{P} = Gal(\overline{K}'/\tilde{K})$ operates on $Y_{\overline{\eta}'}$ over $Y_{\eta'} \times_{\eta'} Spec(\tilde{K})$. For a sheaf $\mathscr{F}$ on $Y_{\eta'} = Y_\eta$ let $\mathscr{G}$ be the pull back of $\mathscr{F}$ to $Y_{\overline{\eta}'}$. Then $h_*(\mathscr{G})^{\tilde{P}}$ is the pull back of $\mathscr{F}$ to $Y_{\overline{\eta}} = Y_{\eta'} \times_{\eta'} Spec(\tilde{K})$. If the characteristic p of $R/\pi R$ is positive, then p is invertible in Λ by assumption. From Lemma D.9 we obtain

Lemma D.10 *Let $\mathscr{F}$ be a sheaf on $Y_\eta = Y_{\eta'}$. Then*

$$R^\nu \Psi_\eta(\mathscr{F}) = \left(R^\nu \Psi_{\eta'}(\mathscr{F})\right)^{\tilde{P}} .$$

In particular this implies: If the sheaves $R^\nu \Psi_{\eta'}(\mathscr{F})$ are constructible, then also the sheaves $R^\nu \Psi_\eta(\mathscr{F})$ are constructible.

Now let $X \to S$ be a finitely generated and affine morphism. Consider a closed embedding into some affine space $\mathbb{A}^m$ over S and a projection onto a factor $\mathbb{A}$ (an affine line over S) of $\mathbb{A}^m$. Then $\mathbb{A} = Spec(R[t])$ in the notations from above

$$\begin{array}{ccc}
X & \hookrightarrow & \mathbb{A}^m \\
& \searrow & \downarrow \\
& & \mathbb{A} \\
& & \downarrow \\
& & S
\end{array}$$

Then we have a diagram

$$\begin{array}{ccccc}
Y = X \times_{\mathbb{A}} S' & \longrightarrow & S' & & \\
\downarrow & & \downarrow & & \\
X & \longrightarrow & \mathbb{A} & \longrightarrow & S .
\end{array}$$

Y is a finitely generated scheme over S' and a – not necessarily finitely generated – scheme over S. Let L be the separable closure of $k(t)$. Then L is the residue class field of $S' = Spec(R')$. We consider the following commutative diagram

$$\begin{array}{ccccc}
Y_{\overline{\eta}} & \longrightarrow & Y & \hookleftarrow & Y_s = X_s \times_{\mathbb{A}_s} Spec(L) \\
\downarrow & & \downarrow \lambda & & \downarrow \lambda_s \\
X_{\overline{\eta}} & \longrightarrow & X & \hookleftarrow & X_s
\end{array}$$

The left square of this diagram is cartesian. The morphism

$$\lambda : Y \to X$$

is a limit of etale affine morphisms. Using the "smooth base change theorem" for such limits, one obtains for a complex $K^\bullet \in D_c^+(X, \Lambda)$

$$\lambda_s^*\big(\mathrm{R}\Psi(K^\bullet)\big) = \mathrm{R}\Psi\big(\lambda^*(K^\bullet)\big) .$$

We now prove Theorem D.7 by induction with respect to the dimension d of the general fiber X_η of X over S. By induction assumption we can assume, that Theorem D.7 is already proven for the complex $\lambda^*(K^\bullet)$ and the morphism $Y \to S'$. By Lemma D.10 therefore

$$\lambda_s^*\big(\mathrm{R}\Psi(K^\bullet)\big) = \mathrm{R}\Psi\big(\lambda^*(K^\bullet)\big)$$

is a constructible complex. Consider the maps

$$X_s \hookrightarrow \mathbb{A}_s^m \longrightarrow \mathbb{A}_s .$$

The complex $\mathrm{R}\Psi(K^\bullet)$ is again constructible over the generic fiber of the morphism

$$X_s \to \mathbb{A}_s .$$

And this is true for every possible projection $X_s \hookrightarrow \mathbb{A}_s^m \longrightarrow \mathbb{A}_s$. Since any sheaf is a direct limit of constructible subsheaves, this implies:

Every cohomology sheaf $\mathrm{R}^\nu\Psi(K^\bullet)$ of $\mathrm{R}\Psi(K^\bullet)$ contains a constructible subsheaf $\mathscr{G}_\nu$, such that the quotient sheaf

$$\mathrm{R}^\nu\Psi(K^\bullet)/\mathscr{G}_\nu$$

is a special skyscraper sheaf in the sense of the first part of this Appendix D.

This is a statement of local nature with respect to X, and holds for any finitely generated scheme X over S with the given dimension d of the generic fiber X_η. Since Theorem D.7 is of local nature with respect to X, it is enough to prove it for affine morphisms $X \to S$. This allows a further modification. Namely by an open dense embedding into a projective scheme over S we can alternatively assume, that $X \to S$ is a projective morphism. In this case one can apply the proper change theorem. It gives

$$\mathrm{R}\Gamma(X_s, \mathrm{R}\Psi K^\bullet) = \mathrm{R}\Gamma(X_{\overline{\eta}}, \overline{j}^*(K^\bullet)) .$$

The finiteness theorem for the proper map $X_{\overline{\eta}} \to \overline{\eta}$ then implies, that the complex $\mathrm{R}\Gamma(X_s, \mathrm{R}\Psi(K^\bullet)) = \mathrm{R}\Gamma(X_{\overline{\eta}}, \overline{j}^*(K^\bullet))$ is constructible. This means, that the cohomology modules of this complex are finite Λ-modules. We have already seen, using the induction hypotheses, that every cohomology sheaf $R^\nu\Psi(K^\bullet)$ contains a constructible subsheaf $\mathcal{G}_\nu$, such that the quotient sheaf $R^\nu\Psi(K^\bullet)/\mathcal{G}_\nu$ is a special skyscraper sheaf. Since for any constructible sheaf $\mathcal{G}$ on X_s also $R\Gamma(X_s, \mathcal{G})$ is constructible, one concludes that $R\Psi(K^\bullet)$ is constructible. See part B of the proof of Theorem D.3. □

Passage to l-adic Cohomology

The prime l is by definition the residue characteristic of the coefficient ring $\mathfrak{o} \subset E$. For $S = Spec(R)$, the spectrum of the strictly Henselian valuation ring R the finiteness conditions of Chap. II, §5 are satisfied. From the first part of this appendix it follows that

$$Ext^\nu_\Lambda(\mathcal{F}, \mathcal{G})$$

is a finite Λ-module for any finitely generated scheme X, any Λ-flat constructible sheaf $\mathcal{F}$ and any constructible sheaf $\mathcal{G}$ on X. More generally for $\mathcal{F}^\bullet, \mathcal{G}^\bullet \in D^b_{ctf}(X, \Lambda)$ the groups $Hom_{D(X,\Lambda)}(\mathcal{F}^\bullet, \mathcal{G}^\bullet)$ are finite groups. One can then define $D(X, \mathfrak{o})$ as projective limit in the sense of Chap. II, §2, i.e. a projective limit of triangulated categories $D^b_{ctf}(X, \mathfrak{o}/\pi^n_E\mathfrak{o})$.

$$D(X, \mathfrak{o}) = \lim_{\substack{\leftarrow \\ n}} D^b_{ctf}(X, \mathfrak{o}/\pi^n\mathfrak{o}) .$$

Now let $X_s \to s = Spec(k)$ be a finitely generated scheme over the separable closed residue field $k = R/\pi R$ of R. Then again

$$D(X_s, \mathfrak{o}) = \lim_{\substack{\leftarrow \\ n}} D^b_{ctf}(X_s, \mathfrak{o}/\pi^n\mathfrak{o}) .$$

In a similar way one can define the category $D(X_s, \mathfrak{o}, \Gamma)$. To do this one has to establish – for every ring $\Lambda = \mathfrak{o}/\pi^n\mathfrak{o}$ – the following finiteness condition: For two sheaves $\mathcal{F}, \mathcal{G}$ of Λ-modules on X_s let the Ext-functors be defined in the category of Λ-module sheaves with continuous Γ-action to be

$$Ext^\nu_{\Lambda[\Gamma]}(\mathcal{F}, \mathcal{G}) .$$

Claim. If $\mathcal{F}$ is a Λ-flat constructible sheaf, $\mathcal{G}$ an arbitrary constructible Λ-sheaf, then the Λ-modules $Ext^\nu_{\Lambda[\Gamma]}(\mathcal{F}, \mathcal{G})$ are finite Λ-modules.

To prove this we consider the spectral sequence

$$H^p(\Gamma, Ext^q_\Lambda(\mathcal{F}, \mathcal{G})) \Longrightarrow Ext^{p+q}_{\Lambda[\Gamma]}(\mathcal{F}, \mathcal{G}) .$$

(To guarantee existence of this spectral sequence we need the Λ-flatness of $\mathcal{F}$)! The Ext-functors $Ext^\nu_\Lambda(\mathcal{F}, \mathcal{G})$ of the category of sheaves of Λ-modules are finite Λ-modules. One has the following

Lemma D.11 *Let $\mathscr{M}$ be a finite Λ-module on which Γ acts continuously. Then the cohomology groups*

$$H^\nu(\Gamma, \mathscr{M})$$

are finite.

Proof. Let now denote $P \subset \Gamma$ the wild ramification group of R. Then we have the exact sequence

$$1 \to P \to \Gamma \to \prod_{p' \neq p} \mathbb{Z}_{p'} \to 0 .$$

Let H be the inverse image in Γ of $\prod_{p' \neq l, p} \mathbb{Z}_{p'}$. Here l denotes the residue characteristic of $\Lambda = \mathfrak{o}/\pi_E^n \mathfrak{o}$. Put $\mathscr{N} = \mathscr{M}^H$. Then $H^\nu(\Gamma, \mathscr{M}) = H^\nu(\mathbb{Z}_l, \mathscr{N})$. The finiteness of $H^\nu(\mathbb{Z}_l, \mathscr{N})$ can be found in [288], II.5.2, proposition 1. Therefore we can construct the derived category $D(X_s, \mathfrak{o}, \Gamma)$ as a projective limit

$$D_c^b(X_s, \mathfrak{o}, \Gamma) = \lim_{\substack{\longleftarrow \\ m}} D_{ctf}^b(X_s, \mathfrak{o}/\pi^m \mathfrak{o})$$

as in Chap. II, §5. So let $X \to S$ be a finitely generated scheme over S. Then we define the functors

$$\mathrm{R}\Psi, \mathrm{R}\Phi : D_c^b(X, \mathfrak{o}) \to D_c^b(X_s, \mathfrak{o}, \Gamma) .$$

To do this we put for a given projective system $K^\bullet = (K_r^\bullet)_{r \geq 1}$ with $K_r^\bullet \in D_c^b(X, \mathfrak{o}/\pi^r \mathfrak{o})$

$$\mathrm{R}\Psi(K^\bullet) = \left(\mathrm{R}\Psi(K_r^\bullet)\right)_{r \geq 1}$$

$$\mathrm{R}\Phi(K^\bullet) = \left(\mathrm{R}\Phi(K_r^\bullet)\right)_{r \geq 1} .$$

By localization, we obtain similar to the cases of the functors already considered the functors

$$\mathrm{R}\Psi, \mathrm{R}\Phi : D_c^b(X, E) \to D_c^b(X_s, E, \Gamma) ,$$

the nearby cycles resp. vanishing cycles functors with coefficients in E. By taking the direct limit over all E such that $\mathbb{Q}_l \subset E \subset \overline{\mathbb{Q}}_l$ we obtain

$$\mathrm{R}\Psi, \mathrm{R}\Phi : D_c^b(X, \overline{\mathbb{Q}}_l) \to D_c^b(X_s, \overline{\mathbb{Q}}_l, \Gamma) .$$

To fill in the details of the arguments is left as an exercise for the reader. □

Bibliography

[BBD] Beilinson A. et al.: Analyse et topologie sur les espaces singuliers. Astérisque 100,1982

[Be1] Beilinson, A. A.: On the derived category of perverse sheaves, K-theory, Arithmetic and Geometry, Ed. Manin, Yu. I., Springer Lecture Notes 1289 (1987), 27–41

[Be2] Beilinson, A.A.: How to glue perverse sheaves, K-theory, Arithmetic and Geometry, Ed. Manin, Yu. I., Springer Lecture Notes 1289 (1987), 42–51

[Del] Deligne P.: La conjecture de Weil II. Inst. Hautes Études Sci., Pul. Math. 52 (1980)

[EGA] Grothendieck A., Dieudonné J.: Elements de Géometrie Algebriques, Parties I–IV. Publ. Math. Iust. des Hautes Études No. 4 (1960), 8 (1961), 11 (1961), 17 (1963), 20 (1964), 24 (1965), 28 (1966), 32 (1967)

[FK] Freitag E., Kiehl R.: Etale Cohomology and the Weil Conjecture. Ergebnisse der Mathematik und ihrer Grenzgebiete, 3. Folge, Band 13 Springer Verlag

[Ga] Popesco N., Gabriel P.: Caractérisation des catégories abéliennes avec générateurs et limites inductives exactes, C. R. Acad. Sc. Paris, t. 258 (27 avril 1964), Groupe 1.

[IL] Illusie, L.: Deligne's l-adic Fourier transform, in Bloch, S.J. (ed.), Algebraic Geometry: Bowdoin 1985, A.M.S., Providence, 1987

[Lau] Laumon G.: Transformation de Fourier, constants d'equations fonctionelles et conjecture de Weil. Publ. Math. No. 65 (1987), Institut des Hautes Études scientifiques

[SGA1] Grothendieck A. et al.: Séminaire Géometrie algébrique. Springer Lecture Notes 224 (1971)

[SGA2] Grothendieck A. et al.: Séminaire Géometrie algébrique, North Holland Pub. Co. (1968)

[SGA3] Grothendieck A. et al.: Séminaire Géometrie algébrique. Springer Lecture Notes 151, 152 and 153 (1970)

[SGA4] Grothendieck A. et al.: Séminaire Géometrie algébrique. Springer Lecture Notes 269, 270 and 305 (1972–73)

[SGA4$\frac{1}{2}$] Deligne P. et al.: Séminaire Géometrie algébrique. Springer Lecture Notes 569 (1977)

[SGA7I] Grothendieck A. et al.: Séminaire Géometrie algébrique. Springer Lecture Notes 288

[SGA7II] Deligne P. et Katz N.: Séminaire Géometrie algébrique. Springer Lecture Notes 340 (1973)

[Spa] Spaltenstein N.: Resolutions of unbounded complexes, Composition Math. 65, (1988), 121–154

[T] Tits J.: Classification of Algebraic Semisimple Groups, in: Proceedings of Symp. in Pure Math., vol. IX, Algebraic Groups and Discontinuous Subgroups, AMS Providence, Rhode Island, 1966

[Tam] Tamme G.: Introduction to Étale Cohomology, Springer-Verlag, Berlin Heidelberg, 1994

[Tar] Tarrio L.A., López A.J., Souto Salorio M.J.: Localization in Categories of Complexes and Unbounded Resolutions, Canad. J. Math. Vol. 52 (2), (2000), 225–247

[Ver] Verdier J.-L.: Des Catégories Dérivées des Catégories Abeliennes. Astérisque 239, 1996

[1] A'Campo N.: Sur la monodromie des singularités isolées d'hypersurfaces complexes. Invent. Math. 20, Fasc. 2 (1973), 147–170

[2] Altman A., Hoobler R., Kleiman S.: A note on the base change map for cohomology. Compositio Math. 27 (1973), 25–38

[3] Altman A., Kleiman S.: Introduction to Grothendieck Duality Theory. Lecture Notes in Math. 146, Springer, Berlin Heidelberg New York, 1970

[4] Amitsur S.: Simple algebras and cohomology groups of arbitrary fields. Trans. Amer. Math. Soc. 9 (1956), 73–112

[5] André M.: Homologie des algèbres commutatives. Springer-Verlag, New York (1974)

[6] Andreotti A., Frankel T.: The Lefschetz theorem on hyperplane sections. Ann. Math. 69, 713–717 (1959)

[7] Anick D.: Recent progress in Hilbert and Poincaré series. Algebraic Topology: Rational Homotopy (Louvain-la-Neuve, 1986), 1–25. Springer Lect. Notes in Math. 1318, Springer-Verlag, New York

[8] Arbarello E., Cornalba M., Griffiths P., Harris J.: Geometry of Algebaic Curves. Springer-Verlag, New York, NY (1985)

[9] Artin E.: The Collected Papers of E. Artin. Ed. S. Lang and J.T. Tate, Addison-Wesley, Reading, MA (1965)

[10] Artin E., Tate J.T.: A note on finite ring extensions. J.Math. Soc. Japan, 3, 74–77 (1951)

[11] Artin E., Tate J.: Class Field Theory, W.A. Benjamin, New York, 1968

[12] Artin M.: Grothendieck Topologies. Lecture Notes, Harvard University Math. Dept. Cambridge, Mass. 1962

[13] Artin M.: Etale coverings of schemes over Hensel rings. Amer. J. Math. 88 (1966), 915–934

[14] Artin M.: The etale topology of schemes. Proc. of Internat. Congr. of Math. (Moscow 1966), edited by I.G. Petrovsky. Izdat "Mir", Moscow, 1968, pp. 44–56

[15] Artin M.: The implicit function theorem in algebraic geometry. Algebraic Geometry (Internat. Colloq. Tata Inst. Fund. Res. Bombay, 1968). Oxford University Press London, 1969, pp. 13–34

[16] Artin M.: Algebraic approximation of structures over complete local rings. Inst. Hautes Etudes sci. Publ. Math. 36 (1969), 23–58

[17] Artin M.: Algebraization of formal moduli, I. In: Global Analysis: Papers in Honor of K. Kodaira, edited by D.C. Spencer and S.E. Iyanaga. University of Tokyo Pres, Tokyo 1969, pp. 21–71

[18] Artin M.: On the joins of Hensel rings. Adv. Math. 7 (1971), 282–296

[19] Artin M.: Algebraic Spaces. Yale Mathematical Monographs 3, Yale University Press, New Haven, 1971

[20] Artin M.: Théorèmes de Représentabilité pour les Espaces Algébriques. Presses de l'Université de Montréal, Montréal 1973

[21] Artin M.: Deformations of Singularities. Tata Inst. Lect. Notes, Bombay, India (1976)

[22] Artin M.: Algebra. Prentice Hall, Englewood Cliffs, NJ (1991)

[23] Artin M., Mazur B.: Etale Homotopy. Lecture Notes in Math. 100, Springer, Berlin Heidelberg New York, 1969

[24] Artin M., Milne J.: Duality in the flat cohomology of curves. Invent. Math. 35 (1976), 111–129

[25] Atiyah M., Macdonald I.: Introduction to Commutative Algebra. Addison-Wesley, Reading, Mass., 1969

[26] A. A. Beilinson, G. Lusztig and R. MacPherson: A geometric setting for the quantum deformation of GL_n. Duke Math. J. 61, No. 2, 655–677 (1990)

[27] Baum P., Fulton W., MacPherson R.: Riemann-Roch and topological K-theory for singular varieties. Acta Math. 143, 155–192 (1979)

[28] Beilinson A., Deligne P.: Interpretation motivique de la conjecture de Zagier reliant polyogarithmes et regulateurs. Jannsen, Uwe (ed.) et al., Motives. Proceedings of the summer research conference on motives, held at the University of Washington, Seattle, WA, USA, July 20–August 2. 1991. Providence, RI: American Mathematical Society. Proc. Symp. Pure Math. 55, Pt. 2, 97–121 (1994)

[29] Beilinson A., Bernstein J., Deligne P.: Faisceaux pervers. Astérisque 100, 172 p. (1982)

[30] Beilinson A., Lusztig G., MacPherson R.: A geometric setting for the quantum deformation of GL_n. Duke Math.J. 61, No. 2, 655–677 (1990)

[31] Berthelot P.: Sur le "Théorème de Lefschetz faible" en cohomologie cristalline. C.R. Acad., Sc. Paris t. 277, Série A p. 955 (1973)

[32] Berthelot P.: Cohomologie cristalline des schémas de caractéristique $p > 0$. Lecture Notes in Math. 407, Springer, Berlin Heidelberg New York, 1974

[33] Berthelot P.: Classes de Chern en cohomologie cristalline. C.R. Acad. Sc. Paris 270, série A, p. 1695 and 1750 (1970)

[34] Berthelot P., Ogus A.: Notes on crystalline cohomology. Math. Notes 21, Princeton University Press (1978)

[35] Bombieri E.: Counting points on curves over finite fields. Sém. Bourb. exp. 430, 1972–1973

[36] Borel A. (ed.): Intersection cohomology. (Notes of a Seminar on Intersection Homology at the University of Bern, Switzerland, Spring 1983). Progres in Mathematics 50, Swiss Seminars. Boston-Basel-Stuttgart: Birkhäuser. X

[37] Borho W., MacPherson R.: Partial resolutions of nilpotent varieties. Astérisque 101/102, 1983 (Analyse et topologie sur les espaces singuliers II–III), 23–74

[38] Borho W., MacPherson R.: Representations de groupes de Weyl et homologie d'intersection pour les varietes nilpotentes. C.R.Acad. Sci., Paris, Sr. I 292, 707–710 (1981)

[39] Borho W., Brylinski J.-L., MacPherson R.: Springer's Weyl group representations through characteristic classes of cone bundles. Math. Ann. 278, 273–289 (1987)

[40] Borho W., MacPherson R.: Partial resolutions of nilpotent varieties. Astérisque 101–102, 23–74 (1983)

[41] Bourbaki N.: Algèbre. Éléments de Math. 4, 6, 7, 11, 14, 23, 24, Hermann, Paris, 1947–59

[42] Bourbaki N.: Algèbre Commutative. Éléments de Math. 27, 28, 30, 31, Hermann, Paris, 1961–65

[43] Bourbaki N.: Algèbre Commutative. Chapters 8–9. Mason, New York (1983)

[44] Bourbaki N.: Commutative Algebra. Chapters 1–7. Springer Verlag, New York (1985)

[45] Bourbaki N.: Algèbre I. Chapters I–III. Hermann, Paris, France (1970), = Algebra I, Chapters I–III. English translation: Springer-Verlag, 1989

[46] Bourbaki N.: Algèbre, Ch. IV–VII. Masson, Paris, France (1981) = Algebra II, Ch. 4–7. English translation: Springer-Verlag, 1990

[47] Brasselet J.P.: Existance des classes de Chern en théorie bivariante. Astérisque 101/102, 1983 (Analyse et topologie sur les espaces singuliers II–III), 7–22

[48] Bredon G.: Sheaf Theory. McGraw-Hill, New York, 1967

[49] Brylinski J.-L.: Transformations canoniques, dualité projective, theorie de Lefschetz, transformations de Fourier et sommes trigonometriques. Geometrie et analyse microlocales, Astérisque 140–141, 3–134 (1986)

[50] Brylinski J.-L.: Transformation de Fourier géometrique. I. C.R. Acad. Sci., Paris, Ser. I 297, 55–58 (1983)

[51] Brylinski J.-L.: Transformation de Fourier géometrique. II. C.R. Acad. Sci., Paris, Ser. I 303, 193–198 (1986)

[52] Brylinski J.-L.: Géometrie et analyse microlocales. Astérisque, 140–141

[53] Brylinski J.-L.: (Co)-homologie d'intersection et faisceaux pervers. Semin. Bourbaki, 34e année, Vol. 1981/92, Exposé 585, Astérisque 92–93, 129–157 (1982) (*)

[54] Brylinski J.-L.: Modules holonomes à singularités régulières et filtration de Hodge II. Astérisque 101/102, 1983 (Analyse et topologie sur les espaces singuliers II–III), 75–117

[55] Bucur I., Deleanu A.: Introduction to the Theory of Categories and Functors. Wiley, London, 1968

[56] Cartan H.: Exposé 7, Sem. Cartan 7 (1954)

[57] Cartan H., Eilenberg S.: Homological Algebra. Princeton University Press 1956

[58] Chase S., Rosenberg A.: Amitsur cohomology and the Brauer group. Mem. Amer. Math. Soc. 52 (1965), 34–79

[59] Cheeger J.: Hodge Theory of complex cones. Astérisque 223, 1994 (Périodes p-adiques), 113–184

[60] Chevalley C.: On the theory of local rings. Ann. Math. 44. 690–708 (1943)

[61] Chevalley C.: Variétés complètes. Fondements de la géométrie algébrique (Chapter IV), Secr. Math. of the Institut Henri Poincaré, Paris, France (1958)

[62] Cornell G., Silverman J. (Eds). Arithmetic Geometry. Springer-Verlag, New York (1986)

[63] Deligne P.: La conjecture de Weil pour les surfaces $K3$. Invent. Math. 15 (1972), 206–226

[64] Deligne P.: Les intersections complètes de niveau de Hodge un. Invent. Math. 15 (1972), 23–250

[65] Deligne P.: Formes modulaires et représentations l-adiques. Sém. Bourb. exp. 355, Lect. Notes in Math. 179 Vol. (1968/69), Springer, Berlin Heidelberg New York, 1971

[66] Deligne P.: Inputs of étale cohomology. Lectures at the 1974 AMS Summer Institute on Algebraic Geometry, Arcata California

[67] Deligne P.: Les constantes des équations fonctionelles de fonctions L. Modular Functions of One Variable, II (Proc. Internat. Summer School, Univ. Antwerp. 1972). Lecture Notes in Math. 349, Springer, Berlin Heidelberg New York, 1973, pp. 501–597

[68] Deligne P.: La conjecture de Weil, I. (in Russian). Usp. Mat. Nauk 30, No. 5 (1985), 159–160 (1975)

[69] Deligne P.: La conjecture de Weil, I. Inst. Hautes Études Sci. Publ. Math. 34 (1974), 274–307

[70] Deligne P.: La conjecture de Weil, II. Inst. Hautes Études Sci. Publ. Math. 52 (1980), 137–252

[71] Deligne P.: Théorie de Hodge I. Actes ICM, Nive, Gauthier-Villars, 1970, t. I, 425–430, II, Publ. Math. IHES 40 (1971), 5–58; III, Publ. Math. IHES 44 (1974), 5–77

[72] Deligne P.: Poids dans la cohomologie des variétés algébriques. Actes ICM, Vancouver, 1974, 79–85

[73] Deligne P.: Extensions centrales de groupes algebriques simplement connexes et cohomologie galoisienne. Publ. Math., Inst. Hautes Etud. Sci. 84, 35–89 (1996)

[74] Deligne P.: Structures de Hodge mixtes réelles. Jannsen, Uwe (ed.) et al., Motives. Proceedings of the summer research conference on motives, held at the University of Washington, Seattle, WA, USA, July 20–August 2, 1991. Providence, RI: American Mathematical Society. Proc. Symp. Pure Math. 55, Pt. 1, 509–514 (1994)

[75] Deligne P.: A quoi servent les motifs? Jannsen, Uwe (ed.) et al., Motives. Proceedings of the summer research conference on motives, held at the University of Washington, Seattle, WA, USA, July 2–August 2, 1991. Providence, RI: American Mathematical Societ. Proc. Symp. Pure Math. 55, Pt. 1, 143–161 (1994)

[76] Deligne P.: Decompositions dans la categorie derivée. Jannsen, Uwe (ed.) et al., Motives. Proceedings of the summer research conference on motives, held at the University of Washington, Seattle, WA, USA, July 20–August 2, 1991. Providence, RI: American Mathematical Society. Proc. Symp. Pure Math. 55, Pt. 1, 115–128 (1994)

[77] Deligne P.: Categories tannakiennes. The Grothendieck Festschrift, Collect. Artic. in Honor of the 60th birthday of A. Grothendieck. Vol. II, Prog. Math. 87, 111–195 (1990)

[78] Deligne P.: Integration sur un cycle evanescent. Invent. Math. 76, 129–143 (1984)

[79] Deligne P.: Applications de la formule des traces aux sommes trigonométriques. Semin. Geom. algebr. Bois-Marie, SGA 4 1/2, Lect. Notes Math. 569, 4–75 (1977)

[80] Deligne P.: Théorèmes de finitude en cohomologie l-adique (with an appendix by L. llusie). Semin. Geom. algebr. Bois-Marie, SGA 4 1/2, Lect. Notes Math. 569, 233–261 (1977)

[81] Deligne P., Lusztig G.: Representations of reductive groups over finite fields. Ann. of Math., II, Ser. 103, 103–161 (1976)

[82] Deligne P., Lusztig G.: Duality for representations of a reductive group over a finite field. II. J. Algebra 81, 540–545 (1983)

[83] Deligne P., Griffiths P., Morgan J., Sullivan D.: Real homotopy theory of Kähler manifolds. Invent. Math. 29 (1975), 245–274

[84] Deligne P., Serre J-P.: Formes modulaires de poids l. Ann. Scient. Ec. Norm. Sup. 4e Série t 7, 507–530 (1974)

[85] Demazure M. Gabriel P.: Groupes Algébriques vol. I, Géometrie algébrique, géneralités, groupes commutatifs. Masson, Paris 1970

[86] Deuring M.: Die Zetafunktion einer algebraischen Kurve vom Geschlechte Eins, Drei Mitteilungen. Nach. Akad. Wiss. Göttingen, 1953, pp. 85–94, 1955, pp. 13–43 and 1956, pp. 37–76

[87] Deuring M.: On the zeta function of an elliptic function field with complex multiplication. Proceedings of the International Symposium on Algebraic Number Theory, Tokyo-Nikko, 1956, pp. 47–50

[88] Deuring M.: Arithmetische Theorie der Korespondenzen algebraischer Funktionenkörper I. J. reine angew. Math. 177 (1937), 161–191

[89] Dieudonné J.: Cours de Géométrie Algébrique I, II. Presses Universitaires de France, Paris, 1974

[90] Dwork B.: On the Rationality of Zeta Functions and L-Series. Proceedings of a Conference on Local Fields. Nuffic Summer School held at Driebergen in 1966, Springer, Berlin Heidelberg New York, 1967

[91] Dwork B.: On the rationality of the zeta function of an algebraic variety. Amer. J. Math. 82 (1960), 631–648

[92] Dwork B.: A deformation theory for the zeta function of a hypersurface. Proc. Int. Cong. Math. Stockholm, 1962, pp. 247–259

[93] Ekedahl T.: On the adic formalism. The Grothendieck Festschrift Vol. II, Birkhäuser Verag 1990

[94] Flicker Y.Z., Kazhdan D.A.: Geometric Ramanujan conjecture and Drinfeld reciprocity law. Number Theory, trace formulas and discrete groups, Symp. in Honor of Atle Selberg, Oslo/Norway 1987, 201–218 (1989)

[95] Fontaine J.-M.: Représentations p-adiques semi-stables. Astérisque 223, 1994 (Périodes p-adiques), 113–184

[96] Fontaine J.-M.: Représentations l-adiques potientiellement semi-stables. Astérisque 223, 1994 (Périodes p-adiques), 321–348

[97] Freyd P.: Abelian Categories; An Introduction to the Theory of Functors. Harper and Row, New York (1964)

[98] Fulton W.: Algebraic Curves; An Introduction to Algebraic Geometry. W.A. Benjamin, New York (1969)

[99] Fulton W.: Introduction to Intersection Theory in Algebraic Geometry. CBMS Lect. Notes 54. Amer. Math. Soc., Providence, RI (1983)

[100] Fulton W.: Introduction to Toric Varieties. Annals of Math Studies 131, Princeton University Press, Princeton, NJ (1993)

[101] Fulton W.: Intersection Theory. Ergebnisse der Mathematik und ihrer Grenzgebiete, 3. Folge, Bd. 2. Springer, Berlin Heidelberg New York Tokio, 1984

[102] Gabriel, categories abeliennes, Bull. Soc. Math. France 90, 1962

[103] Gelfand S., MacPherson E., Vilonen K.: Perverse sheaves and quivers. Duke Math. J. 83, No. 3, 621–643 (1996)

[104] Gelfand S., Manin Y.I.: Methods of homological algebra. Springer-Verlag, Berlin, 1996. English translation of the 1989 Russian original (1989–1996)

[105] Godement R.: Topologie Algébrique et Théorie des Faisceaux. Hermann, Paris, 1958

[106] Goresky M., MacPherson R.: Lefschetz fixed point theorem for intersection homology. Comment. Math. Helv. 60, 366–391 (1985)

[107] Goresky M., MacPherson R.: Morse theory and Intersection Homology theory. Astérisque 101/12, 1983 (Orbits unipotentes et représentations), 135–192

[108] Goresky M., Harder G., MacPherson R.: Weighted cohomology. Invent. Math. 116, No. 1–3, 139–213 (1994)

[109] Goresky M., MacPherson R.: Intersection homology theory. Topology 19, 135–165 (1980)

[110] Goresky M., MacPherson R.: Intersection homology. II. Invent. Math. 72, 77–129 (1983)

[111] Grauert H., Remmert R.: Komplexe Räume, Math. Ann. 136 (1958), 245–318

[112] Griffiths P., Harris J.: Principles of algebraic geometry. John Wiley and Sons, New York, 1978

[113] Grothendieck A.: Sur quelques points d'algèbre homologique. Tôhoku Math. J. 9 (1957), 119–221

[114] Grothendieck A.: La théorie des classes de Chern. Bull. Soc. Math. de France 86 (1958), 137–154

[115] Grothendieck A.: Fondements de la Géométrie Algébrique. (Séminaire Bourbaki 1957–62.) Secrétariat Mathematique, Paris 1962

[116] Grothendieck A.: Le groupe de Brauer. I Algèbres d'Azumaya et interprétations diverses, II. Théorie cohomologique, III. Exemples et compléments. In Dix Exposés sur la Cohomologie des Schémas, North-Holland, Amsterdam, 1968, pp. 46–188

[117] Grothendieck A.: Formule de Lefschetz et rationalité des fonctions L. Séminaire Bourbaki 1964/65. Exposé 279. W.A.Benjamin, New York, 1966

[118] Grothendieck A.: Crystals and the de Rham cohomology of schemes. Dix exposés sur la cohomologie des schémas, North Holland, Amsterdam (1968) 306–350

[119] Grothendieck A.: Éléments de la géométrie algébrique II, Étude globale élémentaire de quelques classes de morphismes. Publ. Math. de l'I.H.E.S 8 (1961a)

[120] Grothendieck A.: Éléments de la géométrie algébrique III, Ëtude globale élémentaire de quelques classes de morphismes. Publ. Math. de l'I.H.E.S 11 (1961b)

[121] Grothendieck A.: Techniques de construction en géometrie analytique IV. Formalisme général des foncteurs représentables. In Séminaire H. Cartan, Familles d'espaces complexes et fondements de la géométrie analytique. Secr. Math. Institut Henri Poincaré, Paris, France (1962)

[122] Grothendieck A.: Éléments de la géométrie algébrique IV, Étude locale des schémas et des morphismes de schémas (premiérre partie). Publ. Math. de l'I.H.E.S 20 (1964)

[123] Grothendieck A.: Éléments de la géométrie algébrique IV, Étude locale des schémas et des morphismes de schémas (deuxième partie). Publ. Math. de l'I.H.E.S. 24 (1965)

[124] Grothendieck A.: Local Cohomology. Springer Lect. Notes in Math. 41, Springer-Verlag, New York (1967)

[125] Grothendieck A.: Revêtements étales et Groupe Fondemental (SGA1). Lect. Notes in Math. 224, Springer-Verlag, New York (1971)

[126] Grothendieck A., Dieudonné J.: Étude globale élémentaire de quelques classes de morphismes. Inst. Hautes études Sci. Publ. Math. 8 (1961)

[127] Grothendieck A., Dieudonné J.: Étude cohomologique des faisceaux cohérents. Ibid. 11 (1961), 17 (1963)

[128] Grothendieck A., Dieudonné J.: Étude locale des schémas et des morphismes de schémas. bid. 20 (1964), 24 (1965), 28 (1966), 32 (1967)

[129] Grothendieck A., Dieudonné J.: Éléments de Géométrie Algébrique. Le langage des Schémas. Springer, Berlin Heidelberg New York, 1971

[130] Grothendieck A. et al.: Séminaire de Géométrie Algébrique Revêtements étales et groupe fondamental (1960–61). Lecture Notes in Math. 224, Springer, Berlin Heidelberg New York, 1971

[131] Grothendieck A. et al. (with M. Artin and J.L. Verdier): Théorie des topos et cohomologie étale des schémas (1963–64). Lecture Notes in Math. 269, 270, 305, Springer, Berlin Heidelberg New York, 1972–73

[132] Grothendieck A. et al. (by P. Deligne with J.F. Boutot, L. Illusie and J.L. Verdier): Cohomologie étale. Lecture Notes in Math. 569, Springer, Berlin Heidelberg New York, 1977

[133] Grothendieck A. et al.: Cohomologie l-adique et fonctions L (1965–66). Lecture Notes in Math. 589, Springer, Berlin Heidelberg New York, 1977

[134] Grothendieck A. et al. (with P. Berthelot and L. Illusie): Théorie des intersections et théorème de Riemann-Roch (1966–67). Lecture Notes in Math. 225, Springer, Berlin Heidelberg New York, 1971

[135] Grothendieck A. et al. (with P. Deligne and N. Katz): Groupes de monodromie en géométrie algébrique (1967–68). Lecture Notes in Math. 288, 340, Springer, Berlin Heidelberg New York, 1971

[136] Grothendieck A. et al. (with J. Murre): The Tame Fundamental Group of a Formal Neighbourhood of a Divisor with Normal Crossings on a Scheme. Lecture Notes in Math. 208, Springer, Berlin Heidelberg New York, 1971

[137] Güemes J.J.: On the homology classes for the components of some fibres of Springer's reduction. Astérisque 173/174, 1989 (Orbites unipotentes et représentations), 257–270

[138] Hardy G.H.: Ramanujan. Twelve lectures on subjects suggested by his life and work. Chelsea, New York: originally published by Cambridge Universiy Press, 1940

[139] Harris J.: Algebraic Geometry. Springer-Verlag, New York (1992)

[140] Hartshorne R.: Residues and Duality. Lecture Notes in Math. 20, Springer, Berlin Heidelberg New York, 1966

[141] Hartshorne R.: Algebraic Geometry. Springer, Berlin Heidelberg New York, 1977

[142] Hartshorne R.: Algebraic vector bundles on projective spaces: A problem list. Topology 18, 117–128 (1979)

[143] Hasse H.: Beweis des Analogons der Riemannschen Vermutung für die Artinschen und F.K. Schmidtschen Kongruenzzetafunktionen in gewissen elliptischen Fällen. Ges. d. Wiss. Nachrichten. Math. Phys. Klasse, 1933, Heft 3, 253–262

[144] Hasse H.: Über die Kongruenz-Zetafunktionen. Sitz.ber. d. Preuss. Akad. d. Wiss., 1934, 250

[145] Hasse H.: Abstrakte Begründung der komplexen Multiplikation und Riemannsche Vermutung in Funktionenkörpern. Abh. Math. Sem. Univ. Hamburg 10 (1934), 325–348

[146] Hasse H.: Zur Theorie der abstrakten elliptischen Funktionenkörper, I, II und III. J. Reine Angew. Math. 175 (1936)

[147] Hasse H.: Über die Riemannsche Vermutung in Funktionenkörpern. Comptes Rendus du congrés international des mathématiciens, Oslo (1930), pp. 189–206

[148] Hasse H.: Modular functions and elliptic curves over finite fields. Simposio Internazionale di Geometria Algebraica, Edizioni, Cremonese, Roma 1967, pp. 248–266

[149] Hironaka H.: Resolution of singularities of an algebraic variety over a field of characteristic zero, I, II. Ann. Math. 79 (1964), 109–326

[150] Hodge W.: The theory and applications of harmonic integrals. Cambridge University Press, Cambridge, 1941

[151] Honda T.: Isogeny classes of abelian varieties over finite fields. J. Math. Soc. Japan 20 (1968), 83–95

[152] Hirzebruch F.: Topological Methods in Algebraic Geometry, 3rd ed. Springer, Berlin Heidelberg New York, 1966

[153] Hyodo O., Kato K.: Semi-stable reduction and crystalline cohomology wtih logarithmic poles. Astérisque 223, 1994 (Périodes p-adiques), 221–268

[154] Ihara Y.: Hecke polynomials as congruence zeta functions in elliptic modular case. Ann. Math., II. Ser. 85 (1967), 267–295

[155] Illusie L.: Exposé I Autour du théorème de monodrome locale. Fontaine, Jean-Marc (ed.), Periodes p-adiques. Séminaire de Bures-sur-Yvette, France, 1988. Paris: Societé Mathématique de France, Astérisque 223, 9–57 (1994)

[156] Illusie L.: Autour du théorème de monodromie locale (with an appendix by P. Colmez: Les nombres algébriques sont denses dans B_{dR}^{+}). Astérisque 223, 1994 (Périodes p-adiques), 59–112

[157] Illusie L.: Logarithmic spaces (according to K. Kato). Cristante, Valentino (ed.) et al., Barsotti symposium in algebraic geometry. Memorial meeting in honor of Iacopo Barsotti, in Abano Terme, Italy, June 24–27, 1991. San Diego, CA: Academic Press. Perspect. Math. 15, 183–203 (1994)

[158] Illusie L.: Crystalline cohomology. Jannsen, Uwe (ed.) et al., Motives. Proceedings of the summer research conference on motives, held at the University of Washington, Seattle, WA, USA, July 20–August 2, 1991. Providence, RI: American Mathematical Society. Proc. Symp. Pure Math. 55, Pt. 1, 43–70 (1994)

[159] Illusie L.: Realisation l-adique de l'accouplement de monodromie d'après A. Grothendieck. Courbes modulaires et courbes de Shimura, C.R. Semin., Orsay/Fr. 1987–88, Astérisque 196–197, 27–44 (1991)

[160] Illusie L.: Cohomologie de Rham et cohomologie étale p-adique (d'après G. Faltings, J.-M. Fontaine et al.) Semin. Bourbaki, Vol. 1989/90, 42ème année, Atérisque 189–190, Exp. No. 726, 325–374 (1990)

[161] Illusie L.: Categories derivées et dualité, travaux de J./L. Verdier. Enseign. Math., II. Ser. 36, No. 3/4, 369–391 (1990)

[162] Iversen B.: Cohomology of Sheaves. Springer-Verlag, New York (1986)
[163] Jensen, : Les Foncteurs Derives de $\lim_{\leftarrow}$ et leurs Applications en Theorie des Modules, Springer Lecture Notes 254 (1972)
[164] Kaplansky I.: Commutative Rings. Allyn and Bacon, Boston, MA. Rev. ed. The University of Chicago Press (1970)
[165] Kaplansky I.: Hilbert's problems, Lecture Notes in Math., Univ. Chicago, Chicago, IL (1977)
[166] Kashiwara M.: Representation theory and D-modules on flag varieties. Astérisque 173/174, 1989 (Orbites unipotentes et représentations), 55–109
[167] Kashiwara M.: Index theorem for constructible sheaves. Systemes differentiels et singularités, Colloq. Luminy/France 1983, Astérisque 130, 193–209 (1985)
[168] Kashiwara M.: The characteristic cycles of holonomic systems on a flag manifold related to the Weyl group algebra. Inent. Mat. 77, 185–198 (1984)
[169] Kashiwara M., Schapira P.: Sheaves on manifolds. With a short history "Les débuts de la theorie des faisceaux" by Christian Houzel. Grundlehren der Mathematischen Wissenschaften, 292. Berlin etc., Springer-Verlag
[170] Kato K.: Semi-stable reduction and p-adic étale cohomology. Astérisque 223, 1994 (Périodes p-adiques), 269–294
[171] Katz N.: Algebraic solutions of differential equations (p-curvature and the Hodge filtration). Invent. Math. 18 (1972), 1–118
[172] Katz N.: An overview of Deligne's proof of the Riemann hypothesis for varieties over finite fields (Hilbert's problem 8). Mathematical Developments Arising from Hilbert Problems, edited by S. Browder, Amer. Math. Soc., Proc. Symp. Pure Math. 28, 1976, pp. 275–306
[173] Katz N.: On a theorem of Ax. Amer. J. Math. 93 (1971), 485–499
[174] Katz N.: Théorème d'uniformité pour la structure cohomologique des sommes exponentielles. Astérisque 79, 1980 (Sommes exponentielles), 83–146
[175] Katz N.: Travaux de Dwork, Séminaire Bourbaki, exp. 409, 71/72, Lecture Notes in Math. 317, Springer, Berlin Heidelberg New York, 1973
[176] Katz N.: Sommes exponentielles. Cours à Orsay, automne 1979. Redigé par Gerard Laumon, preface par Luc Illusie. Astérisque 79, 209 p. (1980)
[177] Katz N., Laumon G.: Transformation de Fourier et majoration de sommes exponentielles. Publ. Math., Inst. Hautes Etud. Sci. 62, 145–202 (1985)
[178] Katz N., Laumon G.: Corrections à "Transformation de Fourier et majoration de sommes exponentielles". Publ. Math., Inst. Hautes Etud. Sci. 69, 233 (1989)
[179] Katz N., Messing W.: Some consequences of the Riemann hypothesis for varieties over finite fields. Invent. Math. 23 (1974), 73–77
[180] Kazhdan D., Laumon G.: Gluing of perverse sheaves and discrete series representation. J. Geom. Phys. 5, No. 1, 63–120 (1988)
[181] Kazhdan D., Lusztig G., Bernstein J.: Fixed point varieties on affine flag manifolds. Isr. J. Math. 62, No. 3, 129–168 (1998)
[182] Kazhdan D., Lusztig G.: A topological approach to Springer's representations. Adv. Math. 38, 222–228 (1980)
[183] Kazhdan D., Lusztig G.: Schubert varieties and Poincaré duality. Geometry of the Laplace operator, Honolulu/Hawai 1979, Proc. Symp. Pure Math., Vol. 36, 185–203 (1980)
[184] Kleiman S.: Algebraic cycles and the Weil conjectures. Dix Exposés sur la Cohomologie des Schémas, North-Holland, Amsterdam, 1968, pp. 359–386
[185] Knutson D.: Algebraic Spaces. Lecture Notes in Math. 203, Springer, Berlin Heidelberg New York, 1971
[186] Koblitz N.: p-adic variation of the zeta-function over families of varieties defined over finite fields. Thesis, Princeton, 1974

[187] Kraft H., Procesi C.: A special decomposition of the nilpotent cone of a classical Lie algebra. Astérisque 173/174, 1989 (Orbites unipotentes et représentation), 193–207

[188] Kunz E.: Introduction to Commutative Algebra and Algebraic Geometry. Birkhäuser, Boston, MA (1985)

[189] Kurke H., Pfise G., Roczen M.: Henselsche Ringe und Algebraische Geometrie. VEB Deutscher Verlag der Wissenschaften, Berlin 1975

[190] Lang S.: Unramified class field theory over function fields in several variables. Ann. Math. 64 (1956), 285–325

[191] Lang S.: Abelian Varieties. Interscience, New York, 1959

[192] Lang S.: Diophantine Geometry. Interscience, New York, 1962

[193] Lang S.: Algebra, (3rd ed). Addison-Wesley, Reading, MA (1993)

[194] Lang S., Weil A.: Number of points of varieties in finite fields. Amer. J. Math. 76 (1954), 819–827

[195] Langlands R.P.: Modular Forms and l-adic Representations, in: Modular Functions of One Variable II, Springer Lecture Notes 349 (1973), p. 361–500

[196] Laumon G.: Cohomology of Drinfeld modular varieties. Part I: Geometry, counting of points and local harmonic analysis. Cambridge Studies in Advanced Mathematics. 41. Cambridge: Cambridge University Press. xiii (1996)

[197] Laumon G.: Cohomology of Drinfeld modular varieties. Part II: Automorphic forms, trace formulas and Langlands correspondence. (With an appendix by Jean-Loup Waldspurger. Cambridge Studies in Advanced Mathematics 56. Cambridge: Cambridge University Press. xi (1997)

[198] Laumon G.: Drinfeld shtukas. Berlin: Springer. Lect. Notes Math. 1649, 50–109 (1997)

[199] Laumon G.: La transformation de Fourier geometrique et ses applications. Proc. Int. Congr. Math., Kyoto/Japan 1990, Vol. I, 437–445 (1991)

[200] Laumon G.: Faisceaux caracteres (d'après Lusztig). Semin. Bourbaki, Vol. 1988/89, 41e année, Exp. No, 709, Astérisque 177–178, 231–260 (1989)

[201] Laumon G.: Transformation de Fourier, constntes d'equations fonctionnelles et conjecture de Weil. Publ. Math., Inst. Hautes Etud. Sci. 65, 131–210 (1987)

[202] Laumon G.: Transformations canoniques et specialisation pour les $\mathscr{D}$-modules filtres. Systemes differentiels et singularites, Colloq. Luminy/France 1983, Astérisque 130, 56–129 (1985)

[203] Laumon G.: Les constants des equations fonctionelles des fonctions L sur un corps global de caractéristique positive. C.R. Acad. Sci., Paris, Ser. I 298, 181–184 (1984)

[204] Laumon G.: Vanishing cycles over a base of dimension ≥ 1. Algebraic geometry, Proc. Jap.-Fr. Conf., Tokyo and Kyoto 1982, Lect. Notes Math. 1016, 143–150 (1983)

[205] Laumon G.: Sur la categorie derivée des $\mathscr{D}$-modules filtres. Algebraic geometry, Proc. Jap.-Fr. Conf., Tokyo and Kyoto 1982, Lect. Notes Math. 1016, 151–237 (1983)

[206] Laumon G.: Caracteristique d'Euler-Poincaré des faisceaux constructibles sur une surface. Astérisque 101–102, 193–207 (1983)

[207] Laumon G.: Majoration de sommes exponentielles attachées aux hypersurfaces diagonales. Ann. Sci. Ec. Norm. Super., IV. Ser. 16, 1–58 (1983)

[208] Laumon G.: Comparaison de caractéristiques d'Euler-Poincaré en cohomologie l-adique. C.R. Acad. Sci., Paris, Ser. I 292, 209–212 (1981)

[209] Lazard D.: Sur les modules plat. C.R. Acad. Sci. Paris 258 (1964), 613–616

[210] Lefschetz S.: L'Analysis Situs et la Géométrie Algébrique. Gauthier-Villars, Paris, 1924

[211] Leray J.: L'anneau spectral et l'anneau filtré d'un espace localement compacte et d'une application continue. J.Math. Pures et Appl. 29, 1–139 (1946)

[212] Lipman J.: Notes on Derived Categories and Derived Functors, Preprint (1995)

[213] Lubkin S.: A p-adic proof of the Weil conjectures, Ann. Math. 87 (1968) 105–194
[214] Lusztig G.: Singularities, character formulas, weight multiplicities. Astérisque 101/102, 1983 (Orbites unipotentes et représentations), 208–229
[215] Lusztig G.: Homology bases arising from reductive groups over a finite field. Carter, R.W. (ed.) et al., Algebraic groups and their representations. Proceedings of the NATO Advanced Study Institute on modular representations and subgroup structure of algebraic groups and related finite groups, Cambridge, UK, June 23–July 4, 1997. Dordreht: Kluwer Academic Publishers. NATO ASI Ser., Ser. C, Math. Phys. Sci. 517, 53–72 (1998)
[216] Lusztig G.: Constructible functions on the Steinberg variety. Adv. Math. 130, No. 2, 287–310 (1997)
[217] Lusztig G.: Affine Weyl groups and conjugacy classes in Weyl groups. Transform. Groups 1, No. 1–2, 83–97 (1996)
[218] Lusztig G.: Classification of unipotent representations of simple p-adic groups. Int. Math. Res. Not. 1995, No. 11, 517–589 (1995)
[219] Lusztig G.: Study of perverse sheaves arising from graded Lie algebras. Adv. Math. 112, No. 2, 147–217 (1995)
[220] Lusztig G.: Appendix. Coxeter groups and unipotent representations. Broue, Michel et al., Representations unipotentes generiques et blocs des groupes reductifs finis avec un appendice de George Lusztig. Montrouge: Societé Mathematique de France, Astérisque 212, 191–203 (1993)
[221] Lusztig G.: Intersection cohomology methods in representation theory. Proc. Int. Congr. Math., Kyoto/Japan 1990, Vol. I, 155–174 (1991)
[222] Lusztig G.: Quivers, perverse sheaves, and quantized enveloping algebras. J. Am. Math. Soc. 4, No. 2, 365–421 (1991)
[223] Lusztig G.: Green functions and character sheaves. Ann. Math., II. Ser. 131, No. 2, 355–408 (1990)
[224] Lusztig G.: Introduction to character sheaves. Representations of finite groups, Proc. Cont., Arcata/Calif. 1986, Pt. 1, Proc. Symp. Pure Math. 47, 165–179 (1987)
[225] Lusztig G.: Character sheaves. I. Adv. Math. 56, 193–237 (1985)
[226] Lusztig G.: Character sheaves. II. Adv. Math. 57, 226–265 (1985)
[227] Lusztig G.: Character sheaves. III. Adv. Math. 57, 266–315 (1985)
[228] Lusztig G.: Character sheaves. IV. Adv. Math. 59, 1–63 (1986)
[229] Lusztig G.: Character sheaves. V. Adv. Math. 61, 103–155 (1986)
[230] G. Lusztig: Erratum to: Character sheaves. V. Adv. Math. 62, 313–314 (1986)
[231] Lusztig G.: On the generalized Springer corespondence for classical groups. Algebraic groups and related topics, Proc. Symp., Kyotoand Nagoya/Jap. 1983, Adv. Stud. Pure Math. 6, 289–316 (1985)
[232] Lusztig G.: Characters of reductive groups over finite fields. Proc. Int. Congr. Math., Warszawa 1983, Vol. 2, 877–880 (1984)
[233] Lusztig G.: Characters of reductive groups over a finite field. Annals of Mathematics Studies, No. 107. Princeton, New Jersey: Princeton University Press. XXI, 384 p (1984)
[234] MacDonald I.G.: Algebraic geometry: Introduction to schemes. New York, Benjamin 1968
[235] MacLane S.: Homology. Springer-Verlag, New York (1963)
[236] MacPherson R., Vilonen K.: Perverse sheaves with singularities along the curve $y^n = x^m$. Comment. Math. Helv. 63, No. 1, 89–102 (1988)
[237] MacPherson R., Vilonen K.: Elementary construction of perverse sheaves. Invent. Math. 84, 403–435 (1986)
[238] Malgrange B.: Rapport sur les theorémes d'inice de Boutet, de Monvel et Kashiwara. Astérisque 101/102, 1983 (Orbites unipotentes et représentations), 243–267

[239] Malgrange B.: Polynômes de Bernstein-Sato et cohomologie évanescente. Astérisque 101/102, 1983 (Orbites unipotentes et représentations), 243–267
[240] Mars J.G.M., Springer T.A.: Character sheaves. Astérisque 173/174, 1989 (Orbites unipotentes et représentations), 111–198
[241] Matsumara H.: Commutative Algebra. W.A. Benjamin, New York, 1970
[242] Mazur B.: Rational points of abelian varieties with values in towers of number fields. Invent. Math. 18 (1972), 183–266
[243] Mazur B.: Notes on étale cohomology of number fields. Ann. Sci. Ecole Norm. Sup. 6 (1973), 521–556
[244] Mazur B.: Frobenius and the Hodge Filtration. B.A.M.S. 78 (1972), 653–667; (estimates) Ann. Math. 98 (1973), 58–95
[245] Mazur B. : Eigenvalues of Frobenius acting on algebraic varieties over finite fields. Proc. of Symp. in Pure Math. Vol. 29, Algebraic Geometry, Arcata 1974
[246] Mazur B., Messing W.: Universal extensions and one dimensional crystalline Cohomology. Lect. Notes in Math. 370, Springer, Berlin Heidelberg New York, 1974
[247] Messing W.: Short sketch of Deligne's proof of the hard Lefschetz theorem. Algebraic Geomety (Arcata 1974), edited by R. Hartshorne. Amer. Math. Soc. Proc. Symp. Pure Math. 29, 1975, pp. 563–580
[248] Miller E.: De Rham cohomology with arbitrary coefficients. Topology 17 (2) (1978), 193–203
[249] Milne J.S.: Étale Cohomology. Princeton Math. Series 33, Princeton University Press 1980
[250] Monsky P.: p-adic analysis and zeta functions. Lectures in Math. Kyoto Univ. pubished by Kinokuininya Book Store Co., Ltd. Tokyo, Japan (1970)
[251] Monsky P., Washnitzer G.: Formal Cohomology I. Ann. Math. 88 (1968), 181–217
[252] Morgan J.: The algebraic topology of smooth algebraic manifolds. Publ. Math. IHEW 48 (1978), 137–204
[253] Mumford D.: Geometric Invariant Theory. Springer, Berlin Heidelberg New York, 1965
[254] Mumford D.: Lectures on Curves on an Algebraic Surface. Annals of Math. Studies 59, Princeton University Press, Princeton, 1966
[255] Mumford D.: Introduction to Algebraic Geometry. Lecture Notes Harvard University Math. Dept., Cambridge, Mass. 1967
[256] Mumford D.: Abelian Varieties. Oxford University Press, Oxford 1970
[257] Mumford D.: Curves and their Jacobians. Univ. of Michigan Press, Ann Arbor, MI (1975)
[258] Mumford D.: Complex Projective Varieties. Springer-Verlag, New York (1976)
[259] Murre J.: Lectures on an Introduction to Grothendieck's Theory of the Fundamental Group. Lecture Notes, Tata Institute of Fundamental Research, Bombay, 1967
[260] Nagata M.: Local Rings. Interscience, New York, 1962
[261] Nagata M.: A generalization of the imbedding problem of an abstract variety in a complete variety. J. Math. Kyoto Univ. 3 (1963), 89–102
[262] Perrin-Riou B.: Représentations p-adiques ordinaires (with an appendix by L. Illusie: Réduction semi-stable ordinaire, cohomologie étale p-adique et cohomologie de Rham après Bloch-Kato et Hyodo). Astérisque 223, 1994 (Périodes p-adiques), 185–220
[263] Peskine C.: An Algebraic Introduction to Complex Projective Geometry, 1. Commutative Algebra. Cambridge University Press, Cambridge, UK (1996)
[264] Pham F.: Structures de Hodge mixtes associées à un germe de fonction à point critique isolé. Astérisque 101/102 (1983) (Orbites unipotentes et représentations), 268–285
[265] Polo P.: Variétés de Schubert et excellentes filtrations. Astérisque 173/174, 1989 (Orbites unipotentes et représentations), 281–312

[266] Popp H.: Fundamentalgruppen algebraischer Mannigfaltigkeiten. Lecture Notes in Math. 176, Springer, Berlin Heidelberg New York, 1970

[267] Rankin R.: Contributions to the theory of Ramanujan's function $\tau(n)$ and similar algebraic functions. Proc. Cambridge Phil. Soc. 35 (1939), 351–372

[268] Raynaud M.: Caractéristique d'Euler-Poincaré d'un faisceau et cohomologie des variétés abéliennes. (Séminaire Bourbaki 1964/65, no. 286.) In Dix Exposés sur la Cohomologie des Schémas, North-Holland, Amsterdam, 1968, pp. 12–30

[269] Raynaud M.: Faisceaux Amples sur les Schémas en Groupes et les Espaces Homogènes. Lecture Notes in Math. 119, Springer, Berlin Heidelberg New York, 1970

[270] Raynaud M.: Anneaux Locaux Henséliens. Lecture Notes in Math. 169, Springer, Berlin Heidelberg New York, 1970

[271] Rossmann W.: Equivariant multiplicities on complex varieties. Astérisque 173/174, 1989 (Orbites unipotentes et représentations), 320–331

[272] Sabbah C.: Morphismes analytiques stratifiés sans éclatement et cycles évonescents. Astérisque 101/102, 1983 (Orbites unipotentes et représentations), 286–319

[273] Saito M.: Supplement to "Gauss-Manin system and mixed Hodge structure". Astérisque 101/102, 1983 (Orbites unipotentes et représentations), 332–364

[274] Schmidt F.K.: Allgemeine Körper im Gebiet der höheren Kongruenzen. Thesis Erlangen (1925)

[275] Schmidt F.K.: Zur Zahlentheorie in Körpern von der Charakteristik p. Sitzungsberichte Erlangen 58/59 (1925)

[276] Schmidt F.K.: Analytische Zahlentheorie in Körpern der Charakteristik p. Math.Z. 33 (1931)

[277] Serre J-P.: Faisceaux algébriques cohérents. Ann. Math. 61 (1955), 197–278

[278] Serre J-P.: Géométrie algébrique et géométrie analytique. Ann. Inst. Fourier 6 (1956), 1–42

[279] Serre J-P.: Critères rationalité pour les surfaces algébriques. (Séminaire Bourbaki 1956/57 no. 146.) Secrétariat Mathématique, Paris

[280] Serre J-P.: Quelques propriétés des variété abéliennes en caractéristique p. Amer. J. Math. LXXX, no. 3 (1958), 715–739

[281] Serre J-P.: Analogues Kählériens de certaines conjectures de Weil. Ann. Math. 71, II. Ser. (1960), 392–394

[282] Serre J-P.: Rationalité des fonctions ζ des variétés algébriques (d'après Bernard Dwork). Séminaire Bourbaki 1959/60, Exposé 198. W.A. Benjamin, New York, 1966

[283] Serre J-P.: Zeta and L-functions in Arithmetic Algebraic Geometry. Harper and Row, New York, 1965

[284] Serre J-P.: Sur la topologie des variétés algébriques en caractéristique p. Symposium internacional de topologia algebraica Universidad Nacional Autónoma de México and UNESCO, Mexico City, 1958, 24–53

[285] Serre J-P.: Espaces fibrés algébriques. (Séminaire Chevalley 2 (1958), Exposé 1.) Secrétariat Mathématique, Paris

[286] Serre J-P.: Groupes Algébriques et Corps de Classes. Hermann, Paris, 1959

[287] Serre J-P.: Corps Locaux. Hermann, Paris, 1962

[288] Serre J-P.: Cohomologie Galoisienne. Lecture Notes in Math. 5, Springer, Berlin Heidelberg New York, 1965

[289] Serre J-P.: Algèbre Locale-Multiplicités. Lecture Notes in Math. 11, Springer, Berlin Heidelberg New York, 1965

[290] Serre J-P.: Abelian l-adic representations and elliptic curves. Benjamin, 1968

[291] Serre J-P.: A Course in Arithmetic, Graduate Texts in Mathematics, Springer Verlag (1973)

[292] Serre J-P.: Représentations Linéaires des Groupes Finis, 2nd ed. Hermann, Paris, 1971
[293] Serre J-P.: Valeurs propres des endomorphismes de Frobenius (d'après P. Deligne). Séminaire Bourbaki 1973/74, Februar 1974, Exposé 446. Lecture Notes in Math. 431, Springer, Berlin Heidelberg New York, 1975
[294] Shafarevich I.R.: Basic Algebraic Geometry. Springer-Verlag, New York (1972)
[295] Shatz S.: The cohomological dimension of certain Grothendieck topologies. Ann. Math. 83 (1966), 572–595
[296] Shatz S.: Profinite Groups, Arithmetic, and Geometry. Annals of Math. Studies 67, Princeton University Press, Princeton, 1972
[297] Shimura G.: Correspondances modulaires et les fonctions L des courbes algébriques. J. Math. Soc. Japan 10 (1958) 1–28
[298] Silverman J.: The Arithmetic of Elliptic Curves. Springer-Verlag, New York (1986)
[299] Steenbrink J.: Limits of Hodge structures, Invent. Math. 31 (1976), 229–257
[300] Steenrod N.: The work and influence of Professor S. Lefschetz in algebraic topology, Algebraic Geometry and Topology: A Symposium in Honor of S. Lefschetz, edited by R.H. Fox, D.C. Spencer, Princeton University Press, Princeton, 1957, pp. 24–43
[301] Stepanov S.A.: The number of points of a hyperelliptic curve over a finite prime field (in Russian). Izv. Akad. Nauk SSSR Ser. Math. 33 (1969), 1171–1181
[302] Sullivan D.: Infinitesimal calculations in topology. Publ. Math. IHES 47 (1977), 269–331
[303] Tate J.: Algebraic cycles and poles of zeta functions. Proceedings of a conference held at Purdue University 1963, Harper and Row, New York, pp. 93–110
[304] Tate J.: Endomorphisms of Abelian Varieties over finite Fields. Invent. Math. 2 (1966), 134–144
[305] Tate J.: Classes d'Isogénie des variétés abéliennes sur un corps fini (d'après T. Honda) Séminaire Bourbaki exp. 352, 1968/1969
[306] Verdier J.-L.: A duality theorem in the etale cohomology of schemes. Poc. Conf. Local Fields (Driebergen 1966), edited by T. Springer, Springer, Berlin Heidelberg New York, 1967, pp. 184–198
[307] Verdier J.-L.: Indépendance par rapport à l des polynomes caractéristiques des endomorphismes de Frobenius de la cohomologie l-adique (d'après P. Deligne), Sém. Bourbaki no. 423, 18 p., Lecture Notes in Math. 383, Springer, Berlin Heidelberg New York, 1974
[308] Verdier J.-L.: The Lefschetz fixed point formula in étale cohomology. Proc. Conf. Local Fields (Driebergen 1966), edited by T. Springer. Springer, Berlin Heidelberg New York, 1967, p. 199–214
[309] Verdier J.-L.: Spécialisation de faisceaux et monodromie modérée. Astérisque 101/102, 1989 (Orbites unipotentes et représentations), 332–364
[310] Weil A.: Number of solutions of equations over finite fields. Bull. Amer. Math. Soc. 55 (1949), 497–508
[311] Weil A.: Sur les fonctions algébriques à corps de constantes finis. C.R. Acad. Sci. Paris 210 (1940), 592–594
[312] Weil A.: On the Riemann Hypothesis in function fields. Proc. Nat. Acad. Sci. U.S.A. 27 (1941), 345–347
[313] Weil A.: Foundations of algebraic geometry. Amer. Math. Soc. Colloq. Pub., Vol. XXIX, New York, 1946
[314] Weil A.: Sur les courbes algébriques et es variétés qui s'en déduisent. Hermann, Paris, 1948
[315] Weil A.: Variétés abeliennes et courbes algébriques. Hermann Paris, 1971
[316] Weil A.: On some exponential sums. Proc. Nat. Acad. Sci. U.S.A. 34 (1948), 204–207

[317] Weil A.: Number of solutions of equations in finite fields. Bull. A.M.S. 55 (1949), 407–508

[318] Weil A.: Fibre Spaces in algebraic geometry. (Notes by A. Wallace) University of Chicago, 1952

[319] Weil A.: Number theory and algebraic geometry. Proceedings of the 1950 International Congress of Mathematicians, Cambridge, vol. 2, pp. 90–100, A.M.S., Providence, 1952

[320] Weil A.: Jacobi sums as Grössencharaktere. Trans. A.M.S. 73 (1952), 487–495

[321] Weil A.: Abstract versus classical algebraic geometry. Proceedings of the 1954 International Congress of Mathematicians, Vol. III, 550–558, North-Holland, Amsterdam, 1956

[322] Wintenberger J.-P.: Théorème de comparaison p-adique pour les schémas abéliens. I: Construction de l'accouplement de périodes. Astérisque 223, 1994 (Périodes l-adiques), 349–397

[323] Zariski O.: Collected Works, vol. II. M.I.T. Press, Cambridge, 1973

[324] Zucker S.: Hodge theory and arithmetic groups. Astérisque 101/102, 1983 (Orbites unipotentes et représentations), 365–381

Glossary

κ 5
$Gal(k/\kappa)$ 5
$W(k/\kappa)$ 5
F 5
F_X 5
Fr_X 5
$|X_0|$ 6
$\kappa(x)$ 6
$N(x)$ 6
$\overline{x}$ 6
$d(x)$ 6
F_x 7
$L(X_0, \mathscr{G}_0, t)$ 7
$L(X_0, \mathscr{G}_0, t)$ 8
$W(X_0, \overline{a})$ 9
$deg(u)$ 9
$\mathscr{L}_\phi$ 10
$\mathscr{L}_b$ 10
$L(X_0, \mathscr{G}_0, t)$ 13
$\tau : \overline{\mathbb{Q}}_l \to \mathbb{C}$ 13
$\tau : \overline{\mathbb{Q}}_l \cong \mathbb{C}$ 14
$w(\mathscr{G}_0)$ 18
$w_{gen}(\mathscr{G}_0) = w(j_0^*(\mathscr{G}_0))$ 19
$f^{\mathscr{G}_0}$ 20
$(f, g)_m$ 20
$\|f\|_m$ 20
$\|\mathscr{G}_0\|$ 21
$\|\mathscr{G}_0\| = w(\mathscr{G}_0)$ 22
$\overline{\mathscr{G}_0}$ 23
G_{geom} 28
$\mathscr{L}_0(\psi)$ 38
T_ψ 40
$f^{\mathscr{G}_0}$ 41
f^K 41
$F_\bullet$ 56
$Gr_i(V)$ 56
$P_i(V)$ 56
I 59
$I^{tame} = I/P$ 59
$X[n]$ 67
$\mathscr{T}^n(f) = f[n]$ 67
$T = (X, Y, Z, u, v, w)$ 67
$rot(T)$ 68
$D(\mathbf{A})$ 72
$Kom(\mathbf{A})$ 72
$K(\mathbf{A})$ 72
$D^+(\mathbf{A})$ 72
$D^-(\mathbf{A})$ 72
$D^b(\mathbf{A})$ 72
$D_c^b(X, \overline{\mathbb{Q}}_l)$ 73
$D^{\leq 0}$ 74
$D^{\geq 0}$ 74
$D^{\leq n}$ 74
$D^{\geq n}$ 74
$\tau_{\leq 0}(E)$ 75
$\tau_{\leq 0}$ 75
$\tau_{\geq 1}$ 75
$ad_{\leq 0}$ 75
$ad_{\geq 1}$ 75
$Core(D)$ 77
$H^n(X)$ 81
$H_0(X)$ 83
$\tau_{[m,n]}$ 86
$D^{[m,n]}$ 86
$D_{ctf}^b(x, \mathfrak{o}_r)$ 88
$D_c^b(X, \mathfrak{o}_r)$ 89
$D_{ctf}^b(X, \mathfrak{o}_r)$ 89
$D_c^b(X, \mathfrak{o}) = \text{“}\lim_r\text{”}\, D_{ctf}^b(x, \mathfrak{o}_r)$ 94

$\tau^{Del}_{\leq n}(K^{\bullet}_{r})$ 100
$Del(\mathscr{G})$ 102
${}^{st}\tau_{\leq n}$ 103
${}^{st}\tau_{\geq n}$ 103
$D^{b}_{c}(X, E)$ 106
$D^{b}_{c}(X, \overline{\mathbb{Q}}_{l})$ 106
$D(X) = D^{b}_{c}(X, \overline{\mathbb{Q}}_{l})$ 106
$f^{!}$ 107
$D_{X}(L)$ 108
$\Gamma_{Y}(\mathscr{G})$ 116
K_{X} 119
$D(L) = D_{X}(L)$ 120
$cl_{X}(\mathscr{L})$ 126
$cl_{X}(\mathscr{L})$ 127
$\phi_{\mathscr{L},G,f}$ 129
$w(B_{0})$ 131
$D_{mixed}(X_{0})$ 131
$D^{b}_{mixed}(X_{0}, \overline{\mathbb{Q}}_{l})$ 131
$D^{b}_{\leq w}$ 131
$D^{b}_{\geq w}$ 131
${}^{p}D^{\leq 0}(X)$ 135
${}^{p}D^{\geq 0}(X)$ 135
$Perv(X)$ 136
${}^{p}H^{\nu}$ 136
${}^{p}\tau$ 136
$i^{!}_{x}L$ 136
$j_{!*}B$ 148
$f^{*}[d]$ 156
$f^{*!}$ 159
$T_{\psi}B$ 160
$(E^{(w_{i})}_{0})$ 162
$f^{K_{0}}$ 173
$f^{K_{0}}_{m}$ 174
Rad 204
$Rad(K)$ 204
$Rad_{!}(K)$ 208
$Rad_{*}(K)$ 208
$\mathscr{P}rim^{n}(K)$ 209
K_{red} 210
$\mathscr{C}o\mathscr{P}rim^{-1}(K)$ 212
$Rad^{0}_{!}$ 214
$\mathscr{P}rim^{0}(K)$ 216
$\mathscr{C}o\mathscr{P}rim^{-1}(K)$ 217
$K^{\vee}$ 222
$\mathbb{A}_{R}$ 226
$\mathbb{A}_{S}$ 226
$\mathrm{Kloos}_{m}(q, \psi, a)$ 226
$\widehat{\mathrm{Kloos}_{m}}(\chi)$ 227
I 231
P 231
I^{tame} 231
$\mathrm{Swan}(M)$ 234
$I_{\overline{s}}$ 235
$\mathrm{Swan}_{s}(\mathscr{G})$ 235
$\mathrm{Swan}_{\overline{s}}(\mathscr{G})$ 235
$\chi(U, \mathscr{G})$ 236
$\chi_{c}(U, \mathscr{G})$ 236
g_{X} 236
$\chi^{tame}_{c}(U, \mathscr{G})$ 237
$\chi^{tame}(U, \mathscr{G})$ 237
$\mathrm{Swan}_{\infty}(\mathscr{L}(\psi))$ 237
$h^{\nu}_{c}(\mathscr{G})$ 238
$rg(\mathscr{G})$ 238
$f(g, \mathbb{F}_{q}, t, \mathbf{a}, \psi, \chi)$ 245
$\mathscr{B}$ 249
$A(u)$ 250
$H^{2\bullet}(\mathscr{B}, \overline{\mathbb{Q}}_{l})$ 254
$\mathscr{U}$ 256
$\mathscr{N}$ 256
σ 257
$\mathscr{B}_{u}$ 258
$\mathfrak{t}$ 261
$\mathfrak{u}$ 261
$\mathfrak{b}$ 261
$\mathfrak{g}$ 261
$\mathfrak{t}_{reg}$ 265
Ψ_{L}, Ψ_{B} 291
Φ_{L}, Φ_{B} 291
$\mathscr{O}_{y}$ 296
$\mathscr{S}(\psi, y)$ 296
$\mathscr{S}(\psi)$ 301
$\mathscr{O}$ 303

Index

N-filtration 56
Λ-dualizing complex 119
π-adic sheaf 87
π-adic sheaves 87
σ-decomposition 55
τ-mixed 13, 61, 129
τ-pure 13, 161
τ-pure of weight w 133
τ-real 19, 34
τ-weights 34
n-th cohomology factor 81

A-R π-adic 88
A-R category 87, 88
adjoint quotient 268
adjoint representation 267
algebraized arithmetic monodromy group 32
arithmetic Frobenius element 5
Artin-Rees 87
Artin-Rees category 88
Artin-Schreier covering 38
Artin-Schreier equation 41
Artin-Schreier extension 39
artinian 150

bad prime 262
biduality 108
biduality map 120
Brylinski action 306
Brylinski-Radon transform 204, 207

Chern class 126
coadjoint version 275
compactifiable IX
constant complex 205
control sequence 287
core 77
core(standard) 103
corestriction 308

Deligne operator 102
Deligne's estimate of Kloosterman sums 231
Deligne's truncation operation 100
derived categories 72
determinant weights 28, 32–34
dimension formula 273
distinguished triangle 68, 69
divisibility 89
dual complex 120
dualizing complex 108
dualizing functor 108

essentially smooth 137
Euler-Poincaré characteristic 236
extensions 76, 211

finiteness theorem 95
flag manifold 249
flat 87
Fourier Inversion 42, 160
Fourier transform 40
Frobenius automorphism 5
Frobenius endomorphism 5

Gabber's theorem 167
Gauss sums 227
generalized Kloosterman sums 226
geometric Frobenius automorphism 5
glued t-structure 140
glueing 139
good prime 253
Green function 315
Grothendieck trace formula 6, 7, 13

Hard Lefschetz Theorem 217
higher ramification groups 232

intermediate extension 148
intermediate pullback 159

Kloosterman manifold 246
Kloosterman sums 225, 230, 247
Künneth type formula 109

L-series 7
L-series of a Weil sheaf 8
largest perverse quotient 146, 171
largest perverse subobject 146
largest perverse subsheaf 171
Lefschetz sequence 209, 212
Lefschetz theory 209
left reduced 214
Lusztig action 303

mapping cone 68
Mellin transform 227
middle perverse t-structure 135
Mittag-Leffler condition 99, 100
mixed 13, 61, 129
modified Radon transform 208

nilpotent variety 256
noetherian 150
null systems 88

octaeder axiom 70, 71
Ogg–Shafarevich–Grothendieck 237
orbit decomposition 279
orthogonality 76
orthogonality relations 42

parity law 205
partial Fourier transform 159
perfect complex 93
perfect constructible complexes 88
perverse IC condition 147
perverse sheaves 136
perverse t-structure 136
Plancherel formula 1, 42, 49, 315
Poincare duality 108
point-wise τ-pure sheaves 133
pointwise pure 13
primitive part 56
pure 133

quasi-unipotent 54
quasiisomorphisms 72
quotient categories 209

radius of convergence 21
Radon inversion formula 206, 210
ramification group 59, 231, 235
Rankin method 37
Rankin-Selberg Method 34

real sheaves 19, 33
reduced 210, 222
regular 265, 269
regular orbit 296
regular representation 254, 272
relative Abhyankar lemma 240
relative Poincare duality 107
restriction isomorphism 307
restriction map 308
restriction sequence 145
rotated triangle 68

scalar product 20
Schur's Lemma 29
sections with support 116
self dual 136
semi-small 277
semicontinuity of weights 18, 161
semisimplicity 168
sheaf of rank 1 25
simple object 149
simple perverse sheaf 149
simultaneous resolution 268
small 274
smooth 87
smooth case 138
smooth complex 137
smooth Weil sheaves 9
spectral sequence of Radon transform 221
Springer fiber 258
Springer representation 251
Springer sheaf 295, 296
standard t-structure 98, 103, 136
Swan conductor 234, 235

t-exact 145
t-left exact 145
t-right exact 145
t-structure 74
tame I-module 232
tame Euler-Poincaré characteristic 236
tame ramification 231
tame ramification group 59
tamely ramified 235
Tate twist 42
trace map 112, 158
triangle 67
triangulated category 68
trigonometric sum 230
truncation 75
truncation factors 76

uniform estimate 228

uniformly constructible 241
unipotent variety 256
universal extensions 155
universal regular orbit 303
universal situation 215, 218
universal Springer sheaf 301
upper numbering 233

very good prime 253

weight 13
weight filtration 162
Weil group 5, 9
Weil sheaf 7, 10, 13
Weil sheaves of rank one 10
Weyl group 267
wild ramification group 59, 231, 232, 235

Printing (Computer to Film): Saladruck Berlin
Binding: Stürtz AG, Würzburg